MATHEMATICS IN ACTION

Algebraic, Graphical, and Trigonometric Problem Solving

FOURTH EDITION

The Consortium for Foundation Mathematics

Ralph Bertelle	*Columbia-Greene Community College*
Judith Bloch	*University of Rochester*
Roy Cameron	*SUNY Cobleskill*
Carolyn Curley	*Erie Community College—South Campus*
Ernie Danforth	*Corning Community College*
Brian Gray	*Howard Community College*
Arlene Kleinstein	*SUNY Farmingdale*
Kathleen Milligan	*Monroe Community College*
Patricia Pacitti	*SUNY Oswego*
Rick Patrick	*Adirondack Community College*
Renan Sezer	*LaGuardia Community College*
Patricia Shuart	*Polk State College—Winter Haven, Florida*
Sylvia Svitak	*Queensborough Community College*
Assad J. Thompson	*LaGuardia Community College*

Addison-Wesley

Boston   Columbus   Indianapolis   New York   San Francisco   Upper Saddle River
Amsterdam   Cape Town   Dubai   London   Madrid   Milan   Munich   Paris   Montréal   Toronto
Delhi   Mexico City   São Paulo   Sydney   Hong Kong   Seoul   Singapore   Taipei   Tokyo

Editorial Director, Mathematics: Christine Hoag
Editor in Chief: Maureen O'Connor
Content Editor: Courtney Slade
Assistant Editor: Mary St. Thomas
Senior Managing Editor: Karen Wernholm
Production Project Manager: Beth Houston
Senior Designer/Cover Designer: Barbara Atkinson
Interior Designer: Studio Montage
Digital Assets Manager: Marianne Groth
Production Coordinator: Katherine Roz
Associate Producer: Christina Maestri
Associate Marketing Manager: Tracy Rabinowitz
Marketing Coordinator: Alicia Frankel
Senior Author Support/Technology Specialist: Joe Vetere
Rights and Permissions Advisor: Michael Joyce
Senior Manufacturing Buyer: Carol Melville
Production Management/Composition: PreMediaGlobal
Cover photo: Eric Michaud/iStockphoto

Many of the designations used by manufacturers and sellers to distinguish their products are claimed as trademarks. Where those designations appear in this book, and Addison-Wesley was aware of a trademark claim, the designations have been printed in initial caps or all caps.

Library of Congress Cataloging-in-Publication Data
Mathematics in action: Algebraic, graphical, and trigonometric problem
    solving / the Consortium for Foundation Mathematics. — 4th ed.
        p. cm.
    Includes index.
    ISBN-13: 978-0-321-69861-2 (student ed.)
    ISBN-10: 0-321-69861-4 (student ed.)
    ISBN-13: 978-0-321-69290-0 (instructor ed.)
    ISBN-10: 0-321-69290-X (instructor ed.)
        1. Algebra—Textbooks. I. Consortium for Foundation Mathematics.
II. Title: Algebraic, graphical, and trigonometric problem solving.

QA152.3.M38 2012
512—dc22   2009052062

Copyright © 2012, 2008, 2004, 2001 Pearson Education, Inc.

All rights reserved. No part of this publication may be reproduced, stored in a retrieval system, or transmitted, in any form or by any means, electronic, mechanical, photocopying, recording, or otherwise, without the prior written permission of the publisher. Printed in the United States of America. For information on obtaining permission for use of material in this work, please submit a written request to Pearson Education, Inc., Rights and Contracts Department, 75 Arlington Street, Suite 300, Boston, MA 02116, fax your request to 617-848-7047, or e-mail at http://www.pearsoned.com/legal/permissions.htm.

Addison-Wesley
is an imprint of

www.pearsonhighered.com   1 2 3 4 5 6 7 8 9 10—EB—14 13 12 11 10

Contents

6. Measure the strength of the correlation (association) by a correlation coefficient.

7. Recognize that a strong correlation does not necessarily imply a linear or a cause-and-effect relationship.

CHAPTER 6   AN INTRODUCTION TO THE TRIGONOMETRIC FUNCTIONS   585

APPENDIXES

Preface

Our Vision

Mathematics in Action: Algebraic, Graphical, and Trigonometric Problem Solving, Fourth Edition, is intended to help college mathematics students gain mathematical literacy in the real world and simultaneously help them build a solid foundation for future study in mathematics and other disciplines.

Our team of fourteen faculty, primarily from the State University of New York and the City University of New York systems, used the AMATYC *Crossroads* standards to develop this three-book series to serve a very large population of college students at the pre-precalculus level. Many of our students have had previous exposure to mathematics at this level. It became apparent to us that teaching the same content in the same way to students who have not previously comprehended it is not effective, and this realization motivated us to develop a new approach.

Mathematics in Action is based on the principle that students learn mathematics best by doing mathematics within a meaningful context. In keeping with this premise, students solve problems in a series of realistic situations from which the crucial need for mathematics arises. *Mathematics in Action* guides students toward developing a sense of independence and taking responsibility for their own learning. Students are encouraged to construct, reflect on, apply, and describe their own mathematical models, which they use to solve meaningful problems. We see this as the key to bridging the gap between abstraction and application and as the basis for transfer learning. Appropriate technology is integrated throughout the books, allowing students to interpret real-life data verbally, numerically, symbolically, and graphically.

We expect that by using the *Mathematics in Action* series, all students will be able to achieve the following goals:

- Develop mathematical intuition and a relevant base of mathematical knowledge.

- Gain experiences that connect classroom learning with real-world applications.

- Prepare effectively for further college work in mathematics and related disciplines.

- Learn to work in groups as well as independently.

- Increase knowledge of mathematics through explorations with appropriate technology.

- Develop a positive attitude about learning and using mathematics.

- Build techniques of reasoning for effective problem solving.

- Learn to apply and display knowledge through alternative means of assessment, such as mathematical portfolios and journal writing.

Our vision for you is to join the growing number of students using our approaches who discover that mathematics is an essential and learnable survival skill for the 21st century.

Pedagogical Features

The pedagogical core of *Mathematics in Action* is a series of guided-discovery activities in which students work in groups to discover mathematical principles embedded in realistic situations. The key principles of each activity are highlighted and summarized at the activity's conclusion. Each activity is followed by exercises that reinforce the concepts and skills revealed in the activity.

The activities are clustered within each chapter. Each cluster contains regular activities along with project activities that relate to particular topics. The lab activities require more than just paper, pencil, and calculator; they also require measurements and data collection and are ideal for in-class group work. The project activities are designed to allow students to explore specific topics in greater depth, either individually or in groups. These activities are usually self-contained and have no accompanying exercises. For specific suggestions on how to use the three types of activities, we strongly encourage instructors to refer to the *Instructor's Resource Manual with Tests* that accompanies this text.

Each cluster concludes with two sections: What Have I Learned? and How Can I Practice? The What Have I Learned? exercises are designed to help students pull together the key concepts of the cluster. The How Can I Practice? exercises are designed primarily to provide additional work with the numeric and algebraic skills of the cluster. Taken as a whole, these exercises give students the tools they need to bridge the gaps between abstraction, skills, and application.

Additionally, each chapter ends with a Summary that contains a brief description of the concepts and skills discussed in the chapter, plus examples illustrating these concepts and skills. The concepts and skills are also cross-referenced to the activity in which they appear, making the format easier to follow for those students who are unfamiliar with our approach. Each chapter also ends with a Gateway Review, providing students with an opportunity to check their understanding of the chapter's concepts and skills.

Changes from the Third Edition

The fourth edition retains all the features of the previous edition, with the following content changes:

- All data-based activities and exercises have been updated to reflect the most recent information and/or replaced with more relevant topics.

- The language in many activities is now clearer and easier to understand.

- Activities 1.1 and 1.2 were expanded to three activities to ensure students get a solid introduction to functions.

- New problem situations were added in Activities 1.9, 1.13, 1.16, 2.2, 4.5, 4.6, and 5.7.

- A new activity, *Activity 1.14: Earth Week*, was added on solving linear systems using matrix methods.

- Chapter 2 was rearranged so that exponents are covered separately from composition and inverse functions. All exponent material is now found in Activities 2.3 and 2.4.

- The coverage of exponential growth and decay was split into two activities. *Activity 3.1: The Summer Job* covers exponential growth, and *Activity 3.2: Half-Life of Medicine* covers exponential decay.

- Chapter 4 was rearranged from three clusters to two. *Activity 4.7: Complex Numbers* is now at the end of Cluster 1.

- The discussion of trigonometric values of special angles is now included in *Activity 6.1: The Leaning Tower of Pisa*.

- Several activities have moved to the *Instructor's Resource Manual with Tests* and MyMathLab to streamline the course without loss of content.

- Several activities have incorporated web-based exercises into the exercise sets.

Supplements

Instructor Supplements

Annotated Instructor's Edition

ISBN-13 978-0-321-69290-0
ISBN-10 0-321-69290-X

This special version of the student text provides answers to all exercises directly beneath each problem.

Instructor's Resource Manual with Tests

ISBN-13 978-0-321-69291-7
ISBN-10 0-321-69291-8

This valuable teaching resource includes the following materials:

- Sample syllabi suggesting ways to structure the course around core and supplemental activities and within different credit-hour options.

- Sample course outlines containing time lines for covering topics.

- Teaching notes for each chapter, specifically for those using the *Mathematics in Action* approach for the first time.

- Skills worksheets for topics with which students typically have difficulty.

- Sample chapter tests and final exams for individual and group assessment.

- Sample journal topics for each chapter.

- Learning in groups with questions and answers for instructors using collaborative learning for the first time.

- Incorporating technology, including sample graphing calculator assignments.

TestGen®

ISBN-13 978-0-321-70560-0
ISBN-10 0-321-70560-2

TestGen enables instructors to build, edit, print, and administer tests using a computerized bank of questions developed to cover all the objectives of the text. TestGen is algorithmically based, allowing instructors to create multiple but equivalent versions of the same question or test with the click of a button. Instructors can also modify test bank questions or add new questions. The software and testbank are available for download from Pearson Education's online catalog.

Instructor's Training Video on CD

ISBN-13 978-0-321-69279-5
ISBN-10 0-321-69279-9

This innovative video discusses effective ways to implement the teaching pedagogy of the *Mathematics in Action* series, focusing on how to make collaborative learning, discovery learning, and alternative means of assessment work in the classroom.

Student Supplements

Worksheets for Classroom or Lab Practice

ISBN-13 978-0-321-73835-6
ISBN-10 0-321-73835-7

- Extra practice exercises for every section of the text with ample space for students to show their work.

- These lab- and classroom-friendly workbooks also list the learning objectives and key vocabulary terms for every text section, along with vocabulary practice problems.

- Concept Connection exercises, similar to the "What Have I Learned?" exercises found in the text, assess students' conceptual understanding of the skills required to complete each worksheet.

MathXL® Tutorials on CD

ISBN-13 978-0-321-69292-4
ISBN-10 0-321-69292-6

This interactive tutorial CD-ROM provides algorithmically generated practice exercises that are correlated at the objective level to the exercises in the textbook. Every practice exercise is accompanied by an example and a guided solution designed to involve students in the solution process. The software provides helpful feedback for incorrect answers and can generate printed summaries of students' progress.

InterAct Math Tutorial Web site www.interactmath.com

Get practice and tutorial help online! This interactive tutorial Web site provides algorithmically generated practice exercises that correlate directly to the exercises in the textbook. Students can retry an exercise as many times as they like with new values each time for unlimited practice and mastery. Every exercise is accompanied by an interactive guided solution that provides helpful feedback for incorrect answers, and students can also view a worked-out sample problem that steps them through an exercise similar to the one they're working on.

Pearson Math Adjunct Support Center

The **Pearson Math Adjunct Support Center** (http://www.pearsontutorservices.com/mathadjunct.html) is staffed by qualified instructors with more than 100 years of combined experience at both the community college and university levels. Assistance is provided for faculty in the following areas:

- Suggested syllabus consultation

- Tips on using materials packed with your book

- Book-specific content assistance

- Teaching suggestions, including advice on classroom strategies

Supplements for Instructors and Students

MathXL® Online Course (access code required)

MathXL® is a powerful online homework, tutorial, and assessment system that accompanies Pearson Education's textbooks in mathematics or statistics. With MathXL, instructors can:

- Create, edit, and assign online homework and tests using algorithmically generated exercises correlated at the objective level to the textbook.

- Create and assign their own online exercises and import TestGen tests for added flexibility.

- Maintain records of all student work tracked in MathXL's online gradebook.

With MathXL, students can:

- Take chapter tests in MathXL and receive personalized study plans and/or personalized homework assignments based on their test results.

- Use the study plan and/or the homework to link directly to tutorial exercises for the objectives they need to study.

- Access supplemental animations and video clips directly from selected exercises.

MathXL is available to qualified adopters. For more information, visit our Web site at www.mathxl.com, or contact your Pearson representative.

MyMathLab® Online Course (access code required)

MyMathLab® is a text-specific, easily customizable online course that integrates interactive multimedia instruction with textbook content. MyMathLab gives you the tools you need to deliver all or a portion of your course online, whether your students are in a lab setting or working from home.

- **Interactive homework exercises**, correlated to your textbook at the objective level, are algorithmically generated for unlimited practice and mastery. Most exercises are free response and provide guided solutions, sample problems, and tutorial learning aids for extra help.

- **Personalized homework** assignments that you can design to meet the needs of your class. MyMathLab tailors the assignment for each student based on their test or quiz scores. Each student receives a homework assignment that contains only the problems he or she still needs to master.

- **Personalized Study Plan**, generated when students complete a test or quiz or homework, indicates which topics have been mastered and links to tutorial exercises for topics students have not mastered. You can customize the Study Plan so that the topics available match your course content.

- **Multimedia learning aids**, such as video lectures and podcasts, animations, and a complete multimedia textbook, help students independently improve their understanding and performance. You can assign these multimedia learning aids as homework to help your students grasp the concepts.

- **Homework and Test Manager** lets you assign homework, quizzes, and tests that are automatically graded. Select just the right mix of questions from the MyMathLab exercise bank, instructor-created custom exercises, and/or TestGen® test items.

- **Gradebook**, designed specifically for mathematics and statistics, automatically tracks students' results, lets you stay on top of student performance, and gives you control over how to calculate final grades. You can also add offline (paper-and-pencil) grades to the gradebook.

- **MathXL Exercise Builder** allows you to create static and algorithmic exercises for your online assignments. You can use the library of sample exercises as an easy starting point, or you can edit any course-related exercise.

- **Pearson Tutor Center** (www.pearsontutorservices.com) access is automatically included with MyMathLab. The Tutor Center is staffed by qualified math instructors who provide textbook-specific tutoring for students via toll-free phone, fax, e-mail, and interactive Web sessions.

Students do their assignments in the Flash®-based MathXL Player, which is compatible with almost any browser (Firefox®, Safari™, or Internet Explorer®) on almost any platform (Macintosh® or Windows®). MyMathLab is powered by CourseCompass™, Pearson Education's online teaching and learning environment, and by MathXL®, our online homework, tutorial, and assessment system. MyMathLab is available to qualified adopters. For more information, visit www.mymathlab.com or contact your Pearson representative.

Acknowledgments

The Consortium would like to acknowledge and thank the following people for their invaluable assistance in reviewing and testing material for this text:

Rogelio Briones, *Los Medanos College*

April D. Strom, *Scottsdale Community College*

Mary Kay Abbey, *Montgomery College*

Barbara Burke, *Hawai'i Pacific University*

Edward (Ted) Coe, *Scottsdale Community College*

Jennifer Dollar, *Grand Rapids Community College*

Irene Duranczyk, *University of Minnesota*

Ernest East, *Northwestern Michigan College*

Maryann B. Faller, *Adirondack Community College*

John R. Furino, *Foothill College*

Linda Green, *Santa Fe Community College*

Teresa Hodge, *University of the Virgin Islands*

Maria Ilia, *Clarke College*

Ashok Kumar, *Valdosta State University*

David Lynch, *Prince George's Community College*

J. Robert Malena, *Community College of Allegheny County—South Campus*

Raquel Mesa, *Xavier University of Louisiana*

Beverly K. Michael, *University of Pittsburgh*

Paula J. Mikowicz, *Howard Community College*

Adam Parr, *University of the Virgin Islands*

Debra Pharo, *Northwestern Michigan College*

Kathy Potter, *St. Ambrose University*

Dennis Risher, *Loras College*

Sandra Spears, *Jefferson Community College*

Christopher Teixeira, *Rhode Island College*

Kurt Verderber, *SUNY Cobleskill*

Lynn Wolfmeyer, *Western Illinois University*

We would also like to thank our accuracy checkers, Shannon d'Hemecourt, Diane E. Cook, Jon Weerts, and James Lapp.

Finally, a special thank-you to our families for their unwavering support and sacrifice, which enabled us to make this text a reality.

The Consortium for Foundation Mathematics

To the Student

The book in your hands is most likely very different from any mathematics textbook you have seen before. In this book, you will take an active role in developing the important ideas of arithmetic and beginning algebra. You will be expected to add your own words to the text. This will be part of your daily work, both in and out of class. It is the belief of the authors that students learn mathematics best when they are actively involved in solving problems that are meaningful to them.

The text is primarily a collection of situations drawn from real life. Each situation leads to one or more problems. By answering a series of questions and solving each part of the problem, you will be using and learning one or more ideas of introductory college mathematics. Sometimes, these will be basic skills that build on your knowledge of arithmetic. Other times, they will be new concepts that are more general and far-reaching. The important point is that you won't be asked to master a skill until you see a real need for that skill as part of solving a realistic application.

Another important aspect of this text and the course you are taking is the benefit gained by collaborating with your classmates. Much of your work in class will result from being a member of a team. Working in small groups, you will help each other work through a problem situation. While you may feel uncomfortable working this way at first, there are several reasons we believe it is appropriate in this course. First, it is part of the learning-by-doing philosophy. You will be talking about mathematics, needing to express your thoughts in words. This is a key to learning. Secondly, you will be developing skills that will be very valuable when you leave the classroom. Currently, many jobs and careers require the ability to collaborate within a team environment. Your instructor will provide you with more specific information about this collaboration.

One more fundamental part of this course is that you will have access to appropriate technology at all times. You will have access to calculators and some form of graphics tool—either a calculator or computer. Technology is a part of our modern world, and learning to use technology goes hand in hand with learning mathematics. Your work in this course will help prepare you for whatever you pursue in your working life.

This course will help you develop both the mathematical and general skills necessary in today's workplace, such as organization, problem solving, communication, and collaborative skills. By keeping up with your work and following the suggested organization of the text, you will gain a valuable resource that will serve you well in the future. With hard work and dedication, you will be ready for the next step.

The Consortium for Foundation Mathematics

Chapter 1

Function Sense

Cluster 1 — Modeling with Functions

Activity 1.1

Parking Problems

Objectives

1. Identify input and output in situations involving two variable quantities.

2. Identify a functional relationship between two variables.

3. Identify the independent and dependent variables.

4. Use a table to numerically represent a functional relationship between two variables.

5. Write a function using function notation.

Introduction

A key step in the problem-solving process is to look for relationships and connections between the variable quantities in a given situation. Problems encountered in the world around us, including the environment, medicine, economics, and the Internet, are often very complicated and contain several variables. In this text, you will primarily deal with situations that contain two variables. In many of these situations, the variables will have a special relationship called a **function**.

Function

Did you have trouble finding a parking space this morning? Was the time that you arrived on campus a factor? As part of a reconstruction project at a small community college, the number of cars in the parking lot was counted each hour from 7:00 A.M. to 10:00 P.M. on a particular day. The results are shown in the following table.

 **Park It**

TIME OF DAY	NUMBER OF CARS
7 A.M.	24
8 A.M.	212
9 A.M.	384
10 A.M.	426
11 A.M.	538
12 P.M.	497
1 P.M.	384
2 P.M.	337
3 P.M.	285
4 P.M.	278
5 P.M.	302
6 P.M.	427
7 P.M.	384
8 P.M.	315
9 P.M.	187
10 P.M.	56

This situation involves two variables, the time of day and the number of cars in the parking lot. A **variable**, usually represented by a letter, is a quantity that may change in value from one particular instance to another. Typically, one variable is designated as the **input** and the other is called the **output**. The input is the value given first, and the output is the value that corresponds to, or is determined by, the given input value.

1. In the parking lot situation, identify the input variable and the output variable.

2. a. For an input of 10:00 A.M., how many cars are in the parking lot (output)?

   b. For an input of 5:00 P.M., how many cars are in the parking lot (output)?

   c. For each value of input (time of day), how many different outputs (number of cars) are there?

The set of data in the Park It table is an example of a mathematical function.

Definition

A **function** is a correspondence between an input variable and an output variable that assigns a single output value to each input value. Therefore, for a function, any given input value has exactly one corresponding output value. If x represents the input variable and y represents the output variable, then the function assigns a single, unique y-value to each x-value.

---

3. Explain how the data in the Park It table fits the description of a function.

A functional relationship is stated as follows: "The output variable is a function of the input variable." Using x for the input variable and y for the output variable, the functional relationship is stated "y is a function of x." Because the input for the parking lot function is the time of day and the output is the number of cars in the lot at that time, you write that the number of cars in the parking lot is a function of the time of day.

Example 1 *Consider the following table listing the official high temperature (in °F) in the village of Lake Placid, New York, during the first week of January. Note that the date has been designated the input and the high temperature on that date the output. Is the high temperature a function of the date?*

Almost Freezing

Date (Input)	1	2	3	4	5	6	7
Temperature (Output)	25	30	32	24	23	27	30

SOLUTION

From this table, you observe that the high temperature is a function of the date. For each date there is exactly one high temperature. The relationships in this example can be visualized as follows:

DATE (INPUT)	TEMPERATURE (OUTPUT)
1 ⟶	25
2 ⟶	30
3 ⟶	32
4 ⟶	24
5 ⟶	23
6 ⟶	27
7 ⟶	30

If d represents the input (date), and T represents the output (temperature), then T is a function of d.

If the input and output in Example 1 are switched (see table below), the daily high temperature becomes the input and the date becomes the output. The date is not a function of the high temperature. The input value 30 has two output values, 2 and 7.

TEMPERATURE (INPUT)	DATE (OUTPUT)
25 ⟶	1
30 ⟶	2
32 ⟶	3
24 ⟶	4
23 ⟶	5
27 ⟶	6
	7

4. Interchange the input and the output in the parking lot situation. Let the number of cars in the lot be the input and the time of day be the output. Is the time of day a function of the number of cars in the lot? Write a sentence explaining why this switch does or does not fit the description of a function.

Example 2   *Determine whether or not the following situation describes a function. Give a reason for your answer.*

The amount of postage for a letter is a function of the weight of the letter.

SOLUTION

Yes, this statement does describe a function. The weight of the letter is the input, and the amount of postage is the output. Each letter has one weight. This weight determines the postage necessary for the letter. There is only one amount of postage for each letter. Therefore, for each value of input (weight of the letter) there is one output (postage). Note that if w represents the input (weight of the letter), and p represents the output (postage), then p is function of w.

5. Determine whether or not each situation describes a function. Give a reason for your answer.

   a. The amount of property tax you have to pay is a function of the assessed value of the house.

   b. The weight of a letter in ounces is a function of the postage paid for mailing the letter.

   c. The speed at which a free-falling baseball strikes the ground is a function of the height from which it was dropped.

   d. The amount of your savings account is a function of your salary.

Definition

If the relationship between two variables is a function, the input variable is called the **independent** variable, and the output variable is called the **dependent** variable. If x represents the input variable, and y represents the output variable, then x is the independent variable and y is the dependent variable.

---

6. The independent variable in Example 1 is the date. The dependent variable is the temperature. Identify the independent and dependent variables in Example 2.

Defining Functions Numerically

The input/output pairing in the parking lot function on page 1 is presented as a **table of matched pairs**. In such a situation, the function is defined **numerically**. Another way to define a function numerically is as a set of ordered pairs.

> **Definition**
>
> An **ordered pair** of numbers consists of two numbers written in the form
>
> (input value, output value).
>
> The order in which they are listed is significant.

Example 3 | *The ordered pair (3, 4) is distinct from the ordered pair (4, 3). In the ordered pair (3, 4), 3 is the input and 4 is the output. In the ordered pair (4, 3), 4 is the input and 3 is the output.*

> **Definition**
>
> A function may be defined **numerically** as a set of ordered pairs in which the first number of each pair represents the input value and the second number represents the corresponding output value. No two ordered pairs have the same input value and different output values.

Example 4 | *(9:00 A.M., 384) or (0900, 384) (using a 24-hour clock) is an ordered pair that is part of the parking lot function.*

7. Using a 24-hour clock, write three other ordered pairs for the parking lot function.

8. In Example 1, the high temperature in Lake Placid is a function of the date.

   Convert to ordered pairs all the values in the Almost Freezing table on page 2.

Function Notation

There is a special notation for functions in which the function itself is represented by a name or letter. For example, the function that relates the time of day to the number of cars can be represented by the letter f. Let t represent time, the input variable, and let c represent the number of cars, the output variable. The following simplification (really an abbreviation) is now possible.

The number of cars in the parking lot **is a function of** the time of day.

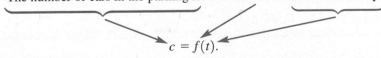

$$c = f(t).$$

The final function notation is read "c equals f of t."

Notice that the output c (the number of cars) is equal to $f(t)$. So $f(t)$ is the output of f when the input is t. For example, $f(1400)$ represents the number of cars (the output of f) when the input is 1400 (at 2 P.M.).

> In general, function notation is written as follows:
>
> *output variable* = name of function (*input variable*).
>
> The input variable or input value is also called the **argument** of the function.

> **Example 5**   *Values from the table or ordered pairs for the parking lot function ca*
> *be written as follows using function notation.*
>
> $$212 = f(800), \qquad 302 = f(1700), \qquad f(2100) = 187$$

9. a. Rewrite the three examples given in Example 5 as three ordered pairs. Pay attention to which is the input value and which is the output value.

b. Write a sentence explaining the meaning of $f(1600) = 278$ in the parking lot situation.

10. a. Referring to the Almost Freezing table in Example 1, determine $g(3)$, where g is the name of the temperature function.

b. Write a sentence explaining the meaning of $g(5) = 23$.

Gross Pay Function

11. If you work for an hourly wage, your gross pay is a function of the number of hours tha you work.

a. Identify the input and output.

b. If you earn $9.50 per hour, complete the following table.

Number of Hours	0	3	5	7	10	12
Gross Pay						

c. Let n represent the number of hours worked and $f(n)$ represent the gross pay. Use the table to determine $f(5)$.

d. Write a sentence explaining the meaning of $f(10) = 95$.

SUMMARY: ACTIVITY 1.1

1. A **variable**, usually represented by a letter, is a quantity that may change in value from one particular instance to another.

2. In a situation involving two variables, one variable is designated the **input** and the other the **output**. The input is the value given first, and the output is the value that corresponds to or is determined by the given input value.

3. A **function** is a rule relating an input variable (sometimes called the argument) and an output variable so that a single output value is assigned to each input value. In such a case, you state that the output variable is a function of the input variable.

4. Independent variable is another name for the input variable of a function.

Note that the graph of the Park It function consists of 16 distinct points that are not connected. The input variable (time of day) is defined only for the car counts in the parking lot for each hour from 7 A.M. to 10 P.M. The Park It function is said to be **discrete** because it is defined only at isolated, distinct input values (practical domain). The function is not defined for input values between these particular values.

Caution. In order to use the graph for a relationship such as the parking lot situation to make predictions or to recognize patterns, it is convenient to connect the points with line segments. This creates a type of continuous graph. This changes the domain shown in the graph from "some values" to "all values." Therefore, you need to be cautious. Connecting data points may cause confusion when working with real-world situations.

3. a. In Example 1 on page 2 of Activity 1.1, the high temperature in Lake Placid is a function of the date. Plot each ordered pair as a point on an appropriately scaled and labeled set of coordinate axes.

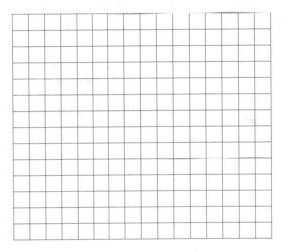

b. Determine the practical domain of the temperature function.

c. Determine the practical range of the function.

d. Is this function discrete?

4. The cost-of-fill-up function in Activity 1.2 was defined by the equation $c = 12.6p$, where c is the cost to fill up your car with 12.6 gallons of gas at a price of p dollars per gallon.

a. What is the practical domain of this function? Refer to Problem 7 in Activity 1.2.

b. List five ordered pairs of the cost-of-fill-up function in the following table.

Price per Gallon, p (dollars)				
Cost of Fill-Up, c				

c. Sketch a graph of the cost-of-fill-up function by first plotting the five points from part b on properly scaled and labeled coordinate axes.

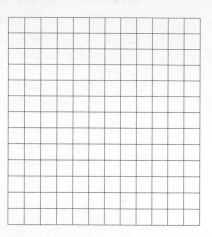

d. Does the graph of the function consist of just the five points from part b? Explain.

e. Describe any patterns or trends in the graph.

The cost-of-fill-up function is defined for all input values in the practical domain in Problem 4a. The five points determined in part b can be connected to form a **continuous** graph. Such a graph is said to be **continuous** over its practical domain.

5. a. Consider the gross pay function defined by $g = 9.50h$ in Problem 11 of Activity 1.2. Plot the ordered pairs determined in Problem 11b, page 14.

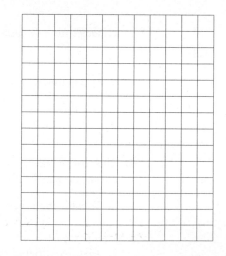

b. Is the gross pay function discrete or continuous? Explain.

Graphing Functions Using Technology

Following is the graph of the cost-of-fill-up function defined by $c = 12.6p$ over its practical domain.

Graph 1

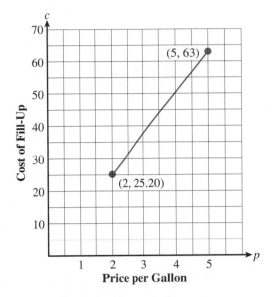

The domain for the general function defined by $c = 12.6p$, with no connection to the context of the situation, is the set of all real numbers, since any real number can be substituted for p in $12.6p$. Following is a graph of $c = 12.6p$ for any real number p.

Graph 2

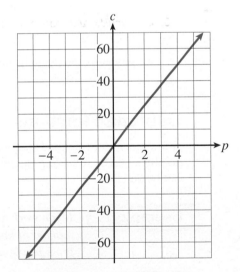

Each of these graphs can be obtained using your graphing calculator as demonstrated in Problems 6 and 7.

Appendix

Recall that the independent variable in your graphing calculator is represented by x and the dependent variable is represented by y. Therefore, the cost-of-fill-up equation $c = 12.6p$ needs to be keyed in the "Y=" menu as $y = 12.6x$. The procedure for graphing a function using the TI-83/TI-84 Plus calculator can be found in Appendix C.

6. a. The viewing window is the portion of the rectangular coordinate system that is displayed when you graph a function. Use the practical domain and practical range of the cost-of-fill-up function to determine X_{min}, X_{max}, Y_{min}, and Y_{max} in the window screen. Key these values into your calculator. Your screens should appear as follows:

 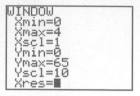

b. Graph the function. Your screen should appear as follows:

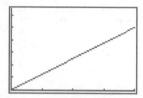

7. a. Determine reasonable window settings to obtain the graph of the general function defined by $c = 12.6p$ for which the domain is any real number.

b. Type in the values determined in part a and graph the function. The screens should appear as follows:

c. How does the graph in part b compare to Graph 2 on the previous page?

Defining Functions: A Summary

A function can be defined by a written statement (verbally), symbolically, numerically, and graphically. The following example illustrates the different ways that the gross pay function can be defined.

Example 1   *If you work for an hourly wage, then your gross pay is a function of the number of hours worked. If you earn $9.50 per hour, then define the gross pay function verbally, symbolically, numerically, and graphically.*

SOLUTION:

Verbal Definition: A Statement of the Definition of the Function:

The gross pay will be the number of hours worked multiplied by $9.50.

Symbolic Definition: If g represents the gross pay and h represents the number of hours worked, then

$$g = 9.50h$$

Definition

A **mathematical model** is an equation or a graph that fits or approximates the actual data. The model can be used to predict output values for input values not in the table.

Note that the points on the graph in Problem 3 do not lie exactly on a specific curve. However, calculators can produce an equation that best models actual data. From the data in the table on page 28, the TI-83/TI-84 Plus can be used to generate the following model:

$$f(x) = -0.00029x^2 + 0.31x + 18.6 \text{ (coefficients are rounded)},$$

where x represents the length of skid distances in feet and $f(x)$ represents the speed in miles per hour. The process for generating these equations is covered in later activities.

Appendix

4. a. Enter the function equation above into your calculator. For help with the TI-83/84 Plus, see Appendix C. The screen should appear as follows:

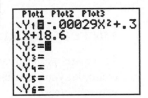

b. The values in the table The Long and Short of It can help to set appropriate window values in your calculator to view the graph. Starting at a minimum skid distance of 0, a reasonable maximum value of the skid distance in this situation would be 450 feet. Beginning at a minimum speed of 0, a reasonable maximum speed value would be 100 mph. Using these settings, the window screen should appear as follows:

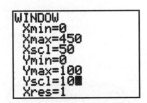

c. Display your graph using the window settings from part b. Your graph should resemble the following:

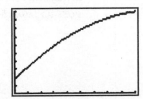

5. a. By pressing the trace button and the left and right arrow leys, you can display the x-y values of points on the graph on the bottom of the display. Use the trace feature of your calculator to approximate the speed of the car when the length of the skid distance is 200 feet. You can obtain the exact value for any x-value between X_{min} and X_{max} by entering the x-value while in trace mode and pressing (ENTER). Your screen should appear as follows:

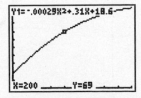

b. Use your calculator to verify the result in part a by evaluating
$-0.00029(200)^2 + 0.31(200) + 18.6$.

There is another way to evaluate $f(200)$ when the function is in the calculator as a Y variable
$(Y1, Y2, Y3, \ldots)$. Since the skid distance function is in Y1, you can evaluate $f(200)$ by typing Y1(200) in the home screen. To place Y1 in the home screen, begin by pressing VARS,
then arrow to Y-VARS, select Option 1: Function, and (ENTER). Follow this with (200), then
(ENTER).

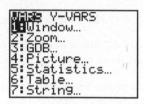 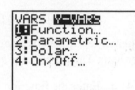 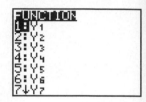

Increasing and Decreasing Functions

There are many advantages to having function models represented in graphical form. For
example, you are often interested in determining how the output values change as the input
values increase.

> **Definition**
>
> A function is **increasing** if its graph goes up to the right, **decreasing** if its graph goes
> down to the right, and **constant** if its graph is horizontal. In each case, you are viewing
> the graph as a point moves along the curve from left to right, that is, as the input values
> increase.

Example 1

a. *The graph of the function $y = 3x + 1$ is*
always increasing. The graph of this function
goes up from left to right.

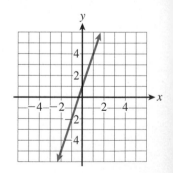

b. *The graph of the function $f(x) = -x + 2$ is*
decreasing because its graph goes down from
left to right.

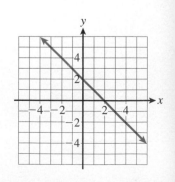

SUMMARY: ACTIVITY 1.4

1. A **mathematical model** is an equation or a graph that fits or approximates the actual data. The model can be used to predict output values for input values not in the table.

2. A function is **increasing** if its graph goes up to the right, **decreasing** if its graph goes down to the right, and **constant** if its graph is horizontal.

3. In the **vertical line test**, a graph defines a function if any vertical line drawn through the graph intersects the graph no more than once.

EXERCISES: ACTIVITY 1.4

1. The following table defines snowfall as a function of elevation for a recent snow storm in upstate New York.

Higher and Higher

ELEVATION (IN FEET), x	SNOWFALL (IN INCHES), $f(x)$
1000	4
2000	6
3000	9
4000	12

A function that closely models the data in the table is

$$f(x) = 0.0027x + 1.$$

Enter this function into a Y variable on your calculator.

a. Complete the following table using the function f. Verify your answers using the table feature of your graphing calculator.

ELEVATION	1000	2000	3000	4000
$f(x)$				

b. Determine $f(2500)$ and explain its practical meaning in this situation.

c. Determine $f(-2000)$. Does this have any meaning in this context?

d. Use the table values to set appropriate window values to view the graph. Graph the function *f* on your calculator. Identify the window you used.

e. Does the graph pass the vertical line test for a function? Explain.

f. Does the graph indicate that the function is increasing, decreasing, or constant?

g. Use the trace function of your calculator to determine $f(2500)$. Compare your answer to your result in part b.

2. The number of new hotel projects in the United States from the fourth quarter of 2007 through the first quarter of 2009 can be modeled by the function

$$h(x) = -7.21x^2 - 3.87x + 447.14,$$

where $x = 0$ is the fourth quarter of 2007, $x = 1$ is the first quarter of 2008, and $h(x)$ is the number of new hotel projects.

a. Using your calculator complete the following table. Round your results to the nearest whole number.

x, Number of Quarters Since Last Quarter of 2007	0	1	2	3	4	5
h(x), The Number of New Hotel Projects						

b. Evaluate $h(2)$ and explain its practical meaning in this situation.

c. Use the table values to set appropriate window values to view the graph of $h(x)$. Graph the function on your calculator. Identify the window you used.

d. Does the graph pass the vertical line test for a function? Explain.

e. Use the trace function on your calculator to determine $h(3)$. Compare your answer to your result in part a.

3. The value of almost everything you own (assets), such as a car, computer, or house, depreciates (goes down) over time. When an asset's value decreases by a fixed amount each year, the depreciation is called straight-line depreciation.

Suppose your truck has an initial value of $12,400 and depreciates $820 per year.

a. State a question that you might want answered in this situation.

b. What two variables are involved in this problem?

c. Which variable can best be designated as the dependent variable? As the independent variable?

d. Complete the following table.

Independent Variable	1	2	3	4	5
Dependent Variable					

e. State in words the relationship between the independent and dependent variables.

f. Use appropriate letters to represent the variables involved and translate the written statement in part e as an equation.

g. If you plan to keep the truck for 7 years, determine the value of the truck at the end of this period. Explain the process you used.

h. What assumption was made regarding the rate of depreciation of the truck? Does this seem reasonable?

4. a. Sketch the graph of a function that is everywhere decreasing.

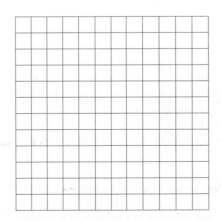

b. What does the graph of a constant function look like?

5. Use the vertical line test to determine whether either graph represents a function.

a.

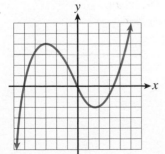

b.

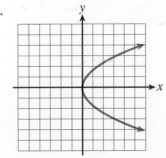

In Exercises 6 and 7, enter the given function into a y variable on your calculator. Match the graph given for the function by determining the appropriate window. (Hint: Use a table to help determine a window.) Give the X_{min}, X_{max}, Y_{min}, and Y_{max} window values that you use.

6. $f(x) = 5400x + 3600$

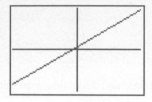

7. $g(x) = 2x^4 - 5x - 2$

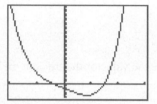

Activity 1.5

Graphs Tell Stories

Objectives

1. Describe in words what a graph tells you about a given situation.

2. Sketch a graph that best represents the situation described in words.

3. Identify increasing, decreasing, and constant parts of a graph.

4. Identify minimum and maximum points on a graph.

The expression "a picture is worth a thousand words" is a cliché, but it is true. Functions are often easier to understand when presented in visual form. To understand such pictures, you need to practice going back and forth between graphs and word descriptions.

Every graph shows how the inputs and outputs change in relation to one another. As you read a graph from left to right, the input variable is increasing in value. The graph indicates the change in the output values (increasing, decreasing, or constant) as the input values increase.

a.

The graph of an increasing function rises to the right.

b.

The graph of a decreasing function falls to the right.

c.

The graph of a constant function remains horizontal.

If a function increases and then decreases, the point where the graph changes from rising to falling is called a **maximum point**. The y-value of this point is called a **local maximum value**. If a function decreases and then increases, the point where the graph changes from falling to rising is called a **minimum point**. The y-value of this point is called a **local minimum value**.

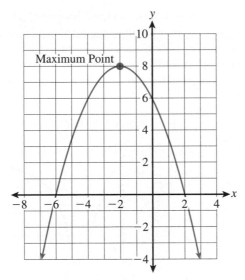

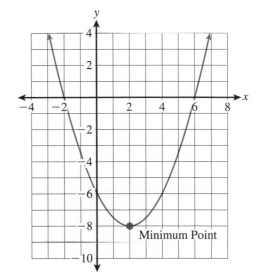

Example 1 *The following graph describes your walk from the parking lot to the library.*

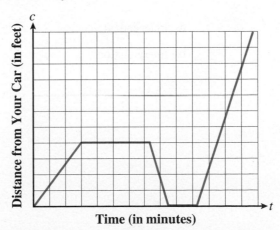

As you read the graph in Example 1 from left to right, it shows how your distance from your car changes as time passes. One possible scenario this graph describes is:

You leave your car and walk at a steady pace toward the library. You meet some friends and stop to chat for a while. You realize that you forgot something and quickly return to your car. After rummaging around for a while, you hurry off to the library.

How did anyone come up with this from the graph? Look at the graph in sections.

a. The first increasing line segment indicates you are moving away from your car because the time and the distance are increasing.

b. The first horizontal section indicates that your distance from the car is constant, so you are standing still.

c. The decreasing line segment indicates your distance from the car is decreasing. When it reaches the horizontal axis, it tells you that you are back at your car.

d. The second horizontal segment indicates you stay at your car for a time.

e. The final increasing segment is steeper and longer than the first, so you are moving away from the car faster and farther than in the first segment.

Graphs to Stories

The graphs in Problems 1–4 present visual images of several situations. Each graph shows how the y-values change in relation to the x-values. In each situation, identify the independent variable and the dependent variable. Then, interpret the situation; that is, describe, in words, what the graph is telling you about the situation. Indicate whether the graph rises, falls, or is constant and whether the graph reaches either a minimum (smallest) or maximum (largest) y-value.

1. A person's core body temperature (°F) in relation to time of day

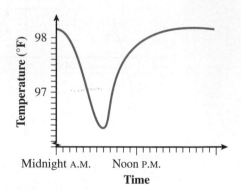

a. Independent: _____ Dependent: _____

b. Interpretation:

8. Hair grows at a steady rate. Suppose you get your hair cut every month. Measuring the longest hair on your head, graph your hair length over the course of 6 months.

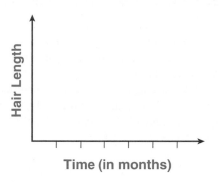

9. The distance traveled is a function of speed in a fixed time interval.

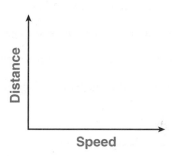

SUMMARY: ACTIVITY 1.5

If a function increases and then decreases, the point where the graph changes from rising to falling is called a **maximum point**. The y-value of this point is called a **local maximum value**. If a function decreases and then increases, the point where the graph changes from falling to rising is called a **minimum point**. The y-value of this point is called a **local minimum value**.

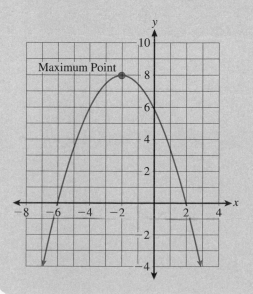

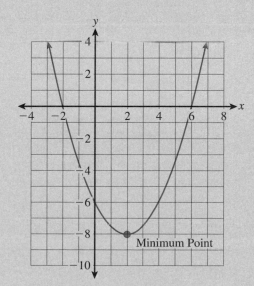

EXERCISES: ACTIVITY 1.5

1. You are a technician at the local power plant, and you have been asked to prepare a report that compares the output and efficiency of the six generators in your sector. Each generator has a graph that shows output of the generator as a function of time over the previous week, Monday through Sunday. You take all the paperwork home for the night (your supervisor wants this report on his desk at 7:00 A.M.), and to your dismay your cat scatters your pile of papers out of the neat order in which you left them. Unfortunately, the graphs for generators A through F were not labeled (you will know better next time!). You recall some information and find evidence elsewhere for the following facts.

- Generators A and D were the only ones that maintained a fairly steady output.

- Generator B was shut down for a little more than two days during midweek.

- Generator C experienced a slow decrease in output during the entire week.

- On Tuesday morning, there was a problem with generator E that was corrected in a few hours.

- Generator D was the most productive over the entire week.

Match each graph with its corresponding generator. Explain in complete sentences how you arrive at your answers.

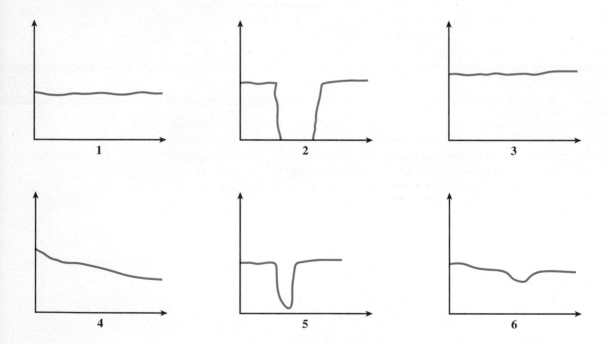

In Exercises 2 and 3, identify the independent variable and the dependent variable. Then interpret the situation being represented. Indicate whether the graph rises, falls, or is constant and whether the graph reaches either a minimum (smallest) or maximum (largest) output value.

2. Time required to complete a task in relation to number of times the task is attempted

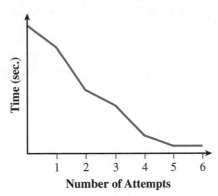

a. Independent: _____ Dependent: _____

b. Interpretation:

3. Number of units sold in relation to selling price

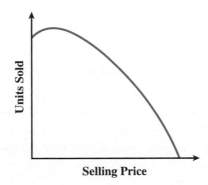

a. Independent: _____ Dependent: _____

b. Interpretation:

In Exercises 4 and 5, sketch a graph that represents the situation. Be sure to label your axes.

4. The sale price of a computer is a function of the percent of discount.

5. The area of a square is a function of the length of one side of the square.

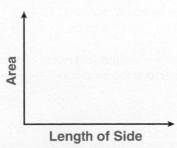

Cluster 1   What Have I Learned?

1. What is the mathematical definition of a function? Give a real-life example, and explain how this example satisfies the definition of a function.

2. Describe how you can tell from a graph when a function is increasing and when it is decreasing.

3. Identify four different ways that a function can be defined. Give an example of each.

4. Explain how to use a vertical line to determine if a graph represents a function, and explain why this works.

5. You know that the point $(1, 2)$ is on the graph of a function f. Give the coordinates of a point that you know is not on the graph of f. Explain.

6. Is it possible for an input/output relationship to represent a function if two different inputs produce the same output? Explain.

7. Is there ever a difference between the domain of a function and practical domain of a function if their symbolic representations are identical? Explain.

8. If $h(5) = 11$ in function h, what is the relationship between the numbers 5 and 11 and the graph of h?

9. If $f(x) = -3x + 2$, explain how to determine $f(4)$.

10. The notation $g(t)$ represents the weight (in grams) of a melting ice cube t minutes after being removed from the freezer. Interpret the meaning of $g(10) = 4$.

11. Use your newspaper to find at least four examples of functions and report your findings back to the class. For each example, you should do the following:

   a. Explain how the example satisfies the definition of a function.

   b. Describe how the function is defined (see Exercise 3).

   c. Identify the independent and dependent variables.

   d. Determine the domain of the function.

Cluster 1   How Can I Practice?

1. The following table shows the total number of points accumulated by each student and the numerical grade in the course.

STUDENT	TOTAL POINTS	NUMERICAL GRADE
Tom	432	86.4
Jen	394	78.8
Kathy	495	99
Michael	330	66
Brady	213	42.6

a. Is the numerical grade a function of the total number of points? Explain.

b. Is the total number of points for these five students a function of the numerical grade? Explain.

c. Using the total points as the input and the numerical grade as the output, write the ordered pairs that represent each student. Call this function f, and write it as a set of ordered pairs.

d. Plot the ordered pairs determined in part c on an appropriately scaled and labeled set of axes.

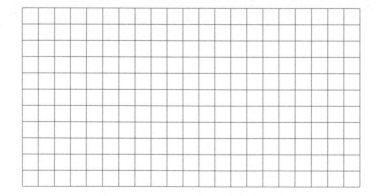

e. What is the value of $f(394)$?

f. What is the practical meaning of $f(394)$?

g. What is the value of $f(213)$?

h. What is the practical meaning of $f(213)$?

i. Determine the numerical value n, given that $f(n) = 66$.

n Exercises 2–10, determine which of the given relationships represent functions.

2. The money you earn at a fixed hourly rate is a function of the number of hours you work.

3. Your heart rate is a function of your level of activity.

4. The cost of daycare depends on the number of hours a child stays at the facility.

5. The number of children in a family is a function of the parents' last name.

6. $\{(2, 3), (4, 3), (5, -5)\}$

7. $\{(-3, 4), (-3, 6), (2, 6)\}$

8.

x	-3	5	7
$f(x)$	0	-5	9

9.

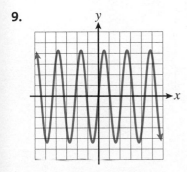

10.

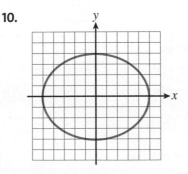

11. a. For a part-time student, the cost of college tuition is a function of the number of hours for which the student is registered. Write an equation to represent the tuition cost, c, if the cost per hour is $120 and h represents the number of hours taken for the current semester.

b. Use $f(h)$ to represent the cost, and rewrite the equation in part a using function notation.

c. Complete the table.

h	2	4	7	8	11
$f(h)$					

d. Evaluate $f(3)$, and write a sentence describing its meaning. Write the results as an ordered pair.

e. Given $f(h) = \$600$, determine the value of h.

f. Which variable is the output? Explain.

g. Which is the independent variable? Explain.

h. Explain (using the table in part c) how you know that the data represents a function.

i. What is the practical domain for this function?

j. Plot the ordered pairs on an appropriately scaled and labeled coordinate system. Which axis represents the input values?

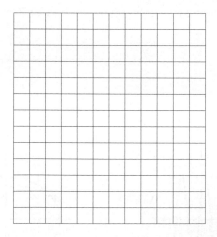

k. Explain from the graph how you know that f is a function.

l. Use your graphing calculator to verify your answers to parts c and j.

m. Use the trace and table features to determine the cost of 9 credit hours.

12. Given $p(x) = 2x + 7$, determine each of the following.

a. $p(3)$

b. $p(-4)$

c. $p\left(\dfrac{1}{2}\right)$

d. $p(0)$

13. Given $t(z) = 2z^2 - 3z - 5$, determine each of the following.

   a. $t(2)$ **b.** $t(-3)$

14. According to the U.S. Centers for Disease Control and Prevention, the average life expectancy from birth for males in the United States may be modeled by the function $f(x) = 0.18x + 64.8$, where x is the number of years since 1950.

   a. Use your calculator to complete the following table. Round your results to the nearest tenth.

YEAR					
1950	**1960**	**1970**	**1985**	**2000**	**2010**
x, Years Since 1950					
0	10	20	35	50	60
$f(x)$, Life Expectancy					

   b. Evaluate $f(35)$, and explain its practical meaning in this situation.

   c. Use the table values to set appropriate window values to view the graph of f. Graph the function on your calculator. Identify the window you used.

   d. Is the graph increasing, decreasing, or constant? Explain.

   e. Use the trace feature of your calculator to determine $f(35)$. Compare your answer to your result in part b.

15. Determine the domain and the range of each of the following functions.

   a. $\{(3, 5), (4, 5), (5, 8), (6, 10)\}$

   b.

CELLULAR PHONE CHARGES (MONTHLY)				
Number of Minutes Input	0	50	100	150	200
Monthly Cost (Dollars) Output	19.95	23.45	26.95	30.45	33.95

c.

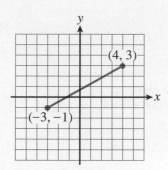

d.

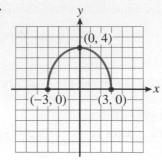

e. $y = 3x$

16. Each of the following graphs shows how the inputs and outputs change in relation to each other. Describe in words what the graph is telling you about the situation. Provide a reasonable explanation for the behavior you describe.

a.

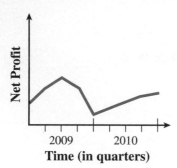

b.

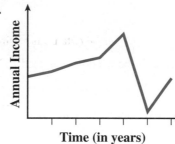

a.

b.

17. Sketch a graph that represents the fact that the hours of daylight depend upon the day of the year.

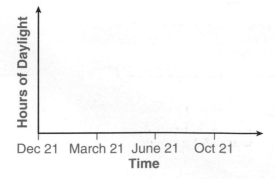

Cluster 2 Linear Functions

Activity 1.6

Walking for Fitness

Objective

1. Determine the average rate of change.

Suppose you are a member of a health-and-fitness club. A special diet and exercise program has been developed for you by your personal trainer. At the beginning of the program, and once a week thereafter, you are tested on the treadmill. The test consists of how many minutes it takes you to walk, jog, or run 3 miles on the treadmill. The following data gives your time, t, over an 8-week period.

End of Week, w	0	1	2	3	4	5	6	7	8
Time, t (in minutes)	45	42	40	39	38	38	37	39	36

Note that $w = 0$ corresponds to the first time on the treadmill, $w = 1$ is the end of the first week, $w = 2$ is the end of the second week, and so on.

1. a. Is time, t, a function of weeks, w? If so, what are the input and output variables?

b. Plot the data points using ordered pairs of the form (w, t).

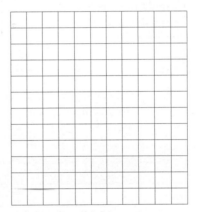

Definition

A set of points in the plane whose coordinate pairs represent input/output pairs of a data set is called a **scatterplot**.

Example 1 *The points plotted in Problem 1b are a scatterplot of the treadmill data. Your graphing calculator can generate a scatterplot of data points. Refer to Appendix C for instructions. The final screen should appear as follows:*

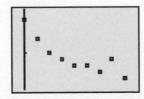

2. a. What was your treadmill time at the beginning of the program?

b. What was your treadmill time at the end of the first week?

An important question that can be asked about this situation is, how did your time change from one week to the next?

3. a. During which week(s) did your time increase?

b. During which week(s) did your time decrease?

c. During which week(s) did your time remain unchanged?

Procedure

Determining Total Change

The change in time, t, is represented by the symbol Δt. The symbol Δ (delta) is used to represent "change in." You generally calculate the change in time, t, from a first (initial) value to a second (final) value of t. The first time is represented by t_1 (read "t sub 1"), and the second time is represented by t_2 (read "t sub 2"). The change in t is then calculated by subtracting the first (initial) value from the second (final) value. This is symbolically represented by

$$\Delta t = t_2 - t_1 \quad \text{or} \quad \Delta t = \text{final time} - \text{initial time}.$$

Because t is the output variable, Δt is the change in output.

Similarly, Δw represents the change in weeks, w; w_1 represents the first (initial) value of w; and w_2 represents the second (final) value of w. Symbolically,

$$\Delta w = w_2 - w_1 \quad \text{or} \quad \Delta w = \text{final week} - \text{initial week}.$$

Because w is the input variable, Δw is the change in input.

---

4. Your time decreased during each week of the first 4 weeks of the program.

a. Determine the total change in time, t, during the first 4 weeks of the program (i.e., from $t = 45$ to $t = 38$). Why should your answer contain a negative sign? Explain.

b. Determine the change in weeks, w, during this period (i.e., from $w = 0$ to $w = 4$).

5. Use the Δ notation to express your results in Problems 4a and 4b.

3. The rate of change measures the change in output for a one-unit change in the input.

4. The line segment connecting the points (x_1, y_1) and (x_2, y_2)

   a. increases from left to right if $\dfrac{\Delta y}{\Delta x} > 0$

   b. decreases from left to right if $\dfrac{\Delta y}{\Delta x} < 0$

   c. remains constant if $\dfrac{\Delta y}{\Delta x} = 0$

5. The average rate of change indicates how much, and in which direction, the output changes when the input increases by a single unit. It measures how the output changes on average.

EXERCISES: ACTIVITY 1.6

The following table of data from the United States Bureau of the Census gives the median age of an American man at the time of his first marriage.

Year	1910	1920	1930	1940	1950	1960	1970	1980	1990	2000	2010
Median Age	25.1	24.6	24.3	24.3	22.8	22.8	23.2	24.7	26.1	26.8	27.5

Use this data to answer Exercises 1–6.

1. a. Determine the average rate of change in median age per year from 1950 to 2010.

   b. Describe what the average rate of change in part a represents in this situation.

2. Determine the average rate of change in median age per year from 1930 to 1960.

3. What is the average rate of change over the 100-year period described in the table?

4. During what 10-year period did the average age increase the most?

5. a. What does it mean in this situation if the average rate of change is negative?

   b. Determine at least one 10-year period when the average rate of change is negative.

   c. What trend would you observe in the graph of median age if the average rate of change were negative? That is, would the graph rise, fall, or remain horizontal?

Exercise numbers appearing in color are answered in the Selected Answers appendix.

6. a. Is the average rate of change zero over any 10-year period? If so, when?

b. What does a rate of change of zero mean in this situation?

c. What trend would you observe in the graph during this period? That is, would the graph rise, fall, or remain horizontal?

7. The following table gives information about new hotel construction in the United States from the fourth quarter of 2007 through the first quarter of 2009.

YEARLY QUARTER	2007 Q4	2008 Q1	2008 Q2	2008 Q3	2008 Q4	2009 Q1
Quarters Since 2007 Q4, t	0	1	2	3	4	5
Number of New Hotel Projects, h	439	459	381	403	289	257

a. Plot the data points using ordered pairs of the form (t, h).

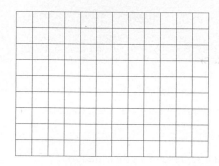

b. Determine the average rate of change of new hotel construction projects from 2007 Q4 to 2008 Q1.

c. Determine the average rate of change of new hotel construction projects from 2008 Q3 to 2009 Q1.

d. Compare the average rate of change from 2007 Q4 to 2008 Q1 with the rate of change from 2008 Q3 to 2009 Q1.

e. When the average rate of change is negative, what trend will you observe in the graph? What does that mean in this situation?

2. Determine whether the following tables contain data that represents a linear function. Assume that the first row of the table is the input and the second row is the output. Explain your reasoning.

a. You owe your grandmother $1000. She paid for your first semester at community college. The conditions of the loan are that you must pay her back the whole amount in one payment using a simple interest rate of 6% per year. She doesn't care in which year you pay her. The table contains input and output values to represent how much money you will owe your grandmother 1, 2, 3, or 4 years later.

Year	1	2	3	4
Amount Owed (in $)	1060	1120	1180	1240

b. You are driving along the highway, and just before you reach the crest of a hill, you notice a sign that indicates the elevation at the crest is 2250 feet. As you proceed down the hill, your elevation is as given by the following table.

Distance traveled from the crest of the hill, in feet	0	1000	3000	6000	10,000
Elevation, in feet	2250	2180	2040	1830	1550

c. You decide to invest $1000 of your 401k funds into an account that pays 5.5% interest compounded continuously. The table contains input and output values that represent the amount an initial investment of $1000 is worth at the end of each year.

Year	1	2	3	4	5
Total Investment (in $)	1057	1116	1179	1246	1317

d. For a fee of $20 per month you may have breakfast (all you can eat) in the college snack bar each day. The table contains input and output values that represent the total number of breakfasts consumed each month and the amount you pay each month.

Number of Breakfasts	10	22	16	13
Cost (in $)	20	20	20	20

3. You belong to a health-and-fitness center. You and your friends are enrolled in the center's weight-loss program. The charts contain input and output values (assume that week is input and weight is output) that represent the weight over a 4-week period for you and your two friends. Determine which charts contain data that is linear and explain why.

a.

Week	1	2	3	4
Weight (in lb.)	150	147	144	141

b.

Week	1	2	3	4
Weight (in lb.)	183	178	174	171

c.

Week	1	2	3	4
Weight (in lb.)	160	160	160	160

4. Consider the equation $y = -2x + 5$.

a. Construct a table of five ordered pairs that satisfy the equation.

x					
y					

b. What is the slope of the line represented by the equation?

c. What is the vertical intercept?

d. What is the horizontal intercept?

e. Sketch a graph of the line using each of the following methods:

Method 1:   Plot the five ordered pairs.

Method 2:   Plot one point and use the slope to obtain additional points on the line.

Method 3:   Plot the intercepts.

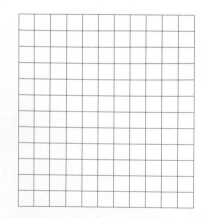
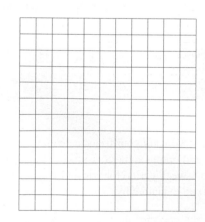
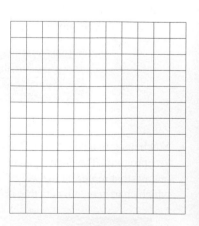

5. a. Determine the slope of the line that goes through the points $(2, -7)$ and $(0, 5)$.

b. Determine the vertical intercept of this line.

c. What is the equation of the line through these points? Write the equation in function notation.

d. What is the horizontal intercept?

6. A car is traveling on a highway. The distance (in miles) from its destination at time t (in hours) is given by the equation $d = 420 - 65t$.

a. What is the vertical intercept of the line?

b. What is the practical meaning of the vertical intercept?

c. What is the slope of the line represented by the equation?

d. What is the practical meaning of the slope determined in part c?

e. What is the horizontal intercept? What is the practical meaning of the horizontal intercept?

f. What is the practical domain of this function?

g. Graph the equation, both by hand and with your graphing calculator, to verify your answers.

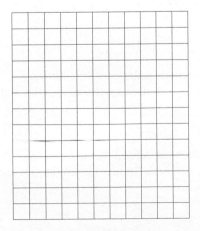

7. The following table gives a jet's height above the ground (in feet) as a function of time (in seconds) as the jet makes its landing approach to the runway.

t (IN SECONDS)	h (IN FEET)
0	3500
20	3000
40	2500
60	2000
80	1500
100	1000

a. Is this function linear? Explain.

b. Calculate the slope using the formula $m = \dfrac{\Delta h}{\Delta t}$.

c. What is the significance of the sign of the slope in part b?

d. Determine where the graph crosses the vertical axis.

e. Write the equation in slope-intercept form.

f. Determine the horizontal intercept. What is its significance in this situation?

8. Determine the horizontal intercept of the line whose equation is given.

a. $y = 4x + 2$

b. $y = \dfrac{x}{2} - 3$

9. a. Use your graphing calculator to graph the linear functions defined by the following equations: $y = 2x - 3$, $y = 2x$, $y = 2x + 2$, $y = 2x + 5$. Discuss the similarities and the differences of the graphs.

b. Use your graphing calculator to graph the linear functions defined by the following equations: $y = x - 2$, $y = 2x - 2$, $y = -x - 2$, $y = -2$. Discuss the similarities and the differences in the graphs.

c. Use your graphing calculator to graph the linear functions defined by the following equations: $y = 3x$, $y = -2x$, $y = \frac{1}{2}x$, $y = -5x$. Discuss the similarities and differences in the graphs.

Activity 1.8

A New Computer

Objectives

1. Write a linear equation in the slope-intercept form, given the initial value and the average rate of change.

2. Write a linear equation given two points, one of which is the vertical intercept.

3. Use the point-slope form to write a linear equation given two points, neither of which is the vertical intercept.

4. Compare slopes of parallel lines.

The new iBook computers are now available at a cost of $1200. There is a special promotion for students that allows you to make monthly payments for 2 years at 0% interest. You are required to make a 20% down payment, with the balance to be paid in 24 equal monthly payments. This is an opportunity to get the computer you really want, so you investigate to see if you can afford to take advantage of this promotion.

1. a. Determine the amount of the down payment required.

b. Determine the amount owed after the down payment.

c. What are the monthly payments?

A monthly payment of $40 is something you can afford, so you decide to go for it.

At any time during the next 2 years, the total amount, A, paid is a function of the number, n, of payments made.

2. a. What is the amount A when $n = 0$? What does this value of A represent in this situation?

b. What is the total amount paid, A, when $n = 1$ (after the first payment)?

c. What is the total amount paid, A, when $n = 2$?

d. Record the results from parts 2a through 2c to complete the following table.

PAYMENT NUMBER, n	AMOUNT PAID, A ($)
0	
1	
2	

e. What is the average rate of change from $n = 0$ to $n = 1$?

f. What is the average rate of change from $n = 1$ to $n = 2$?

g. What type of function is this? Explain.

To write an equation of a linear function in the slope-intercept form, $y = mx + b$, you need the values of the slope, m, and the vertical intercept, b. The slope is the average rate of change, which is constant for a linear function. The vertical intercept is the output when the input is zero. This output value is often referred to as the initial value.

3. a. What is the slope, m, of the payment function?

b. What is the initial value, b, of the payment function?

c. Write the linear equation that gives the amount paid, A, as a function of the number of payments made, n. Note that the equation will have the general form $A = mn + b$.

d. Graph the equation from part c.

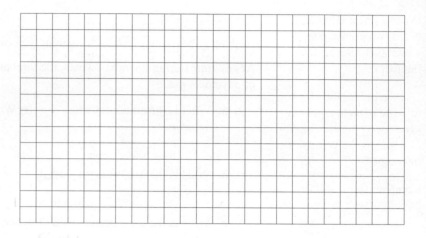

e. What is the slope of the line graphed in part b? What is the practical meaning of the slope in this situation?

f. What is the vertical intercept of the line graphed in part b (write the answer as an ordered pair)? What is the practical meaning of the vertical intercept in this situation?

4. The number of students at Columbia–Greene Community College (CGCC) was 1100 in 1985 and 1940 in 2009. The number of students increased at a constant rate over this 24-year period. Let t represent the number of years after 1985 and N represent the number of students at CGCC in a given year.

a. Use the given enrollment figures at CGCC to write two ordered pairs of the form (t, N). Note 1985 is $t = 0$.

b. Use the result from part a to complete the following table.

YEARS AFTER 1985, t	NUMBER OF STUDENTS, N
0	
24	

c. What is the initial value of this function? (Remember, the initial value is the output when the input is zero.)

c. Repeat Problems 7a and 7b using the ordered pair $(9, 1130)$ for (n_0, C_0).

Problem 7 demonstrates that you obtain the same equation of the line regardless of which point you use for (n_0, C_0).

8. Follow the steps below to write the point-slope form of the equation of the line containing the points $(2, 4)$ and $(5, -2)$.

    a. Use the given points to determine the slope, m.

    b. Use the point $(2, 4)$ for (x_0, y_0) to write the point-slope form of the equation.

Parallel Lines

Recall in Problems 1 through 3 you developed the equation $A = 40n + 240$, where A represented the amount paid toward the computer after n monthly payments. Suppose you are considering ordering the computer with extra memory (RAM). The cost for the memory is $160. You do not want to increase your monthly payments, so you want to pay for the extra memory by adding its cost to the down payment.

9. a. What will be the amount of the down payment if you add the cost of the memory?

    b. Write the linear equation that gives the amount paid, A, as a function of the number of payments made, n, with the new down payment amount. Remember the payment amount is unchanged at $40 per month.

    c. Graph the equation from part b on the grid from Problem 3b copied below.

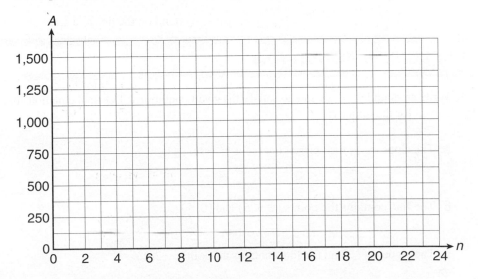

d. What can you say about the graphs of the two lines?

e. Compare the equations of the two lines. What is the same and what is different?

> ### Definition
> Two lines are **parallel** if the lines have equal slopes but different vertical intercepts.

10. Write an equation of a line parallel to the line whose equation is $y = 0.5x + 4$.

SUMMARY: ACTIVITY 1.8

1. To determine the equation of a line when two points on the line are known:

Step 1. Determine the **slope** of the line: $m = \dfrac{\Delta \text{ output}}{\Delta \text{ input}}$.

Step 2. If the input value of one of the points is zero, then b is the output value and you can write the equation in **slope-intercept form,** $y = mx + b$.

Step 3. If neither point has an input value of zero, then choose one of the points as (x_0, y_0) and write the equation in **point-slope form,** $y = y_0 + m(x - x_0)$.

2. Parallel lines have the same slopes with different y-intercepts.

EXERCISES: ACTIVITY 1.8

1. Write an equation of the line to satisfy the given conditions. The final equation should be solved for the ouput y.

a. The slope $= \dfrac{1}{2}$, the vertical intercept $= -1$.

b. The line contains the points $(-3, 5)$ and $(0, 1)$.

c. The slope is -3 and the line contains the point $(-2, 1)$.

d. The line contains the points $(-4, -3)$ and $(2, 6)$.

e. The line is parallel to the line $y = 3x - 7$ and passes through the point $(2, -5)$. Recall that parallel lines have the same slope.

2. The following table gives the cost, c, of a car rental as a function of miles driven, x.

x (IN MILES)	c (IN DOLLARS)
0	35
100	40

Assume that the function is linear and that you want to determine the equation of the line from the table.

a. What is the vertical intercept? How do you know this?

b. Calculate the slope $m = \dfrac{\Delta c}{\Delta x}$ from the data in the table. What is the practical meaning of this slope?

c. Use your results from parts a and b to write the equation of the line in slope-intercept form.

3. The following table gives the distance, d, in miles of a boat from a marina as a linear function of time, t, in hours.

t (IN HOURS)	d (IN MILES)
2	75
4	145

a. Determine the slope of the line. What is the practical meaning of slope in this situation?

b. Write the equation of the line in slope-intercept form.

4. For each graph below,

i. Determine the slope of the graph.

ii. Determine the y-intercept of each graph.

iii. Write the equation of the line whose graph is given.

Each tick mark denotes 1 unit.

a.

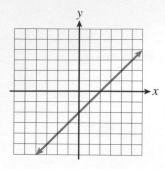

b.

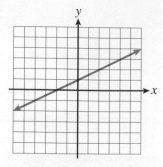

c.

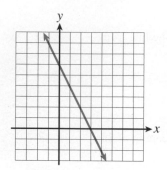

d.

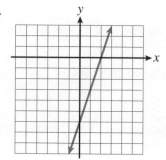

5. The following graph represents the distance from home of a car as a function of time (in hours).

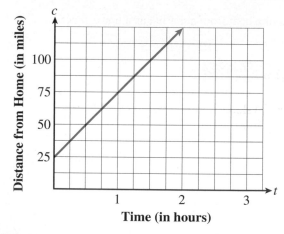

Time (in hours)

a. How fast is the car traveling?

b. Determine the vertical intercept.

c. Write the equation of the line in slope-intercept form.

6. a. Graph a line with a slope of 4 that goes through the point (1, 5). Write the equation of the line.

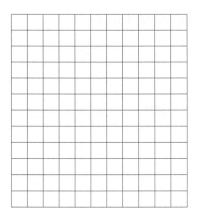

b. Graph a line with a slope of $-\frac{1}{2}$ that goes through the point $(-2, 3)$. Write the equation of the line.

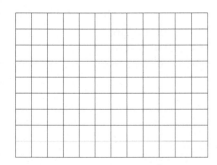

7. The data in the following table shows the circumference of a circle as a function of its radius.

r (radius in feet)	0	5	10	15
C (circumference in feet)	0	31.42	62.83	94.25

a. Assuming that this function is linear, determine the slope to the nearest hundredth place.

b. From the table, determine the vertical intercept.

c. Write the equation of the linear function.

d. What is the formula of the circumference of a circle?

e. Does your equation approximate this formula? Explain.

8. You own a kayak company and open only during the summer months. You discover that if you sell a certain type of kayak for $400, your sales per day average $5200. If you raise the price of the kayak to $450, the sales fall to approximately $3600 per day.

   a. Assume that the sales per day is a function of the price of the kayak. Write two ordered pairs that describe this situation.

   b. Assume that the sales per day is a linear function of the price of the kayak. Write an equation describing this relationship.

   c. You cannot make enough profit if you sell the kayak for less than $375. What would be the average sales per day if you change the price to $375?

9. Your architect will charge you a flat fee of $2000 for the plans for your home. The cost of your home is estimated by the square footage. The following table gives the total estimated cost of your home, c, including the architect's fees, as a function of the square footage, h. Assume that the total cost is a linear function of square footage.

TOTAL SQUARE FEET, h	TOTAL COST, c ($)
0	2000
3000	311,000

   a. What is the vertical intercept of the line containing these points? Explain how you determined this intercept.

   b. Using the data in the table and the formula $m = \dfrac{\Delta c}{\Delta h}$, calculate the slope.

   What is the practical meaning of the slope in this situation?

   c. Use the results from parts a and b to write the equation of the line in slope-intercept form that can be used to determine the cost for any given square footage.

   d. You decide that you cannot afford a house with 3000 square feet. Using the equation from part c, determine the cost of your home if you decrease its size to 2500 square feet.

    c. Describe the graph in words.

    d. Choose two ordered pairs from part a and determine the slope, m, of the line. What is the practical meaning of the slope in this situation?

    e. What are the intercepts (vertical and horizontal) of the line (if they exist)?

    f. Write the equation of this line in slope-intercept form, $y = mx + b$.

Definition

A graph in which the output y is a constant or, equivalently, $f(x)$ is a constant, is a **horizontal line**. The equation of a horizontal line is $y = c$ or $f(x) = c$, where c is some fixed real number. The slope of any horizontal line is zero.

---

7. Determine three ordered pairs that satisfy each of the following equations, and then sketch each graph on the same coordinate axes.

    a. $y = -2$ **b.** $f(x) = 1$

    c. $g(x) = \frac{5}{2}$

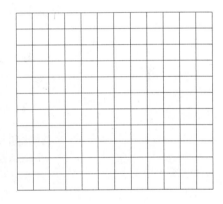

Equation of a Vertical Line

8. You apply for a part-time job at the skateboarding rink to help cover your weekly expenses while going to college. The following table gives your weekly salary, x, and corresponding weekly expenses, y, for a typical month.

x, Weekly Salary in Dollars	70	70	70	70
y, Weekly Expenses in Dollars	45	35	50	60

a. Sketch a graph of the given data points.

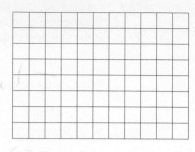

b. Describe the graph in words.

c. Choose two ordered pairs from the table and determine the slope, m.

d. What are the intercepts (vertical and horizontal) of the line (if they exist)?

e. Is y a function of x? Explain using the definition of a function.

f. Do the ordered pairs in the table satisfy the equation $x = 70$? Explain why or why not.

g. Can you graph the equation $x = 70$ using your graphing calculator? Explain.

Definition

A graph in which x is a constant is a **vertical line**. The equation of a vertical line is $x = a$, where a is some fixed real number. The slope of a vertical line is undefined.

---

9. Determine three ordered pairs that satisfy each of the following equations, and then sketch each graph.

    a. $x = -2$ **b.** $x = 4$ **c.** $x = \dfrac{5}{2}$

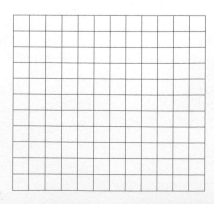

SUMMARY: ACTIVITY 1.9

1. The **standard form** of a linear equation is $Ax + By = C$, where A, B, and C are constants, and A and B are not both zero.

2. The graph of $y = c$ or $f(x) = c$ is a **horizontal line**. In this case, f is called a constant function. Every point on this line has a y-coordinate equal to c. A horizontal line has slope of zero.

3. The graph of $x = a$ is a **vertical line**. Every point on this line has an x-coordinate equal to a. The slope of a vertical line is undefined.

EXERCISES: ACTIVITY 1.9

1. Write the following linear equations in slope-intercept form. Determine the slope and vertical intercept of each line.

   a. $2x - y = 3$

   b. $x + y = -2$

   c. $2x - 3y = 7$

   d. $-x + 2y = 4$

   e. $0x + 3y = 12$

2. a. Sketch the graph of the horizontal line through the point $(-2, 3)$.

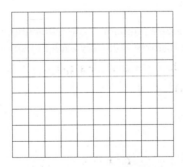

   b. Write the equation of a horizontal line through the point $(-2, 3)$.

   c. What is the slope of the line?

   d. What are the vertical and horizontal intercepts of the line?

   e. Does the graph represent a function? Explain.

3. a. Sketch the graph of the vertical line through $(-2, 3)$.

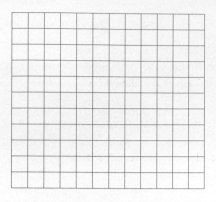

b. Does the graph represent a function? Explain.

c. Write the equation of a vertical line through the point $(-2, 3)$.

d. What is the slope of the line?

e. What are the vertical and horizontal intercepts of the line?

4. Explain the difference between a line with a zero slope and a line with an undefined slope.

5. You are retained as a consultant for a major computer company. You receive $2000 per month as a fee no matter how many hours you work.

a. Using x to represent the number of hours you work each month, write a function, f, in symbolic form to represent the total amount received from the company each month.

b. Complete the following table of values.

Hours Worked per Month	15	25	35
Fee per Month (in $)			

c. Use your graphing calculator to sketch the graph of this function.

d. Use your equation from part c to predict the value of y when $x = 10$.

e. Use your equation from part c to predict the value of y when $x = 25$.

f. Which prediction, $f(10)$ or $f(25)$, would be more accurate? Explain.

2. Public debt in your state increased at a relatively constant rate from 2002 to 2007. The following table gives the average debt per capita (in thousands of dollars) for years since 2002, where $t = 0$ corresponds to the year 2002.

t (in years since 2002)	0	1	2	3	4	5
d (in thousands of dollars)	8.77	10.53	13.00	15.85	18.02	19.81

a. Plot the data.

b. Use your graphing calculator to determine the equation of the regression line. Write the result here.

c. What is the slope of the line? What is the practical meaning of *slope* in this situation?

d. Use your regression line to determine the average debt of an individual in 2004 $(x = 2)$. Compare your result with the actual value of $13,000 in 2004.

e. Use the regression equation to predict the average debt of an individual in 2015.

f. What is the process called that you used to make the prediction in part e?

3. In 1966, the U.S. Surgeon General's health warnings began appearing on cigarette packages. The following data seems to demonstrate that public awareness of the health hazards of smoking has had some effect on consumption of cigarettes.

	YEAR							
	1997	1998	1999	2000	2001	2002	2003	2004
% of Total Population 18 and Older Who Smoke	24.7	24.1	23.5	23.2	22.7	22.4	21.6	20.9

Source: U.S. National Center for Health Statistics.

a. Plot the given data as ordered pairs of the form (t, P), where t is the number of years since 1997 and P is the percent of the total population (18 and older) who smoke. Appropriately scale and label the coordinate axes.

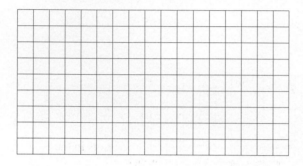

b. Determine the equation of the regression line that best represents the data.

c. Use the equation to predict the percent of the total population 18 and older that will smoke in 2010.

4. The number of Internet users in the United States has increased steadily over the past several years, as indicated in the following table.

YEAR	NUMBER OF INTERNET USERS IN U.S. (in millions)
2000	121
2001	127
2002	140
2003	146
2004	156
2005	163

a. Plot the data points on appropriately scaled and labeled coordinate axes. Let x represent the number of years since 2000.

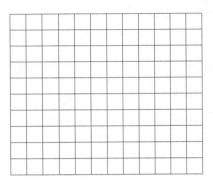

b. Use your graphing calculator's statistics menu (STAT) to determine the equation of the regression line.

c. What is the slope of the line in part b? What is the practical meaning of the slope?

d. Use the linear model from part b to predict when the number of Internet users in the U.S. will reach 200 million. How confident are you in this prediction?

Collecting and Analyzing Data

5. One measure of how well the American economy is doing is the number of new houses built in a given year. Your economics teacher has asked the class to investigate trends in new home construction in recent years and to predict the number of new homes that will be constructed in 2020.

Use the U.S. Bureau of the Census[*] as a resource to obtain the latest data available on new home construction from 2000 to 2007. Work with a group and be prepared to give an oral presentation of your findings to the class. The presentation should include visual displays showing any tables, scatterplots, regression equations, and graphs used in the problem-solving process.

[*]The U.S. Bureau of the Census gathers large amounts of data about the United States and its population. This information is published in the Statistical Abstract of the United States. The publication is available on the Internet as well as in print form at most libraries. This is an excellent source of data on variety of topics.

Each member of the group should submit a report of the group's findings. The report should contain the following information:

a. The source(s) of the data.

b. A description of the dependent and independent variables.

c. A table and scatterplot of the data.

d. Identification of any patterns in the scatterplot.

e. A linear regression equation and correlation coefficient for the data.

f. A prediction of the number of new homes to be constructed 2020 and a description of the level of confidence in this prediction.

g. State a conclusion of the data analysis process.

Cluster 2   What Have I Learned?

1. Describe how you recognize that a function is linear when it is given

  a. graphically.

  b. as an equation involving x and y.

  c. numerically in a table.

2. The coordinates of all points (x, y) on the line with slope m and vertical intercept $(0, b)$ satisfy the equation

$$y = mx + b.$$

  a. The point $(2, 3)$ is on the line whose equation is $y = 4x - 5$. Show how the coordinates of the point $(2, 3)$ satisfy the equation $y = 4x - 5$.

  b. Determine another point on the line whose coordinates satisfy the equation $y = 4x - 5$.

3. Given an input/output table that represents a function, how can you determine if the function is linear?

4. Write a procedure for determining the horizontal intercept of a graph, given its equation.

5. Consider two linear functions $f(x) = m_1 x + b_1$ and $g(x) = m_2 x + b_2$, whose graphs are distinct and parallel. What can you say about m_1 and m_2 and about b_1 and b_2? Explain.

6. If the graph of $Ax + By = C$ is a vertical line, what can you conclude about the values of A and B? Explain.

Cluster 2   How Can I Practice?

1. Identify the independent and dependent variables, and write a linear equation model for each of the following situations. Then give the practical meaning of the slope and vertical intercept in each situation.

   a. You make a down payment of $50 and pay $10 per month for your new computer.

   b. You pay $16,000 for a new car whose value decreases by $1500 each year.

2. Match each of the following functions or equations with its corresponding graph below. Use your graphing calculator for checking purposes only. Each unit equals 1.

   $f(x) = -2x + 3$ $g(x) = 2x - 3$ $y = 2$
   $h(x) = -2x - 3$ $y = -2x$ $x = 2$

   a.
   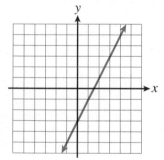

   b.
   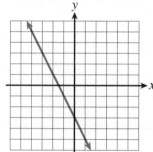

   c.
   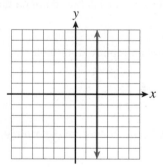

   d.
   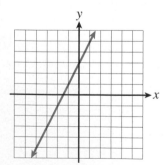

   e.
   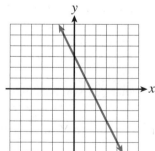

   f.
   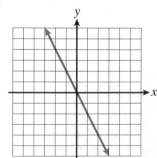

g.

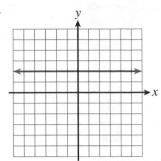

h.

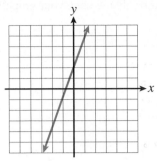

i.

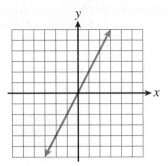

3. The cost of renting a graphing calculator from the bookstore is $20, plus $4 per month for as long as you wish to rent.

  a. Complete the table.

Months, *m*	2	5	8	10	12
Cost In $, *c*					

  b. Is the cost, *c*, a linear function of months, *m*?

  c. What is the slope?

  d. Write the equation for the function.

  e. Graph this function and compare the results from your graphing calculator.

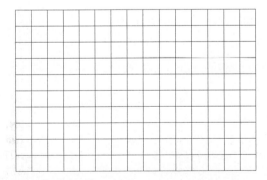

  f. What is the practical meaning of the slope in this situation?

  g. What is the vertical intercept of the graph of the function? What is the practical meaning of the vertical intercept in this situation?

  h. What is the horizontal intercept of the graph of the function? What practical meaning does this have in this situation?

    i. Approximately how many months will you be able to keep your graphing calculator if you have $65 budgeted for this expense?

4. Given the table of values:

t	10	20	40
s(t)	0	15	45

    a. Determine the average rate of change from $t = 10$ to $t = 20$.

    b. Determine the average rate of change from $t = 20$ to $t = 40$.

    c. Is s a linear function? Explain.

5. Determine the slope of the line through the points $(3, 8)$ and $(-5, 12)$.

6. Determine the slope of the line represented by the equation $y = -4x + 2$.

7. Determine the slope of the line having equation $2x - 5y = 9$.

8. Write an equation of a line with a slope of -7 and a vertical intercept of 4.

9. Write an equation of a line that has a slope of 2 and goes through the point $(0, 10)$.

10. Write an equation of a line that has a slope of 0 and goes through the point $(-4, 5)$.

11. Write an equation of a vertical line that goes through the point $(-3, -5)$.

12. Write an equation of a line that goes through the points $(2, -3)$ and $(-4, 0)$.

13. Write an equation of a line parallel to $y = \frac{1}{3}x - 7$ and through the point $(6, -1)$.

14. Sketch a graph of the line through $(2, -3)$ with slope of $\frac{1}{2}$.

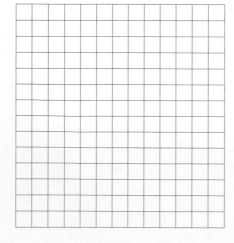

Exercises 15–17, graph each of the functions by hand. Then compare your results using your graph-
ing calculator.

15. $y = \frac{4}{5}x - 3$ **16.** $4x - 5y = 20$ **17.** $y = -4$

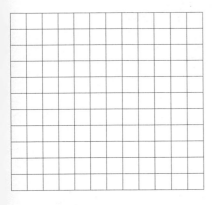   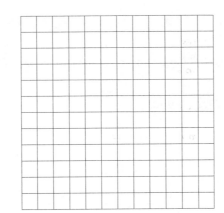   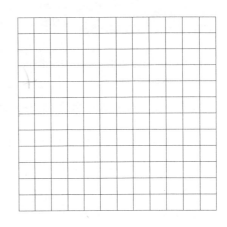

18. a. Plot the data on the following grid using an appropriate scale.

x	0	2	4	6	8	10	12
y	23.76	24.78	25.93	26.24	26.93	27.04	27.93

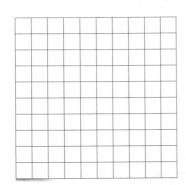

b. Use your graphing calculator to determine the equation of the regression line. Write the result.

c. Use the result from part b to predict the value of y when $x = 9$.

d. Use the result from part b to predict the value of y when $x = 20$.

19. The following table shows the percentage of elected female state executives from the years 1977 to 2004, as tabulated by the Center for American Women and Politics.

Year	1977	1981	1985	1989	1993	1997	2001	2004
Percentage of Elected Women Executives	9.9	10.5	13.3	14.3	22.2	25.4	27.6	25.4

Let x represent the number of years since 1977 and y represent the percentage of women state executives.

a. Plot the data from the table on the following grid.

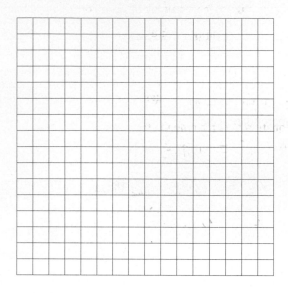

b. Use your graphing calculator to produce the linear regression equation for the data in the graph, and record it here.

c. Use the equation from part b to estimate the percentage of women state executives in 1986 and 2010.

d. In which of the values found in part c do you have more confidence? Explain.

oblem?

ces, called
ndle contains
is modeled by

ct. Write an
of pavers.

n the practical

he practical mean-

will break even
nt from the graph.
of the break-even

---

c. $y = 5x - 3$

$y = 5x + 7$

Numerically

x	y_1	y_2
0		
1		
2		
3		
4		

Graphically

Algebraically (substitution method)

2. Two companies sell software products. In 2010, company A had total sales of $17.2 million. Its marketing department projects that sales will increase by $1.5 million per year for the next several years. Company B had total sales of $9.6 million of software products in 2010 and projects that its sales will increase by $2.3 million each year.

Let n represent the number of years since 2010.

a. Write an equation that represents the total sales, s, of company A since 2010.

b. Write an equation that represents the total sales, s, of company B since 2010.

c. The two equations in parts a and b form a system. Solve this system to determine the year in which the total sales of both companies will be the same.

3. You are considering installing a security system in your new house. You gather the following information from two local home security dealers for similar security systems:

Dealer 1: $3560 to install and $15 per month monitoring fee

Dealer 2: $2850 to install and $28 per month monitoring fee

Although the initial fee of dealer 1 is much higher than that of dealer 2, dealer 1's monitoring fee is lower.

Let n represent the number of months you have the security system.

a. Write an equation that represents the total cost, c, of the system with dealer 1.

b. Write an equation that represents the total cost, c, of the system with dealer 2.

---

c. Try to solve the system algebraically using the substitution method. What type of equation do you obtain?

The linear system in Problem 9 is said to be **inconsistent**. There is no solution because the lines never intersect. Graphically, the slopes of the lines are equal, but the vertical intercepts are different. Therefore, the graphs are parallel lines. Solving such an equation algebraically results in a false equation such as $30 = 0$.

10. a. You and your friend are taking one more trip. This time she does not have a head start. You both leave from your house, both travel in the same direction, and both travel at 60 miles per hour. When will you both be at the same point?

b. The system for this situation is

$$d = 60t$$
$$d = 60t.$$

Try to solve this system graphically. Do the lines intersect? What is the solution to the system?

c. Try to solve this system algebraically. What type of equation do you obtain?

The system in Problem 10 is an example of a **dependent** system. Graphically, in such a system, both equations represent the same line. The system has an infinite number of solutions. Solving a dependent system algebraically results in an equation that is true for all t, such as $0 = 0$.

SUMMARY: ACTIVITY 1.11

1. A 2×2 system of linear equations consists of two equations with two variables. The graph of each equation is a line.

2. A **solution** to a 2×2 system of equations is an ordered pair that satisfies both equations of the system.

3. Solutions can be found in three different ways:

- **numerically**, by examining tables of values for both functions.

- **graphically**, by graphing each equation and finding the point of intersection.

- **algebraically**, by combining the two equations to form a single equation in one variable, which can then be solved. This is called the substitution method.

4. A linear system is **consistent** if there is at least one solution, the points of intersection of the graphs.

5. A linear system is **inconsistent** if there is no solution; the lines are parallel.

6. A linear system is **dependent** if there are infinitely many solutions. The equations represent the same line.

EXERCISES: ACTIVITY 1.11

Solve the following systems of linear equations numerically, graphically, and algebraically (substitution method).

1. a. $y = 2x + 3$

$y = -x + 6$

Numerically

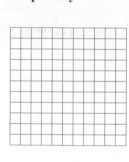

x	y_1	y_2
−2		
−1		
0		
1		
2		
3		

Graphically

Algebraically (substitution method)

b. $y = 0.5x + 9$

$y = 4.5x + 17$

Numerically

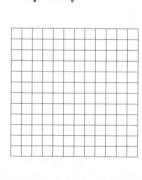

x	y_1	y_2
−2		
−1		
0		
1		
2		
3		

Graphically

Algebraically (substitution method)

c. Solve the system of equations that results from parts a and b to determine the number of months for which the total cost of the systems will be equal.

d. If you plan to live in the house and use the system for 10 years, which system would be less expensive?

4. You can run a 400-meter race at an average rate of 6 meters per second. Your friend can run the race at a rate of 5 meters per second. You give your friend a 40-meter head start. She then runs 360 meters.

a. Write an equation for your distance, in meters, from the starting point as a function of time, in seconds.

b. Write an equation for your friend's distance, in meters, from the starting point as a function of time, in seconds.

c. How long does it take you to catch up with your friend?

d. How far from the finish line do you meet?

5. For many years, the life expectancy for women has been longer than the life expectancy for men. In the past few years, the life expectancy for men has been increasing at a faster rate than that for women. Using data from the Centers for Disease Control and Prevention and the National Center for Health Statistics, the life expectancy, E, for men and women in the United States can be modeled by

Women: $E = 0.115x + 77.42$

Men: $E = 0.212x + 69.80$;

where t represents the number of years since 1980.

a. Solve the system numerically by completing the following table. Approximate the value of t to the nearest year. Use your calculator.

t, NUMBER OF YEARS SINCE 1980	LIFE EXPECTANCY FOR WOMEN	LIFE EXPECTANCY FOR MEN
0		
25		

e. Determine the exact break-even point imate your answer from the graph in p

f. How many bundles of pavers does th

g. What is the total cost to the company values are equal at the break-even po

h. For what values of x will your reven

i. As manager, what factors do you ha

j. If you knew you could sell only 30 much it would cost you and how m

Collecting and Analyzing

7. The women's world record time in th men's record times. A similar occure women's Olympic 500-meter speed s

Choose one of these competitions a event and for the men's event. Use t equal the men's record time.

Work in a group and be prepared to include visual displays showing the t process. Be sure to identify the sour

Healthy Lifestyle

Objectives

1. Solve a 2 × 2 linear system algebraically using the substitution method and the addition method.

2. Solve equations containing parentheses.

You are trying to maintain a healthy lifestyle. You eat a well-balanced diet and follow a regular schedule of exercise. One of your favorite exercise activities is a combination of walking and jogging in your nearby park.

On one particular day, it takes you 1.3 hours to walk and jog a total of 5.5 miles in the park. You are curious about the amount of time you spent walking and the amount of time you spent jogging during the workout.

Let x represent the time you walked and y represent the time you jogged.

1. Write an equation, using x and y, for the total time of your walk/jog workout in the park.

2. a. If you walk at 3 miles per hour, write an expression that represents the distance you walked.

b. If you jog at 5 miles per hour, write an expression that represents the distance you jogged.

c. Write an equation for the total distance you walked/jogged in the park.

The situation just described can be represented by the following system.

$$x + y = 1.3$$
$$3x + 5y = 5.5$$

Note that each equation in this system is in standard form. One approach to solving this system is to solve each equation for one variable in terms of the other and then use the substitution method.

3. a. Solve each of the equations in the system above for y.

b. Solve the system in part a using the substitution method.

c. Check your answer graphically using your graphing calculator. You may want to use the window Xmin = −2.5, Xmax = 2.5, Ymin = −2.5, and Ymax = 2.5.

Addition Method

Sometimes it is more convenient to leave each equation in the linear system in standard form ($Ax + By = C$) rather than solving for one variable in terms of the other. Look again at the original system.

$$x + y = 1.3 \qquad (1)$$
$$3x + 5y = 5.5 \qquad (2)$$

If you apply the addition principle of algebra by adding the two equations (left side to left side and right side to right side), you may be able to obtain a single equation containing only one variable.

$$x + y = 1.3 \qquad (1)$$
$$\underline{3x + 5y = 5.5} \qquad (2)$$
$$4x + 6y = 6.8$$

In this case, adding the equations does not eliminate a variable. But if you multiply both sides of equation (1) by -5, the coefficients of y will be opposite, and the variable y can be eliminated.

$$-5(x + y) = -5(1.3) \longrightarrow -5x - 5y = -6.5 \qquad \text{Add the corresponding sides of the}$$
$$3x + 5y = 5.5 \qquad\qquad \underline{3x + 5y = 5.5} \qquad \text{equation.}$$
$$-2x + 0 \;\; = -1$$

Solving the resulting equation, $-2x = -1$, for x, you have

$$-2x = -1$$
$$x = \tfrac{1}{2} = 0.5.$$

Substituting for x in $x + y = 1.3$, you have

$$0.5 + y = 1.3$$
$$\text{or } y = 0.8.$$

This method of solving systems algebraically is called the **addition method**.

> To solve a 2 × 2 linear system by the addition method,
>
> **Step 1.** line up the like terms in each equation vertically;
>
> **Step 2.** if necessary, multiply one or both equations by constants so that the coefficients of one of the variables are opposite;
>
> **Step 3.** add the corresponding sides of the two equations.

4. Solve the following system again using the addition method. Multiply the appropriate equation by the appropriate factor to eliminate x and solve for y first.

$$x + y = 1.3$$
$$3x + 5y = 5.5$$

Not all systems will have convenient coefficients, and you may need to multiply one or both equations by a factor that will produce coefficients of the same variable that are additive inverses, or opposites.

5. Consider solving the following system with the addition method.

$$-2x + 5y = -16$$
$$3x + 2y = 5$$

a. Identify which variable you wish to eliminate. Multiply the appropriate equation by the appropriate factor so that the coefficients of your chosen variable are opposite.

Show the two equations after you multiply by the factor. (Remember to multiply both sides of the equation by the factor.)

b. Add the two equations to eliminate the chosen variable.

c. Solve the resulting linear equation.

d. Determine the complete solution. Remember to check by substituting into both of the original equations.

Substitution Method Revisited

The substitution method of solving a 2×2 system of linear equations is generally used when each equation in the system is solved for one variable in terms of the other. However, it can be convenient to use the substitution method when only one of the equations is solved for a variable.

Note: As you solve a system using the substitution method, you will encounter an equation involving parentheses. It would be an equation like

$$2(3x - 7) + 9 = -x + 2.$$

To solve this equation, you first simplify the expression on the left-hand side.

$$2(3x - 7) + 9 = -x + 2$$
$$6x - 14 + 9 = -x + 2$$
$$6x - 5 = -x + 2$$

Using the addition principle to move all x terms to one side,

$$6x - 5 = -x + 2$$
$$\underline{+x \qquad\quad +x}$$
$$7x - 5 = \qquad 2.$$

Using the addition principle again to isolate the x term,

$$7x - 5 = \quad 2$$
$$\underline{+ 5 \quad\; + 5}$$
$$7x \quad = \quad 7$$
$$x = 1.$$

Example 1   *In the walk/jog system,*

$$x + y = 1.3 \qquad \textbf{(1)}$$
$$3x + 5y = 5.5 \qquad \textbf{(2)}$$

you could solve equation (1) for y and then substitute for y in equation (2) as follows.

Step 1. Solve $x + y = 1.3$ for y.

$$y = 1.3 - x$$

Step 2. Substitute $1.3 - x$ for y in equation (2).

$$3x + 5(1.3 - x) = 5.5$$

Step 3. Solve the resulting equation for x.

$3x + 5(1.3 - x) = 5.5$	Remove the parentheses by applying the distributive property.
$3x + 6.5 - 5x = 5.5$	Collect like terms on the same side.
$-2x + 6.5 = 5.5$	
$\underline{\quad -6.5 \quad -6.5 \quad}$	Add the opposite of 6.5 to both sides.
$-2x = -1$	Divide each side by -2.
$\frac{-2x}{-2} = \frac{-1}{-2}$	
$x = 0.5$	

Step 4. From equation (1), $y = 0.8$ as before.

6. a. Solve the following linear system using the substitution method in which you solve only one of the equations for a variable.

$$x - y = 5$$
$$4x + 5y = -7$$

b. Check your answer in part a by solving the system using the addition method.

SUMMARY: ACTIVITY 1.12

1. There are two methods for solving a 2×2 system of linear equations algebraically:

   a. the substitution method and

   b. the addition method.

2. To solve linear systems by **substitution**,

   Step 1. Solve one or both equations for a variable;

   Step 2. Substitute the expression that represents the variable in one equation for that variable in the other equation;

   Step 3. Solve the resulting equation for the remaining variable;

   Step 4. Substitute the value from step 3 into one of the original equations, and solve for the other variable.

3. To solve linear systems by **addition** with equations written in the form $Ax + By = C$,

   Step 1. Multiply one equation or both equations by the number(s) that will make the coefficients of one of the variables opposites;

   Step 2. Add the two equations to eliminate one variable and solve the resulting equation;

   Step 3. Substitute the value from step 2 into one of the original equations, and solve for the other variable.

EXERCISES: ACTIVITY 1.12

1. Solve each of the following equations.

   a. $5(x + 3) + 4 = 6x - 1 - 5x$

   b. $-3(1 - 2x) - 3(x - 4) = -5 - 4x$

   c. $2(x + 3) - 4x = 5x + 2$

2. Solve the following systems algebraically using the substitution method.

a. $y = 3x + 1$

$\quad y = 6x - 0.5$

b. $y = 3x + 7$

$\quad 2x - 5y = 4$

c. $2x + 3y = 5$

$\quad -2x + y = -9$

d. $4x + y = 10$

$\quad 2x + 3y = -5$

3. Solve the following systems algebraically using the addition method.

a. $y = 2x + 3$

$\quad y = -x + 6$

b. $2x + y = 1$

$\quad -x + y = -5$

4. Solve the system both graphically and algebraically.

$$3x + y = -18$$
$$5x - 2y = -8$$

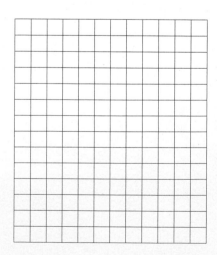

5. A catering service placed an order for eight centerpieces and five glasses, and the bill was $106. For the wedding reception, they were short one centerpiece and six glasses and had to reorder. This order came to $24. Let x represent the cost of one centerpiece, and let y represent the cost of one glass.

a. Write a system of equations that represents both orders.

b. Solve the system using the substitution method. Interpret your solution.

c. Check your result in part b using the addition method.

d. Use your graphing calculator to solve the system.

Activity 1.13

Manufacturing Cell Phones

Objective

1. Solve a 3 × 3 linear system of equations.

You recently started your own company to manufacture and sell your latest improvements i the cell phone market. You have decided to start with three models: a *basic model* for thos people with not a lot of disposable income, a *standard model* that has most of the modern ap plications built in, and a *deluxe version* that has all of the latest technology and is expandable.

You have hired and trained your employees to perform all of the basis tasks: the assembly the testing, and the packaging of each phone. You have determined that you have a total of 260 hours for assembly during the week, 170 hours available for testing, and 120 hours avail able for packing your phones.

The time allotted for each task on each type of phone is summarized in the following table.

	BASIC MODEL	STANDARD MODEL	DELUXE MODEL
Assembly	1 hour	3 hours	4 hours
Testing	1 hour	2 hours	2 hours
Packaging	1 hour	1 hour	2 hours

Use x, y, and z to represent the number of each type of phone you are to build each week, with x the number of basic phones, y the number of standard phones, and z the number of deluxe phones.

1. Write an equation for the total hours spent on the assembly of the phones each week.

$$x + 3y + 4z = 260$$

2. Write an equation for the total hours spent on testing phones each week.

$$x + 2y + 2z = 170$$

3. Write an equation for the total hours spent on packaging phones each week.

$$x + y + 2z = 120$$

Taken together, the equations in Problems 1–3 form a **3 × 3 system of linear equations.**

The solution to this system is the ordered triple of numbers (x, y, z) that satisfies all three equations. The strategy for solving such a system is typically to reduce the system to a 2 × 2 linear system and then proceed to solve this smaller system.

4. Select two equations from Problems 1–3 and use substitution or addition to eliminate one of the variables.

$$\begin{aligned} &① \quad x + 3y + 4z = 260 \\ &② \quad x + 2y + 2z = 170 \\ &\overline{\quad 2x + 5y + 6z = 430} \end{aligned}$$

5. Select a different pair of equations from Problems 1–3 and eliminate the same variable.

6. The equations from Problems 4 and 5 form a 2 × 2 system. Now solve this new 2 × 2 system.

7. Substitute the solutions from Problem 6 into one of the original three equations. Now solve for the third varible.

8. How many of each type of phone should you manufacture each week to make optimal use of your available times? Check to make sure your solution agrees with each of the three original assumptions.

9. Explain why it is not possible to solve this 3 × 3 system by graphing on your calculator.

All of the equations in the 3 × 3 systems in this activity are called *linear equations*, even though they cannot all be graphed as single lines. In this case, linearity refers to each variable being linear, that is, raised to the first power.

10. Solve the following 3 × 3 linear system.

$$x - 2y + z = -5 \qquad (1)$$
$$2x + y - z = 6 \qquad (2)$$
$$3x + 3y - z = 11 \qquad (3)$$

If you are not sure where to start, follow these steps.

Step 1. Is it possible to add two of the equations (right side to right side and left side to left side) so that one of the variables is eliminated? (Add equation (1) to equation (2).)

Step 2. Is it possible to add a different pair of equations to eliminate the same variable? (Add equation (1) to equation (3).)

Step 3. Notice that your equations from parts a and b form a 2 × 2 linear system. Solve this 2 × 2 system.

Step 4. Substitute your solution from step 3 into any one of the three original equations, and solve the resulting equation for the remaining variable.

Step 5. The final step is to substitute your potential solution into each of the three original equations. This is the only way you can be confident that your solution is correct.

Check: Equation 1:

        Equation 2:

        Equation 3:

Most 3 × 3 systems will not have coefficients that are quite so convenient as the ones you just encountered. The following application provides a case in point.

11. In your job as buyer for Sam's Café, a nationwide coffee bar, you need to buy three grades of coffee bean, to be blended with various flavors to make Sam's well-known coffee drinks. This week the three grades of beans are selling for $0.80, $1.20, and $1.80 per pound. Determine the equation that corresponds to each of the following assumptions.

   a. The total weight of beans needed is 11,400 pounds.

   $$11,400 = x + y + z$$
   $$T = .8x + 1.2y + 1.8z$$

   b. You can only spend $13,010.

   $$13010 = .8x + 1.2y + 1.8z$$

   c. You need 500 more pounds of the least expensive grade than the other two grades combined.

   $$11400 = x + y + z + 500$$

12. Solve the system you determined in Problem 11. State your solution in terms of the application. Verify that all three assumptions are satisfied.

Appendix

Further examples and practice in solving 3 × 3 linear systems of equations can be found in Appendix A.

SUMMARY: ACTIVITY 1.13

- A **3 × 3 system of linear equations** consists of three equations involving three variables.

- A **linear equation in the three variables** x, y, and z is of the form $Ax + By + Cz = D$, where A, B, C, and D are any constants. The equation is linear because the variables are all raised to the first power.

- To solve a 3 × 3 system algebraically,

  1. reduce the system in size to a 2 × 2 system

  2. solve the 2 × 2 system

  3. substitute the 2 × 2 solution into any original equation to solve for the third unknown

  4. check the solution by substituting into all three of the original equations

- Note that 3 × 3 systems may also be **inconsistent** (have no solution) or **dependent** (have infinitely many solutions).

EXERCISES: ACTIVITY 1.13

In Exercises 1–4, solve the 3 × 3 linear systems. Be sure to check your solution in all three of the original equations.

1. $x + y - z = -8$
$-x + y + z = 2$
$2x - y + z = 8$

2. $2x - 3y + z = 7$
$x + 2y - 2z = -5$
$-2x + y + z = -1$

3. $x + 2y - z = 0$
$3x + 2y + z = -8$
$2x + 3y + z = 0$

4. $x + 2y - 3z = 5$
$-x + y + 2z = 0$
$2x - y + z = -1$

5. Recall that some 2×2 linear systems do not have unique solutions. Try to solve these 3×3 linear systems. Identify each system as either dependent or inconsistent.

a. $x + 2y + z = 4$
$\qquad 2x - y + 3z = 2$
$\qquad 3x + y + 4z = 6$

b. $2x - y + 3z = 3$
$\qquad -x + 2y - z = 1$
$\qquad x + y + 2z = 2$

6. You are responsible for buying parts from a wholesale distributor. There are three types of comparable switches that are needed. The cost per switch is \$1.20, \$1.90, and \$2.30. You need all three types and will place your order as dictated by the following facts.

i. You need a total of 12,000 switches in this order.

ii. Your budget will allow an expenditure of \$23,400.

iii. You need three times as many of the most expensive switches as the least expensive switches.

Let x, y, and z represent the number of the first type, second type, and third type of switch, respectively.

a. Write a 3×3 linear system of equations to model this problem.

b. Solve the system. Be sure to check your solution.

Earth Day in your community has been expanded to Earth Week. As part of the festivities, groups of local middle school, high school, and community college students will all participate in three projects: cleaning up roadsides, planting trees, and helping to collect electronic equipment for recycling.

For the week, a total of 2150 pounds of trash was collected along roadsides, 525 new trees were planted in the community, and 9750 pounds of electronic equipment was recycled. The average production of the three groups is summarized in the following table.

	MIDDLE SCHOOL	HIGH SCHOOL	COMMUNITY COLLEGE
Highway Cleanup	10 lb/student	20 lb/student	30 lb/student
Trees Planted	3 trees/student	4 trees/student	7 trees/student
Electronics Recycled	50 lb/student	75 lb/student	150 lb/student

Activity 1.14

Earth Week

Objective

1. Solve a linear system of equations using matrices.

1. Let x represent the number of middle school student participants, y represent the number of high school student participants, and z the number of community college student participants. Write a 3×3 linear system that describes this situation.

Suppose you solve this system algebraically. Do the variables actually enter into the calculations? The answer is, "not really." The variables are important in setting up (placement or alignment) the system. But once the system is set up, the variables represent a position within the system. It is the coefficients and the constants that are important.

Using just the coefficients and constants, any system of linear equations can be represented by a matrix.

Definition

Any rectangular array of numbers or symbols is called a **matrix**. That is, a matrix is a systematic arrangement of numbers or symbols in rows and columns.

Matrices are very useful for displaying information. The following example demonstrates how a system of linear equations can be represented by a matrix.

Example 1 *Consider the 3×3 linear system*

$$3x + 2y - 3z = -2$$
$$2x - 5y + 2z = -2$$
$$4x - 3y + 4z = 10.$$

Note that the variables x, y, and z hold a particular position in the system of equations. The system can be viewed as an array of numbers.

The matrix

$$\begin{bmatrix} 3 & 2 & -3 \\ 2 & -5 & 2 \\ 4 & -3 & 4 \end{bmatrix}$$

derived from the linear system is called the **matrix of coefficients**. Note that the first column contains the coefficients of the variable x in the system. The second and third columns are the coefficients of the y and z variables, respectively.

The matrix

$$\begin{bmatrix} 3 & 2 & -3 & -2 \\ 2 & -5 & 2 & -2 \\ 4 & -3 & 4 & 10 \end{bmatrix},$$

also derived from the same linear system, is called the **augmented matrix**. The fourth column contains the constants on the right side of the system.

Now, to solve a system using matrices, you use a combination of operations on the rows of the augmented matrix to obtain new matrices that give the same solution as the original system.

Elementary Row Operations

a. Interchange two rows.

b. Multiply one row by a nonzero constant.

c. Add a multiple of one row to another row and replace the second row.

Note that these row operations are the same operations you use on a system of linear equations to eliminate variables and reduce the system to a simpler system. For a 3×3 linear system, your goal is to apply a sequence of row operations to obtain an augmented matrix of the form

$$\begin{bmatrix} 1 & 0 & 0 & c_1 \\ 0 & 1 & 0 & c_2 \\ 0 & 0 & 1 & c_3 \end{bmatrix}.$$

Such a matrix is said to be in reduced row echelon form.

Definition

A matrix is said to be in **reduced row echelon form** if the matrix has 1s down the main diagonal and 0s above and below each 1.

The following example demonstrates how to solve a 2×2 linear system using matrices. In this situation, your goal is to obtain a matrix in the reduced row echelon form

$$\begin{bmatrix} 1 & 0 & c_1 \\ 0 & 1 & c_2 \end{bmatrix}.$$

Example 2   *Solve the following system by using an augmented matrix and row operations.*

$$8x + 3y = 23$$
$$3x - 9y = 12$$

SOLUTION

Step 1.   Write the system as an augmented matrix.

The augmented matrix for this system is

$$\begin{bmatrix} 8 & 3 & 23 \\ 3 & -9 & 12 \end{bmatrix}.$$

Step 2. Use row operations to reduce the augmented matrix to the reduced row echelon form

$$\begin{bmatrix} 1 & 0 & c_1 \\ 0 & 1 & c_2 \end{bmatrix}.$$

In general, your strategy is first to change the number in the first column, first row, to a 1.

1. Interchange the rows to obtain

$$\begin{bmatrix} 3 & -9 & 12 \\ 8 & 3 & 23 \end{bmatrix}.$$

2. Divide the first row by 3.

$$\begin{bmatrix} 1 & -3 & 4 \\ 8 & 3 & 23 \end{bmatrix}$$

Next, you want to change the number in the first column, second row, to a 0.

3. Multiply the first row by -8 and add it to the second row to obtain

$$\begin{bmatrix} 1 & -3 & 4 \\ 0 & 27 & -9 \end{bmatrix}.$$

Now, change the number in the second column, second row, to a 1.

4. Divide the second row by 27.

$$\begin{bmatrix} 1 & -3 & 4 \\ 0 & 1 & -\frac{1}{3} \end{bmatrix}$$

Last, change the number in the second column, first row, to a 0.

5. Multiply the second row by 3 and add to the first.

$$\begin{bmatrix} 1 & 0 & 3 \\ 0 & 1 & -\frac{1}{3} \end{bmatrix}$$

This augmented matrix is in reduced row echelon form and is equivalent to the system

$$\begin{aligned} 1x + 0y &= 3 \quad \text{or} \quad x = 3 \\ 0x + 1y &= -\tfrac{1}{3} \qquad\quad y = -\tfrac{1}{3} \end{aligned}$$

Therefore, the solution to the system is $x = 3$ and $y = -\frac{1}{3}$.

2. Why would you want to rewrite an augmented matrix in reduced row echelon form?

You can use your calculator to determine reduced row echelon form. Problem 3 demonstrates the steps in solving the system in Example 2 using the TI-83/84 plus.

3. a. Choose the MATRIX option and choose the EDIT menu.

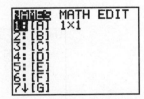
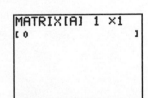

b. For the system

$$\begin{aligned} 8x + 3y &= 23 \\ 3x - 9y &= 12 \end{aligned}$$

enter 2 × 3 for the dimension and enter the elements of the matrix into MATRIX [A]

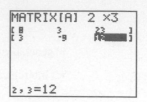

c. Return to the Home screen; then choose the MATRIX option again. This time select the MATH menu.

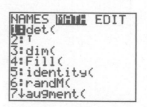

d. Choose option B, the reduced row echelon form command rref(, and follow this with MATRIX and 1: for matrix [A].

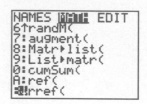

e. Close the parentheses and (ENTER).

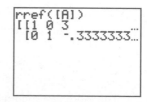

f. Read the solution to the system from the matrix.

4. Solve the 3 × 3 linear system from Example 1

$$3x + 2y - 3z = -2$$
$$2x - 5y + 2z = -2$$
$$4x - 3y + 4z = 10$$

using your calculator to determine the reduced row echelon form. Your screens should appear as follows:

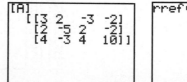
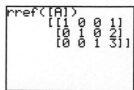

5. a. Use matrices to solve the system from Problem 1.

b. Interpret your solution in part a in the context of the original Earth Day situation.

SUMMARY: ACTIVITY 1.14

1. Any rectangular array of numbers or symbols is called a **matrix**.

2. The **matrix of coefficients of a linear system** is the matrix that contains only the coefficients of the variables of the system. The variables must be in the same order in each equation. For any missing term, a 0 is entered for the coefficient.

3. The **augmented matrix** is the matrix of coefficients with one column added to the right-hand end of the matrix. This column contains the constant terms from each equation in the system.

4. Matrix **elementary row operations** are as follows:
a. Interchange two equations.
b. Multiply one equation by a nonzero constant.
c. Add a multiple of one row to another equation and replace the second equation.

5. To solve a system, use elementary row operations to rewrite the augmented matrix in **reduced row echelon form.**

6. You can use your calculator to determine reduced row echelon form.

Step 1 Choose the MATRIX option and choose the EDIT menu.

Step 2. Enter the dimension and enter the elements of the matrix into MATRIX [A].

Step 3. Return to the Home screen; then choose the MATRIX option again. This time, select the MATH menu.

Step 4. Choose option B, the reduced row echelon form command rref(, and follow this with MATRIX and 1: for matrix [A].

Step 5. Close the parentheses and click ENTER.

Step 6. Read the solution to the system from the matrix.

EXERCISES: ACTIVITY 1.14

Write the augmented matrix for each system of linear equations.

1. $4x + 3y = -1$
   $2x - y = -13$

2. $3x - 2y + 5z = 31$
   $x + 3y - 3z = -12$
   $-2x - 5y + 3z = 11$

3. $4x - 2y + z = 15$
   $3x + 2y - 2z = -4$
   $x + z = 5$

4. $x - 2y + 3z = 9$
   $y + 3z = 5$
   $z = 2$

Write the system of linear equations represented by the augmented matrix. Use variables of x, y, and z when appropriate.

5. $\begin{bmatrix} 4 & 3 & 15 \\ 2 & -5 & 1 \end{bmatrix}$

6. $\begin{bmatrix} 2 & -6 & 4 & 10 \\ 1 & 5 & -5 & 0 \\ 3 & 0 & 4 & 7 \end{bmatrix}$

7. The given system of equations

$$2x - y + 3z = 9$$
$$x + 2y - z = -3$$
$$3x + y + z = 4$$

has the following row-reduced augmented matrix (use your calculator to check).

$$\begin{bmatrix} 1 & 0 & 0 & 1 \\ 0 & 1 & 0 & -1 \\ 0 & 0 & 1 & 2 \end{bmatrix}$$

What is the solution to the system?

8. Your senior class will sell pizzas as their fund-raiser this year. The medium cheese and pepperoni pizza will sell for $10.95 and the large for $14.95. On Valentine's Day, a total of 49 pizzas were sold for a total of $620.55. Your class president asked your class treasurer how many medium and how many large pizzas were sold. Can you help the treasurer figure this out using matrices?

Exercise numbers appearing in color are answered in the Selected Answers appendix.

9. In a recent game at your community college, the men's basketball team made 45 baskets from the field; some were 2-pointers and some were 3-pointers. In total, 101 points were made from the field. The coach wanted to know how many were 2-pointers and how many were 3-pointers. Use matrices to determine the answer, and give the information to your coach.

10. Your college offers a degree in cosmetology. You are interested in the number of men (over the age of 25), young adults (25 and under), and women (over the age of 25), who take advantage of this school. For 1 week you keep track of the number of hours spent working in each area, as shown in the table below. Students spend a maximum of 31 hours doing haircuts, 19 hours giving permanents, and 25 hours applying haircolor. If all of the available time must be used, how many young adults, men, and women will be serviced at your school during that week?

	MEN	YOUNG ADULTS	WOMEN
Haircuts	3 hours	1 hour	2 hours
Permanents	1 hour	1 hour	2 hours
Haircolor	1 hour	3 hours	2 hours

11. A car dealer needs to buy luxury sedans, small hatchbacks, and hybrid models. The dealer needs a total of 28 cars and has a budget of $692,000. Luxury sedans cost $36,000 each, small hatchbacks cost $18,000 each, and hybrids cost $28,000 each. The dealer needs as many small hatchbacks as the number of luxury sedans and hybrids combined. How many of each model should the dealer buy?

12. Your doctor tells you that your diet should consist of 2100 calories per day. Your doctor also recommends that you consume a total of 437.5 grams of protein, carbohydrates, and fat. Protein and carbohydrates have 4 calories per gram, and fat has 9 calories per gram. To maintain the correct proportions, the number of grams of carbohydrates is 140 grams more than the sum of the grams of protein and fat. How many grams of protein, carbohydrates, and fat should you consume daily?

Activity 1.15

How Long Can You Live?

Objectives

1. Solve linear inequalities in one variable numerically and graphically.

2. Use properties of inequalities to solve linear inequalities in one variable algebraically.

3. Solve compound inequalities algebraically.

4. Use interval notation to represent a set of real numbers described by an inequality.

Life expectancy in the United States is steadily increasing, and the number of American aged 100 or older will exceed 850,000 by the middle of this century. Medical advancemen have been a primary reason for Americans living longer. Another factor has been the in creased awareness of maintaining a healthy lifestyle.

The life expectancies at birth for men and women born after 1980 in the United States can b modeled by the following functions.

$$W(x) = 0.115x + 77.42$$
$$M(x) = 0.212x + 69.80,$$

where $W(x)$ represents the life expectancy for women, $M(x)$ represents the life expectanc for men, and x represents the number of years since 1980 that the person was born. That i: $x = 0$ corresponds to the year 1980, $x = 5$ corresponds to 1985, and so forth.

1. a. Complete the following table.

 Counting the Years

	YEAR					
	1980	**1985**	**1990**	**1995**	**2000**	**2005**
x, Years Since 1980	0	5	10	15	20	25
W(x)						
M(x)						

b. For people born between 1980 and 2005, do men or women have the greater life expectancy?

c. Is the life expectancy of men or women increasing more rapidly? Explain using slope.

You would like to determine in what birth years the life expectancy of men is greater than that of women. The phrase "greater than" indicates a mathematical relationship called an **inequality**. Symbolically, the relationship can be represented by

$$\underbrace{M(x)}_{\substack{\text{Life expectancy} \\ \text{for men}}} \quad \underbrace{>}_{\text{is greater than}} \quad \underbrace{W(x)}_{\substack{\text{life expectancy} \\ \text{for women.}}}$$

Other commonly used phrases that indicate inequalities are given in the following example.

Example 1

STATEMENT, WHERE x REPRESENTS A REAL NUMBER	TRANSLATION TO AN INEQUALITY
x is greater than 10	$x > 10$ or $10 < x$
x is less than 10	$x < 10$ or $10 > x$
x is at least 10	$x \geq 10$ (also read, "x is greater than or equal to 10")
x is at most 10	$x \leq 10$ (also read, "x is less than or equal to 10")

2. Substitute the appropriate expressions for $M(x)$ and $W(x)$ to obtain an inequality involving x that can be used to determine the birth years for which the life expectancy of men is greater than that of women.

Solving Inequalities in One Variable Numerically and Graphically

--- **Definition** ---

Solving an inequality in one variable is the process of determining the values of the variable that make the inequality a true statement. These values are called the **solutions** of the inequality.

3. Solve the inequality in Problem 2 numerically. That is, continue to construct a table of values (see Problem 1) until you determine the values of the years x (inputs) for which $0.212x + 69.80 > 0.115x + 77.42$. Use the Table feature of your graphing calculator.

Therefore, if the trends given by the equations for $M(x)$ and $W(x)$ continue, the approximate solution to the inequality $M(x) > W(x)$ is $x > 79$. That is, according to the models, after the year 2059, men will live longer than women.

4. Now, solve the inequality $0.212x + 68.80 > 0.115x + 77.42$ graphically.

   a. Use your graphing calculator to sketch a graph of $M(x) = 0.212x + 68.80$ and $W(x) = 0.115x + 77.42$ on the same coordinate axis.

   b. Determine the point of intersection of the two graphs using the intersect feature of your graphing calculator. What does the point represent in this situation?

To solve the inequality $M(x) > W(x)$ graphically, you need to determine the values of x for which the graph of $M(x) = 0.212x + 68.80$ is above the graph of

$$W(x) = 0.115x + 77.42.$$

   c. Use the graph to solve $M(x) > W(x)$. How does your solution compare to the solution in Problem 3?

5. a. Write an inequality to determine the birth years of women whose life expectancy is at least 85.

b. Solve the inequality numerically, using the Table feature of your graphing calculator.

c. Use your graphing calculator to solve this inequality graphically.

Solving Inequalities in One Variable Algebraically

The process of solving an inequality in one variable algebraically is very similar to solving an equation in one variable algebraically. Your goal is to isolate the variable on one side of the inequality symbol. You isolate the variable in an **equation** by performing the same operations to both sides of the equation so as not to upset the **balance**. You isolate the variable in an **inequality** by performing the same operations to both sides so as not to upset the **imbalance**.

6. a. Write the statement "15 is greater than 6" as an inequality.

b. Add 5 to each side of $15 > 6$. Is the resulting inequality a true statement? (That is, is the left side still greater than the right side?)

c. Subtract 10 from each side of $15 > 6$. Is the resulting inequality a true statement?

d. Multiply each side of $15 > 6$ by 4. Is the resulting inequality true?

e. Multiply each side of $15 > 6$ by -2. Is the left side still greater than the right side?

f. Reverse the direction of the inequality symbol in part e. Is the new inequality a true statement?

Problem 6 demonstrates two very important properties of inequalities.

> **Property 1** If $a < b$ represents a true inequality, then if
>
> **i.** the same quantity is added to or subtracted from both sides or
>
> **ii.** both sides are multiplied or divided by the same *positive number,* then the resulting inequality remains a true statement and the direction of the inequality symbol remains the same

For example, because $-4 < 10$, then

i. $-4 + 5 < 10 + 5$ or $1 < 15$ is true;

$-4 - 3 < 10 - 3$ or $-7 < 7$ is true.

ii. $-4(6) < 10(6)$ or $-24 < 60$ is true;

$\frac{-4}{2} < \frac{10}{2}$ or $-2 < 5$ is true.

> **Property 2**
>
> If $a < b$ represents a true inequality, then if both sides are multiplied or divided by the same *negative number*, then the inequality symbol in the resulting inequality statement must be reversed ($<$ to $>$ or $>$ to $<$) in order for the resulting statement to be true.

For example, because $-4 < 10$, then $-4(-5) > 10(-5)$ or $20 > -50$. Because $-4 < 10$, then $\dfrac{-4}{-2} > \dfrac{10}{-2}$ or $2 > -5$.

Properties 1 and 2 will be true if $a < b$ is replaced by $a \le b$, $a > b$, or $a \ge b$.

The following example demonstrates how properties of inequalities can be used to solve an inequality algebraically.

Example 2	*Solve* $3(x - 4) > 5(x - 2) - 8.$

SOLUTION

$3(x - 4) > 5(x - 2) - 8$ Apply the distributive property.

$3x - 12 > 5x - 10 - 8$ Combine like terms on the right side.

$3x - 12 > 5x - 18$

$\underline{-5x \qquad\quad -5x}$

$-2x - 12 > -18$ Subtract $5x$ from both sides; the direction of the inequality symbol remains the same.

$\underline{+12 \quad\; +12}$

$\dfrac{-2x}{-2} > \dfrac{-6}{-2}$ Add 12 to both sides; the direction of the inequality does not change.

$x < 3$ Divide both sides by -2; the direction is reversed!

Therefore, from Example 2, any number less than 3 is a solution to the inequalit $3(x - 4) > 5(x - 2) - 8$. The solution set can be represented on a number line by shadin all points to the left of 3:

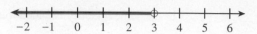

The open circle at 3 indicates that 3 is *not* a solution. A closed circle indicates that the num ber beneath the closed circle *is* a solution. The arrow shows that the solutions extend indefi nitely to the left.

7. Solve the inequality $0.212x + 69.80 > 0.115x + 77.42$ algebraically to determine th birth years in which men will be expected to live longer than women. How does you solution compare to the solutions determined numerically and graphically in Problems and 4c?

Compound Inequality

You have joined a health-and-fitness club. Your aerobics instructor recommends that t achieve the most cardiovascular benefit from your workout, you should maintain your puls rate between a lower and upper range of values. These values depend on your age.

8. If the variable a represents your age, then the lower and upper values for your pulse rate are determined by the following.

$$\text{lower value: } 0.72(220 - a)$$
$$\text{upper value: } 0.87(220 - a).$$

a. Determine your lower value.

b. Determine your upper value.

For the most cardiovascular benefit, a 20-year-old's pulse rate should be between 144 and 174. The phrase "between 144 and 174" means the pulse rate should be greater than 144 *and* less than 174. Symbolically, this combination or **compound inequality** is written as

$$144 < \text{pulse rate } and \text{ pulse rate} < 174.$$

This statement is written more compactly as $144 <$ pulse rate < 174.

The numbers that satisfy this compound inequality can be represented on a number line as

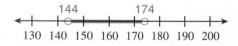

Other commonly used phrases that indicate compound inequalities involving the word *and* are given in the following example.

Example 3

STATEMENT, WHERE x REPRESENTS A REAL NUMBER	TRANSLATION TO A COMPOUND INEQUALITY
x is greater than or equal to 10 and less than 20	$10 \leq x < 20$
x is greater than 10 and less than or equal to 20	$10 < x \leq 20$
x is from 10 to 20 inclusive	$10 \leq x \leq 20$

9. Recall that the life expectancy for men is given by the expression $0.212x + 69.80$, where x represents the number of years since 1980. Use this expression to write a compound inequality that can be used to determine in what birth years men will be expected to live into their 80s.

The following example demonstrates how to solve a compound linear inequality algebraically and graphically.

Example 4   *Solve* $-4 < 3x + 5 \leq 11$ *using an algebraic approach.*

SOLUTION

Note that the compound inequality has three parts: left: -4, middle: $3x + 5$, and right: 11. To solve this inequality, isolate the variable in the middle part.

$$-4 < 3x + 5 \leq 11$$
$$\underline{-5 \qquad\quad -5 \quad -5} \qquad \text{Subtract 5 from each part.}$$
$$-9 < 3x \leq 6$$
$$-\frac{9}{3} < \frac{3x}{3} \leq \frac{6}{3} \qquad \text{Divide each part by 3.}$$
$$-3 < x \leq 2$$

The solution can be represented on a number line as follows.

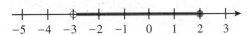

10. Solve the compound inequality $80 \leq 0.212x + 69.80 < 90$ from Problem 9 to determine in what birth years men will be expected to live into their 80s.

Interval Notation

Interval notation is an alternative method to represent a set of real numbers described by an inequality. The **closed interval** $[-3, 4]$ represents all real numbers x for which $-3 \leq x \leq 4$. The square brackets [ ] indicate that the endpoints of the interval are

included. The **open interval** $(-3, 4)$ represents all real numbers x for which $-3 < x <$
Note that the parentheses $(\ )$ indicate that the endpoints are not included. The interv.
$(-3, 4]$ is said to be **half-open** or **half-closed**. The interval is open at -3 (endpoint not i.
cluded) and closed at 4 (endpoint included).

Suppose you want to represent the set of real numbers x for which x is greater than 3. Th
symbol $+\infty$ (positive infinity) is used to indicate **unboundedness** in the positive directio.
Therefore, the interval $(3, +\infty)$ represents all real numbers x for which $x > 3$. Note th.
$+\infty$ is always open.

The symbol $-\infty$ (negative infinity) is used to represent unboundedness in the negativ
direction. Therefore, the interval $(-\infty, 5]$ represents all real numbers x for which $x \leq 5$.

11. In parts a–d, express each inequality in interval notation.

a. $-5 \leq x \leq 10$　　**b.** $4 \leq x < 8.5$　　**c.** $x > -2$　　**d.** $x \leq 3.75$

In parts e–h, express each of the following using inequalities.

e. $(-6, 4]$　　　　**f.** $(-\infty, 1.5]$　　**g.** $(-2, 2)$　　　**h.** $(-3, +\infty)$

SUMMARY: ACTIVITY 1.15

1. The **solution set** of an inequality is the set of all values of the variable that satisfy the inequality.

2. The direction of an inequality is not changed when

　i. the same quantity is added to or subtracted from both sides of the inequality. Stated symbolically,

　　if $a < b$, then $a + c < b + c$ and $a - c < b - c$.

　ii. the same positive quantity is multiplied or divided on both sides of the inequality. Stated symbolically.

　　If $a < b$, then $ac < bc$, and $\frac{a}{c} < \frac{b}{c}$, whenever $c > 0$.

3. The direction of an inequality is reversed if both sides of an inequality are multiplied by or divided by the same negative number. These properties can be written symbolically as

　i. if $a < b$, then $ac > bc$, where $c < 0$;

　ii. if $a < b$, then $\frac{a}{c} > \frac{b}{c}$, where $c < 0$.

The two properties of inequalities above (items 2 and 3) will still be true if $a < b$ is replaced
by $a \leq b, a > b$, or $a \geq b$.

4. Inequalities such as $f(x) < g(x)$ can be solved using three different methods:

　i. a **numerical approach**, in which a table of input/output pairs is used to determine values of
　　x for which $f(x) < g(x)$;

　ii. a **graphical approach**, in which values of x are located so that the graph of f is below the
　　graph of g;

　iii. an **algebraic approach**, in which the properties of inequalities are used to isolate the variable.

Similar statements can be made for solving inequalities of the form $f(x) \leq g(x), f(x) > g(x)$,
and $f(x) \geq g(x)$.

EXERCISES: ACTIVITY 1.15

Exercises 1–6, translate the given statement into an algebraic inequality or compound inequality.

1. To avoid an additional charge, the sum of the length, l, width, w, and depth, d, of a piece of luggage to be checked on a commercial airline can be at most 61 inches.

2. A PG-13 movie rating means that your age, a, must be at least 13 years for you to view the movie.

3. The cost, $C(A)$, of renting a car from company A is less expensive than the cost, $C(B)$, of renting from company B.

4. The label on a bottle of film developer states that the temperature, t, of the contents must be kept between 68° and 77° Fahrenheit.

5. You are in a certain tax bracket if your taxable income, i, is over $24,650, but not over $59,750.

6. The range of temperature, t, on the surface of Mars is from 28°C to 140°C.

Solve Exercises 7–14 graphically and algebraically.

7. $3x > -6$

8. $3 - 2x \leq 5$

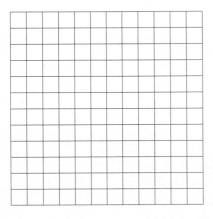
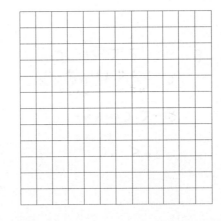

9. $x + 2 > 3x - 8$

10. $5x - 1 < 2x + 11$

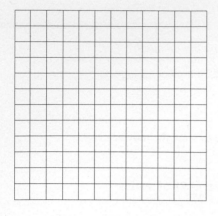

11. $8 - x \geq 5(8 - x)$

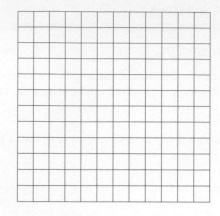

12. $5 - x < 2(x - 3) + 5$

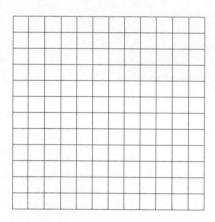

13. $\frac{x}{2} + 1 \leq 3x + 2$

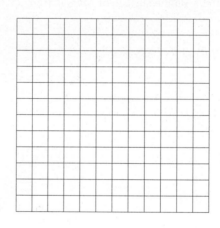

14. $0.5x + 3 \geq 2x - 1.5$

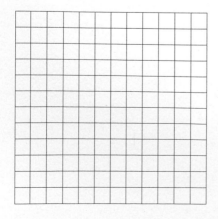

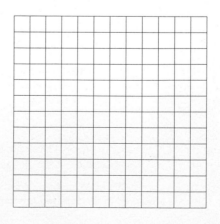

olve Exercises 15–16 algebraically.

15. $1 < 3x - 2 < 4$

16. $-2 < \frac{x}{3} + 1 < 5$

17. The consumption of cigarettes in the United States is declining. If t represents the number of years since 1990, then the consumption, C, is modeled by

$$C = -9.90t + 529.54,$$

where C represents the number of billions of cigarettes smoked per year.

a. Write an inequality that can be used to determine the first year in which cigarette consumption is less than 200 billion cigarettes per year.

b. Solve the inequality in part a using an algebraic as well as a graphical approach.

18. In Activity 1.11, Moving Out, you contacted two local rental companies and obtained the following information for the 1-day cost of renting a truck.

   Company 1: $19.99 per day plus $0.79 per mile

   Company 2: $29.99 per day plus $0.59 per mile

Let n represent the total number of miles driven in 1 day.

a. Write an expression to determine the total cost, C, of renting a truck for 1 day from company 1.

b. Write an expression to determine the total cost, C, of renting a truck for 1 day from company 2.

c. Use the expressions in parts a and b to write an inequality that can be used to determine for what number of miles it is less expensive to rent the truck from company 2.

d. Solve the inequality.

19. The sign on the elevator in a seven-story building on campus states that the maximum weight it can carry is 1200 pounds. As part of your work-study program, you need to move a large shipment of books to the sixth floor. Each box weighs 60 pounds.

a. Let n represent the number of boxes placed in the elevator. If you weigh 150 pounds, write an expression that represents the total weight in the elevator. Assume that only you and the boxes are in the elevator.

b. Using the expression in part a, write an inequality that can be used to determine the maximum number of boxes that you can place in the elevator at one time.

c. Solve the inequality.

20. The following equation is used in meteorology to determine the temperature humidity index T:

$$T = \tfrac{2}{5}(w + 80) + 15,$$

where w represent the wet-bulb thermometer reading. For what values of w would T range from 70 to 75?

21. The temperature readings in the United States have ranged from a record low of $-79.8°F$ (Alaska, January 23, 1971) to a record high of $134°F$ (California, July 10, 1913).

   a. If F represents the Fahrenheit temperature, write a compound inequality that represents the interval of temperatures (in °F) in the United States.

   b. Recall that Fahrenheit and Celsius temperatures are related by the formula

   $$F = 1.8C + 32.$$

   Rewrite the compound inequality in part a to determine the temperature range in degrees Celsius.

   c. Solve the compound inequality.

22. You are enrolled in a wellness course at your college. You achieved grades of 70, 86, 81, and 83 on the first four exams. The final exam counts the same as an exam given during the semester.

   a. If x represents the grade on the final exam, write an expression that represents your course average (arithmetic mean).

   b. If your average is greater than or equal to 80 and less than 90, you will earn a B in the course. Using the expression from part a for your course average, write a compound inequality that must be satisfied to earn a B.

   c. Solve the inequality.

bjectives

. Graph a piecewise linear function.

. Write a piecewise linear function to represent a given situation.

. Graph a function defined by $y = |x - c|$.

As an incentive to obtain more sales, a salesperson is often paid on a commission basis. A common commission scheme is the **accumulative plateau method**. Using this method, a salesperson is paid commission at a greater rate for a greater level of sales.

A particular company pays commission as follows:

> 5% on sales up to $10,000
>
> 7.5% on the next $10,000 in sales
>
> 10% on all sales over $20,000

For example,

i. The commission on $8000 in sales is 5% of $8000:

$$0.05 \cdot 8000 = \$400.$$

ii. The commission on $12,000 in sales is 5% of the first $10,000 plus 7.5% of the sales over $10,000:

$$0.05 \cdot 10,000 + 0.75 \cdot 2000 = 500 + 150 = \$650.$$

iii. The commission on $27,000 in sales is 5% of the first $10,000 plus 7.5% of the second $10,000 plus 10% of the sales over $20,000:

$$500 + 750 + 0.10(7000) = 1250 + 700 = \$1950.$$

1. a. How much will a salesperson earn in commission on $6000 in sales?

b. How much will a salesperson earn in commission on $15,000 in sales?
Note: The 7.5% rate only applies to sales above $10,000.

c. How much will a salesperson earn in commission on $24,000 in sales?
Note: The 10% rate only applies to sales above $20,000.

2. Is the commission earned a function of the amount of the sales? Explain.

Note that when using the accumulation plateau method, the commission earned is calculated differently depending on the level (plateau) of sales.

> When the output value of a function is calculated differently depending on the input value, the function is called a **piecewise function.**

3. a. Refer to Problem 1a and write an equation that gives the commission, C, as a function of sales, s, if the sales amount is less than or equal to $10,000.

b. Refer to Problem 1b and write an equation that gives the commission, C, as a function of sales, s, if the sales amount is greater than $10,000 but less than or equal to $20,000.

c. Refer to Problem 1c and write an equation that gives the commission, C, as a function of sales, s, if the sales amount is greater than $20,000.

The equations in Problem 3 are all part of one piecewise function, the commission functio⋯ This piecewise function is written in the following form:

$$C = f(s) = \begin{cases} 0.05s & \text{if } 0 \le s \le 10{,}000 \\ 500 + 0.075(s - 10{,}000) & \text{if } 10{,}000 < s \le 20{,}000. \\ 1250 + 0.10(s - 20{,}000) & \text{if } s > 20{,}000 \end{cases}$$

Note that in the form a piecewise function is written, each row represents one piece of the fun⋯ tion. The equation for one of the function pieces is followed by the interval over which th⋯ equation applies. All the pieces are joined together by the brace to the left of the equations.

To determine the commission earned, $f(s)$, for a particular sales amount, s, you first look ⋯ each interval in the definition to determine the interval in which the value of s is containe⋯ Then use the expression in that row to calculate the commission, $f(s)$.

4. a. Which equation would be used to calculate $f(18{,}000)$?

b. Calculate $f(18{,}000)$ and interpret its meaning in this situation.

5. Sketch a graph of the commission function over the domain 0 to 30,000. Be sure to us⋯ an appropriate scale for the horizontal and vertical axes.

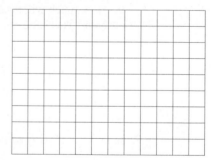

Cell Phone Minutes

A major wireless company offers a plan that includes 450 anytime minutes for $39.99 pe⋯ month. This is the least expensive plan, and many people choose it for that reason. However⋯ this plan can become expensive for someone who goes over the 450 minutes in 1 month⋯ There is a $0.45 charge for each minute over 450 in 1 month.

6. Suppose you have signed up for this plan. Complete the following table.

Minutes Used in One Month	200	300	400	500
Cost (in dollars)				

The method used to determine the cost for 500 minutes is different from the method used to⋯ determine the charge for 200, 300, or 400 minutes. This is another example of a piecewise⋯ function. The input for this function is the total number of anytime minutes used in 1 month⋯ and the output is the cost for that month.

7. Let m represent the total anytime minutes used in 1 month and let C represent the cost for that month. Therefore, $C = f(m)$.

   a. Write an equation for $f(m)$ if m is less than or equal to 450 minutes.

   b. Write an equation for $f(m)$ if m is greater than 450 minutes. *Note*: Check your equation by replacing m with 500 and confirm that you get the same result as in Problem 6.

   c. Combine the results from parts a and b to write C as a piecewise function of m.

8. a. Use the function from Problem 7c to determine the charge for 600 anytime minutes with this plan.

   b. Sketch a graph of the cell phone cost function over the domain $0 \le m \le 600$.

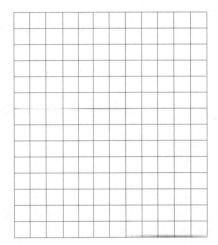

   c. Would you suggest this plan to a friend whose monthly cell phone use is regularly in excess of 450 minutes? Explain.

Overtime Function

According to federal overtime pay law, overtime pay is extra cash compensation for the number of hours nonexempt (eligible) employees work in excess of 40 in one workweek. The current overtime pay rate for eligible employees is one and one-half (1.5) times their regular rates of pay (also known as time-and-a-half).

9. a. If your regular hourly rate of pay is $9.00 per hour, then what is your overtime rate?

   b. If your regular hourly rate of pay is $14.00 per hour, then what is your overtime rate?

10. a. Suppose your regular hourly rate of pay is $12.00 per hour. Complete the following table.

Number of Hours Worked	25	35	45
Gross Pay			

b. Is gross pay a piecewise function of the number of hours worked? Explain.

The method used to determine your gross pay for 45 hours is different from the method used for 25 or 35 hours. Gross pay is a piecewise function of number of hours worked in 1 week.

11. Let h represent the number of hours worked in 1 week and let P represent the gross pay. Assume your regular hourly rate of pay is $12.00 per hour.

a. Write an equation for P if h is less than or equal to 40 hours.

b. Write an equation for P if h is greater than 40 hours.

c. Combine the equations from parts a and b to write P as a piecewise function of h.

d. Sketch a graph of the gross pay function.

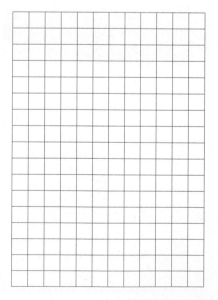

e. Use the gross pay piecewise function to calculate the gross pay for 45 hours, and compare the result with the answer to Problem 10a.

Absolute Value Function

12. Consider the piecewise function defined by

$$f(x) = \begin{cases} -x & \text{if} \quad x < 0 \\ x & \text{if} \quad x \geq 0 \end{cases}.$$

a. What is the domain of the function f?

b. Complete the following table.

x	−4	−3	−2	−1	0	1	2	3	4
f(x)									

c. Sketch a graph of the function f. Use your graphing calculator to verify the graph.

d. Describe the shape of the graph of function f.

e. What is the range of the function?

The function f defined in Problem 12 may look familiar. This function is the piecewise definition of the absolute value function defined by $f(x) = |x|$.

13. a. Use your graphing calculator to obtain the graph of $y = |x|$. The absolute value function is located in the Math menu under the Num submenu.

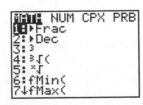 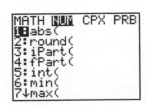 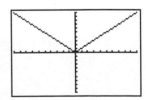

b. How does the graph compare to the graph in Problem 12?

14. a. Use your graphing calculator to sketch a graph of $y = |x - 2|$.

b. Describe how the graph of $y = |x - 2|$ can be obtained from the graph of $y = |x|$

c. Describe how the graph of $y = |x| + 2$ can be obtained from the graph of $y = |x|$

SUMMARY: ACTIVITY 1.16

1. A **piecewise** function is a function that is defined differently for certain "pieces" of its domain.

2. The **absolute value** function is a special piecewise function defined by

$$|x| = \begin{cases} x & \text{if} \quad x \geq 0 \\ -x & \text{if} \quad x < 0 \end{cases}.$$

3. The absolute value of the linear function, $g(x) = |x - c|$, always has a V-shaped graph with a vertex at $(c, 0)$.

EXERCISES: ACTIVITY 1.16

1. To travel outside the city limits, a certain taxicab company charges $1.20 for the first mile or less of travel. After the first mile, the charge is an additional $0.90 per mile.

a. Complete the following table.

x, Number of Miles Outside City Limits	0	0.5	1	2	5	10
C, Total Cost (in $)						

b. Write an equation for the total cost, C, of the first mile or less of travel.

c. Write an equation that gives the total cost, C, if you travel more than 1 mile, all outside city limits.

d. Write a piecewise function for the total cost, C.

e. Sketch a graph of the cost function over the domain 0 to 10 miles.

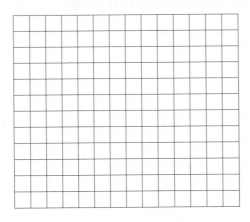

f. Determine the cost of a 12-mile taxi ride, all outside the city limits.

2. Sketch the graph of the piecewise function.

$$H(x) = \begin{cases} -2x + 3 & \text{if} \quad x < -2 \\ 4 & \text{if} \quad -2 \le x < 1 \\ x - 1 & \text{if} \quad x \ge 1 \end{cases}$$

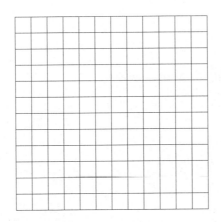

3. You are an author about to publish your first novel, *Hockey in the 90s*. The book will sell for $25. You will be paid royalties of 10% on the first 15,000 copies sold, 12% on the next 6000 copies, and 16% on any additional copies.

a. Write a piecewise function, f, that specifies the total royalties if x copies are sold.

b. Use your function from part a to determine if the royalty is $46,500 when 18,000 books are sold. Is the royalty $71,500 when 25,000 copies are sold? If you do not obtain these answers when substituting into the function, then discover where you went wrong.

c. With the royalties from your book, you would like to pay for your advanced degree in journalism. It will cost $65,000. How many books must sell to cover the cost of the degree?

4. You receive a bill each month for your credit-card use. The bill indicates the minimum amount that is due by a certain date. The minimum amount due depends upon your unpaid balance. One credit-card company uses the following criteria to determine your bill.

- The entire amount is due if the balance is less than $10.

- A minimum of $10 is due if the balance is $10 or more but less than $500.

- A minimum of $30 is due if the balance is $500 or more but less than $1000.

- A minimum of $50 is due if the balance is $1000 or more but less than $1500.

- A minimum of $70 is due if the balance is $1500 or more.

a. Let x represent the dollar amount of the unpaid balance. Complete the following table.

Where Credit Is Due

Unpaid Balance (in $)	0	5	10	100	300	500	700	1200	1700
Minimum Amount Due, M (in $)									

b. Write a piecewise function for the minimum payment due.

c. Sketch a graph of the minimum payment function.

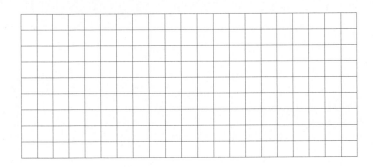

5. a. Complete the following table, where $f(x) = |x - 3|$ and $g(x) = |x + 3|$.

x	−5	−4	−3	−2	−1	0	1	2	3	4	5
f(x)											
g(x)											

b. Sketch a graph of each of the functions f and g. Verify using your graphing calculator.

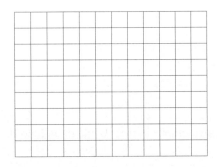

c. Describe how the graphs of f and g can be obtained from the graph of $y = |x|$.

d. Write $f(x) = |x - 3|$ as a piecewise function.

e. What is the domain of $f(x)$? What is the domain of $g(x)$?

f. What is the range of $f(x)$? What is the range of $g(x)$?

6. a. Without using a table of values or a graphing calculator, sketch the graph of $f(x) = |x - 5|$.

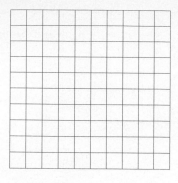

b. Describe the location and shape of the graph of $g(x) = |x + 5|$.

c. Sketch the graph of $h(x) = |x| + 5$ without a graphing calculator or a table of values.

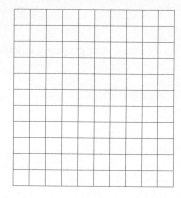

d. Describe the similarities and differences in the two graphs in parts b and c. Include shape, location, and intercepts.

Cluster 3 — What Have I Learned?

1. You are given two linear equations in slope-intercept form. How can you tell by inspection if the system is consistent, inconsistent, or dependent? Give examples.

2. In this cluster, you solved 2×2 linear systems four ways. List them. Give an advantage of each approach.

3. Describe a procedure that will combine the following two linear equations in three variables into a single linear equation in two variables.

$$2x + 3y - 5z = 10$$
$$3x - 2y + 2z = 4$$

4. What number is its own opposite?

5. The graphs of the absolute value functions in this cluster look like a V. What are the coordinates of the point of the V of the graph of $f(x) = |x - 10|$?

6. Explain when the addition method would be more efficient to use than the substitution method as you solve a system of linear equations algebraically.

7. In solving an inequality, explain when you would change the direction of the inequality symbol.

Cluster 3 How Can I Practice?

1. Solve the following systems both graphically and algebraically.

a. $x + y = -3$

$y = x - 5$

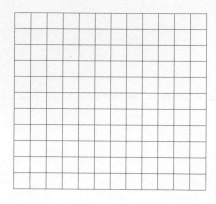

b. $x - 2y = -1$

$4x - 3y = 6$

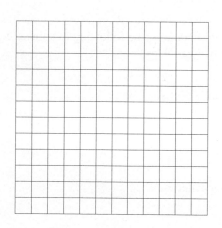

c. $2x - 3y = 7$

$5x - 4y = 0$

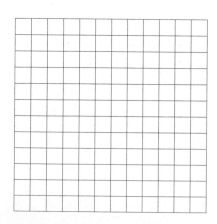

d. $x - y = 6$

$y = x + 2$

2. Rewrite the systems in Exercise 1 in the form

$$y = ax + b$$
$$y = cx + d$$

and check your solutions numerically using the table feature of your graphing calculator.

a.   **b.**   **c.**   **d.**

3. Solve the 3 × 3 system algebraically, or using matrix techniques.

$2x - y + z = -5$

$x - 2y + 2z = -13$

$3x + y - 2z = 12$

4. Solve the following inequalities algebraically. Check your solutions graphically in parts a and b.

a. $2.5x + 9.8 \geq 14.3$ **b.** $-3x + 14 < 32$ **c.** $-5 \leq 3x - 8 < 7$

5. You are going to help your grandmother create a garden of tulips and daffodils. She has space for approximately 80 bulbs. The florist tells you that tulips cost $0.50 per bulb and daffodils cost $0.75 per bulb. How many of each can you purchase if her budget is $52?

a. Write the system of equations.

b. Solve the system algebraically.

c. Check your solution graphically using your graphing calculator.

6. You need some repair work done on your truck. Towne Truck charges $80 just to examine the truck and $30 per hour for labor costs. World Transport Co. charges $50 for the initial exam and $40 per hour for the labor.

a. Write a cost equation for each company. Use *y* to represent the total cost of doing the work and *x* to represent the number of hours of labor.

b. Complete the table of values for the cost functions.

x (NUMBER OF HOURS)	y, TOWNE TRUCK COST ($)	y, WORLD TRANSPORT COST ($)
2		
4		
6		
8		

c. Graph the functions.

d. From the graph, determine after how many hours the costs will be equal. What will be the total cost?

e. Check your solution in part d by solving the system algebraically.

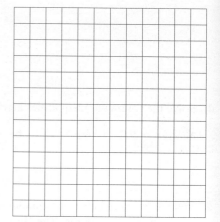

f. You think that you have a transmission problem that will take approximately 6 hours to fix. Determine from the graph which company you will hire for this job. Explain.

7. Your friend tells you that he has 27 coins. Some coins are nickels, some are dimes, and the rest are quarters. The total value is $3.25. When your friend gives you the last clue by saying he has twice as many dimes as nickels, you can easily solve the system to tell him how many of each coin he has.

a. Write the system of equations.

b. Solve the system algebraically.

c. Check your solution.

8. Translate each of the following into an inequality statement.

   a. x is greater than -5 and at most 6.

   b. x is greater than or equal to -3 and less than 4.

9. You own a hot dog cart in New York City. Your monthly profit is determined from the expression $1.50x - 50$, where x represents the number of hot dogs sold each month. The number 1.50 in the expression is the profit for each hot dog. The cost of leasing the hot dog stand is the number 50 in the expression $1.50x - 50$.

   a. To ensure a profit of at least $2000 per month, approximately how many hot dogs do you have to sell? Write the inequality and solve.

   b. Your profit has been fluctuating between $1500 and $2200 per month. Determine approximately between what two values your hot dog sales have to be to realize this range of profit. Write the inequality and solve.

Chapter 1 Summary

The bracketed numbers following each concept indicate the activity in which the concept is discussed.

CONCEPT/SKILL	DESCRIPTION	EXAMPLE
Variable [1.1]	A variable, usually represented by a letter, is a quantity or quality that may change in value from one particular instance to another.	In a survey of your class, an individual's height, weight, and gender are all variables.
Input variable [1.1]	The input variable is the value given first in a relationship.	In the relationship between the perimeter and the side of a square, $P = 4s$, s is the input variable.
Output variable [1.1]	The output is the value that corresponds to or is determined by the given input value.	In the relationship between the perimeter and the side of a square, $P = 4s$, P is the output variable.
Function [1.1]	A function is a correspondence between an input variable and an output variable that assigns a single output value to each input value.	See Example 1 in Activity 1.1 (pages 2 and 3).
Ordered pair [1.1]	An ordered pair of numbers consists of two numbers written in the form (input value, output value). The order in which they are listed is significant.	(2, 3) is an ordered pair. In this pair, 2 is the input and 3 is the output.
Verbally defined function [1.1]	A function is defined verbally when it is defined using words.	The high temperature in Albany, New York, is a function of the day of the year, because for each day there is one high temperature.
Numerically defined function [1.1]	A function is defined numerically using ordered pairs.	The Park It table (page 1) in Activity 1.1 is a numerically defined function because for each hour there is only one value for the number of cars in the parking lot.
Function notation [1.1]	Output variable = name of function(input variable). $y = f(t)$ is read "y equals f of t."	See Example 5 of Activity 1.1 (page 6).
Independent variable [1.1]	*Independent variable* is another name for the input variable of a function.	See Problem 6 of Activity 1.1 (page 4).
Dependent variable [1.1]	*Dependent variable* is another name for the output variable of a function.	See Problem 6 of Activity 1.1 (page 4).

CONCEPT/SKILL	DESCRIPTION	EXAMPLE
Domain of a function [1.2]	The domain of the function is the collection of all replacement values for the independent or input variable.	See Example 3 in Activity 1.2 (page 13).
Practical domain of a function [1.2]	The practical domain is the collection of replacement values of the input variable that make practical sense in the context of the situation.	See Example 3 in Activity 1.2 (page 13).
Range of a function [1.2]	The range of a function is the collection of all output values of a function.	See Problem 8 in Activity 1.2 (page 13).
Practical range of a function [1.2]	The practical range is the collection of all output values that make practical sense in the context of the situation.	See Problem 8 in Activity 1.2 (page 13).
Graphically defined function [1.3]	A function is defined graphically when the input variable is represented on the horizontal axis and the output variable on the vertical axis.	The graph drawn on page 18 in Problem 1 of Activity 1.3 defines a function.
Mathematical model [1.4]	A function can be used as a mathematical model that best fits the actual data and can be used to predict output values for input values not in the table.	See Problem 4 in Activity 1.4 (page 29).
Increasing function [1.4]	A function is increasing if its graph goes up to the right.	See Example 1 in Activity 1.4 (page 30).
Decreasing function [1.4]	A function is decreasing if its graph goes down to the right.	See Example 1 in Activity 1.4 (page 30).
Constant function [1.4]	A function is constant if its graph is horizontal.	See Example 1 in Activity 1.4 (page 30).
Vertical line test [1.4]	A graph defines a function if any vertical line intersects the graph no more than once. This is called the vertical line test.	No circle can represent a function because a vertical line through the center will pass through the circle twice, indicating that there is at least one input value paired with two different output values.
Average rate of change [1.6]	The average rate of change of a function over a specified input interval is the ratio $$\frac{\text{change in output}}{\text{change in input}}.$$	See Problem 8 in Activity 1.6 (page 53).

CONCEPT/SKILL	DESCRIPTION	EXAMPLE
Linear function [1.7]	A function for which the rate of change between any pair of points remains constant is called a linear function.	$f(x) = 2x + 1$ defines a linear function.
Slope [1.7]	The slope of a line segment joining two points (x_1, y_1) and (x_2, y_2) is denoted by m and defined by $m = \dfrac{y_2 - y_1}{x_2 - x_1}$.	The slope of the line segment joining $(2, -1)$ and $(5, 2)$ is given by $m = \dfrac{2 - (-1)}{5 - 2} = \dfrac{3}{3} = 1$.
Vertical intercept [1.7]	The vertical intercept $(0, b)$ of a graph is the point where the graph crosses the vertical axis.	The vertical intercept of $f(x) = 2x + 1$ is $(0, 1)$.
Horizontal intercept [1.7]	The horizontal intercept $(a, 0)$ of a graph is the point where the graph crosses the horizontal axis.	The horizontal intercept of $f(x) = 2x + 1$ is $\left(-\frac{1}{2}, 0\right)$.
Slope-intercept form [1.7]	The slope-intercept form of the equation of a line is $f(x) = mx + b$.	$f(x) = 2x + 1$ is a linear function in slope-intercept form.
Parallel lines [1.8]	The graphs of linear functions with the same slope but different y-intercepts are parallel lines.	The graphs of $f(x) = 2x + 1$ and $g(x) = 2x - 3$ are parallel lines.
Point-slope form [1.8]	The point slope form of the equation of a line is $y = y_0 + m(x - x_0)$	The point slope form of the equation of the line through $(2, -1)$ with slope 4 is $y = -1 + (x - 2)$
Standard form of a linear equation [1.9]	A linear function whose equation is in the form $Ax + By = C$, where A, B, and C are constants, is said to be written in standard form.	$2x + 3y = 6$ is an equation of a linear function written in standard form.
Horizontal line [1.9]	The graph of $y = c$ or $f(x) = c$ is a horizontal line.	The graph of $y = 3$ is a horizontal line 3 units above the x-axis.
Vertical line [1.9]	The graph of $x = a$ is a vertical line.	The graph of $x = 2$ is a vertical line 2 units to the right of the y-axis.
Linear regression equation [1.10]	The linear regression equation is the linear equation that best fits a set of data.	See Problem 4 in Activity 1.10 (page 88).
Interpolation [1.10]	Interpolation is the process of using a regression equation to predict a value of output for an input value that lies within the range of the original input data.	See Problem 6 in Activity 1.10 (page 89).
Extrapolation [1.10]	Extrapolation is the process of using a regression equation to predict a value of output for an input value that lies outside the range of the original input data.	See Problem 6 in Activity 1.10 (page 89).

CONCEPT/SKILL	DESCRIPTION	EXAMPLE
2×2 system of linear equations [1.11]	A 2×2 system of linear equations consists of two linear equations with two variables.	$y = 3x - 10$ $y = 5x + 14$
Solution to a 2×2 linear system [1.11]	A solution to a 2×2 linear system is an ordered pair that solves both equations of the system.	$(-12, -46)$ is a solution to $y = 3x - 10$ $y = 5x + 14.$
Consistent system [1.11]	A linear system is consistent if there is at least one solution.	$y = 3x - 10$ $y = 5x + 14$ is a consistent system.
Inconsistent system [1.11]	A linear system is inconsistent if there is no solution. The lines are parallel.	$y = 2x + 1$ $y = 2x - 3$
Dependent system [1.11]	A linear system is dependent if there are infinitely many solutions. The equations represent the same line.	$2x - 3y = 6$ $4x - 6y = 12$
Linear equation in three variables [1.13]	A linear equation in three variables x, y, and z is of the form $Ax + By + Cz = D$, where A, B, C, and D are any constants.	$2x + 3y - 7z = 23$ is a linear equation in three variables.
3×3 system of linear equations [1.13]	A 3×3 system of linear equations consists of three equations with a total of three variables.	$x + y - z = 8$ $-x + y + z = 2$ $2x - y + z = 8$ is a system of three linear equations.
Matrix [1.14]	Any rectangular array of numbers or symbols is called a **matrix**. That is, a matrix is a systematic arrangement of numbers or symbols in rows and columns.	$\begin{bmatrix} 1 & 2 & 3 \\ -1 & 5 & 0 \end{bmatrix}$ is an example of a 2×3 matrix; two rows and three columns.
Matrix of coefficients [1.14]	A matrix of coefficients of a linear system is a matrix containing the coefficients of the variables of the system.	For the system $2x + 3y = 1$ $-x - 2y = 4,$ $\begin{bmatrix} 2 & 3 \\ -1 & -2 \end{bmatrix}$ is the matrix of coefficients.
Augmented matrix [1.14]	An augmented matrix of a linear system is the matrix of coefficients with the constant terms added.	For the system $2x + 3y = 1$ $-x - 2y = 4,$ $\begin{bmatrix} 2 & 3 & 1 \\ -1 & -2 & 4 \end{bmatrix}$ is the augmented matrix

CONCEPT/SKILL	DESCRIPTION	EXAMPLE
Reduced row echelon form [1.14]	A matrix is said to be in **reduced row echelon form** if the matrix has 1s down the main diagonal and 0s above and below each 1.	The matrix $$\begin{bmatrix} 1 & 0 & 3 \\ 0 & 1 & -2 \end{bmatrix}$$ is in reduced row echelon form.
Interval notation [1.15]	Interval notation is an alternative method to represent a set of real numbers described by an inequality.	See Problem 11 of Activity 1.15 (page 142).
A compound inequality [1.15]	A compound inequality is a statement that involves more than one inequality symbol: $<, >, \leq, \geq,$ or $\neq$.	$-3 < x + 7 \leq 10$
Piecewise function [1.16]	A piecewise function is a function that is defined differently for certain "pieces" of its domain.	$f(x) = \begin{cases} x & \text{if} \quad x \leq 2 \\ -x + 1 & \text{if} \quad x > 2 \end{cases}$
Absolute value function [1.16]	The absolute value function is the function defined by $$|x| = \begin{cases} x & \text{if} \quad x \geq 0 \\ -x & \text{if} \quad x < 0 \end{cases}.$$	$|x| = f(x) = \begin{cases} x & \text{if} \quad x \geq 0 \\ -x & \text{if} \quad x < 0 \end{cases}$
The graph of the absolute value of a linear function defined by $g(x) = |x - c|$ [1.16]	The absolute value of a linear function $g(x) = |x - c|$ has a V-shaped graph with a vertex at $(c, 0)$.	$g(x) = |x + 2|$ has a vertex at $(-2, 0)$.

1. Determine whether each of the following is a function.

    a. The loudness of an iPod is a function of the position of the volume dial.

    b. $\{(2, 9), (3, 10), (2, -9)\}$

    c.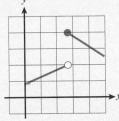

2. For a certain yard, the fertilizer costs $20. You charge $8 per hour to do yard work. If x represents the number of hours worked on the yard and $f(x)$ represents the total cost, including fertilizer, complete the following table.

x	0	2	3	5	7
f(x)					

    a. Is the total cost a function of the hours worked? Explain.

    b. Which variable represents the input?

    c. Which is the dependent variable?

    d. Which value(s) of the domain would not be realistic for this situation? Explain.

    e. What is the average rate of change from 0 to 3?

    f. What is the average rate of change from 5 to 7?

    g. What can you say about the rate of change between any two of the points?

    h. What kind of relationship exists between the two variables?

    i. Write this relationship in the form $f(x) = mx + b$.

Answers to all Gateway exercises are included in the Selected Answers appendix.

j. What is the practical meaning of the slope in this situation?

k. What is the vertical intercept? What is the practical meaning of this point?

l. Determine $f(4)$.

m. For what value(s) of x does $f(x) = 92$? Interpret your answer in the context of the situation.

3. Let $f(x) = x^2 - 5x$ and let $g(x) = -3x + 4$. Evaluate each of the following.

a. $f(-2)$ and $g(-2)$

b. $f(3) + g(3)$

c. $f(-3) - g(-4)$

d. $f(-4) \cdot g(2)$

4. Which of the following sets of data represent a linear function?

a.

x	0	2	4	6	8
f(x)	14	22	30	38	46

b.

x	5	10	15	20	25
y	4	2	0	−2	−4

c.

x	1	3	4	6	7
g(x)	10	20	30	40	50

d.

t	0	10	20	30	40
d	143	250	357	464	571

5. a. Determine the slope of the line through the points $(5, -3)$ and $(-4, 9)$.

b. From the equation $3x - 7y = 21$, determine the slope.

c. Determine the slope of the line from its graph.

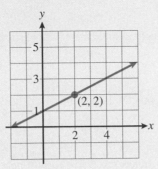

6. Write the equation of the line described in each of the following.

a. A slope of 0 and passing through the point $(2, 4)$

b. A slope of 2 and a vertical intercept of $(0, 5)$

c. A slope of -3 and passing through the point $(6, -14)$

d. A slope of 2 and passing through the point $(7, -2)$

e. A line with no slope passing through the point $(2, -3)$

f. A slope of -5 and a horizontal intercept of $(4, 0)$

g. A line passing through the points $(-3, -4)$ and $(2, 16)$

h. A line parallel to $y = -\frac{1}{2}x$ and passing through the point $(0, 5)$

7. Given the following graph of the linear function, determine the equation of the line.

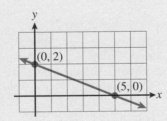

8. a. The building where your computer graphics store is located is 10 years old and has a value of $200,000. When the building was 1 year old, its value was $290,000. Assuming that the building's depreciation is linear, express the value of the building as a function, f, of its age, x, in years.

b. What is the slope of the line? What is the practical meaning of the slope in this situation?

c. What is the vertical intercept? What is the practical meaning of the vertical intercept in this situation?

d. What is the horizontal intercept? What is the practical meaning of the horizontal intercept in this situation?

9. Determine the vertical intercept of the following functions. Solve for y if necessary.

  a. $y = 2x - 3$ 　　　　　　　**b.** $y = -3$ 　　　　　　　**c.** $x - y = 3$

  d. What relationship do the graphs of these functions have to one another?

  e. Use your graphing calculator to graph the functions in parts a–c on the same coordinate axes. Compare your results with part d.

10. Determine the slopes and y-intercepts of each of the following functions. Solve for y if necessary.

  a. $y = -2x + 1$ 　　　　　　**b.** $2x + y = -1$ 　　　　　　**c.** $-4x - 2y = 6$

  d. What relationship do the graphs of these functions have to one another?

  e. Use your graphing calculator to graph the functions in parts a–c on the same coordinate axes. Compare your results with part d.

11. Determine the slopes and y-intercepts of each of the following functions. Solve for y if necessary.

  a. $y = -3x + 2$ 　　　　　　**b.** $3x + y = 2$ 　　　　　　**c.** $6x + 2y = 4$

  d. What relationship do the graphs of these functions have to one another?

  e. For two lines to be parallel to each other, what has to be the same?

  f. For two lines to lie on top of each other (coincide), what has to be the same?

  g. Use your graphing calculator to graph the functions in parts a–c on the same coordinate axes. Compare your results with part d.

12. a. Graph the function defined by $y = -2x + 150$. Indicate the vertical and horizontal intercepts. Make sure to include some negative values of x.

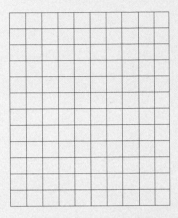

b. Using your graphing calculator, verify the graph you have drawn in part a.

c. Using the graph, determine the domain and range of the function.

d. Assume that a 150-pound person starts a diet and loses 2 pounds per week for 15 weeks. Write the equation modeling this situation.

e. Compare the equation you found in part d with the one given in part a.

f. What is the practical meaning of the vertical and horizontal intercepts you found in part a?

g. What is the practical domain and range of this function for the situation given in part d?

13. a. You pay a flat fee of $25 per month for your trash to be picked up, and it doesn't matter how many bags of trash you have. Use x to represent the number of bags of trash, and write a function, f, in symbolic form to represent the total cost of your trash for the month.

b. Sketch the graph of this function.

c. What is the slope of the line?

14. You work as a special events salesperson for a golf course owned by your city. Your salary is based on the following. You receive a flat salary of $1500 per month for sales of $10,000 or less; for the next $30,000 of sales, you receive your salary plus 2% of the sales over $10,000 and up to $40,000; and for any sales exceeding $40,000, you receive your salary and commission of 4% of sales over $40,000.

a. Write a piecewise function, f, that specifies the total monthly salary when x represents the amount of sales for the month.

b. Graph the function.

c. What is your salary if your sales are $25,000?

d. You need to make $3150 to cover your expenses this month. What will your sales have to be for your salary to be that amount?

15. During the years 2006–2010, the number of finishers in a large marathon increased. The following table gives the total number of finishers (to the nearest hundred) each year, where t represents the number of years after 2006.

Years After 2006, t	0	1	2	3	4
Number of Finishers, n	7800	9100	10,000	10,900	12,100

a. Enter the data from your table into your calculator. Determine the linear regression equation model, and write the result.

b. What is the slope of the regression line? What is the practical meaning of the slope in this situation?

c. What is the vertical intercept? What is the practical meaning of the vertical intercept in this situation?

d. Use your graphing calculator to graph the regression line in the same screen as the scatterplot. How well do you think the line fits the data?

e. Use your regression model to determine the number of finishers in 2012.

f. Did you use interpolation or extrapolation to determine your result in part e? Explain.

g. Do you think that the prediction for the year 2024 will be as accurate as that in 2012? Explain.

16. Solve the systems of equations. Solve at least one algebraically and at least one graphically.

a. $3x - y = 10$
$5x + 2y = 13$

b. $4x + 2y = 8$
$x - 3y = -19$

c. $2x + y = 10$
$y = -2x + 13$

d. $2x + 6y = 4$
$x + 3y = 2$

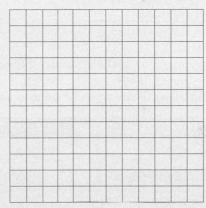

17. The employees of a beauty salon order lunch 2 days in a row from the corner deli. Lunch on the first day consists of five small pizzas and six cookies for a total of $27. On the second day, eight pizzas are ordered along with four cookies, totaling $39. To know how much money to pay, employees have to determine how much each pizza and each cookie cost. How much does the deli charge for each pizza and each cookie?

18. Solve these 3×3 systems of linear equations algebraically or use matrices.

a. $x + y + z = 3$
$2x - y + 2z = 3$
$3x + 2y - z = 0$

b. $2x + 4y - z = -2$
$x - 2y + z = -5$
$-2x + y + 2z = 7$

c. $4x - 2y + z = 1$
$2x + 6y - z = 3$
$-3x + 4y + z = -1$

d. $3x - 4y - z = -1$
$4x + y - 5z = 10$
$-x + 2y + z = 11$

19. Your favorite photography studio advertises a family portrait special in the newspaper. There are three different print sizes available: small (3 inches $\times$ 5 inches), medium (5 inches $\times$ 7 inches), and large (8 inches $\times$ 10 inches). There are three different packages that can be ordered.

PACKAGE	SMALL	MEDIUM	LARGE	TOTAL COST
A	4	2	1	$6
B	6	4	2	$11
C	10	6	4	$19

a. Normally, small prints sell for $0.65 per print, medium for $1.10, and large for $2.50. Decide whether you are getting a good deal by determining the cost per print of each size print.

b. Will you take advantage of the special? Explain.

20. a. Sketch the graph of $f(x) = |x + 2|$.

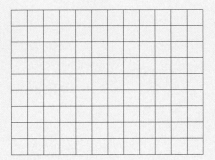

b. Determine the interval over which the function is increasing. Over which interval is the function decreasing?

c. Determine the domain of the function.

d. Determine the range of the function.

e. Graph $g(x) = -|x + 2|$. How does the graph compare with the graph in part a?

f. Let $h(x) = |x| + 2$. Explain what makes the functions $f(x)$ and $h(x)$ similar. How are the functions different?

The Algebra of Functions

chapter

2

Cluster 1

Addition, Subtraction, and Multiplication of Polynomial Functions

Activity 2.1

Spending and Earning Money

Objectives

1. Identify a polynomial expression.

2. Identify a polynomial function.

3. Add and subtract polynomial expressions.

4. Add and subtract polynomial functions.

You are planning a trip with friends and are going to rent a van. The van you want rents for $75 per day. You are given 100 free miles each day and are charged $0.20 per mile for extra miles. The dealer claims that you can expect to average 25 miles per gallon. The Auto Club says you can expect to pay an average of $2.50 per gallon for gas on your trip. You are planning to be gone for 8 days, and you know that you will be traveling at least 1000 miles on the trip. You have been assigned the job of estimating the total cost of operating the van for the trip.

1. The cost of renting the van is a function of the total number of miles driven on the trip. Note that with 100 free miles each day, you will not have to pay for 800 miles of the trip over the 8 days.

    a. Complete the following table. $R(m)$ represents the cost of renting the van for a given total number of miles, m, driven in the 8 days.

On the Road Again

Total Miles Driven in 8 Days, m	1000	1200	1400	1600
Cost of Renting the Van, $R(m)$ ($)				

    b. Write an equation for the cost, $R(m)$, of renting the van in terms of the total number of miles, m, traveled in the 8 days.

2. The cost of the fuel is also a function of the total number of miles, m, traveled on the trip. The rental agency will start you with a full tank but expects a full tank when you return.

    a. Complete the following table, where $F(m)$ represents the cost of the fuel for a given number of miles, m, driven in the 8 days.

m	1000	1200	1400	1600
$F(m)$				

177

b. Write an equation for the cost, $F(m)$, of fuel in terms of m.

3. The total cost of using the van for the 8-day period is a function of the total number of miles, m, traveled.

a. Complete the following table, where $C(m)$ represents the total cost of renting. The entries for $R(m)$ and $F(m)$ were determined in Problems 1 and 2.

m, TOTAL MILES	R(m), RENTAL COST ($)	F(m), FUEL COST ($)	C(m), TOTAL COST ($)
1000	640	100	
1200	680	120	
1400	720	140	
1600	760	160	

b. What is the relationship among the rental cost, $R(m)$, the fuel cost, $F(m)$, and the total cost, $C(m)$?

c. Add the equations in Problems 1 and 2 to determine an equation to define the total cost, $C(m)$, of using the van, as a function of the total miles, m, traveled on the trip.

Definition

The rental van situation involves the addition of functions. The total cost function determined in Problem 3 is called the **sum function.** The notation is

$$C(m) = R(m) + F(m).$$

Example 1   *Functions f and g are defined by the following tables.*

x	−2	0	2	4	6	8
f(x)	0	2	6	20	42	72

x	−2	0	2	4	6	8
g(x)	7	3	7	19	39	67

Complete the following table for the sum of f and g.

SOLUTION

For any given x, when functions are added, the outputs are added.

x	−2	0	2	4	6	8
f(x) + g(x)	0 + 7 = 7	2 + 3 = 5	6 + 7 = 13	20 + 19 = 39	42 + 39 = 81	72 + 67 = 1

4. Enter the rental function, R, the fuel function, F, and the total cost function, C, into your graphing calculator. Use the table feature, in ASK mode, to complete the following table for four input values.

m	$R(m)$ ($)	$F(m)$ ($)	$C(m)$ ($)
1250			
904			
1303			
1675			

Note: The outputs of the sum function, C, are sums of the outputs of the functions R and F.

Subtraction of Functions

You have returned from your trip, and now it's back to work. You are the owner of a small pet kennel. Your kennel can accommodate at most 20 dogs. Your current charge for boarding a dog is $12 per day. Utility bills are approximately $15 per day. The cost of feeding each dog, cleaning its stall, and exercising it is approximately $7.15 per day.

5. a. Suppose you let the input variable d represent the number of dogs boarding on a given day. Determine an equation that expresses the total revenue, $R(d)$, as a function of the number of dogs, d, boarding on a given day.

b. Complete the following input/output table for the revenue function R.

d	0	5	10	15	20
$R(d)$ ($)					

6. a. The total daily cost of operating the kennel is a function of the number of dogs boarding on a given day. If $C(d)$ represents the cost, write an equation for $C(d)$ in terms of d.

b. Complete the following table for the daily cost function, C.

d	0	5	10	15	20
$C(d)$ ($)					

7. The results from Problems 5 and 6 can be used to determine the profit, $P(d)$, in terms of the number of boarded dogs, d.

a. Use the output values from Problems 5b and 6b to complete the following table for the profit. Recall that profit = revenue − cost.

d	0	5	10	15	20
$P(d)$ ($)					

b. Using the equations for revenue, $R(d)$, and cost, $C(d)$, in Problems 5a and 6a, determine an equation for the profit, $P(d)$, as a function of d. Use the new equation to verify some of the entries in the table in part a.

Definition

The kennel situation involves the subtraction of functions. The profit function determined in Problem 7b is called the **difference function**. The notation is

$$P(d) = R(d) - C(d).$$

Example 2   *Functions f and g are defined by the following tables.*

x	−2	0	2	4	6	8
f(x)	2	0	6	20	42	72

x	−2	0	2	4	6	8
g(x)	7	3	7	19	39	67

Complete the following table for the difference of f and g.

SOLUTION

For any given x, when functions are subtracted, the outputs are subtracted.

x	−2	0	2	4	6	8
f(x) − g(x)	2 − 7 = −5	0 − 3 = −3	6 − 7 = −1	20 − 19 = 1	42 − 39 = 3	72 − 67 = 5

8. a. Sketch the graphs of the revenue function R, the cost function C, and the profit function P on the following grid. Label the axes with the appropriate scales.

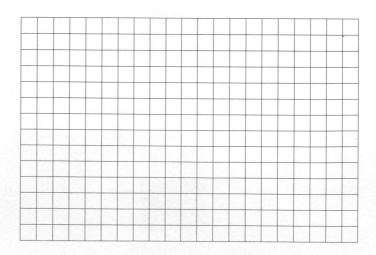

Note: The practical domain of these functions consists of whole numbers. While it is common practice to draw the graphs with continuous lines, the functions are *discrete*.

b. Use the graphs from part a to determine the break-even point for your pet-kennel business. That is, determine the number of dogs necessary for the profit to equal $0. Explain how you determined your answer.

c. Write and solve an equation to determine the break-even point.

d. For you to obtain a profit of at least $50, how many dogs must board?

Polynomial Expressions, Functions, and Terminology

The functions encountered in this activity, such as the total cost of van rental function defined by $C(m) = 0.2m + 440$, are examples of a special category of functions called **polynomial functions**. Such functions are defined by equations of the form $y = p(x)$, where x is the input variable and $p(x)$ is a polynomial expression involving the input variable. Therefore, to identify a polynomial function, you must be able to identify a polynomial expression.

Definition

Any expression that is formed by adding terms of the form ax^n, where a is a real number and n is a nonnegative integer, is called a **polynomial expression** in x.

Example 3

POLYNOMIAL EXPRESSIONS	EXPRESSIONS THAT ARE NOT POLYNOMIALS
10, $5x$, $-3x^2 + 2$, $4x^3 + 7x^2 - 3$, and $\frac{5}{3}x^4$	$\frac{3}{x}$, $\sqrt{x}$, and $2x^3 + \frac{1}{x^3}$

The polynomial expression $4x^3 + 7x^2 - 3$ is said to be written in **descending order** because the term having the largest exponent is written first, the term having the next largest exponent is written second, and so on. The same polynomial written in **ascending order** is

$$-3 + 7x^2 + 4x^3.$$

The expressions $\frac{3}{x}$ and $\sqrt{x}$ are not polynomial expressions because a variable appears in denominator and under a radical, respectively. Later in this chapter, you will learn tha $\frac{3}{x} = 3x^{-1}$ and $\sqrt{x} = x^{1/2}$. Variables in polynomial expressions can only have nonnegativ integer exponents.

Polynomial expressions can be classified by the number of terms that are contained in the expression.

Terminology

- A **monomial** is a single term that consists of a constant or a constant times a variable or variables raised to nonnegative integers, such as -3, $2x^4$, and $\frac{1}{2}s^2$.
- A **binomial** is a polynomial that has two terms, such as $4x^3 + 2x$ and $3t - 4$.
- A **trinomial** is a polynomial that has three terms, such as $3x^4 - 5x^2 + 10$ and $5x^2 + 3x - 4$.

---

9. a. Write "yes" if the expression is a polynomial in x. Write "no" if the expression is not a polynomial.

$3x - 5$ _____ $5x^3 - 2x + 7$ _____

$\frac{5x}{3}$ _____ $\sqrt{x} + 10$ _____

$\dfrac{2}{x^2} - 8$ _____

b. In part a, classify any polynomial expressions as a monomial, binomial, or trinomial.

Example 4   *Examples of polynomial and nonpolynomial functions are given in the following table.*

POLYNOMIAL FUNCTIONS	NONPOLYNOMIAL FUNCTIONS
$y = 10$, $y = 5x$	$f(x) = \frac{3}{x}$
$f(x) = -3x^2 + 2$	$g(x) = \sqrt{x}$
$y = 4x^3 + 7x - 3$	$h(x) = 2x^3 + \frac{1}{x^3}$
$g(x) = \frac{5}{3}x^4$	$y = 3x^{-2}$

---

10. Write "yes" if the equation defines a polynomial function. Write "no" if it does not.

a. $y = 5x^2 + 2x - 1$ **b.** $f(x) = 3x + \frac{1}{x}$

c. $g(m) = 1.75m - 7$ **d.** $R(t) = \sqrt{t} + 7$

Addition and Subtraction of Polynomial Expressions

Operations with polynomial functions (such as addition and subtraction) involve operations with polynomial expressions. Example 5 demonstrates how to perform these operations.

> **Example 5** *Consider the polynomial functions f and g defined by*
>
> $$f(x) = 2x^2 + 3x - 5 \quad \text{and} \quad g(x) = -x^2 + 5x + 1.$$
>
> Determine each of the following.
>
> **a.** $f(x) + g(x) = (2x^2 + 3x - 5) + (-x^2 + 5x + 1)$ Remove parentheses.
>
> $\qquad\qquad\quad = 2x^2 + 3x - 5 - x^2 + 5x + 1$ Combine like terms.
>
> $\qquad\qquad\quad = x^2 + 8x - 4$
>
> **b.** $f(x) - g(x) = (2x^2 + 3x - 5) - (-x^2 + 5x + 1)$ Change the sign of each term of
> the polynomial being subtracted.
>
> $\qquad\qquad\quad = 2x^2 + 3x - 5 + x^2 - 5x - 1$
>
> $\qquad\qquad\quad = 3x^2 - 2x - 6$
>
> **c.** $-5f(x) = -5(2x^2 + 3x - 5)$ Apply the distributive property.
>
> $\qquad\quad\;\; = -10x^2 - 15x + 25$

11. Given the polynomial functions g and h, defined by

$$g(x) = 4x^2 - 3x + 10, \qquad h(x) = -3x^2 + 5x - 2,$$

determine each of the following:

a. $g(x) + h(x)$

b. $g(x) - h(x)$

c. $-2g(x)$

SUMMARY: ACTIVITY 2.1

1. Given two functions, f and g, the **sum function**, s, is defined by

$$s(x) = f(x) + g(x)$$

and the **difference function**, d, is defined by

$$d(x) = f(x) - g(x).$$

2. Any expression that is formed by adding terms of the form ax^n, where a is a real number and n is a nonnegative integer, is called a **polynomial expression** in x.

3. A **monomial** is a polynomial with one term. A **binomial** is a polynomial with two terms. A **trinomial** is a polynomial with three terms. Polynomials having more than three terms are not given special names.

4. A **polynomial function** is any function defined by an equation of the form $y = f(x)$, where $f(x)$ is a polynomial expression in x. For example,

$$y = \underbrace{2x^3 + 5x^2 - x + 1}_{f(x),\ \text{a polynomial}}.$$

EXERCISES: ACTIVITY 2.1

1. You are a financial planner. In an effort to attract new customers, you sponsor a dinner at a local restaurant. The restaurant will charge $100 for the banquet room, plus $12.50 per person for each meal. You will pay these expenses yourself. From past experience, you can expect to make sales to about 15% of the people attending. You also know that the average in sales you can expect from each new client is $750, for which you receive a 13% commission. It is clear that your personal financial success from the event depends on how many people you can attract to this dinner.

a. Complete the following table.

Food for Thought

x, NUMBER OF PEOPLE ATTENDING	COST OF THE BANQUET HALL ($)	TOTAL MEAL COST ($)	TOTAL COST ($)
20	100		
40			
60			
80			
100			

b. Determine a formula for the total cost of restaurant expenses as a function of x, the number of attendees. Represent the total cost by $C(x)$.

c. If 20 people attend, determine the number of new customers.

d. Determine the total sales of the new customers in part c.

e. What is your commission on the sales in part d?

f. Complete the following table. The results from parts c, d, and e are recorded.

x, NUMBER OF PEOPLE ATTENDING	NUMBER OF NEW CUSTOMERS	TOTAL SALES ($)	YOUR COMMISSION ($)
20	3	2,250	292.50
40			
60			
80			
100			

Exercise numbers appearing in color are answered in the Selected Answers appendix.

g. Determine a formula for the total commission that you can expect to generate from this dinner as a function of x, the number of attendees. Represent the total commission by $T(x)$.

h. Combine the formulas in parts b and g to define a new function for the profit, P, that you can expect from your dinner. A basic business equation is

$$\text{profit} = \text{revenue} - \text{cost}.$$

In this situation, the revenue is the total commission.

i. What is the practical domain of this new function?

j. Use your new function to determine how many people must attend for you to break even. Explain how you arrive at your decision.

k. You hope to make a profit of $500 on the dinner. How many people must attend for you to meet this goal? Explain. Write an equation that can be solved to answer this question. Then show how to solve the equation.

2. Write "yes" if the expression is a polynomial in x. Write "no" if the expression is not a polynomial. If the expression is a polynomial, classify it as a monomial, binomial, or trinomial.

a. $5x^{-3} + 4$

b. $-3x^{10} - 2x^2 - 1$

c. $x^{1/2}$

d. x

e. $\frac{5}{4x} - 8$

3. Suppose that f and g are defined by the following tables.

x	0	2	4	6	8	10
f(x)	3	-5	0	7	-1	4

x	0	2	4	6	8	10
g(x)	1	-1	1	-1	3	4

Complete the following table.

x	0	2	4	6	8	10
$f(x) + g(x)$						
$f(x) - g(x)$						

4. a. Suppose that $f(x) = 4x + 1$ and $g(x) = -2x + 4$. Determine an algebraic expression for $f(x) - g(x)$ by subtracting $g(x)$ from $f(x)$ and combining like terms.

b. Complete the following table, using f and g from part a.

x	$f(x)$	$g(x)$	$f(x) - g(x)$
0			
2			
4			
6			

c. Use the table feature of your graphing calculator to complete the following table for four input values not used in part b.

x	$f(x)$	$g(x)$	$f(x) - g(x)$

d. Do the results in part c agree with your understanding of the difference of two functions? Explain.

5. The algebraic skills necessary for determining the algebraic form of the new functions are those of simplifying expressions and combining like terms. Simplify the following.

a. $(2x + 3) + (3x - 5)$

b. $(2x^2 - 3x + 1) - (x^2 - 6x + 9)$

c. $2(x + 9) - 3(x - 4)$

d. $14x - 9 - 3(x^2 + 2x - 2)$

e. $4(3x - 2) - (7 - 3x)$

f. $6x + 5 + 3(2 - 2x)$

g. $2x^2 + 5x - 3(3 - x^2)$

h. $(5x - 2) - 2(3x^2 - 5x + 1)$

i. $2x + 5 - [3x - 4(5 - x)]$

j. $7x + 2[3x - 2(4 - 5x)] + 6$

6. Given $f(x) = 3x - 5$ and $g(x) = -x^2 + 2x - 3$, determine a formula, in simplest form, for each of the following.

a. $f(x) + g(x)$

b. $f(x) - g(x)$

c. $2f(x) + 3g(x)$

d. $f(x) - 2g(x)$

7. Given $h(x) = 6$, $p(x) = 3 - 4x$, and $r(x) = 4x^2 - x - 6$, determine an expression, in simplest form, for each of the following.

a. $r(x) + h(x)$

b. $p(x) + r(x) - h(x)$

c. $h(x) - p(x)$

d. $r(x) + p(x) + h(x)$

8. Given $f(x) = x^2 - 2$ and $g(x) = x + 4$, determine a value for each of the following.

a. $f(2) + g(2)$

b. $g(3) - f(3)$

c. $f(-5) + g(-5)$

d. $f(-2) - g(-2)$

9. Suppose f and g are defined by the following tables.

x	−6	−4	−2	0	2	4
f(x)	50	16	−2	−4	10	40

x	−6	−4	−2	0	2	4
g(x)	49	25	9	1	1	9

a. Complete the following table.

x	−6	−4	−2	0	2	4
f(x) − g(x)						

b. If $f(x) = 2x^2 + 3x - 4$ and $g(x) = x^2 - 2x + 1$, determine an algebraic expression for $f(x) - g(x)$.

c. Check your answers in the table in part a by using the function you determined in part b.

Activity 2.2

The Dormitory Parking Lot

Objectives

1. Multiply two binomials using the FOIL method.

2. Multiply two polynomial functions.

3. Apply the property of exponents to multiply powers having the same base.

Your dormitory has recently been renovated to bring it up to twenty-first-century standards, but the parking lot is still very inadequate. The current lot has two rows with six parking spaces in each row.

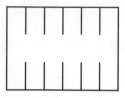

1. What is the current number of parking spaces?

Fortunately, the college has decided to expand this lot. It has also been determined that in order to maintain the relative shape of the lot, for every four cars added to a row, the college will add another row to the lot.

Starting with a geometric representation of the original lot, extend the length by four cars and the width by one row to obtain a geometric model of the new parking lot.

Number of Parking Spaces per Row

6   +   4

Number of Rows

2

+

1

2. Determine the number of parking spaces in each of the four sections of this parking lot diagram; then record the total in the appropriate place in the diagram above.

3. Determine the total number of parking spaces in two different ways.

a. Sum the parking spaces of the sections.

b. Multiply the number of cars in each row by the total number of rows.

4. Let x represent the number of new rows added to the original parking lot.

a. Determine an expression for the total number of rows when x new rows are added.

b. Determine an expression for the number of cars in each row when x new rows are added. Remember, each new row requires an additional four cars to the length of each row.

5. Starting with a geometric representation of the current parking lot, extend the parking lot by x rows to obtain a geometric model of the new lot.

6. **a.** The number of parking spaces in each row in Problem 5 is a function of x. Write an equation for the number of parking spaces per row, $s(x)$, as a function of x.

   b. The number of rows in the parking lot is also a function of x. Write an equation for the number of rows, $n(x)$, in terms of x.

   c. The total number of parking spaces is a function of x. If $T(x)$ represents the total number of parking spaces in the new lot, use the results in parts a and b to write an equation that defines $T(x)$ as a function of x. Do not simplify.

Definition

The total parking spaces function T in Problem 6c is called a **product function** since $T(x)$ is determined by the multiplication or product of two functions. The notation for a product function is
$$T(x) = s(x) \cdot n(x).$$

---

   d. Complete the following table using the results from parts a–c.

x	s(x)	n(x)	T(x)
2			
5			
8			

Multiplication of Binomials

The total parking space function is defined by $T(x) = s(x) \cdot n(x) = (6 + 4x)(2 + x)$. The product of the binomials $6 + 4x$ and $2 + x$ can be determined using a geometric model.

7. **a.** Return to the geometric model of the parking lot from Problem 5. Determine the number of parking spaces in each section, and fill in that number in each section of the geometric model.

Number of Parking Spaces per Row

b. Sum the parking spaces in all of the sections in the geometric model and simplify the expression. What does this algebraic expression represent?

As a result of Problem 7, you now know that

$$T(x) = (6 + 4x)(2 + x) = 12 + 14x + 4x^2.$$

The geometric model can be used to develop an algorithm (process or procedure) for determining the product of two binomials such as $6 + 4x$ and $2 + x$. The sum of the parking spaces in Problem 7 is

$$2 \cdot 6 + 2 \cdot 4x + x \cdot 6 + x \cdot 4x.$$

This sum can be obtained from the terms of the binomial factors $6 + 4x$ and $2 + x$ as follows:

$$(6 + 4x) \cdot (2 + x) = \overbrace{6 \cdot 2}^{F} + \overbrace{6 \cdot x}^{O} + \overbrace{4x \cdot 2}^{I} + \overbrace{4x \cdot x}^{L}$$

$$= 12 + 6x + 8x + 4x^2$$

Combining like terms, you have $12 + 14x + 4x^2$. This procedure is called the FOIL method. FOIL is an acronym for the sum of the products of the first, outer, inner, and last terms of the binomials. Essentially, you multiply each term of the first binomial by each term of the second binomial.

Example 1   *If $f(x) = 3x + 2$ and $g(x) = 2x - 5$, determine $f(x) \cdot g(x)$.*

SOLUTION

$$f(x) \cdot g(x) = (3x + 2)(2x - 5)$$

$$= 3x \cdot 2x + 3x(-5) + 2 \cdot 2x + 2(-5)$$

$$= 6x^2 - 15x + 4x - 10$$

$$= 6x^2 - 11x - 10$$

Note that the product $(3x + 2)(2x - 5)$ can be represented by the following diagram.

	$2x$	-5
$3x$	$6x^2$	$-15x$
$+2$	$4x$	-10

Therefore, $(3x + 2)(2x - 5) = 6x^2 - 15x + 4x - 10 = 6x^2 - 11x - 10.$

8. a. Given $f(x) = x + 7$ and $g(x) = x + 5$, determine a single polynomial expression for $f(x) \cdot g(x)$ by multiplying $(x + 7)(x + 5)$. Write your answer as a sum of terms.

b. Using the functions defined in part a, complete the following table.

x	f(x)	g(x)	f(x)·g(x)
0			
1			
2			
3			
4			

c. Use the table feature of your graphing calculator to complete the following table for four input values not used in part b.

x	f(x)	g(x)	f(x)·g(x)

d. Are the results found in part c consistent with your algebraic solutions in part b? Explain.

e. When you determine the product function, what do you multiply: domain values, range values, or both? Explain.

Multiply Powers Having the Same Base

The volume, v (in cubic feet), of a partially cylindrical storage tank of liquid fertilizer is represented by the formula

$$v(r) = r^2(4.2r + 37.7),$$

where r is the radius (in feet) of the cylindrical part of the tank.

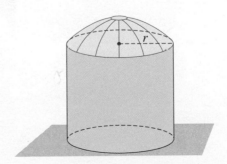

9. Determine the volume of the tank if its radius is 3 feet.

Suppose you were asked to write the expression $r^2(4.2r + 37.7)$ as an equivalent expression without parentheses. Using the distributive property, you would multiply each term within the parentheses by r^2. The first product is $r^2(4.2r)$. What is r^2 times r?

Recall that in the expression r^2, the exponent 2 tells you that the base r is used as a factor two times. In the expression r, the exponent, 1, tells you that the base r is used as a factor once.

$$r^2 \cdot r = \underbrace{(r \cdot r) \cdot r}_{\substack{\text{base } r \text{ is used as} \\ \text{a factor three times.}}} = r^3$$

10. a. Complete the following table.

INPUT r	OUTPUT FOR $r^2 \cdot r$	OUTPUT FOR r^3
2		
4		
5		

b. How does the table demonstrate that $r^2 \cdot r$ is equivalent to r^3?

c. Consider the following products.

   i. $x \cdot x^4 = x^5$　　　　**ii.** $w^2 \cdot w^5 = w^7$　　　　**iii.** $a^2 \cdot a^3 \cdot a^4 = a^9$

What pattern do you observe?

Multiplication Property of Exponents

Let m and n be rational numbers.

To multiply powers of the same base, keep the base and add the exponents.

$$a^m \cdot a^n = a^{m+n}$$

---

11. Expressions for $p(x)$ and $q(x)$ are given in the following table. Fill in the last column of the table with a single power of x.

$p(x)$	$q(x)$	$p(x) \cdot q(x)$
x^2	x^4	
$2x^3$	x	
$-3x^5$	$4x^2$	
$5x^4$	$3x^4$	

12. Multiply $(-2a^5)(8b^3)(3a^2b)$. Explain the steps you used to determine this product.

13. If $f(x) = 2x^2 + 3$ and $g(x) = 5x^3 - 2$, determine $f(x) \cdot g(x)$.

When multiplying polynomials with more than two terms, FOIL cannot be applied. However, the geometric principles behind the FOIL method can still be applied, or you can simply multiply each term of the first polynomial by each term of the second and collect like terms.

14. a. Multiply $(4x + 2)(x^2 - 4x + 3)$. Determine the appropriate products to complete the chart below. Combine like terms, and write the final answer for the product in descending order of the exponents.

	x^2	$-4x$	$+3$
$4x$			
2			

b. Multiply $(4x + 2)(x^2 - 4x + 3)$ by multiplying each term of the first polynomial by each term of the second. Combine like terms, and write the final answer for the product in descending order of the exponents.

c. How do the final answers in parts a and b compare?

Special Products

15. Use FOIL or the rectangle method to determine the product $(x + 4)(x - 4)$.

Notice that the outer and inner products subtract out. The general form for this product is the following identity:

$$(a + b)(a - b) = a^2 - b^2.$$

This expression, $a^2 - b^2$, is the **difference of the squares** of the binomial terms. For Problem 15, $a = x$ and $b = 4$ so $a^2 = x^2$, and $b^2 = 4^2 = 16$. Then $a^2 - b^2 = x^2 - 16$.

16. To use the identity $(a + b)(a - b) = a^2 - b^2$ to determine the product $(2x + 3)(2x - 3)$, complete the following steps.

a. Identify a and b.   **b.** Identify a^2 and b^2.   **c.** Write the product as $a^2 - b^2$.

17. Use FOIL or the rectangle method to determine the product $(x - 4)^2 = (x - 4)(x - 4)$.

Notice that the outer and inner products are equal. The general form for this product is the following identity:

$$(a + b)^2 = a^2 + 2ab + b^2.$$

The expression $(a + b)^2$ is the **square of a binomial**. For Problem 17, $a = x$ and $b = -$ so $a^2 = x^2$, and $b^2 = (-4)^2 = 16$. Then $(a + b)^2 = a^2 + 2ab + b^2 = x^2 + 2 \cdot x \cdot (-$ $+ 4^2 = x^2 - 8x + 16$.

18. To use the identity $(a + b)^2 = a^2 + 2ab + b^2$ to determine the square of the binomia $(2x + 3)^2$, complete the following steps.

    a. Identify a and b.   **b.** Identify a^2 and b^2.

    c. Write the product as $a^2 + 2ab + b^2$.

19. Use the identity $(a + b)(a - b) = a^2 - b^2$ or $(a + b)^2 = a^2 + 2ab + b^2$ to deter mine the following products.

    a. $(x + 6)(x - 6)$ **b.** $(x - 6)^2$

    c. $(4x + 1)(4x - 1)$ **d.** $(3x + 5)^2$

SUMMARY: ACTIVITY 2.2

1. To multiply any two polynomials, multiply each term of the first by each term of the second.

2. A common method to multiply two binomials is the **FOIL method**.

    Step 1. Multiply the FIRST terms in each binomial.

    Step 2. Multiply the OUTER terms.

    Step 3. Multiply the INNER terms.

    Step 4. Multiply the LAST terms.

    Step 5. Sum the products in steps 1–4.

3. Given two functions, f and g, the **product function** is defined by $y = f(x) \cdot g(x)$.

4. To multiply powers of the same base, keep the base and add the exponents. Symbolically, this property of exponents is written as $a^m \cdot a^n = a^{m+n}$, where m and n are rational numbers.

5. Special products:

    a. $(a + b)(a - b) = a^2 - b^2$

    b. $(a + b)^2 = a^2 + 2ab + b^2$

EXERCISES: ACTIVITY 2.2

1. a. You are drawing up plans to enlarge your square patio. You want to triple the length of one side and double the length of the other side. If x represents a side of your square patio, write an expression for the new area in terms of x.

b. You discover from the plan that after doubling one side of the patio, you must cut off 3 feet from that side to clear a shrub. Write an expression in terms of x to represent the length of this side.

c. Use the result from part b to write an expression without parentheses to represent the new area of the patio. Remember that the length of the other side of the original square patio was tripled.

2. A rectangular bin has the following dimensions.

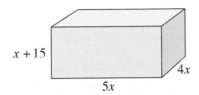

a. Write an expression that represents the area of the base of the bin.

b. Using the result from part a, write an expression that represents the volume of the bin. (Note that the volume is computed by multiplying the area of the base by the height.)

3. You are working for a concert promoter and she has assigned you the task of setting the ticket prices. She knows from experience that you will sell 3000 tickets if you price them at $40 each. She also knows that you will sell 100 more tickets for every dollar that you reduce the ticket price. Your job is to determine the ticket price that will maximize the revenue for the concert.

a. Let x represent the number of $1 reductions in the price of the tickets. Write an equation for the price of each ticket, $P(x)$.

b. Write an equation for the number of tickets sold, $N(x)$.

c. The revenue is the total amount of money collected. In this case the revenue will be determined by multiplying the number of tickets sold by the cost of each ticket. Determine an equation for the total revenue, $R(x)$, as a function of the number of $1 reductions in price, x.

d. What is the domain of the revenue function if the promoter has informed you that she will not sell the tickets for less than $30 each?

e. Complete the following table.

Number of $1 Reductions, x	0	2	4	6	8	10
Price per Ticket, $P(x)$ ($)						
Number of Tickets Sold, $N(x)$						
Total Revenue, $R(x)$ ($)						

f. How do the values in the fourth row of the table in part e relate to the values in the second and third rows?

g. Using the table in part e, determine an appropriate window and graph the total revenue function on your calculator.

h. Rewrite the cost function by multiplying the factors and then combining like terms.

i. Graph the new function from part h, in the same window as part g. What do you see? Compare this graph to the graph in part g.

4. Use the property of exponents $a^m \cdot a^n = a^{m+n}$ to determine the following products.

a. $3^5 \cdot 3^7$

b. $t^4 \cdot t$

c. $x^2 y^5$

d. $(2z^4)(3z^8)$

e. $(-2x)(3x^2)(-5x^3)$

f. $(a^2b^2)(a^3b^4)$

g. $x^{2n} \cdot x^n$

5. Multiply $(x + 3)(x^2 + 3x - 5)$. Determine the appropriate products to complete the chart. Combine like terms, and write the final answer for this multiplication in descending order of the exponents.

	x^2	$3x$	-5
x	x^3	$3x^2$	
3			-15

6. Multiply $(x^2 + 2x - 3)(2x^2 + 3x - 4)$. Determine the appropriate products to complete the chart. Combine like terms, and write the final answer for this multiplication in descending order of the exponents.

	$2x^2$	$3x$	-4
x^2	$2x^4$	$3x^3$	
$2x$		$6x^2$	$-8x$
-3			12

7. Determine each product, and simplify the result.

a. $(3x + 2)(2x + 5)$

b. $(3x - 2)(2x-5)$

c. $(x + 2)(4x - 3)$

d. $(x - 2)(4x + 3)$

8. Determine the following products, and simplify the results.

a. $(2x + 5)(x - 3)$

b. $(4x + 3)(3x - 2)$

c. $(x + 2)(x^2 + 4x - 3)$

d. $(4 - 3x + x^2)(2x^2 + x)$

e. $(x - 3)(2x^2 - 5x + 1)$

f. $(x - 4)(4 - x^2)$

g. $(x^2 - 3x + 1)(3x^2 - 5x + 2)$

h. $(2x^2 + 5x)(6 - 2x)$

9. a. Expand $(2x + 3)^2$.

b. Multiply $(3x - 2)^2$.

c. Multiply $(5x + 2)(5x - 2)$.

d. Multiply $(x^2 + 5)(x^2 - 5)$.

e. After simplifying in parts c and d, the product contains only two terms. Explain why.
(*Hint*: Compare the first terms to each other and the second terms to each other.)

10. a. Given $f(x) = x + 1$ and $g(x) = 2x - 3$, determine $f(x) \cdot g(x)$ by multiplying and combining like terms.

b. Use f and g as defined in part a to complete the following table.

x	f(x)	g(x)	f(x) · g(x)
0			
1			
2			
3			
4			

c. Use your graphing calculator to plot all three functions, f, g, and the product of f and g. Use the trace or table feature to complete the following table for four input values not used in part b.

x	f(x)	g(x)	f(x) · g(x)

Activity 2.3

Stargazing

Objectives

1. Convert scientific notation to decimal notation.

2. Convert decimal notation to scientific notation.

3. Apply the property of exponents to divide powers having the same base.

4. Apply the definition of exponents $a^0 = 1$, where $a \neq 0$.

5. Apply the definition of exponents $a^{-n} = \frac{1}{a^n}$, where $a \neq 0$ and n is any real number.

On any clear evening, the sky is filled with millions of stars. Some of these are closer to Earth than others. Some are large. Some are small. All of them send light to us. The speed of light through the universe is constant. Light travels at a speed of 300,000 kilometers per second.

1. How many kilometers does light travel in 1 minute?

2. How many kilometers does light travel in 1 hour?

3. How many kilometers does light travel in 1 day?

The result of Problem 3 displayed on the TI-83/TI-84 Plus is 2.592E10. This is the way your calculator displays a very large number in **scientific notation**.

Definition

Scientific notation is a convenient way to write a very large (or small) number. A positive number is written in scientific notation as a number (the base) between 1 and 10 times a power of 10. A negative number is written as a number (the base) between -10 and -1 times a power of 10.

Example 1 *Convert 3.2×10^3 and -9.8×10^7 from scientific notation to decimal notation.*

SOLUTION

$3.2 \times 10^3 = 3.2 \times 1000 = 3200$

$-9.8 \times 10^7 = -9.8 \times 10,000,000 = -98,000,000$

Notice that to convert a number whose absolute value is greater than 1 from scientific notation to decimal notation, you move the decimal point of the base to the right, the number of decimal places being indicated by the exponent of the power of 10.

4. Convert the following numbers from scientific notation to decimal notation.

   a. $2.23 \times 10^4 =$

   b. $-4.78 \times 10^6 =$

   c. $8.37 \times 10^{12} =$

You can use the TI-83/TI-84 Plus to check your answers. To input 2.23×10^4, type in the base, 2.23, and then press $\boxed{\text{2nd}}$ $\boxed{,}$ to access the EE command, then $\boxed{4}$ $\boxed{\text{ENTER}}$. Your display should look like this:

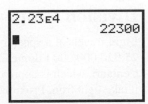

The EE button, which displays as E on the screen, is followed by the exponent of the pow
of 10.

The calculator display for converting -4.78×10^6 is

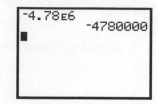

Because the calculator displays at most 10 digits, it will not convert 8.37×10^{12} to a decim
number. Try it.

Decimal Notation to Scientific Notation

In English units, light travels approximately 5,880,000,000,000 miles in 1 year. This distanc
is called the light-year. If you enter 5,880,000,000,000 into the calculator, it converts th
number to scientific notation and returns 5.88E12.

5. a. Write the number 5,880,000,000,000 in scientific notation.

    b. Describe the process the calculator used to convert 5,880,000,000,000 into 5.88E12.

In general, to convert a number whose absolute value is greater than 1 from decimal notation into scientific notation, move the decimal point to the immediate right of the first
nonzero digit. Count the decimal places moved. This number is the exponent of the
power of 10.

Example 2   *Convert 345,000,000 into scientific notation.*

SOLUTION

Because the first nonzero digit is 3, the decimal part of the number is 3.45. Because the deci
mal point needs to be moved eight places to produce 3.45, the exponent on the 10 is 8
Therefore, $345,000,000 = 3.45 \times 10^8$ in scientific notation.

6. Convert the following decimal numbers into scientific notation.

    a. 7,605,000,000,000   **b.** −98,300,000

Division Property of Exponents

Light travels at a speed of approximately 300,000 kilometers per second or approximately
25,920,000,000 kilometers per day. The nearest star to Earth other than the Sun is Proxima
Centauri, which is approximately 39,740,000,000,000 kilometers from Earth. How long does
it take light from Proxima Centauri to reach Earth?

The answer to the question is determined by dividing 39,740,000,000,000 kilometers by 25,920,000,000 kilometers per day. If you do this with your calculator, it produces the following screen.

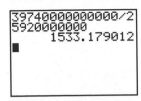

7. Change the mode of the calculator to Scientific Notation (Sci).

Now divide 39,740,000,000,000 kilometers per day by 25,920,000,000 kilometers by entering both numbers into your calculator in scientific notation. How does your result compare to the result performed in Normal mode?

8. Look at this calculation in scientific notation more closely:

$$\frac{3.974 \times 10^{13}}{2.592 \times 10^{10}} = 1.533179012 \times 10^{3}.$$

a. The 1.533179012 is a result of dividing 3.974 by 2.592. Check this result with your calculator.

b. The 10^3 is a result of dividing 10^{13} by 10^{10}. This means that $\frac{10^{13}}{10^{10}} = 10^3$. Note that

$$\frac{10^{13}}{10^{10}} = \frac{10 \times 10 \times 10 \times 10 \times 10 \times 10 \times 10 \times 10 \times 10 \times 10 \times 10 \times 10 \times 10}{10 \times 10 \times 10 \times 10 \times 10 \times 10 \times 10 \times 10 \times 10 \times 10} = 10^3.$$

Rather than expanding the powers and dividing out the common factors, you can obtain the exponent 3 from the exponents of the powers you are dividing. You simply subtract the exponent in the denominator, 10, from the exponent in the numerator, 13.

Problem 8b demonstrates another important property of exponents.

Division Property of Exponents

Let m and n be rational numbers. To divide powers of the same base, keep the base and subtract the exponents

$$\frac{a^m}{a^n} = a^{m-n}, \text{ where } a \neq 0.$$

Example 3

a. $\dfrac{3^5}{3^2} = 3^{5-2} = 3^3$

b. $\dfrac{x^{15}}{x^9} = x^{15-9} = x^6$

c. $\dfrac{15t^6}{3t^2} = \dfrac{15}{3} \cdot \dfrac{t^6}{t^2} = 5 \cdot t^{6-2} = 5t^4$

d. $\dfrac{-4a^7}{2a^5} = -2a^2$

e. $\dfrac{x^9}{y^5}$ *cannot be simplified because the bases x and y are different.*

9. Use the division property to simplify the following expressions.

a. $\dfrac{8^9}{8^4} =$ **b.** $\dfrac{x^6}{x} =$ **c.** $\dfrac{10w^8}{4w^5} =$

d. $\dfrac{6t^{13}}{2t^7} =$ **e.** $\dfrac{5^2}{5^2} =$

The result of Problem 9e using the division property is $\dfrac{5^2}{5^2} = 5^{2-2} = 5^0$. If you did the sam~ problem by first writing 5^2 as 25, the result would be $\dfrac{5^2}{5^2} = \dfrac{25}{25} = 1$. Therefore, it must be tr~ that $5^0 = 1$. In the same way it can be shown that $2^0 = 1$, $10^0 = 1$, and so on. This leads ~ the following definition.

Definition

Zero Exponents

$a^0 = 1$ if $a \neq 0$.

Example 4

a. $16^0 = 1$ **b.** $\left(\dfrac{3}{x}\right)^0 = 1, x \neq 0$

c. $(3x)^0 = 1$ provided that $x \neq 0$. Note that $3x^0 = 3 \cdot 1 = 3$.

d. $5(x + 3)^0 = 5, x \neq -3$

10. Simplify the following expressions. Assume $x \neq 0$.

a. 7^0 **b.** $2x^0$ **c.** $(5x)^0$

d. $\left(\dfrac{4}{x}\right)^0$ **e.** $-3(x^2 + 4)^0$

Negative Integer Exponents

Light travels at a rate of 25,920,000,000 kilometers per day. The Sun is 149,600,000 kilome~ ters from Earth. You have determined that it takes 1533 days for light to travel from the sec~ ond-nearest star, Proxima Centauri, to Earth. How many days does it take for light from the nearest star, the Sun, to travel to Earth?

To answer this question, divide 149,600,000 kilometers by 25,920,000,000 kilometers pe~ day. If you convert both numbers to scientific notation and work in Sci mode on your calcu~ lator, your results should resemble the following.

```
1.496E8/2.592E10
     5.771604938E-3
```

This says that light travels from the Sun to Earth in 5.772×10^{-3} days. If you perform the same calculation in Normal mode, your results should resemble the following.

```
1.496E8/2.592E10
      5.771604938E-3
1.496E8/2.592E10
         .0057716049
■
```

Therefore, you know that 5.772×10^{-3} days = 0.005772 days.

11. Describe in your own words how to convert a number such as 5.772×10^{-3} (written in scientific notation) to its equivalent representation 0.005772 (written in decimal notation).

Therefore, to convert a number in scientific notation with a negative exponent, n, to decimal notation, move the decimal point $|n|$ places to the left.

Example 5 *Convert 6.3×10^{-3} and -17.7×10^{-4} to decimal notation.*

SOLUTION

 a. $6.3 \times 10^{-3} = 0.0063$

 b. $-17.7 \times 10^{-4} = -0.00177$

You can check these results by entering the numbers into your calculator in scientific notation as long as your calculator is in Normal mode.

12. Convert the following to decimal notation.

 a. $5.61 \times 10^{-5} =$

 b. $9.071 \times 10^{-7} =$

Look again at 5.772×10^{-3} days = 0.005772 days. You know that

$$0.005772 = 5.772 \times 0.001$$

$$= 5.772 \times \frac{1}{1000}$$

$$= 5.772 \times \frac{1}{10^3}.$$

Therefore, $5.772 \times 10^{-3} = 5.772 \times \frac{1}{10^3}$. It follows that $10^{-3} = \frac{1}{10^3}$. This leads to the following definition of negative exponents.

Definition

Negative Exponents

If $a \neq 0$ and n is a rational number, then $a^{-n} = \left(\frac{1}{a}\right)^n = \frac{1}{a^n}$.

Example 6 *Rewrite the following expressions using only positive exponents.*

a. $3^{-4} = \dfrac{1}{3^4} = \dfrac{1}{81}$

b. $(2x)^{-3} = \dfrac{1}{(2x)^3} = \dfrac{1}{8x^3}$

c. $x^{-1} = \dfrac{1}{x}$

d. $\dfrac{1}{x^{-4}} = x^4$

e. $3y^{-2} = \dfrac{3}{y^2}$

f. $\dfrac{-2a^{-3}}{b^{-2}} = \dfrac{-2b^2}{a^3}$

g. $x^{-3} \cdot x^{-5}$

Method 1. Apply the multiplication property of exponents first.

$x^{-3} \cdot x^{-5} = x^{-8} = \dfrac{1}{x^8}$

Method 2. Apply the definition of negative exponents first.

$x^{-3} \cdot x^{-5} = \dfrac{1}{x^3} \cdot \dfrac{1}{x^5} = \dfrac{1}{x^8}$

h. $\dfrac{x^{-3}}{x^4}$

Method 1. Apply the division property of exponents first.

$\dfrac{x^{-3}}{x^4} = x^{-3-4} = x^{-7} = \dfrac{1}{x^7}$

Method 2. Apply the definition of negative exponents first.

$\dfrac{x^{-3}}{x^4} = \dfrac{1}{x^3} \cdot \dfrac{1}{x^4} = \dfrac{1}{x^7}$

13. Rewrite the following expressions using positive exponents only.

a. 5^{-3}

b. $(2z)^{-4}$

c. $6y^{-5}$

d. $\dfrac{4}{x^{-1}}$

e. $\left(\dfrac{x}{y}\right)^{-3}$

f. $\dfrac{x^3}{y^{-4}}$

g. $x^{-4} \cdot x^{-2}$

h. $\dfrac{a^{-2}}{a^{-5}}$

Appendix

Additional examples and exercises involving properties of exponents in this Activity are given in Appendix A.

SUMMARY: ACTIVITY 2.3

1. In **scientific notation**, a positive number is written as a number (the base) between 1 and 10 times a power of 10. A negative number is written as a number (the base) between -10 and -1 times a power of 10.

2. To convert a number whose absolute value is greater than 1 from scientific notation to decimal notation, you move the decimal point of the base to the right, the number of decimal places being indicated by the exponent of the power of 10.

3. To convert a number whose absolute value is greater than 1 from decimal notation into scientific notation, move the decimal point to the immediate right of the first nonzero digit. Count the decimal places moved. This number is the exponent of the power of 10.

4. To divide powers of the same base, keep the base and subtract the exponents: $\dfrac{a^m}{a^n} = a^{m-n}$,

   where $a \neq 0$, and m and n are rational numbers.

5. $a^0 = 1$, where $a \neq 0$.

6. To convert a number in scientific notation with a negative exponent, n, to decimal notation, move the decimal point $|n|$ places to the left.

7. If $a \neq 0$ and n is a real number, then $a^{-n} = \left(\dfrac{1}{a}\right)^n = \dfrac{1}{a^n}$.

EXERCISES: ACTIVITY 2.3

1. According to the United States Department of the Treasury, in December 2008 the total currency estimated in circulation in the United States was \$853,200,000,000. Write this number using scientific notation. Check your result by entering this number into your calculator.

2. According to CTIA, the International Wireless Association, in 2008, there were an estimated 2.6×10^8 cellular phone subscribers in the United States.

   a. Write this number in standard notation.

   b. According to CITA, in 1999 the estimated number of cell phone subscribers in the United States was 86,000,000. Write the number of cell phone subscribers in the United States at that time in scientific notation.

   c. Show by using the properties of exponents that the number of cell phone subscribers in the United States has more than tripled from 1999 to 2008.

3. a. A sextillion has 21 zeros. Write 3 sextillion in scientific notation.

   b. The number 45,000,000,000,000,000 is read 45 quadrillion. Write this number in scientific notation.

   c. Write the number 9×10^{27} in standard notation. The number will be read 9 octillion.

   d. Use scientific notation to divide 9 octillion by 45 quadrillion. Use the properties of exponents.

Exercise numbers appearing in color are answered in the Selected Answers appendix.

4. a. One square inch is equivalent to approximately 0.000000159423 acre. Write this number in scientific notation.

   b. 5.78704×10^{-4} cubic feet is equivalent to 1 cubic inch. Write this number in standard notation.

5. The amount of federal acreage in the United States is approximately 6.35×10^8. The total acreage in the United States is approximately 2.27×10^9.

   a. Use scientific notation to write the ratio of the federal acreage to the total acreage.

   b. Use the properties of exponents to simplify your answer.

In Exercises 6–20, use the properties of exponents to simplify the following, where $x \neq 0$. Write your results with positive exponents only.

6. $\dfrac{3^5}{3^2}$

7. $\left(\dfrac{6}{x}\right)^0$

8. 2^{-5}

9. $10x^0$

10. $\left(\dfrac{1}{x}\right)^{-2}$

11. $4x^{-4}$

12. $(2x)^{-3}$

13. $\dfrac{6x^8}{3x}$

14. $\dfrac{9x^8}{3x^{12}}$

15. $\dfrac{6x^3y^5z^2}{10x^7yz^2}$

16. $x^{-3} \cdot x$

17. $(3x^{-2})(x^{-3})x^2$

18. $\dfrac{10x^4}{5x^{-3}}$

19. $\dfrac{4a^0b^{-4}}{-8a^2b^{-1}}$

20. $a^{-3}(4a^{-1})(-5a^7)$

Activity 2.4

The Cube of a Square

Objectives

1. Apply the property of exponents to simplify an expression involving a power to a power.

2. Apply the property of exponents to expand the power of a product.

3. Determine the nth root of a real number.

4. Write a radical as a power having a rational exponent and write a base to a rational exponent as a radical.

Suppose that you are in charge of decorations for a Monte Carlo Night benefit for the math club. Some of the decorations will be in the form of dice (cubes) that you will make out of foam material. You decide that if the length of each edge of the small die is x units, you want the edges of each large die to be x^2 units.

1. a. Determine the volume V of a die for the edge lengths given in the following table. Also evaluate the expression in the fourth column for the given lengths. The first row is done for you.

LENGTH OF EDGE x INCHES	SMALL DIE $V = x^3$ CUBIC INCHES	LARGE DIE $V = (x^2)^3$ CUBIC INCHES	x^6
2	8	$(2^2)^3 = 4^3 = 64$	$2^6 = 64$
3			
4			

b. The volume of the large die is the cube of a square function defined by $f(x) = (x^2)^3$. Use your graphing calculator to sketch a graph of $f(x) = (x^2)^3$ and $g(x) = x^6$.

c. How do these graphs compare?

d. What do the results of parts a and c demonstrate about the relationship between $(x^2)^3$ and x^6?

The results of Problem 1 suggest that the expression $(x^2)^3$ can be written as x^6. Note also that $(x^2)^3$ indicates the base x^2 is used as a factor three times. Therefore,

$$(x^2)^3 = \underbrace{x^2 \cdot x^2 \cdot x^2}_{\text{The base } x^2 \text{ is used as a factor 3 times.}} = \underbrace{x^{2+2+2}}_{\text{Property 1 of exponents}} = (x^{2\cdot3}) = x^6$$

The simplified expression x^6 can be obtained from $(x^2)^3$ by multiplying the exponents 2 and 3. This leads to the following property of exponents.

Property of Exponents (Power to a Power)

If a is a real number and m and n are rational numbers, then

$$(a^n)^m = (a^m)^n = a^{mn}.$$

Example 1

   a. $(3^2)^3 = 3^{2 \cdot 3} = 3^6 = 729$; $(3^3)^2 = 3^{2 \cdot 3} = 3^6 = 729$

   b. $(x^3)^5 = x^{3 \cdot 5} = x^{15}$; $(x^5)^3 = x^{5 \cdot 3} = x^{15}$

   c. $(a^{-2})^3 = a^{-6} = \dfrac{1}{a^6}$

2. Use the properties of exponents to simplify each of the following.

   a. $(t^3)^5$

   b. $(y^2)^4$

   c. $(3^2)^4$

   d. $2(a^5)^3$

   e. $x(x^2)^3$

   f. $-3(t^2)^4$

   g. $(5xy^2)(3x^4y^5)$

Example 2   *Multiply the series of factors $3x^4(x^2)^3 2x^3$.*

SOLUTION

$3x^4(x^2)^3 \cdot 2x^3$

$= 3x^4x^6 \cdot 2x^3$ Remove parentheses by applying a property of exponents (power to power).

$= 6x^4x^6x^3$ Multiply the numerical coefficients.

$= 6x^{13}$ Apply the multiplicative property of exponents to variable factors that have the same base.

3. Multiply the series of factors $2a^3(a^2)^4 7a^5$.

The Power of a Product

Consider the square of the expression $4x^3$, written as $(4x^3)^2$. Since the base for the squaring is $4x^3$, you can write $(4x^3)^2$ as

$$(4x^3)^2 = 4x^3 \cdot 4x^3 = 16x^6. \qquad (1)$$
$$\uparrow$$
$$\text{base}$$

Note that $4x^3 \cdot 4x^3$ can be written equivalently as

$$4x^3 \cdot 4x^3 = 4 \cdot 4 \cdot x^3 \cdot x^3 = 4^2(x^3)^2. \qquad (2)$$

Comparing the results on lines 1 and 2, the expression $(4x^3)^2$ can be written as

$$(4x^3)^2 = 4^2 \cdot (x^3)^2.$$

Note that each factor in the base $4x^3$ (namely 4 and x^3) can be raised to the second power. This illustrates another important property of exponents.

Property of Exponents (Power of a Product)

If a and b are real numbers, and n is a rational number, then
$$(ab)^n = a^n b^n.$$

Example 3

a. $(x^2 y^3)^4 = (x^2)^4 (y^3)^4 = x^8 y^{12}$

b. $(-2a^5)^3 = (-2)^3 (a^5)^3 = -8a^{15}$

c. $(3x^{-3})^4 = 3^4 (x^{-3})^4 = 81x^{-12} = \dfrac{81}{x^{12}}$

4. Use the properties of exponents to simplify the following.

a. $(2x^4)^3$

b. $(-3x^3)^2$

c. $(a^3 b^{-3})^2$

d. $x^2(x^4 x^5)^2$

e. $3x(-2x^3)^4$

5. Apply the properties of exponents to simplify the following.

i. $y = (5x^3)^2$

ii. $y = (-x^4)^3$

Fractional Exponents

What does $a^{1/2}$ represent? The properties of exponents allow you to adopt a reasonable definition for rational (fractional) exponents such as $\frac{1}{2}$. Let us begin by reviewing the definition of square root.

Definition

Let a represent a nonnegative real number, symbolically written as $a \geq 0$. The principal **square root** of a, denoted by $\sqrt{a}$, is defined as the nonnegative number that when squared produces a.

Example 4

a. $\sqrt{9} = 3$ *because* $3^2 = 9$

b. $\sqrt{100} = 10$ *because* $10^2 = 100$

Note that because $\sqrt{9} = 3$ *and* $3^2 = 9$, *it follows that* $\left(\sqrt{9}\right)^2 = 9$.

In general, $\left(\sqrt{a}\right)^2 = a$ if $a \geq 0$.

Is there a relationship between $\sqrt{a}$ and $a^{1/2}$? To answer this question, you need to assum there is an exponent, m, such that

$$a^m = \sqrt{a}, \qquad a \geq 0.$$

Squaring both sides of this equation, you have

$$(a^m)^2 = (\sqrt{a})^2$$
$$a^{2m} = a^1.$$

Because the bases are equal, the exponents must be equal. Therefore,

$$2m = 1$$
$$m = \tfrac{1}{2}.$$

From this result, the following definition is obtained.

Definition

$a^{1/2} = \sqrt{a}$, where a is a real number, and $a \geq 0$.

---

6. Evaluate each of the following, if possible, and check the answer using your graphing calculator.

a. $36^{1/2}$ **b.** $-9^{1/2}$ **c.** $(-9)^{1/2}$ **d.** $0^{1/2}$

Cube Roots

7. The volume of the following cube is 64 cubic inches.

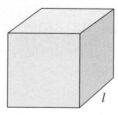

$V = 64$ cubic inches

a. Determine the length, l, of one side (edge) of this cube.

b. Explain how you obtained your answer.

The answer in Problem 7a is called the **cube root** of 64.

Definition

The **cube root** of any real number a, denoted by $\sqrt[3]{a}$, is defined as the number that when cubed gives a.

Example 5

a. $\sqrt[3]{8} = 2$ *since* $2^3 = 8$

b. $\sqrt[3]{125} = 5$ *since* $5^3 = 125$

c. $\sqrt[3]{-1000} = -10$ *since* $(-10)^3 = -1000$

Numbers such as **8, 125,** *and* **−1000** *that have exact cube roots are called perfect cubes.*

Appendix

8. Evaluate each of the following, and check the answer using your graphing calculator. See Appendix C for help in determining cube roots on the TI-83/TI-84 Plus.

a. $\sqrt[3]{1000}$ b. $\sqrt[3]{0}$ c. $\sqrt[3]{-8}$ d. $\sqrt[3]{100}$
                                                                                                 (nearest tenth)

Just as $\sqrt{a}$, where $a \geq 0$, can be written equivalently as $a^{1/2}$, the cube root of a real number a can be written as $a^{1/3}$.

Consider the following argument. If $\sqrt[3]{a}$ can be written in exponential form,

$a^m = \sqrt[3]{a}$	Raise both sides to the 3rd power.
$(a^m)^3 = \left(\sqrt[3]{a}\right)^3$	Simplify both sides.
$a^{3m} = a^1$	Since the bases are equal, the exponents must be equal as well.
$3m = 1$	
$m = \dfrac{1}{3}$	

Therefore, $\sqrt[3]{a} = a^{1/3}$, the cube root of a real number a

Similarly, $\sqrt[4]{a} = a^{1/4}$, the fourth root of $a \geq 0$, *and*

$$\sqrt[5]{a} = a^{1/5}, \text{ the fifth root of a real number } a$$

Definition

In general, $\sqrt[n]{a} = a^{1/n}$, the nth root of a. The number a, called the **radicand**, must be nonnegative if n, called the **index**, is even.

Appendix

9. Calculate each of the following, and then verify your answer using your graphing calculator. See Appendix C for determining nth roots on the TI-83/TI-84 Plus.

a. $\sqrt[4]{81}$ b. $32^{1/5}$ c. $\sqrt[5]{-32}$ d. $-225^{1/4}$

10. a. Try to compute $\sqrt[4]{-81}$ using your graphing calculator. Explain what happens.

b. Explain why the value of a in $\sqrt[n]{a}$, where n is even, cannot be negative.

11. Yachts that compete in the America's Cup must satisfy the International America's Cup Class rule that requires

$$L + 1.25\sqrt{S} - 9.8\sqrt[3]{D} = 16.296 \text{ meters,}$$

where L represents the yacht's length in meters,

S represents the rated sail area, in square meters, and

D represents the water displacement, in cubic meters.

a. Is a yacht having length 21.85 meters, sail area 305.5 square meters, and displacement 21.85 cubic meters eligible to compete? Explain.

b. Explain why the units of your numerical answer in part a are meters.

Rational Exponents

The properties of exponents can be expanded to include rational exponents where the numerator is different from 1. For example,

$$8^{2/3} = 8^{2 \cdot (1/3)} \qquad \text{Apply the property of exponents.}$$
$$= (8^2)^{1/3} \qquad \text{Apply the definition of the 1/3 exponent.}$$
$$= \sqrt[3]{8^2}$$
$$= \sqrt[3]{64}$$
$$= 4$$

In a similar fashion,

$$8^{2/3} = 8^{(1/3) \cdot 2} = (8^{1/3})^2 = \left(\sqrt[3]{8}\right)^2 = 2^2 = 4.$$

Therefore, $8^{2/3}$ can be written equivalently as $\sqrt[3]{8^2}$ or $\left(\sqrt[3]{8}\right)^2$. Note that 3, the denominator of the rational exponent 2/3, is the index. The numerator 2 indicates the power.

Definition

$a^{p/q} = \sqrt[q]{a^p}$ or $a^{p/q} = \left(\sqrt[q]{a}\right)^p$, where $a \geq 0$ if q is even and p is an integer.

Example 6

a. $(-27)^{2/3} = \left(\sqrt[3]{-27}\right)^2 = (-3)^2 = 9$

b. $16^{3/4} = \left(\sqrt[4]{16}\right)^3 = 2^3 = 8$

c. $8^{-2/3} = \dfrac{1}{\left(8^{\frac{1}{3}}\right)^2} = \dfrac{1}{\left(\sqrt[3]{8}\right)^2} = \dfrac{1}{2^2} = \dfrac{1}{4}$

12. Compute each of the following, and then verify the answer using your graphing calculator.

a. $25^{3/2}$

b. $(-8)^{2/3}$

c. $32^{4/5}$

d. $-16^{3/4}$

e. $243^{2/5}$

f. $(-16)^{3/4}$

13. Compute $7^{2/3}$ on your calculator, and explain why your answer is reasonable.

14. Write each of the following using fractional exponents.

a. $\sqrt[3]{x}$ b. $\sqrt[5]{x^3}$

c. $\sqrt[4]{x + 1}$ d. $\sqrt{xy}$

15. Perform the indicated operations by applying the appropriate property of exponents.

a. $x^{1/2}x^{2/3}$ b. $\left(x^3\right)^{3/4}$

c. $\dfrac{x^3}{x^{1/3}}$

16. Determine the domain of each of the following.

a. $f(x) = \sqrt{x}$ b. $g(x) = \sqrt[3]{x}$

17. If $f(x) = \sqrt{x + 2}$, then determine each of the following.

a. $f(-2)$ b. $f(0)$

c. $f(7)$ d. $f(-6)$

18. If $g(x) = \sqrt[3]{x - 5}$, then determine each of the following.

a. $g(5)$ b. $g(13)$ c. $g(-3)$

19. The area of the base of a cube is related to the volume of the cube by the formula

$$A = V^{2/3}.$$

Determine the base area of a cube having volume 216 cubic inches.

Appendix

Additional examples and exercises involving properties of exponents are given in Appendix A.

SUMMARY: ACTIVITY 2.4

1. If a is a real number and m and n are rational numbers, then $(a^m)^n = a^{mn}$.

2. If a and b are real numbers and n is a rational number, then $(ab)^n = a^n b^n$.

3. Let a represent a nonnegative number, symbolically written as $a \geq 0$. The **principal square root** of a, denoted by $\sqrt{a}$, is defined as the nonnegative number that, when squared, produces a.

4. $\left(\sqrt{a}\right)^2 = a$, if $a \geq 0$.

5. $a^{1/2} = \sqrt{a}$, where $a \geq 0$.

6. $\sqrt[n]{a} = a^{1/n}$, the nth root of a. The number a, called the **radicand**, must be nonnegative if n, called the **index**, is even.

7. $a^{p/q} = \sqrt[q]{a^p}$ or $a^{p/q} = \left(\sqrt[q]{a}\right)^p$, where $a \geq 0$ if q is even and p is a positive integer.

EXERCISES: ACTIVITY 2.4

1. Simplify each expression by applying the properties of exponents. Write your results with positive exponents only.

   a. $(x^3)^6$

   b. $(2x^5)^2$

   c. $(-3x^2)^3$

   d. $(x^4)^{-3}$

   e. $(-4a^{-5})^2$

   f. $(a^2b^{-2}c^3)^{-4}$

   g. $-(x^6)^3$

   h. $(-x^3)^6$

2. Compute each of the following quantities.

   a. $100^{1/2}$ **b.** $144^{1/2}$ **c.** $64^{1/3}$ **d.** $64^{4/3}$

   e. $5^{2/5}$ **f.** $(-8)^{2/3}$ **g.** $25^{-1/2}$ **h.** $27^{-2/3}$

3. Simplify the following.

   a. $x^{1/3} \cdot x^{3/4}$

   b. $(x^{-1/3})^{-1/2}$

   c. $\dfrac{x^{4/5}}{x^{1/3}}$

4. Write each of the following using fractional exponents.

   a. $\sqrt{x}$

   b. $\sqrt[4]{x^3}$

   c. $\sqrt[3]{x + y}$

   d. $\sqrt[5]{a^2 b^3}$

5. If $f(x) = \sqrt{x} - 3$, determine each of the following.

   a. $f(28)$

   b. $f(3)$

   c. $f(-1)$

6. If $g(x) = \sqrt[3]{x} + 10$, determine each of the following.

   a. $g(-74)$

   b. $g(-10)$

   c. $g(17)$

7. The length of time, t (in seconds), it takes the pendulum of a clock to swing through one complete cycle is a function of the length of the pendulum in feet. The relationship is defined by

$$t = f(L) = 2\pi \sqrt{\tfrac{L}{32}},$$

where L is the length of the pendulum.

   a. Rewrite the formula using fractional exponents.

   b. Determine the length of the cycle in time if the pendulum is 4 feet long.

Cluster 1   What Have I Learned?

1. For any two functions f and g,

    I. determine which of the following equations are true and explain why.

    II. determine which of the following equations are not always true and give an example to show why not.

    a. $f(x) + g(x) = g(x) + f(x)$

    b. $f(x) - g(x) = g(x) - f(x)$

    c. $f(x) \cdot g(x) = g(x) \cdot f(x)$

2. Given $f(x) = 2x - 3$ and $g(x) = 4$, for what values of x is $f(x) + g(x) = f(x) \cdot g(x)$?

3. Given the defining equations for two functions, describe how to determine the output of the product function for a particular input value.

4. Explain the difference between 3^4 and 3^{-4}.

5. What will be the sign of the answer if you raise a positive base to a negative exponent?

Cluster 1   How Can I Practice?

1. a. The College International Club is planning a holiday dinner. A banquet room has been reserved, and catering arrangements have been made for a total of $600, a fixed fee independent of the number of persons who attend. The planning committee has decided that a price of $20 per couple is the most it will charge for tickets. What is the *least* number of tickets the committee needs to sell to break even?

b. The committee also decides that once it has met its expenses, it will reduce the ticket charge by $0.50 per couple for each additional ticket (couple) above the break-even point you determined in part a. Let t represent the *additional* tickets sold. Write an equation to show the total number, N, of couples attending the banquet (i.e., the number of tickets sold) as a function of t. Call this function f.

c. The charge per ticket can be represented by the function $C = g(t)$. Write an equation for the charge per ticket, C.

d. Complete the following table for each of the functions $N = f(t)$ and $C = g(t)$.

t	0	2	4	6	8	10
$N = f(t)$						
$C = g(t)$						

e. The total revenue obtained from the ticket sales is the total number of tickets sold multiplied by the charge per ticket. Use the output values from part d to complete the following table for the total revenue function, $R(t) = N \cdot C = f(t) \cdot g(t)$.

t	0	2	4	6	8	10
$R(t)$						

f. Use the results from parts b and c to determine a symbolic rule for $R(t)$.

g. Use your graphing calculator to graph the total revenue function, $R(t)$, on the accompanying grid. What window values—Xmin, Xmax, Ymin, Ymax—do you use?

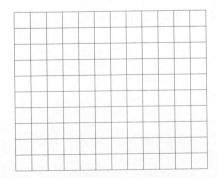

Answers to all How Can I Practice exercises are included in the Selected Answers appendix.

h. Determine the maximum revenue that can be obtained.

i. What is the total number of tickets that must be sold to obtain this maximum revenue? (Be careful.)

2. If $f(x) = x + 2$ and $g(x) = 2x - 3$, determine the following.

a. $f(x) + g(x)$

b. $f(x) - g(x)$

c. $f(x) \cdot g(x)$

d. $f(3) - g(3)$

e. $f(-2) \cdot g(-2)$

f. $3 \cdot f(x)$

3. If $f(x) = x^2 - 2x + 1$ and $g(x) = x^2 + x - 4$, determine each of the following.

a. $f(x) - g(x)$

b. $f(x) \cdot g(x)$

c. $f(-1) + g(-1)$

d. $2f(x) - 3g(x)$

4. Perform the indicated operations and simplify.

a. $(4x + 5) + (x - 7)$

b. $(x^2 - 3x + 1) + (x^2 + x - 9)$

c. $(x + 4) - (3x - 8)$

d. $(3x^2 - 4x - 5) - (x^2 + 9x + 3)$

e. $(5x^2 + 6x - 1) - (7x - 3)$

5. Perform the indicated operations and simplify.

a. $x \cdot x^3$

b. $x^4 \cdot x^5$

c. $(2x^6)(3x^2)$

d. $(xy^3)(x^4y^3z)$

e. $(2x^4y^5z^7)(5x^2z)$

f. $(-3a^2b)(-2a)(-5a^2b^2)$

6. Perform the indicated operations and simplify.

a. $(x - 2)(x - 5)$

b. $(4x - 3)(x + 7)$

c. $(2x - 3)(2x + 3)$

d. $(x - 2)(x^2 + 3x - 5)$

e. $(2x + 1)(x^2 - x + 2)$

f. $3(x - 7) - 2(x^2 + 4x)$

g. $2x(x + 5) - 3x(4 - 3x)$

h. $3x^2 - (x^3 + 1) - x(x^4 - 2)$

i. $(3x + 5)^2$

j. $(2x - 7)^2$

k. $(x + 4)^3$

l. $(5x - 7)(5x + 7)$

7. The tuition at a local state college was \$3350 in 2005. Since then, tuition has increased approximately \$140 per year. Let t represent the years since 2005 and $f(t)$ represent the tuition for any given year. At this same college, the cost for room and board was \$5250 in 2005, and it has increased at a rate of \$350 per year since 2005. Let $g(t)$ represent the cost of room and board for any given year since 2005. The college fees have not changed over the years. They remain \$400. Let $h(t)$ represent the college fees. Assume that the rate of increase in tuition and in room and board continues and that the college fees remain at \$400.

a. Complete the following input/output table for the tuition function, $f(t)$.

t, Years Since 2005	0	5	10	15	20
$f(t)$, Cost of Tuition (\$)					

b. Write an equation that will give the cost of tuition for any year beginning in 2005.

c. Complete the following input/output table for the room and board, $g(t)$.

t, Years Since 2005	0	5	10	15	20
$g(t)$, Cost of Room and Board (\$)					

d. Write an equation that will give the cost of room and board for any year beginning in 2005.

e. Complete the following input/output table for the fees, $h(t)$.

t, Years Since 2005	0	5	10	15	20
$h(t)$, Cost of Fees (\$)					

f. Write an equation that will give the cost of fees for any year beginning in 2005.

g. Use the equations in parts b, d, and f to determine a function, k, that will give the total cost of attending this college for any year after 2005. Write the function in simplest form.

h. Use your graphing calculator to graph functions f, g, h, and k. Use the trace or table feature to complete the following table for four input values.

t, Years since 2005	f(t)	g(t)	h(t)	k(t)
3				
12				
18				
25				

i. If the increase continues at the same rate, in what year will the total cost first equal or exceed \$25,000?

8. Functions f and g are defined by the following tables.

x	−4	−2	0	2	4	6
f(x)	49	9	−7	1	33	89

x	−4	−2	0	2	4	6
g(x)	32	12	0	−4	0	12

a. Complete the following table.

x	−4	−2	0	2	4	6
f(x) − g(x)						

b. If $f(x) = 3x^2 - 2x - 7$ and $g(x) = x^2 - 4x$, determine an algebraic expression for $f(x) - g(x)$.

c. Check your answers in the table in part a by using the function you found in part b.

9. Use the properties of exponents to simplify the following, where $x \neq 0$. Write your results with positive exponents only.

a. $2x^{-3}$

b. $(-3x)^2$

c. 3^{-4}

d. $\left(\dfrac{x}{5}\right)^0$

e. $\dfrac{4x^9}{2x^2}$

f. $(-2x^3y^6)(3xy^2)$

g. $4x^0$

h. $\dfrac{8x}{12x^5}$

i. $\dfrac{10xy^3z^4}{2x^5yz^4}$

j. $(3x^{-2})(-5x^{-4})(x)$ 　　　　**k.** $\dfrac{-4x^3}{2x^{-3}}$ 　　　　**l.** $\dfrac{a^4b^{-2}c^{-5}}{a^0b^2}$

10. Simplify the following by applying the appropriate properties of exponents.

a. $(x^2)^4$ 　　　　　　**b.** $(xy)^3$ 　　　　　　**c.** $(2x^4y)^5$

d. $(3x^2)^3(2xy)^4$ 　　　**e.** $(-5)^2$ 　　　　　　**f.** -5^2

g. $(-2x)^2$ 　　　　　**h.** $(-2x)^3$ 　　　　　**i.** $(x^2)^5(-2x^4)^3$

j. $25^{3/2}$ 　　　　　　**k.** $64^{2/2}$ 　　　　　　**l.** $(-27)^{1/3}$

m. $\left(\dfrac{9}{16}\right)^{3/2}$ 　　　　**n.** $4^{1/3}$ 　　　　　　**o.** $(x^3y^6)^{2/3}$

p. $x^{2/3}x^{1/2}$ 　　　　　**q.** $(8xy)^{1/3}$

11. a. The *Exxon Valdez* spilled oil in Prince William Sound, Alaska, when it ran aground in March 1989. It is estimated that 1.008×10^7 gallons of oil spilled. Write this number in standard form.

b. For insurance purposes you need to determine how many square miles of hunting land you own. One square foot is equivalent to 0.00000003587006 square miles. If your hunting land measures approximately 6,000,000 square feet, how many square miles is this? Use scientific notation and the rules of exponents to determine how many square miles of land you own.

c. The mean radius of the largest planet, Jupiter, is 43,441 miles. The formula for determining the volume of a sphere, V, is

$$V = \tfrac{4}{3}\pi r^3.$$

Determine the approximate volume, in cubic miles, of Jupiter. Express your answer in scientific notation rounded to four decimal places.

12. According to the United States population clock at the United States Census Bureau, the population estimate for January 2010 is 310,233,000. CTIA estimates there will be 280 million cell phone subscribers by January 2010. Write both of these numbers in scientific notation.

13. According to their respective Web sites in June 2009, Yale University's endowment wa approximately $22,600,000,000 and the University of Michigan's endowment wa $7,570,000,000. Show that Yale's endowment is three times the endowment of the University of Michigan by using scientific notation. Which property of exponents di you use in your explanation?

14. You have planned to put in a rectangular patio that measures 5 feet by 7 feet. However you neglected to include enough seating room around your patio table. Let x be the num-ber of additional feet you will extend your plan in each direction.

a. Determine a formula for the area of the extended patio.

b. If x is 4 feet, by how much have you increased the area of the patio from that of the original plan?

c. Your patio table is round with a radius of r feet. You need to purchase an umbrella for the table with an overhang of 2 feet all around. Write an expression for the area that the umbrella will cover.

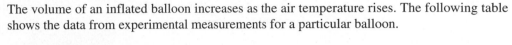

Cluster 2 | Composition and Inverses of Functions

Activity 2.5

Inflated Balloons

Objectives

1. Determine the composition of two functions.

2. Explore the relationship between $f(g(x))$ and $g(f(x))$.

The volume of an inflated balloon increases as the air temperature rises. The following table shows the data from experimental measurements for a particular balloon.

Temperature (in °F)	32	39	42	45	50	58	63	68
Volume (in cu. in.)	35.1	36.5	37.1	37.7	38.7	40.3	41.3	42.3

1. Treating volume as a function of temperature, is this relationship a linear function? Describe how you determined your answer.

Appendix

2. Use your graphing calculator to sketch a scatterplot of these data points. Refer to Appendix C for procedures to plot a set of data on the TI-83/TI-84 Plus. Your screen should resemble the following.

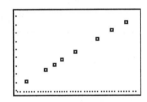

Does your plot verify your answer to Problem 1?

3. a. Use the table of values and/or your graphing calculator to determine the equation for this function. Use V for the output variable and F for the input variable. Call the function g so that $V = g(F)$.

b. Use your graphing calculator to sketch the graph of the volume function, g. Explain how you can be reasonably sure that your function is correct.

c. Use the volume function, g, to determine the volume of the balloon when the temperature is 55°F.

4. Suppose you only have a Celsius thermometer and you want to know the volume of the balloon when the temperature is 10°C. To use the volume function from Problem 3, you must first convert degrees Celsius to degrees Fahrenheit.

a. The formula $F = 32 + 1.8C$ is used to convert from degrees Celsius, C, to degrees Fahrenheit, F. Note that the formula defines F as a function of C. Therefore, F can be written as $h(C)$, where h is the name of the function. Use the given formula to determine the Fahrenheit temperature equivalent to 10°C. That is, determine $h(10)$.

b. Now, use the result from part a to determine the volume for 10°C.

5. You have two functions. For the first function defined by $V = g(F) = 0.2F + 28.7$ F is the input, and $V = g(F)$ is the output.
In the second function defined by $F = h(C) = 32 + 1.8C$, C is the input and $F = h(C)$ is the output.

Complete the following table using a combination of these functions. (Note that the tem perature is given in Celsius units.)

Temperature (in °C)	0	10	20	30	40
Volume (in cu. in.)					

In calculating the volumes in Problem 5, you followed a two-step calculation.

Step 1. First you used the degrees Celsius, C, as the input to $F = g(C) = 32 + 1.8C$, to convert the temperature to degrees Fahrenheit.

Step 2. Then you used this output, degrees Fahrenheit, as the input to the **function** $V = f(F) = 0.2F + 28.7$ to obtain the volume, V.

To shorten this calculation, you can combine these two functions in a special way as described in Problem 6.

6. a. Since $F = 32 + 1.8C$, substitute the expression $32 + 1.8C$ for F in $V = 0.2F + 28.7$, and simplify. You have just determined an equation for V as a function of C.

   b. Use the equation in part a and your graphing calculator to verify the table of values in Problem 5.

Using function notation to describe the procedures in Problem 6a, you started with $F = h(C)$. Then, substituting $h(C)$ for F in the second function, $V = g(F)$, you have

$$V = g(F) = g(h(C)).$$

This function of a function is called the **composition of g and h**.

Example 1   *Given $f(x) = 2x + 3$ and $g(x) = 4x - 1$, determine $g(f(x))$.*

SOLUTION

Substitute $2x + 3$ for $f(x)$ in $g(f(x))$.

$g(f(x)) = g(2x + 3)$ Replace x in the function rule for $g(x)$ with the expression $2x + 3$.

$\quad\quad = 4(2x + 3) - 1$

$\quad\quad = 8x + 12 - 1$ Simplify.

$\quad\quad = 8x + 11$

Therefore, $g(f(x)) = 8x + 11$.

7. If $f(x) = 2x + 5$, determine each of the following.

   a. $f(3)$

   b. $f(a)$

   c. $f(\text{KATE})$

   d. $f(\square)$

   e. $f(g(x))$

8. If $g(x) = 3x^2 + 2x - 4$, then determine each of the following.

   a. $g(2)$

   b. $g(\text{KATE})$

   c. $g(h(x))$

9. a. If $f(x) = 2x + 1$ and $g(x) = 4x - 3$, determine an equation for $f(g(x))$.

   b. Use the result from part a to determine $f(g(3))$.

   c. Determine an equation for $g(f(x))$.

   d. Does $f(g(x)) = g(f(x))$?

SUMMARY: ACTIVITY 2.5

1. If x is the input of a function g, the output is $g(x)$. If $g(x)$ is then used as the input of a function f, the output is $f(g(x))$. The result is a function h, defined by $h(x) = f(g(x))$. The function h is the **composition** of the functions f and g.

2. In general, $f(g(x)) \neq g(f(x))$.

EXERCISES: ACTIVITY 2.5

1. Oil is leaking from a tanker and is spreading outward in the shape of a circle. The area, A, of the oil slick is a function of radius, r (in feet), and is given by $A = f(r) = \pi r^2$. The input, r, for the area function is itself a function of time, t (in hours), since the oil began leaking.

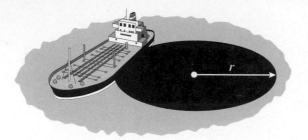

a. If $r = g(t) = 100t$, determine $g(2)$, and interpret its meaning in this situation.

b. Determine $f(g(2))$, and interpret its meaning in this situation.

c. Determine $f(g(10))$, and interpret its meaning in this situation.

d. Determine a general expression for $f(g(t))$.

e. Determine the area of the circular oil spill after 10 hours using the new composite area function found in part d, and compare the result with your answer to part c.

2. If $A = g(r) = 4\pi r^2$, and $r = f(t) = t + 1$, determine $g(f(t))$.

3. If $s = u(t) = -2t^2 + 2t + 1$ and $t = v(x) = 3x - 1$, determine each of the following.

a. $u(v(x))$

b. $v(u(t))$

4. In parts a and b, use the first two tables to complete the third.

a.

x	1	2	3	4	5	6
$f(x)$	2	−1	5	7	19	4

x	−1	2	4	5	7	19
$g(x)$	0	−3	4	1	5	12

x	1	2	3	4	5	6
$g(f(x))$						

b.

x	−3	−2	0	1	2	3
$g(x)$	4	−3	9	6	−1	8

x	−3	−1	4	6	8	9
$h(x)$	5	0	13	−4	−2	1

x	−3	−2	0	1	2	3
$h(g(x))$						

5. You read about the safety features on your brand-new car. According to *Consumer Reports*, in a 30-mile-per-hour collision, the seat belt locks properly about 99% of the time. In approximately 90% of such collisions, the air bag will successfully deploy. Let x represent the number of 30-mile-per-hour collisions in cars of the same year and model that you purchased.

a. Write an equation for $L(x)$ that represents the number of collisions in which the seat belt locks.

b. Write an equation for $D(x)$ that represents the number of collisions in which the air bag deploys.

c. If you suppose that every occupant is uninjured only when everything works as it should in the 30-mile-per-hour collision—that is, the seat belts lock and air bags deploy—write an equation for $S(x)$ that represents the number of times no one is injured out of x collisions. (*Hint:* Use the results of parts a and b to write a composite function.)

d. Evaluate $S(500)$.

e. Examine $L(x)$ and $D(x)$, and decide which of the safety features should be improved immediately to increase the number of survivors, $S(x)$.

6. A quality-control inspector at a bottler of carbon-filtered drinking water notes that the first six bottles processed each day are not acceptable because they are not properly labeled. After those first six bottles, the labeler is warmed up and works just fine for the rest of the shifts. After labeling, the bottles are filled and caps are put in place. Caps are properly applied approximately 99% of the time. After capping, the bottles are inspected. Let x represent the number of bottles processed in a day.

a. Write an equation for $f(x)$ that represents the number of bottles that are properly labeled.

b. Write an equation for $g(x)$ that represents the number of bottles that are satisfactorily capped.

c. Now determine $g(f(10,000))$, and interpret its meaning in this situation.

Activity 2.6

Finding a Bargain

Objective

1. Solve problems using the composition of functions.

You have been waiting for the best price for a winter coat. You see the following advertisement

SUPER SUNDAY
60–70% OFF
original price
when you take
an additional 40% off
already reduced prices

This is it! The time is right! Last week the coat was on sale for 25% off the original price and now you can get the coat for 40% less than last week's sale price.

1. a. Complete the following table.

ORIGINAL COST ($) OF THE COAT, x	LAST WEEK'S PRICE ($)	TODAY'S PRICE ($)
80	60	36
100		
120		
140		

b. What percent of the original price would you have paid for the coat during last week's sale?

c. Write an equation that gives last week's sale price, y, as a function of the original price, x. Therefore, y represents the sale price after the first reduction.

2. The ad indicates that you can now save an additional 40%.

a. What percent of last week's sale price would you now have to pay?

b. Write an equation that gives this week's sale price, z, as a function of last week's sale price, y. Therefore, z represents the sale price after the second reduction.

3. Determining the final sale price of the coat requires a sequence of two calculations:

First: $y = f(x) = 0.75x$ (With 25% off, you pay 75% of the cost.)

Second: $z = g(y) = 0.60y$ (With 40% off, you now pay 60% of the reduced price.)

a. Substitute the expression $0.75x$ for y into the second equation to determine the single-step equation for this composition function.

b. Your answer to part a should imply a savings of 55% from the original price. Does it? Explain.

c. Using function notation, you can determine the equation of the composition function as follows.

$$z = g(y) = g(f(x))$$

Determine $g(f(x))$. How does this compare with the final equation in part a?

d. Use the single-step equation to determine the final price of a coat having the original price

i. $80   **ii.** $100   **iii.** $120   **iv.** $140

SUMMARY: ACTIVITY 2.6

Taking a discount of a discount is an example of **composition of functions** denoted by $g(f(x))$, where

1. f is the function that calculates the first reduction

2. x is the original price and $f(x)$ is the price after the first reduction

3. g is the function that calculates the second reduction

4. $g(f(x))$ is the price after the second reduction has been applied to the reduced price, $f(x)$

EXERCISES: ACTIVITY 2.6

1. In this Activity, the coat was on sale for 25% off the original price, followed by an additional 40% off the sale price. Suppose the sales clerk entered the discount as a 65% reduction.

a. Why do you think the clerk took a 65% reduction?

b. Is this okay with you? Explain.

c. Would this be okay with the manager of the department? Explain.

2. Function f gives the approximate percent increase in harmful ultraviolet rays for an x percent decrease in the thickness of the ozone layer.

x	0	1	2	3	4	5	6
f(x)	0	1.5	3.0	4.5	6.0	7.5	9.0

Exercise numbers appearing in color are answered in the Selected Answers appendix.

Function g gives the expected percent increase in cases of skin cancer for a p percent increase in ultraviolet radiation.

p	0	1.5	3.0	4.5	6.0	7.5	9.0
$g(p)$	0	5.25	10.5	15.75	21.0	26.75	31.5

In Exercises a–d, determine the output value, and interpret your answer.

a. $f(3)$

b. $g(4.5)$

c. $g(f(3))$

d. $g(f(6))$

e. What does $g(f(x))$ determine? That is, what does the output for $g(f(x))$ represent in this situation?

f. Complete the following table.

x	0	1	2	3	4	5	6
$g(f(x))$							

3. A car dealership advertises a factory rebate of $1500, followed by a 10% discount.

a. Let x represent the price of the car. Let $f(x)$ represent the price of the car after the rebate. Determine a rule for $f(x)$.

b. If $g(x)$ represents the price of the car after the 10% discount, determine a rule for $g(x)$.

c. Determine $g(f(20,000))$, and interpret your answer.

d. Suppose the price of a car is $20,000. Determine $f(g(20,000))$, and interpret your answer.

e. Compare the sale price obtained by subtracting the rebate first and then taking the discount, with the sale price obtained by taking the discount first and then subtracting the rebate.

4. You drop a pebble off a bridge. Ripples move out from the point of impact as concentric circles. The radius (in feet) of the outer ripple is given by

$$R = f(t) = 0.5t,$$

where t is the number of seconds after the pebble hits the water. The area, A, of a circle is a function of its radius and is given by

$$A = g(r) = \pi r^2.$$

a. Determine a formula for $g(f(t))$.

b. What are the input and output for the function defined in part a?

5. The following table gives some conversions from U.S. dollars to Canadian dollars in June 2009.

U.S. Dollars	50	100	250	500	1000
Canadian Dollars	56.37	112.75	281.87	563.74	1127.48

The next table gives some conversion from Canadian dollars to Euros on the same day.

Canadian Dollars	56.37	112.75	281.87	563.74	1127.48
Euros	35.89	71.78	179.44	358.89	717.77

Use the two previous tables to complete the following table for converting U.S. dollars to Euros on the same day in June 2009.

U.S. Dollars	50	100	250	500	1000
Euros					

Activity 2.7

Study Time

Objectives

1. Determine the inverse of a function represented by a table of values.

2. Use the notation f^{-1} to represent an inverse function.

3. Use the property
$f(f^{-1}(x)) =$
$f^{-1}(f(x)) = x$
to recognize inverse functions.

4. Determine the domain and range of a function and its inverse.

You are interested in taking 16 credits this semester to complete your program, but you are concerned about the amount of time you will need for studying. The Academic Advising Center provides you with the information listed in the following table.

CREDITS TAKEN (INPUT)	HOURS OF STUDY PER WEEK (OUTPUT)
12	22
13	24
14	26
15	28
16	30
17	32
18	34

Notice that the number of hours of study time is a function of the number of credits taken. Call this function h.

1. Determine $h(14)$ and explain its meaning in practical terms.

2. Use the information from the preceding table to construct another table in which the number of hours of study time is the input and the number of credits taken is the output.

HOURS OF STUDY PER WEEK (INPUT)	CREDITS TAKEN (OUTPUT)

3. a. Using the input/output values from the second table, determine if the number of credits taken is a function of the number of hours of study time. Explain.

b. Call this new function c, and determine $c(30)$. Explain its meaning in practical terms.

4. a. What are the domain and the range of function h, where the number of credits is the input and the number of hours of study per week is the output?

b. What are the domain and the range of function c, where the number of hours of study time is the input and the number of credits taken is the output?

c. How are the domain and range of function h related to the domain and range of function c?

5. Determine each of the following. Refer to the appropriate table.

a. $h(15)$ **b.** $c(28)$

Notice that in Problem 5, the output of h—namely, 28—in part a was used as the input of function c in part b. Recall that this is the composition of h and c and can be written as $c(h(15))$.

6. Determine each of the following. Use the preceding tables given at the beginning of this activity.

a. $c(h(18))$ **b.** $h(c(34))$

c. $c(h(13))$ **d.** $h(c(32))$

7. a. Let x represent the input, number of credits taken, for the function h. What is the result of the composition $c(h(x))$?

b. Let x represent the input, number of hours of study, for the function c. What is the result of the composition $h(c(x))$?

Inverse Functions

Definition

When the output of the composition of two functions is always the same as the input of the composition, the two functions are **inverses** of one another. Each function "undoes" the other. Symbolically, if functions f and g are inverses, then $f(g(x)) = x$ and $g(f(x)) = x$.

Example 1   *Show that f and g, defined by $f(x) = 3x - 1$ and $g(x) = \frac{x+1}{3}$, are inverses.*

SOLUTION

First check the composition $f(g(x))$.

$$f(g(x)) = f\left(\tfrac{x+1}{3}\right) = 3\left(\tfrac{x+1}{3}\right) - 1 = (x + 1) - 1 = x$$

This verifies the first equation. Now check $g(f(x))$.

$$g(f(x)) = g(3x - 1) = \frac{(3x - 1) + 1}{3} = \frac{3x}{3} = x$$

Since $f(g(x)) = g(f(x)) = x, f$, and g are inverses.

8. Are the functions h and c in Problem 7 inverses? Explain.

In general, the inverse of a function f is written f^{-1}. Using this notation, $h = c^{-1}$ and $c = h^{-1}$

Inverse Functions

- The two functions f and g are inverses if $f(g(x)) = x$ and $g(f(x)) = x$.
- The notation for the inverse of f is f^{-1}.
- The domain of f is the range of f^{-1}, and the range of f is the domain of f^{-1}.

Important Note About Notation

The notation for the inverse function of f is potentially confusing. The -1 in $f^{-1}(x)$ is not an exponent! The notation is derived from the fact that the inverse function undoes the arithmetic operations of the original function. You may have seen -1 used as an exponent to denote the reciprocal of a number: $2^{-1} = \frac{1}{2}$; however, $f^{-1}(x) \neq \frac{1}{f(x)}$.

An important property of inverse functions is that the domain and range values are interchanged. For example, from the first table in this section,

$$h = \{(12, 22), (13, 24), (14, 26), \text{etc.}\}.$$

If the input and output values of each ordered pair are interchanged, you have

$$\{(22, 12), (24, 13), (26, 14), \text{etc.}\}.$$

These ordered pairs match the (input, output) pairs of the function c, the inverse of h. See the second table in this section.

9. Given the function $p = \{(2, 4), (-5, 6), (0, 1), (7, 8)\}$, determine the following.

   a. The inverse function, p^{-1}

   b. $p(2)$ **c.** $p^{-1}(4)$ **d.** $p(p^{-1}(4))$

   e. $p^{-1}(p(2))$ **f.** $p(p^{-1}(x))$ **g.** $p^{-1}(p(x))$

10. The function q is defined by the following table, where $s = q(t)$.

t (INPUT)	s (OUTPUT)
1	2
2	3
3	5
4	3

a. Interchange the input and output values, and record the results in the following table.

s (INPUT)	t (OUTPUT)

b. Does the table in part a represent a function? Explain.

c. As a result of part b, the function q does not have an inverse. Could you have predicted that the function q does not have an inverse from the original table? Explain.

SUMMARY: ACTIVITY 2.7

1. The two functions f and g are **inverses** if $f(g(x)) = x$ and $g(f(x)) = x$.

2. The notation for the inverse of f is f^{-1}.

3. The domain of f is the range of f^{-1}, and the range of f is the domain of f^{-1}.

4. $f^{-1}(x) \neq \frac{1}{f(x)}$.

EXERCISES: ACTIVITY 2.7

1. The functions f and g are defined by the following tables.

x	y = f(x)
2	3
4	5
6	7
8	9

x	y = g(x)
3	2
5	4
7	6
9	8

Determine each of the following.

a. $f(g(7))$ **b.** $g(f(1))$ **c.** $f(g(x))$ **d.** $g(f(x))$

2. The function h is defined by the following set of ordered pairs.

$$\{(2, 3), (3, 4), (4, 5), (5, 6)\}$$

a. Write h^{-1} as a set of ordered pairs.

b. Determine $h(3)$ and $h^{-1}(h(3))$.   **c.** Determine $h^{-1}(5)$ and $h(h^{-1}(5))$.

3. The function r is defined by the following table.

Does the function r have an inverse that is also a function? Explain.

x	r(x)
0	2
1	3
2	4
3	2

4. You are planning a trip to Canada and you want to exchange some U.S. currency for Canadian money. The following table will help you with this conversion.

Function, f

Amount in U.S. Dollars (Input)	50	100	250	500	1000
Amount in Canadian Money (Output)	56.37	112.75	281.87	563.74	1127.48

Source: Exchange Calculator, June 2009.

a. Determine $f(100)$ and determine its meaning in practical terms.

b. Use the information from the preceding table to construct another table in which the amount of Canadian money is the input and the U.S. dollar amount is the output.

Amount in Canadian Money (Input)					
Amount in U.S. Dollars (Output)					

c. Use the information from the table in part b to determine if the amount in U.S. dollars is a function of the amount in Canadian money. Explain.

d. Let g represent the function in the table in part b. Determine $g(281.87)$. Explain its meaning in this practical situation.

e. Determine the domain of f and the range of f.

f. Determine the domain of g and the range of g.

g. How are the domain and range of f related to the domain and range of g?

h. Determine $g(f(1000))$ by using the two preceding tables. Explain your answer.

i. Determine $f(g(56.37))$ by using the two preceding tables.

5. You are now familiar with the conversion tables in Exercise 4, but you would like to be able to take any amount of money and convert between U.S. and Canadian currencies.

First verify that the functions f and g in Exercise 4 are linear. Then, refer to the data in the tables to answer the following.

a. Determine the average rate of change (slope) of function f. Round to four decimal places.

b. What is the practical meaning of the slope in this situation?

c. Use x to represent the input (amount in U.S. dollars) to write the linear equation of the function f.

d. Use the equation to determine $f(3000)$. Explain the practical meaning in this situation.

e. Determine the average rate of change (slope) for function g. Round to four decimal places.

f. What is the practical meaning of the slope in function g?

g. Use x to represent the input to write the linear equation of the function g.

h. Use the equation to determine $g(6000)$. Explain the meaning in this situation.

i. Prove that the two functions are inverse functions by showing that $f(g(x)) = x$ and $g(f(x)) = x$. Round your answer to the nearest whole number.

Activity 2.8

Temperature Conversions

Objectives

1. Determine the equation of the inverse of a function represented by an equation.

2. Describe the relationship between the graphs of inverse functions.

3. Determine the graph of the inverse of a function represented by a graph.

4. Use the graphing calculator to produce graphs of an inverse function.

In Activity 2.5, Inflated Balloons, you used the function defined by $F = 32 + 1.8C$ to convert from a temperature measured in degrees Celsius to a temperature measured in degrees Fahrenheit. Call this function T.

1. **a.** Identify the input and output variables for the conversion function T.

   b. Determine $T(-5)$ and explain its meaning in this situation.

2. When the temperature is 70°F, what is the temperature in degrees Celsius?

3. If you need to determine the temperature in Celsius for several Fahrenheit temperatures, it is easier to have a single formula (in fact, a new function) in which Celsius is the output and Fahrenheit is the input. Solve $F = 32 + 1.8C$ for C to determine this new function. Call this function H.

4. **a.** Identify the input and output variables for the function H.

   b. Determine $H(62)$, and explain its meaning in this situation.

5. Determine each of the following.

   a. $T(10)$ **b.** $H(T(10))$

6. Determine each of the following.

   a. $T(H(95))$ **b.** $H(T(0))$ **c.** $T(H(212))$

7. **a.** Write a general expression for $H(T(C))$ and simplify the result.

   b. Write a general expression for $T(H(F))$ and simplify the result.

   c. Based on the results of parts a and b, what can you conclude about the functions H and T?

Inverse Function Algorithm

Given y as a function of x follow the steps below to determine the inverse function.

Step 1. Write the equation for y in terms of x.

Step 2. Solve for x to get $x = f^{-1}(y)$.

Step 3. If f is a noncontextual (abstract) function, you can interchange x and y to get
$y = f^{-1}(x)$.

Example 1 *Given $f(x) = 7 + 5x$, determine the equation for the inverse function f^{-1}.*

SOLUTION

Step 1. Write the equation for y in terms of x: $y = 7 + 5x$.

Step 2. Solve for x to get $x = f^{-1}(y)$: $5x = y - 7$, $x = \dfrac{y - 7}{5}$, $f^{-1}(y) = \dfrac{y - 7}{5}$.

Step 3. Since f is a noncontextual function, we can interchange x and y to get
$$y = f^{-1}(x) = \dfrac{x - 7}{5}.$$

8. Given $y = f(x) = 5 - 3x$, determine an equation of the inverse function f^{-1}.

Graphs of Inverse Functions

9. **a.** Consider the graphs of the functions $f(x) = 7 + 5x$ and $f^{-1}(x) = \dfrac{x - 7}{5}$ from Example 1. Graph the two functions on the same coordinate system.

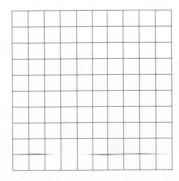

b. Draw the graph of $y = x$ on the grid in part a. Describe the relationship among the graphs of $f(x) = 7 + 5x$, $f^{-1}(x) = \dfrac{x - 7}{5}$, and $y = x$.

You can draw the inverse of a function with your TI-83/TI-84 Plus without determinin the equation. Turn off the graph of f^{-1} from Problem 9, and keep the graphs of f and $y =$ Use the window Xmin $= -9$; Xmax $= 9$, Ymin $= -6$; and Ymax $= 6$. Your screen shoul resemble the following.

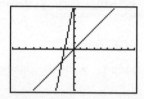

Access the Draw menu by pressing (2nd) (PRGM). Toggle down to option 8, DrawInv.

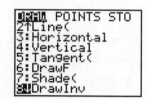

Pressing (ENTER) will place the DrawInv command into the Home screen. Now press (VARS and then (Y-VARS) and (1: Functions) to access the following screen.

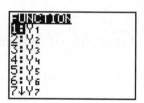

Choose the variable that corresponds to your function T, most likely Y1, and press (ENTER again. Your screen should resemble the following.

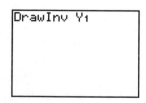

Press (ENTER). This adds the graph of the inverse to the graph screen.

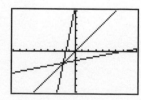

How does this compare with the graph of f^{-1} in Problem 9a? Note that the graph of f^{-1} in the graph screen of the TI-83/TI-84 Plus is not active. It cannot be traced, nor can values be viewed in the table.

10. Determine the point of intersection of the graphs of f and f^{-1}. What is the significance of this point in this situation?

11. a. Complete the following table for the graphs of f and f^{-1} given in Problem 9.

	VERTICAL INTERCEPT	HORIZONTAL INTERCEPT
$f(x) = 7 + 5x$		
$f^{-1}(x) = \dfrac{x - 7}{5}$		

b. Write a sentence that describes the relationship of the intercepts of the graphs of f and f^{-1}.

12. a. Determine the slopes of the graphs of f and f^{-1} given in Problem 9.

b. Write a sentence that describes the relationship of the slope of f and the slope of f^{-1}.

> **Graphs of Inverse Functions**
>
> The graphs of inverse functions are reflections about the line $y = x$.

13. Sketch a graph of f^{-1} on the same coordinate system as the graph of f for the functions that follow.

a.

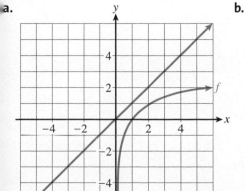

b.

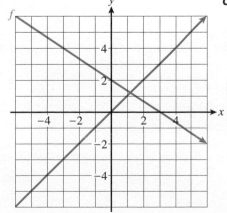

c.

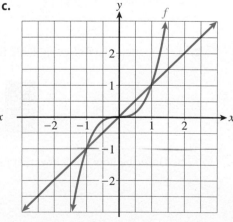

SUMMARY: ACTIVITY 2.8

1. To determine the equation of the inverse of a function defined by $y = f(x)$,

   Step 1. Write the equation for y in terms of x.

   Step 2. Solve for x to get $x = f^{-1}(y)$.

   Step 3. If f is noncontextual, interchange x and y to get $y = f^{-1}(x)$.

2. The graphs of inverse functions are reflections about the line $y = x$.

3. If two linear functions are inverses, the slopes of the graphs of the lines are reciprocals.

EXERCISES: ACTIVITY 2.8

1. As a sales representative, you are paid a base salary of $250 a week, plus a 5% commission.

   a. Determine a function, f, for your weekly gross pay, P (before taxes), as a function of S, your weekly sales in dollars, $P = f(S)$.

   b. Determine $f(6000)$, and interpret its meaning in this situation.

   c. Solve the equation $P = 0.05S + 250$ for S to determine the equation for a new function, $g(P)$, whose input is P and whose output is S.

   d. Determine $g(400)$, and interpret its meaning in this situation.

   e. Determine $g(f(8000))$.

2. **a.** Complete the following table for the function $y = f(x) = 3x^2 - 2$.

x	-2	-1	0	1	2
y					

   b. Describe how you know that f is a function by examining the table in part a.

c. Use the table of values from part a to determine f^{-1} if possible. If it is not possible, explain why.

3. Determine the equation of the inverse of the given function.

a. $y = f(x) = 3x - 4$

b. $w = g(z) = \dfrac{z - 4}{2}$

c. $s = \dfrac{5}{t}$

4. a. Given $y = f(x) = 2x + 6$, determine f^{-1}.

b. Sketch the graphs f and f^{-1} on the same coordinate system.

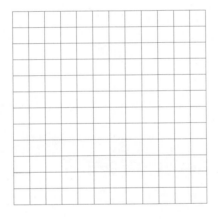

c. Add the sketch of the graph of $y = x$ to the sketches of f and f^{-1} in part b. Describe any symmetry in the graphs of f and f^{-1}.

d. Determine the horizontal and vertical intercepts of the graphs of f and of f^{-1}. Explain the relationship between the intercepts of f and the intercepts of f^{-1}.

e. Determine the slope of the graph of f and the slope of the graph of f^{-1}. What is the relationship between the slope of f and the slope of f^{-1}?

5. Consider the graphs of functions g and h. Are the functions inverses of each other? Explain, using a symmetry argument.

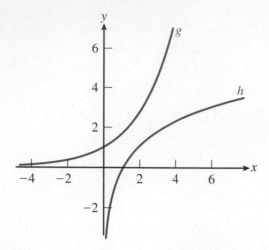

6. Consider the functions $f(x) = 3x + 6$ and $g(x) = \frac{1}{3}x - 2$.

a. Determine $f(g(x))$.

b. Determine $g(f(x))$.

c. Are f and g inverse functions? Explain.

d. Complete the following tables for the given functions.

x	f(x)
−2	
0	
2	
4	

x	g(x)
0	
6	
12	
18	

e. Do you notice anything about the ordered pairs of f and g?

f. Graph f and g on the following grid.

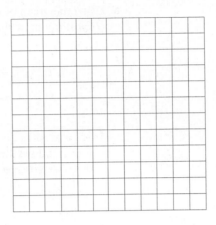

g. What can you say about the graphs of f and g with respect to the graph of $y = x$?

7. a. Given $g(x) = \dfrac{6 + 4x}{3}$, determine an equation for $g^{-1}(x)$.

b. Sketch the graphs of g and g^{-1} on the following grid.

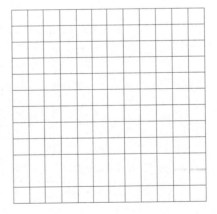

c. Do you believe that your equation for g^{-1} in part a is correct? Explain.

d. Determine $g^{-1}(g(x))$. Does your result support your answer in part c? Explain.

The functions in Problems 8 and 9 model contextual situations. The input and output repre-
sent real-world quantities. In general, you should not graph the function and its inverse on
the same set of axes as the input and output quantities are different. However, we do here to
emphasize the geometric relationship between the graph of a function and its inverse.

8. One inch measures 2.54 centimeters. To convert inches to centimeters, use the equation $C(x) = 2.54x$, where x is the number of inches.

a. Using the graph of C, draw the graph of the inverse function C^{-1}.
(*Hint:* First draw the graph of $y = x$ on the same axes.)

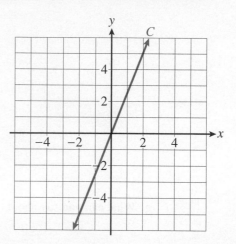

b. Locate $C^{-1}(5)$ on the graph, and interpret its meaning in this situation.

9. Your new kitchen has a square island that needs a granite top. The side of the largest square top you can purchase measures $S(a) = \sqrt{a}$, where S is the length of the square in feet and a is the area of the square.

a. Using the graph of S, sketch S^{-1}.

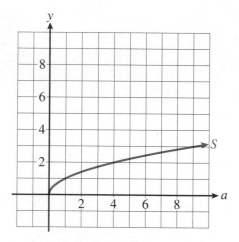

b. Check your graph of S^{-1} by using the DrawInv command.

c. What are input and output variables for S?

d. What are the input and output variables for S^{-1}?

e. Would S^{-1} ever be useful in this situation? Explain.

Cluster 2 · **What Have I Learned?**

1. For any two functions f and g,

   i. determine which of the following equations are true and explain why.

   ii. determine which of the following equations are not always true and give an example to show why not.

   a. $f(g(x)) = g(f(x))$

   b. $f(f^{-1}(x)) = f^{-1}(f(x))$

2. Describe how you would determine whether two functions are inverses of each other.

3. Suppose f is a nonconstant linear function.

   a. Will f always have an inverse function?

   h. What will always be true about the slopes of the lines representing the graphs of f and f^{-1}?

4. If $f(g(x)) = x^6$, determine at least three different ways to define f and g.

5. It is possible to compose more than two functions. If $f(x) = 2x - 1$, $g(x) = 5 - 3x$, and $h(x) = 2x^2$, determine each of the following.

   a. $f(g(h(1)))$ **b.** $g(f(h(1)))$

   c. $h(g(f(1)))$ **d.** $h(f(g(1)))$

Cluster 2   How Can I Practice?

1. Given $f(x) = x^2 - 4$ and $g(x) = x + 2$, determine each of the following.

a. $f(g(-3))$ **b.** $g(f(-3))$

c. $f(g(x))$ **d.** $g(f(x))$

e. $f(f(x))$ **f.** $g^{-1}(x)$

2. Given $f(x) = x - 4$ and $g(x) = 4 + x - x^2$, determine each of the following.

a. $f(g(3))$ **b.** $g(f(3))$

c. $g(g(3))$ **d.** $f(g(x))$

e. $g(f(x))$ **f.** $f^{-1}(x)$

3. Given the following two tables, complete the third table.

x	0	1	2	3	4
f(x)	6	−1	−3	0	2

x	−3	−1	0	2	6
g(x)	2	3	1	0	−1

x	0	1	2	3	4
g(f(x))					

4. Given $f(x) = 3x^2$ and $g(x) = -2x^3$, determine each of the following.

a. $g(f(x))$ **b.** $f(g(x))$

c. $g(f(-4))$

5. Given $s(x) = x^2 + 4x - 1$ and $t(x) = 4x - 1$, determine each of the following.

   a. $s(t(x))$

   b. $t(s(x))$

   c. $t^{-1}(x)$

6. Given $p(x) = \dfrac{1}{x}$ and $c(x) = \sqrt{x + 2}$, determine each of the following.

   a. $p(c(x))$ **b.** $c(p(x))$ **c.** $p^{-1}(x)$

7. The function q is defined by the following set of ordered pairs:
$$\{(4, 6), (7, -9), (-2, 1), (0, 0)\}.$$
Determine q^{-1} as a set of ordered pairs.

8. Show that $f(x) = 2x - 3$ and $g(x) = \dfrac{x + 3}{2}$ are inverses of each other.

9. a. Determine the equation of the inverse function of $f(x) = 4x + 3$.

   b. Sketch the graph of the function f and its inverse, f^{-1}, on the same coordinate system.

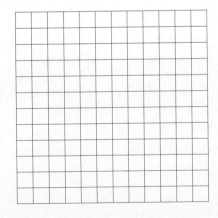

   c. Determine the horizontal and vertical intercepts of the graph of f and of f^{-1}.

d. Determine the slopes of the graphs of f and of f^{-1}.

e. Sketch a graph of the line $y = x$ on the same coordinate system.

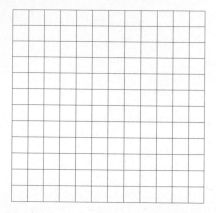

10. The 1990–2000 population boom in U.S. public schools from kindergarten through grade 12 can be modeled by the function $f(x) = 0.6x + 41$, where x represents the years after 1990 and $f(x)$ represents the student population in millions. (These figures are based on fall enrollments.) *Source:* U.S. Department of Education.

a. Complete the table using the given function model.

x, Years Since 1990	0	3	6	10
f(x), Student Population in Millions				

b. Interchange the input and output data.

x, Student Population in Millions				
f(x), Years Since 1990				

c. Determine the equation of the inverse function, and use this equation to check the table values in part b.

d. Graph the function and its inverse using your graphing calculator. Use the window Xmin = −20, Xmax = 20, Ymin = −100, and Ymax = 100. Use Zoom, ZSQR to see the graphs in a square window. Using the trace feature and the appropriate graph, determine when the population will be 46 million. Round your answer to the nearest year.

e. Add the graph of $y = x$ to the graphs of the function and its inverse. Describe the symmetry in the graphs.

f. Determine the horizontal and vertical intercepts of the graphs of the function and its inverse. Explain the relationship between the intercepts of the two functions.

	HORIZONTAL INTERCEPT	VERTICAL INTERCEPT
$f(x) = 0.6x + 41$		
$f^{-1}(x) = \dfrac{x - 41}{0.6}$		

g. Determine the slopes of the graphs of the function and its inverse. What is the relationship between the slope of the function and the slope of its inverse?

h. Predict the year in which the school population will be 50 million.

11. A Japanese student is coming to the United States as an exchange student. She has saved 60,000 yen for the trip. She realizes that because she is going to stop briefly in Europe, she can convert her money to the euro. After looking up the currency conversion factor on the Internet for June 21, 2009, she determines that the function that will convert her Japanese money is $f(x) = 0.007416x$, with x representing the amount in Japanese currency and $f(x)$ representing the amount in euros.

a. Use the function to determine how many euros she will receive for her 60,000 yen.

b. She then comes to the United States with her euros. To convert her money to U.S. dollars, she finds another conversion factor on the Internet. The function is $g(E) = 1.39320E$, with E representing the amount in euros and $g(E)$ representing the U.S. dollar amount. How many dollars will she have if she has 350 euros?

c. If the Japanese student does not go to Europe, but flies directly to the United States, write a new function that would tell her how much her 60,000 yen would be in U.S. dollars. (*Hint:* Determine $g(f(x))$.)

d. Use the function you found in part c to determine what her 60,000 yen would be worth in U.S. dollars.

12. Given $f(x) = 4x - 9$ and $g(x) = 10 - 3x$, what input to the composition function $f(g(x))$ will result in an output of 15?

13. The volume of a rectangular box is equal to the area of the base times the height. Suppose a particular box has a height of 10 inches and a square base.

    a. Using V for the volume of the box, b for the area of the base, and x for the length of one side of the square base, express b as a function of x and V as a function of b.

    b. Show how the volume is the composition of these two functions.

The bracketed numbers following each concept indicate the activity in which the concept is discussed.

CONCEPT/SKILL	DESCRIPTION	EXAMPLE
The sum function $f + g$ [2.1]	Given two functions, f and g, the sum function is defined by $y = f(x) + g(x)$.	See Example 1, Activity 2.1.
The difference function, $f - g$ [2.1]	Given two functions, f and g, the difference function is defined by $y = f(x) - g(x)$.	See Example 2, Activity 2.1.
Polynomial expression [2.1]	Any expression that is formed by adding or subtracting terms of the form ax^n, where a is a real number and n is a nonnegative integer, is called a **polynomial expression** in x.	$3x^3 - 2x^2 + 6x - 7$
Monomial [2.1]	A **monomial** is a polynomial with one term.	$13x^5$
Binomial [2.1]	A **binomial** is a polynomial with two terms.	$14x^4 - 3x$
Trinomial [2.1]	A **trinomial** is a polynomial with three terms.	$5x^4 - 7x + 13$
Polynomial function [2.1]	A polynomial function is any function defined by an equation of the form $y = f(x)$, where $f(x)$ is a polynomial expression.	$f(x) = 3x^3 - 2x - 7$
FOIL [2.2]	FOIL is a common method used to multiply two binomials.	See Example 1, Activity 2.2.
The product of two polynomials [2.2]	To multiply any two polynomials, multiply each term of the first by each term of the second.	$(x^2 + 1)(3x^2 + 6x - 2)$ $= 3x^4 + 6x^3 - 2x^2 + 3x^2 + 6x - 2$ $= 3x^4 + 6x^3 + x^2 + 6x - 2$
The product of two functions [2.2]	Given two functions, f and g, the product function is defined by $y = f(x) \cdot g(x)$.	See Exercise 8, Activity 2.2.
$a^m \cdot a^n$ [2.2]	To multiply powers of the same base, keep the base and add the exponents. Symbolically, this property of exponents is written as $a^m \cdot a^n = a^{m+n}$.	$x^4 \cdot x^3 = x^{4+3} = x^7$
Zero exponents [2.3]	$a^0 = 1$ where $a \neq 0$	$\left(\dfrac{2x}{4y}\right)^0 = 1$ x and $y \neq 0$
Negative exponents [2.3]	$a^{-n} = \dfrac{1}{a^n}$ if $a \neq 0$ and n is a real number	$\left(\dfrac{2}{3}\right)^{-3} = \left(\dfrac{3}{2}\right)^3 = \dfrac{27}{8}$

CONCEPT/SKILL	DESCRIPTION	EXAMPLE
$(a^m)^n$ [2.4]	If a is a real number and m and n are rational numbers, then $(a^m)^n = a^{mn}$.	$(x^2)^5 = x^{2 \cdot 5} = x^{10}$
$(ab)^n$ [2.4]	If a and b are real numbers and n is a rational number, then $(ab)^n = a^n b^n$.	$(2x)^4 = 2^4 x^4 = 16x^4$
The principal square root of a, $\sqrt{a}$ [2.4]	Let a represent a nonnegative real number, symbolically written as $a \geq 0$. The principal square root of a, denoted by $\sqrt{a}$, is defined as the nonnegative real number that, when squared, produces a.	$\sqrt{16} = 4$ because $4^2 = 16$
Fractional exponents [2.4]	$a^{1/2} = \sqrt{a}$, where $a \geq 0$	$13^{1/2} = \sqrt{13}$
$\sqrt[n]{a}$ [2.4]	$\sqrt[n]{a} = a^{1/n}$, the nth root of a. The real number a, called the **radicand**, must be nonnegative if n, called the **index**, is even.	$\sqrt[3]{36} = 36^{1/3}$
$a^{p/q}$ [2.4]	$a^{p/q} = \sqrt[q]{a^p}$ or $a^{p/q} = \left(\sqrt[q]{a}\right)^p$, where $a \geq 0$ if q is even, and p is a positive integer.	$16^{3/4} = \left(\sqrt[4]{16}\right)^3 = 2^3 = 8$
Composition of functions [2.5]	The **composition** of the functions f and g is a function, h, defined by $h(x) = f(g(x))$.	If $f(x) = 2x + 1$ and $g(x) = 3x - 2$, $f(g(x)) = f(3x - 2)$ $= 2(3x - 2) + 1 = 6x - 3$
Noncommutativity of composition [2.5]	In general, $f(g(x)) \neq g(f(x))$.	If $f(x) = 2x + 1$ and $g(x) = 3x - 2$, $f(g(x)) = 6x - 3$, $g(f(x)) = g(2x + 1)$ $= 3(2x + 1) - 2 = 6x + 1$
Inverse functions [2.7]	The two functions f and g are inverses if $f(g(x)) = x$ and $g(f(x)) = x$.	See Example 1, Activity 2.7.
Domain and range of inverse functions [2.7]	The domain of f is the range of f^{-1} and the range of f is the domain of f^{-1}.	See the tables at Problems 1 and 2, Activity 2.7.
Graphs of inverse functions [2.8]	The graphs of inverse functions are reflections about the line $y = x$.	See the graph before Problem 9, Activity 2.8.
Slopes of inverse linear functions [2.8]	If two linear functions are inverses, the slopes of the graphs of the lines are reciprocals.	The slope of $f(x) = 3x + 1$ is 3. The inverse, defined by $f^{-1}(x) = \frac{1}{3}x - \frac{1}{3}$, has a slope of $\frac{1}{3}$.

1. Simplify the following.

a. $(x + 6) + (2x^2 - 3x - 7)$

b. $(x^2 + 4x - 3) - (2x^2 - x + 1)$

c. $(x - 3)(4x - 1)$

d. $(x - 5)(x^2 - 2x + 3)$

e. $4(x + 2) - 3(5x - 1)$

f. $(2x^2 + x - 1)(x^2 - 3x + 4)$

2. Simplify the following. Write all of your results with positive exponents only. Assume all variables in denominators do not equal 0.

a. $(3x^3)(2x^5)$

b. $(4x^3y)^2$

c. $(xy)^2(-2x^3y)$

d. $(5x^3y^4z)(-2x^2yz^3)$

e. $(3x^2y)^0(3x^3y)^2$

f. $(-5xy)^3$

g. $\dfrac{6x^4}{3x}$

h. $2x^0$

i. $\dfrac{3^3}{3^3}$

j. $\dfrac{6xy^4z^2}{4xyz^5}$

k. $(-5x^{-3})(x^{-5})$

l. $\dfrac{8x^{-4}}{-2x^{-6}}$

m. $(-5x^{-3})^3$

n. $x^{4/5} \cdot x^{1/2}$

o. $(x^{2/3})^3$

3. Given $f(x) = 6x - 2$ and $g(x) = -2x + 3$, determine each of the following.

a. $f(-3)$

b. $f(x) + g(x)$

c. $f(3) - g(3)$

d. $f(x) \cdot g(x)$

e. $f(g(x))$

f. $g(f(2))$

g. f^{-1} (Determine the inverse of f.)

Answers to all Gateway exercises are included in the Selected Answers appendix.

4. Given $f(x) = x^2 - x + 3$ and $g(x) = 3x - 2$, determine each of the following.

a. $f(x) - g(x)$ **b.** $f(x) \cdot g(x)$

c. $f(g(x))$ **d.** $g(f(2))$

5. Determine the value of each of the following.

a. $49^{1/2}$ **b.** $32^{2/5}$ **c.** $(-27)^{4/3}$ **d.** $7^{3/5}$

e. $\sqrt[3]{27^2}$ **f.** $\sqrt[4]{16^5}$ **g.** 4^{-2}

6. To ship the mail-order ceramic figures that you produce, you need to make square-bottomed boxes. For the size of the box to be proportional to the figurines, the height of the box must always be three times longer than the width. The cost of the material to make the top and bottom of the box molded to fit the figurine is $0.01 per square inch, and the cost of the material for the sides of the box sells for $0.004 per square inch.

a. Write a function, f, to represent the cost of producing the top and the bottom of a box. Use x to represent the width of the bottom of the box in inches.

b. Write a function, g, to represent the cost of producing the sides of the box.

c. Combine the functions in parts a and b to write one function, T, that represents the total cost of making the box. Write the equation in simplest form.

d. Using $f(x)$, $g(x)$, and $T(x) = f(x) + g(x)$ as defined in parts a, b, and c, complete the following table.

x (in.)	f(x)	g(x)	T(x) = f(x) + g(x)
2			
4			
6			
8			
10			

7. You have a knitting machine in your home and your business is making ski hats. The fixed cost to run your knitting company is $300 per month, and the cost to produce each hat averages approximately $12. The hats will sell for $25.95.

a. Write a function C to represent the cost of making the hats. Use x to represent the number of hats made per month.

b. Write a function R to represent the revenue from the sale of the hats.

c. Write a function p to represent the profit for the month. Express this function in simplest form.

d. Graph the three functions on your graphing calculator. How many hats must be sold in 1 month to break even?

e. Determine the value of $C(50)$, $R(50)$, and $p(50)$. Explain the practical meaning of the values that you find.

f. Explain how the difference function $y = f(x) - g(x)$ pertains to this situation.

8. The manufacturer of a certain brand of computer printer sells her printers at a wholesale price of $110 per printer, based on selling 60 printers. Because the warehouse is overstocked and new high-speed printers are arriving, the manufacturer is offering a one-time-only offer. The price of each printer ordered in addition to the basic 60 will be reduced by $2 per extra printer ordered. You may not, however, purchase more than 90 printers.

a. Let x represent the number of printers in excess of 60 that you will purchase. Write a function for the total cost, C, of the printers as a function of the number of printers, x. Call this function f.

b. Rewrite the cost function by multiplying the factors and combining like terms.

c. What is the domain of the cost function?

d. If you decide that it is to your advantage to purchase 75 printers rather than 60 printers, explain the cost savings to you and your business.

9. Functions f and g are defined by the following tables.

x	− 1	0	1	2	4	5	8
f(x)	8	5	2	− 1	− 7	− 10	− 19

x	− 1	0	1	2	4	5	8
g(x)	5	1	− 1	− 1	5	11	41

Determine the values for each of the following.

a. $f(g(4))$

b. $g(f(-1))$

c. $f(g(0))$

d. $g(f(2))$

10. You manufacture snowboards. You cannot produce more than 30 boards per day. The cost, C, of producing x boards is represented by the function

$$C = f(x) = 150x - 0.9x^2.$$

a. What is the practical domain of the function?

b. What is the cost if 22 boards are produced?

c. The number of snowboards that can be produced in t hours is represented by the function $g(t) = 3.75t$. Determine $f(g(t))$.

d. What is the input variable in the composition of the functions in part c?

e. Because of a blizzard, your employees work only 4 hours on a certain day. Determine the production cost for that day.

f. The company prefers to keep production costs at approximately $3500. How many hours each day does the company have to operate to maintain this production cost? Your employees do not work more than 8 hours per day.

11. a. Determine the equation of the inverse of the function $f(x) = \dfrac{2x - 3}{5}$.

b. What is the slope of the line for each function? What is the relationship between the slopes of the two functions?

12. a. Show that $f(x) = -2x + 1$ and $g(x) = \dfrac{1 - x}{2}$ are inverse functions of each other.

b. Sketch the graphs of the functions in part a on the same axis, and check the result with your graphing calculator.

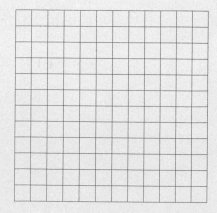

c. What can you say about the graphs of f and g with respect to the graph of $y = x$?

13. You work in the box office of a movie theater. The system for purchasing tickets is automated. You enter the number of tickets you need for each category (adult, child, senior), and the output is the total cost. The following table is a sample from a computer screen that shows the cost for adult tickets only. The number of tickets represents the input, and the total cost represents the output. Let c represent the total cost of the tickets purchased and let n represent the number of tickets purchased.

NUMBER OF ADULT TICKETS (n)	TOTAL COST (c)
2	$11.00
5	$27.50
7	$38.50
12	$66.00

a. Does this table represent a linear function? Explain.

b. Write a function, f, to represent the total cost, c, as a function of the number of tickets, n, that are purchased.

c. Determine the cost of one ticket. What does this value represent in your function?

Often the customer approaches the window with the exact amount of cash for the tickets. In that case, you enter the total amount into the computer and press adult ticket, and out comes the number of tickets. In this case, the total cost is the input and the number of tickets is the output.

d. Fill in the table showing this situation. (Hint: consider the table above.)

TOTAL COST	NUMBER OF TICKETS

e. Write a function g that represents the total number of tickets, n, purchased as a function of the total cost, c.

f. What is the slope of this line? What is the relationship between the slopes of the two functions f and g?

g. Determine $f(g(c))$ and $g(f(n))$. Are the functions inverses of each other? Explain.

Exponential and Logarithmic Functions

Cluster 1   Exponential Functions

Activity 3.1

The Summer Job

Objectives

1. Determine the growth factor of an exponential function.

2. Identify the properties of the graph of an exponential function defined by $y = b^x$, where $b > 1$.

3. Graph an increasing exponential function.

Your neighbor's son will be attending college in the fall, majoring in mathematics. On July 1, he comes to your house looking for summer work to help pay for college expenses. You are interested since you need some odd jobs done, but you don't have a lot of extra money to pay him. He can start right away and will work all day, July 1 for 2 cents. This gets your attention, but you wonder if there is a catch. He says that he will work July 2 for 4 cents, July 3 for 8 cents, July 4 for 16 cents, and so on for *every* day of the month of July.

1. Do you hire him? Justify your answer.

For Problems 2–8, assume that you do hire him.

2. How much will he earn on July 5? July 6?

3. What will be his total pay for the first week of July (July 1 through July 7)?

4. a. Complete the following table.

DAY IN JULY (INPUT)	PAY IN CENTS (OUTPUT)
1	
2	
3	
4	
5	
6	
7	
8	

b. Do you notice a pattern in the output values? Describe how you can obtain the pay on a given day knowing the pay on the previous day.

    c. Use what you discovered in part b to determine the pay on July 9.

5. a. The pay on any given day can be written as a power of 2. Write each pay entry in the output column of the table in Problem 4 as a power of 2. For example, $2 = 2^1, 4 = 2^2$.

    b. Let n represent the number of days worked. Write an equation for the daily pay, $P(n)$ (in cents) as a function of n, the number of days worked. Note that the number of days worked is the same as the July date.

    c. Use the equation from part b to determine how much your neighbor's son will earn on July 20. That is, determine the value of $P(n)$ when $n = 20$. What are the units of measurement of your answer?

    d. How much will he earn on July 31? Be sure to indicate the units of your answer.

    e. Was it a good idea to hire him?

6. a. Determine the average rate of change of $P(n)$ as n increases from $n = 3$ to $n = 4$. What are the units of measurement of your answer?

    b. Determine the average rate of change of $P(n)$ as n increases from $n = 7$ to $n = 8$. Include units in your answer.

    c. Is the function linear? Explain.

7. a. What is the practical domain (inputs that make sense for the situation) of the function defined by $P(n) = 2^n$?

    b. Sketch a scatterplot of ordered pairs of the form $(n, P(n))$ from July 1 to July 10 on an appropriately scaled and labeled axes.

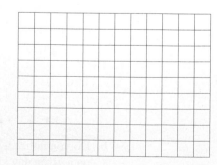

c. Is the function discrete or continuous?

The function defined by $P(n) = 2^n$ gives the relationship between the pay $P(n)$ (in cents) and the given July date, n, worked. This function belongs to a family of functions called **exponential functions**.

> Some **exponential functions** can be defined by equations of the form $y = b^x$, where the base b is a constant such that $b > 1$. Such functions are called **exponential** functions because the independent variable (input) x is the exponent.

Example 1 *Examples of exponential functions are*

i. $g(x) = 10^x$, where $b = 10$.

ii. $h(x) = (1.08)^x$, where $b = 1.08$.

Graphs of Increasing Exponential Functions

Because n in $P(n) = 2^n$ (the summer job situation) represents a given day in July, the practical domain (whole numbers from 1 to 31) limits the investigation of the exponential function.

8. a. Consider the general function defined by $f(x) = 2^x$. Use your graphing calculator to sketch a graph of this function. Use the window Xmin $= -10$, Xmax $= 10$, Ymin $= -2$, and Ymax $= 10$.

b. Because the graph of the general function $f(x) = 2^x$ is continuous (it has no holes or breaks), what appears to be the domain of the function f? What is the range of the function f?

c. Determine the y-intercept of the graph of f by substituting 0 for x in the equation $y = 2^x$ and solving for y.

d. Is the function f increasing or decreasing?

Definition

If the base b of an exponential function defined by $y = b^x$ is greater than 1, then b is the **growth factor**. The graph of $y = b^x$ is increasing if $b > 1$. For each increase of 1 of the value of the input, the output increases by a factor of b.

Example 2   *The base 2 of $f(x) = 2^x$ is the growth factor because each time the input, x, is increased by 1, the output is multiplied by 2.*

9. Identify the growth factor, if any, for the function defined by the given equation.

a. $g(x) = 10^x$ *growing*

b. $y = 1.08^x$ *growing*

c. $y = \left(\frac{4}{3}\right)^x$ *growing*

d. $h(x) = 0.8^x$ *decaying*

e. $y = 8x$

This is a line, not exponential

There is a special relationship between the graph of $f(x) = 2^x$ and the x-axis when the input x becomes more negative. Problem 10 investigates this relationship.

10. Return to the graph of $f(x) = 2^x$.

a. Does the graph of $f(x) = 2^x$ appear to have an x-intercept?

b. Use your calculator to complete the following table.

x	−1	−2	−4	−6	−8	−10
$f(x) = 2^x$						

Note: $2^{-10} = \frac{1}{2^{10}} \approx 0.000977$.

c. As the values of the input variable x decrease (become more negative), what happens to the output values?

d. Use the trace feature of your graphing calculator to trace the graph of $f(x) = 2^x$ for $x < 0$. What appears to be the relationship between the graph of $y = 2^x$ and the x-axis when x decreases (becomes more negative)?

Definition

A horizontal axis having equation $y = 0$ is called a **horizontal asymptote** of the graph of a function defined by $y = b^x$, where $b > 1$. The graph of the function gets closer and closer to the x-axis ($y = 0$) as the input gets farther from the origin, in the negative direction.

Example 3   *The x-axis is the horizontal asymptote of y = 7ˣ because, as x gets*
more negative, the graph gets closer and closer to the x-axis. See the
graph that follows.

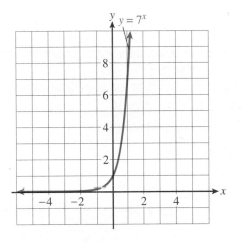

11. a. Complete the following table.

x	−3	−2	−1	0	1	2	3	4	5
$f(x) = 2^x$									
$g(x) = 10^x$									

b. Sketch the graph of the functions f and g on your graphing calculator. Use the window Xmin = −5, Xmax = 5, Ymin = −2, and Ymax = 9.

c. Use the results from parts a and b to describe how the graphs of $f(x) = 2^x$ and $g(x) = 10^x$ are similar and how they are different. Be sure to include domain, growth factor, x- and y-intercepts, and horizontal asymptotes. Also discuss whether the graph of g increases faster or slower than the graph of f.

12. Examine the output pattern to determine which of the following data sets is linear and which is exponential. For the linear set, determine the slope. For the exponential set, determine the growth factor.

a.

x	−2	−1	0	1	2	3	4
y	−8	−4	0	4	8	12	16

b.

x	−2	−1	0	1	2	3	4
y	$\frac{1}{16}$	$\frac{1}{4}$	1	4	16	64	256

SUMMARY: ACTIVITY 3.1

Functions defined by equations of the form $y = b^x$, where $b > 1$, are called **exponential functions** and have the following properties.

1. The domain is all real numbers.

2. The range is $y > 0$.

3. If $b > 1$, the function is increasing and has the following general shape.

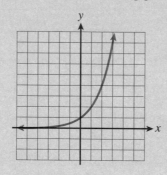

In this case, b is called the **growth factor**.

4. The vertical intercept (y-intercept) is (0, 1).

5. The graph does not intersect the horizontal axis. There is no x-intercept.

6. The line $y = 0$ (the x-axis) is a **horizontal asymptote**.

7. The function is continuous.

EXERCISES: ACTIVITY 3.1

1. a. Complete the following tables.

x	−3	−2	−1	0	1	2	3
$h(x) = 5^x$	$\frac{1}{125}$	$\frac{1}{25}$	$\frac{1}{5}$	1	5	25	125

x	−3	−2	−1	0	1	2	3
$g(x) = 2.65^x$				1	2.65		

b. Sketch graphs of h and g on the following grid.

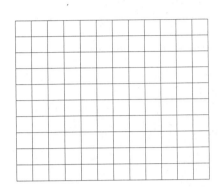

$h(x) = 5x$

x	y
3	125
2	25
1	5
0	1
-1	$\frac{1}{5}$
-2	$\frac{1}{25}$
-3	$\frac{1}{125}$

c. Use the tables and graphs in parts a and b to complete the following table.

FUNCTION	BASE, b	GROWTH FACTOR	x-INTERCEPT	y-INTERCEPT	HORIZONTAL ASYMPTOTE	INCREASING OR DECREASING
$h(x) = 5^x$	5	5	none	(0, 1)	y = 0	Inc
$g(x) = 2.65^x$	2.65	2.65	none	(0, 1)	x = 0	Inc

2. a. Complete the following table.

x	−3	−2	−1	0	1	2	3
$f(x) = 3^x$	.037			1	3	9	27
$g(x) = x^3$	-27						27
$h(x) = 3x$	-9	-6	-3	0	3	6	9

b. Sketch a graph of each of the given functions f, g, and h.

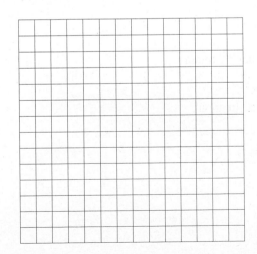

c. Describe any similarities or differences that you observe in the graphs.

3. Determine which of the following data sets are linear and which are exponential. For the linear sets, determine the slope. For the exponential sets, determine the growth factor.

a.

x	−2	−1	0	1	2	3	4
y	$\frac{1}{9}$	$\frac{1}{3}$	1	3	9	27	81

b.

x	−2	−1	0	1	2	3	4
y	2	2.5	3	3.5	4	4.5	5

c.

x	−2	−1	0	1	2	3	4
y	0.75	1.5	3	6	12	24	48

4. Would you expect $f(x) = 3^x$ to increase faster or slower than $g(x) = 2.5^x$ for $x > 0$? Explain. (*Hint:* You may want to use your graphing calculator for help.)

5. Take a piece of paper from your notebook. Let x represent the number of times you fold the paper in half and $f(x)$ represent the number of sections the paper is divided into after the folding.

a. Complete the table of values.

x	0	1	2	3	4	5
f(x)	1					

b. If you could fold the paper eight times, how many individual sections will there be on the paper?

c. Does this data represent an exponential function? Explain.

d. What is the practical domain and range in this situation?

Activity 3.2

Half-Life of Medicine

Objectives

1. Determine the decay factor of an exponential function.

2. Graph a decreasing exponential function.

3. Identify the properties of an exponential functions defined by $y = b^x$, where $b > 0$ and $b \neq 1$.

When you take medicine, your body metabolizes and eliminates the medication until there is none left in your body. The half-life of a medication is the time is takes for your body to eliminate one-half of the amount present. For many people the half-life of the medicine Prozac is 1 day.

1. **a.** What fraction of a dose is left in your body after 1 day?

   b. What fraction of the dose is left in your body after 2 days? (This is one-half of the result from part a.)

   c. Complete the following table. Let t represent the number of days since a dose of Prozac is taken and let Q represent the fraction of the dose of Prozac remaining in your body.

t, days	Q
0	1
1	
2	
3	
4	

2. **a.** The values of the fraction of the dosage, Q, can be written as powers of $\frac{1}{2}$. For example, $1 = \left(\frac{1}{2}\right)^0, \frac{1}{2} = \left(\frac{1}{2}\right)^1$, and so on. Complete the following table by writing each value of Q in the above table as a power of $\frac{1}{2}$. The values for 1 and $\frac{1}{2}$ have already been entered.

t, days	Q
0	$\left(\frac{1}{2}\right)^0$
1	$\left(\frac{1}{2}\right)^1$
2	
3	
4	

   b. Use the result of part a to write an equation for Q in terms of t.

   c. What is the practical domain of the half-life function?

    d. Sketch a scatterplot of the data in part a on an appropriately scaled and labeled coordinate axis.

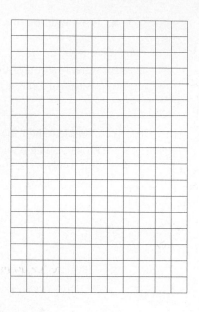

    e. Is this function discrete or continuous?

Notice that the equation $Q = \left(\frac{1}{2}\right)^t$ fits the equation form of an exponential function, $y = b^x$, given in the previous activity. However, the value of b for this exponential function is $\frac{1}{2}$, which is not greater than 1. Therefore, the base $\frac{1}{2}$ is not a growth factor.

When the base, b, of an exponential function is between 0 and 1, the base is called a **decay factor**. The result is a **decreasing exponential function**.

Graphs of Decreasing Exponential Functions

Because t in the equation $Q = \left(\frac{1}{2}\right)^t$ represents time after the drug has been taken, the function is not defined for negative values of t and therefore limits the investigation of the function. Therefore, consider the general function defined by $g(x) = \left(\frac{1}{2}\right)^x$.

3. a. Complete the following table.

x	−3	−2	−1	0	1	2	3	4	5
$g(x) = \left(\frac{1}{2}\right)^x$									

    b. Describe how you can obtain the output value for $x = 6$, using the output value for $x = 5$.

c. Sketch the graph of $g(x) = \left(\frac{1}{2}\right)^x$. Verify your sketch using your graphing calculator.

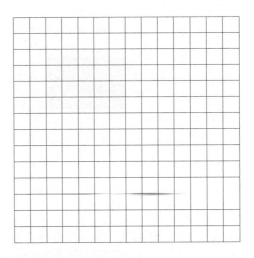

d. What are the domain and range of the function g?

e. Determine the vertical intercept of the graph of g.

f. Is the function g increasing or decreasing?

Definition

If the base b of an exponential function $y = b^x$ is between 0 and 1, then b is the **decay factor**. The graph of $y = b^x$ is decreasing if $0 < b < 1$. For each increase of 1 in the value of the input, the output decreases by a factor of b.

Example 1   *The base $\frac{1}{2}$ in the function $g(x) = \left(\frac{1}{2}\right)^x$ is the decay factor because each time x is increased by 1, the output value is multiplied by $\frac{1}{2}$.*

4. Identify the decay factor, if any, for the given function.

a. $g(x) = \left(\frac{2}{7}\right)^x$

b. $y = 0.98^x$

c. $h(x) = 1.8^x$

d. $y = 0.8x$

5. Return to the graph of $g(x) = \left(\frac{1}{2}\right)^x$.

  a. Does the graph of $g(x) = \left(\frac{1}{2}\right)^x$ have an x-intercept?

  b. Complete the following table.

x	1	3	5	7	10
$g(x) = \left(\frac{1}{2}\right)^x$					

  c. As the values of the input variable x get larger, what happens to the output values?

  d. Does the graph of g have a horizontal asymptote? Explain.

6. a. For each of the following exponential functions, identify the base, b, and determine whether the base is a growth or decay factor. Graph each function on your graphing calculator, and complete the table below.

FUNCTION	BASE, b	GROWTH OR DECAY FACTOR	x-INTERCEPT	y-INTERCEPT	HORIZONTAL ASYMPTOTE	INCREASING OR DECREASING
$h(x) = (1.08)^x$						
$T(x) = (0.75)^x$						
$f(x) = (3.2)^x$						
$r(x) = \left(\frac{1}{4}\right)^x$						

  b. Without graphing, how might you determine which of the functions in part a increase and which decrease? Explain.

7. Examine the output pattern to determine which of the following data sets is linear and which is exponential. For the linear set, determine the slope. For the exponential set, determine the growth or decay factor.

  a.

x	-2	-1	0	1	2	3	4
y	-6	-3	0	3	6	9	12

  b.

x	-2	-1	0	1	2	3	4
y	9	3	1	$\frac{1}{3}$	$\frac{1}{9}$	$\frac{1}{27}$	$\frac{1}{81}$

8. Determine the decay factor of the function represented by the data, and complete the table.

x	−2	−1	0	1	2
f(x)	16	4			

SUMMARY: ACTIVITY 3.2

Functions defined by equations of the form $y = b^x$, where $b > 0$ and $b \neq 1$, are called **exponential functions** and have the following properties.

1. The domain is all real numbers.

2. The range is $y > 0$.

3. If $0 < b < 1$, the function is decreasing and has the following general shape.

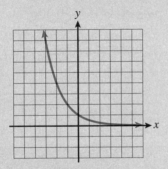

In this case, b is called the **decay factor**.

4. If $b > 1$, the function is increasing and has the following general shape.

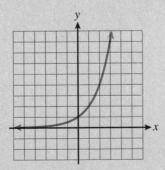

In this case, b is called the **growth factor**.

5. The vertical intercept (y-intercept) is $(0, 1)$.

6. The graph does not intersect the horizontal axis. There is no x-intercept.

7. The line $y = 0$ (the x-axis) is a **horizontal asymptote**.

8. The function is continuous.

EXERCISES: ACTIVITY 3.2

1. a. Complete the following tables.

x	−3	−2	−1	0	1	2	3
$h(x) = (0.35)^x$							

x	−3	−2	−1	0	1	2	3
$g(x) = \left(\frac{1}{5}\right)^x$							

b. Sketch graphs of *h* and *g* on the following grid.

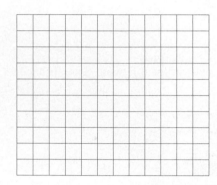

c. Use the tables and graphs in parts a and b to complete the following table.

FUNCTION	BASE, *b*	GROWTH OR DECAY FACTOR	*x*-INTERCEPT	*y*-INTERCEPT	HORIZONTAL ASYMPTOTE	INCREASING OR DECREASING
$h(x) = (0.35)^x$						
$g(x) = \left(\frac{1}{5}\right)^x$						

2. Using your graphing calculator, investigate the graphs of the following families (groups) of functions. Describe any relationships within each family, including domain and range, growth or decay factors, vertical and horizontal intercepts, and asymptotes. Identify the functions as increasing or decreasing.

a. $f(x) = \left(\frac{3}{4}\right)^x$, $g(x) = \left(\frac{4}{3}\right)^x$

b. $f(x) = 10^x$, $g(x) = -10^x$

c. $f(x) = 3^x$, $g(x) = \left(\frac{1}{3}\right)^x$

3. Determine which of the following data sets are linear and which are exponential. For the linear sets, determine the slope. For the exponential sets, determine the growth factor or the decay factor.

a.

x	-2	-1	0	1	2	3	4
y	0.5	1	1.5	2	2.5	3	3.5

b.

x	-2	-1	0	1	2	3	4
y	0.50	2	8	32	128	512	2048

c.

x	-2	-1	0	1	2	3	4
y	6.25	2.5	1	0.4	0.16	0.064	0.0256

4. Assume that y is an exponential function of x.

   a. If the growth factor is 1.25, then complete the following table.

x	0	1	2	3
y	10.5			

   b. If the decay factor is 0.75, then complete the following table.

x	0	1	2	3
y	10			

5. a. Would you expect $f(x) = 5^x$ to increase faster or slower than $g(x) = 7^x$ for $x > 0$? Explain. (*Hint:* You may want to use your graphing calculator for help.)

   b. Would you expect $f(x) = \left(\frac{1}{2}\right)^x$ to decrease faster or slower than $g(x) = (0.70)^x$ for $x > 0$? Explain.

6. Determine the domain and range of each of the following functions.

   a.

   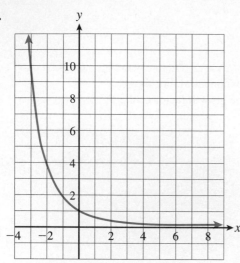

   b.

   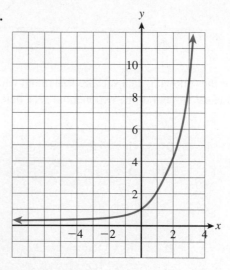

Activity 3.3

Cell Phones

Objectives

1. Determine the growth and decay factor for an exponential function represented by a table of values or an equation.

2. Graph exponential functions defined by $y = ab^x$, where $b > 0$ and $b \neq 1$, $a \neq 0$.

3. Determine the doubling and halving time.

During a meeting, you hear the familiar ring of a cell phone. Without hesitation, several of your colleagues reach into their jacket pockets, briefcases, and purses to receive the anticipated call. Although sometimes annoying, cell phones have become part of our way of life.

The following table shows the rapid increase in the number of cell phones (figures are approximate) in the years 2000–2006. Note that the input variable (year) increases in steps of 1 unit (year).

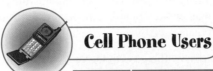

Cell Phone Users

YEAR	NUMBER OF CELL PHONE USERS AS OF JAN. 1 (IN MILLIONS)
2000	109.48
2001	128.37
2002	140.77
2003	156.72
2004	182.14
2005	207.90
2006	233.00

1. Is this a linear function? How do you know?

2. a. Evaluate the indicated ratios to complete the following table.

NO. OF PHONE USERS IN 2001 / NO. OF PHONE USERS IN 2000	NO. OF PHONE USERS IN 2002 / NO. OF PHONE USERS IN 2001	NO. OF PHONE USERS IN 2003 / NO. OF PHONE USERS IN 2002	NO. OF PHONE USERS IN 2004 / NO. OF PHONE USERS IN 2003	NO. OF PHONE USERS IN 2005 / NO. OF PHONE USERS IN 2004	NO. OF PHONE USERS IN 2006 / NO. OF PHONE USERS IN 2005

b. What do you notice about the values of the table?

In an exponential function with base b, equally spaced input values yield output values whose successive ratios are constant. If the input values increase by increments of 1, the common ratio is the base b. If $b > 1$, b is the growth factor; if $0 < b < 1$, b is the decay factor.

3. a. Can the relationship in the table preceding Problem 1 be modeled by an exponential function? Explain.

b. What is the growth factor?

c. As a consequence of the result found in part b, you can start with the number of cell phone users in 2000, and obtain the number of cell phone users in 2001 by multiplying by the growth factor, b. You can then determine the number of cell phone users in 2002 by multiplying the number of cell phone users in 2001 by b, and so on. Verify this with your calculator. Note that because the exponential function is a mathematical model, the results will vary slightly from the actual number of cell phone users given in the table preceding Problem 1.

Once you know the growth factor (b), you can determine the equation that gives the number of cell phone users as a function of t, the number of years since 2000. Note that $t = 0$ corresponds to 2000, $t = 1$ to 2001, and so on.

4. a. Complete the table.

t	CALCULATION FOR THE NUMBER OF CELL PHONE USERS	EXPONENTIAL FORM	NUMBER OF CELL PHONE USERS (IN MILLIONS)
0	109.48	$109.48(1.13)^0$	109.48
1	(109.48)1.13		
2			
3			

b. Use the pattern in the preceding table to help you write the equation of the form $N(t) = a \cdot b^t$, where $N(t)$ represents the number of cell phone users (in millions) at time t, the number of years since 2000.

c. What is the practical domain of the function N?

d. Graph the function N on your graphing calculator, and then sketch the result below on an appropriately scaled and labeled axis.

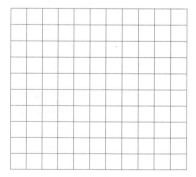

e. Determine the vertical intercept of the graph of N by substituting 0 for the input, t. What is the practical meaning of the vertical intercept in this situation?

f. What does the value of a in the equation in part b represent?

Definition

Many exponential functions can be represented symbolically by $f(t) = a \cdot b^t$, where a is the value of f when $t = 0$ and b is the growth or decay factor. If the input, t, of $y = a \cdot b^t$ represents time, then the coefficient a is called the initial value.

Example 1 *The exponential function defined by $f(x) = 5 \cdot 2^x$ has y-intercept $(0, 5)$ and growth factor $b = 2$. The exponential function defined by $h(x) = \frac{1}{2}(0.75)^x$ has y-intercept $\left(0, \frac{1}{2}\right)$ and decay factor $b = 0.75$.*

5. Use the function defined by $N(t) = 109.48(1.13)^t$ to estimate the number of cell phone users in 2010. Do you think this is a good estimate? Explain.

6. a. Use the graph of the exponential function $N(t) = 109.48(1.13)^t$ and the graph or table feature of your graphing calculator to estimate the number of years it takes for the number of cell phone users to double from 109.48 million to 218.96 million.

b. Estimate the time necessary for the number of cell phone users to double from 218.96 million to 437.92 million. Verify your estimate using your calculator.

c. How long will it take for any given number of cell phone users to double?

Definition

The **doubling time** of an exponential function is the time it takes for an output to double. The doubling time is determined by the growth factor and remains the same for all output values.

Example 2 *The balance $B(t)$, in dollars, of an investment account is defined by $B(t) = 5500(1.12)^t$, where t is the number of years. The initial value for this function is $5500. Determine the value of t when the balance is doubled or equal to $11,000.*

SOLUTION

If you use the table feature of your calculator, the doubling time is estimated at 6.1 years (se the following calculator graphic). The intersect feature on the graphing calculator shows th doubling time to be 6.12 years to the nearest hundredth.

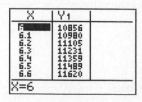

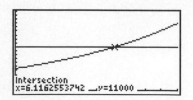

Decreasing Exponential Functions, Decay Factor, and Halving Time

You have just purchased a new automobile for $22,000. Much to your dismay, you have ju: learned that you should expect the value of your car to depreciate by 30% per year! The fo: lowing table shows the book value of the car for the next several years, where V is the valu in thousands of dollars.

DEPRECIATION: TAKING ITS TOLL

t (year)	0	1	2	3	4
V(t) (in thousands of dollars)	22	15.4	10.8	7.5	5.3

When a quantity is increased or decreased by a constant percent rate, it can be modeled by a exponential function. In this situation, the car value is decreased by 30% per year. A decreas ing exponential function has a decay factor, b, with $0 < b < 1$. For consecutive values o the input, an output value is determined by multiplying the previous output value by b.

7. As the input, t, increases from 0 to 1, the output, V, decreases from 22 to 15.4 (in $1000).

   a. Determine the value that 22 is multiplied by to get 15.4.

   b. Use the result from part a to complete the following table.

t	CALCULATION OF THE VALUE OF THE CAR	EXPONENTIAL FORM	VALUE (IN $1000)
0	22	$22(0.7)^0$	22
1	22(0.7)	$22(0.7)^1$	15.4
2	22(0.7)(0.7)	$22(0.7)^2$	
3			
4			

c. Use the pattern in the preceding table to write an equation in the form $V = a \cdot b^t$ that gives the car value, V, as a function years, t.

d. Input the function into Y_1 on your calculator and graph it in an appropriate window.

e. Determine the value of the vertical intercept of the graph. Write the result as an ordered pair.

f. What is the practical meaning of the vertical intercept in this situation?

8. a. Estimate the number of years, t, it takes for the value of the car to be $11,000, half the original value. (*Hint:* Put 11000 in Y_2 and find the point of intersection.)

b. How many years will it take for the car value to be halved again, that is from $11,000 to $5500?

Definition

The **half-life** of an exponential function is the time it takes for an output to decay by one-half. The half-life is determined by the decay factor and remains the same for all output values.

Example 3   *The population of Buffalo, New York, can be modeled by the equation $B(t) = 1102(0.995)^t$, with $t = 0$ representing the year 1970 and $B(t)$ representing the population in thousands. If the population of Buffalo continues to decline at the same rate, determine the number of years it will take for the population of Buffalo to be one half of the 1970 population.*

SOLUTION

The equation indicates that the population of Buffalo, NY, in 1970 ($t = 0$) is $B(0) = 1102(0.995)^0 = 1102$ thousand people. Therefore, half of that is 551 thousand people. Use the table feature of your graphing calculator; the halving time is estimated at 138 years.

X	Y₁	Y₂
135	560.14	551
136	557.34	551
137	554.55	551
138	551.78	551
139	549.02	551
140	546.28	551
141	543.55	551

X=138

Use the intersect feature of your graphing calculator and you will also determine the halving time to be 138 years rounded to the nearest year.

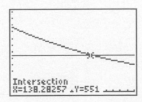

Intersection
X=138.28257 Y=551

9. Inflation means that a current dollar will buy less in the future. According to the U.S Consumer Price Index, the inflation rate for 2005 was 4%. This means that a 1-pound loa of white bread that cost a dollar in January 2005 cost $1.04 in January 2006. The chang in price is usually expressed as an annual percentage rate, known as the inflation rate.

a. If you assume that inflation increases somewhat and remains 5% per year for the next decade, you can calculate the cost of a currently priced $8 pizza for each year over the next 10 years. Complete the following table. Round to the nearest cent.

Years from now, t	0	1	2	3	4	5	6	7	8	9	
Cost of Pizza, $c(t)$, $											

b. Determine an exponential function defined by $c(t) = a \cdot b^t$ to represent the cost of an $8 pizza t years from now.

c. Explain the meaning of a and b in this situation.

d. Use the cost function to determine the cost of a pizza after 5 years.

e. A graph of the function $c(t) = 8(1.05)^t$ is shown below in a window for t between -20 and 20 and $c(t)$ between 0 and 20. Use the same window to graph this function on a graphing calculator.

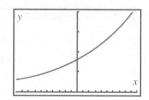

f. Is the entire graph in part e relevant to the original problem? Explain.

g. Resize your window to include only the first quadrant from $x = 0$ to $x = 20$ and $y = 0$ to $y = 20$. Regraph the function.

h. How many years will it take for the price of an $8 pizza to double? Explain how you determined your answer.

SUMMARY: ACTIVITY 3.3

1. For **exponential functions** defined by $f(x) = ab^x$, a is the value of f when $x = 0$ (sometimes called the initial value), and b is the growth or decay factor.

2. The vertical intercept of these functions is $(0, a)$.

3. In an exponential function, equally spaced input values yield output values whose successive ratios are constant. If the input values increase by 1 unit, then

  a. the constant ratio is the **growth factor** if the output values are increasing

  b. the constant ratio is the **decay factor** if the output values are decreasing

4. The **doubling time** of an increasing exponential function is the time it takes for an output to double. The doubling time is set by the growth factor and remains the same for all output values.

5. The **half-life** of a decreasing exponential function is the time it takes for an output to decay by one-half. The half-life is determined by the decay factor and remains the same for all output values.

EXERCISES: ACTIVITY 3.3

1. The population of Russia in selected years can be approximated by the following table.

Year	1995	1996	1997	2000	2006	2009
Population (in Millions)	148.0	147.6	146.9	146.0	143.0	142.0

  a. Let 1995 correspond to $t = 0$. Let b be the ratio between the population of Russia in 1996 and 1995. Determine an exponential function of the form $y = a \cdot b^t$ to represent the population of Russia symbolically. Round to three decimal places.

  b. Does the function in part a give an accurate value of the population of Russia in 2009? Explain.

  c. Use your model in part a to predict the population of Russia in 2015.

2. Without using your graphing calculator, match each graph with its equation. Then check your answer using your graphing calculator.

a. $f(x) = 0.5(0.73)^x$

b. $g(x) = 3(1.73)^x$

i.

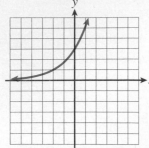

ii.

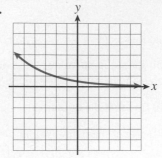

3. Which of the following tables represent exponential functions? Indicate the growth or decay factor for the data that is exponential.

a.

x	0	1	2	3	4
y	0	2	16	54	128

b.

x	0	1	2	3	4
y	1	4	16	64	256

c.

x	1	2	3	4	5
y	1750	858	420	206	101

4. An exponential function may be increasing or decreasing. Determine which is the case for each of the following functions. Explain how you determined each answer.

a. $y = 5^x$

b. $y = \left(\frac{1}{2}\right)^x$

c. $y = 1.5^t$

d. $y = 0.2^p$

5. a. Evaluate the functions in the following table for the input values, x:

Input x	0	1	2	3	4	5
g(x) = 3x						
f(x) = 3^x						

b. Compare the functions $f(x) = 3^x$ and $g(x) = 3x$ from $x = 0$ to $x = 5$ by comparing the output values in the table.

c. Compare the graphs of the functions f and g that are shown in the following window. Approximate the interval in which the exponential function f grows slower than the linear function g and the interval where it grows faster.

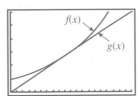

d. Compare the rate of increase of the function $f(x) = 3^x$ and $g(x) = 3x$ from $x = 0$ to $x = 5$ by calculating the average rate of change for each function from $x = 0$ to $x = 5$. Determine which function grows faster on average in the given interval.

6. If $f(x) = 3 \cdot 4^x$, determine the exact value of each of the following, when possible. Otherwise, use your calculator to approximate the value to the nearest hundredth.

a. $f(-2)$

b. $f\left(\frac{1}{2}\right)$

c. $f(2)$

d. $f(1.3)$

7. According to the U.S. Bureau of Justice Statistics, the state and federal prison population in the state of Georgia since December 31, 1980 can be modeled by $P(t) = 12.45 \cdot 1.057^t$, where t represents the number of years since December 31, 1980 and $P(t)$ represents the state and federal prison population in the state of Georgia measured in thousands.

a. Complete the following table.

t, Number of Years Since 1980	0	10	20	25	26	27
P(t), the Georgia State and Federal Prison Population (in Thousands of Prisoners)						

b. Determine the growth factor for the Georgia prison population.

c. Sketch a graph of this exponential equation. Use $0 \leq t \leq 30$ and $0 \leq P(t) \leq 70$.

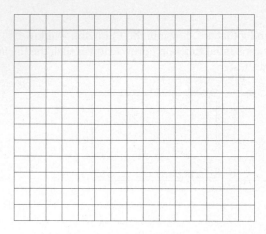

d. Use the equation to determine the state and federal prison population in Georgia in the year 2015.

e. Use your graphing calculator to approximate the year in which the state and federal prison population in Georgia will exceed 100 thousand prisoners.

8. Chlorine is used to disinfect swimming pools. The chlorine concentration should be between 1.5 and 2.5 parts per million (ppm). On sunny, hot days, 30% of the chlorine dissipates into the air or combines with other chemicals. Therefore, chlorine concentration, $A(x)$ (in parts per million) in a pool after x sunny days can be modeled by

$$A(x) = 2.5(0.7)^x.$$

a. What is the initial concentration of chlorine in the pool?

b. Complete the following table.

x	0	1	2	3	4	5
A(x)						

c. Sketch the graph of the chlorine function.

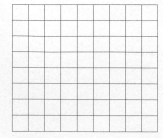

d. What is the chlorine concentration in the pool after 3 days?

e. Approximate graphically and numerically the number of days before chlorine should be added.

9. The population of Las Vegas, Nevada from 2000 to 2006 is approximated in the following table.

Year	2000	2001	2002	2003	2004	2005	2006
Population (in Thousands)	478.4	490.0	502.2	514.1	526.6	539.1	552.5

a. Does the relationship in the table represent an exponential function? Explain.

b. What is the growth factor?

c. Determine the equation that gives the population N, in thousands, of Las Vegas as a function of t, the number of years since 2000. Note that $t = 0$ corresponds to 2000, and so on.

d. Graph the function.

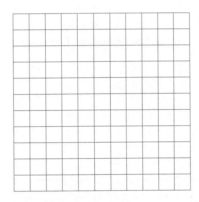

e. What is the vertical intercept? What is the practical meaning of this intercept in this situation?

f. Use the equation to estimate the population of Las Vegas in the year 2020. Do you think this is a good estimate? Explain.

g. Use the graph of the exponential function and the graph or table feature of your graphing calculator to estimate the number of years it takes for the population of Las Vegas to double from 478.4 thousand people to 956.8 thousand people.

10. As a radiology specialist, you use the radioactive substance iodine-131 to diagnose conditions of the thyroid gland. Your hospital currently has a 20-gram supply of iodine-131. The following table gives the number of grams remaining after a specified number of days.

t (Number of Days Starting from A 20-Gram Supply of Iodine-131)	0	1	2	3	4	5	6
N, Number of Grams of Iodine-131 Remaining from A 20-Gram Supply	20.00	18.34	16.82	15.42	14.14	12.97	11.89

a. Does the relationship represent an exponential function? Explain.

b. What is the decay factor?

c. Write an exponential decay formula for N, the number of grams of iodine-131 remaining, in terms of t, the number of days from the current supply of 20 grams.

d. Determine the number of grams of iodine-131 remaining from a 20-gram supply after 2 months (60 days).

e. Graph the decay formula for iodine-131, $N = 20(0.917)^t$, as a function of the time t (days). Use appropriate scales and labels on the following grid or an appropriate window on a graphing calculator.

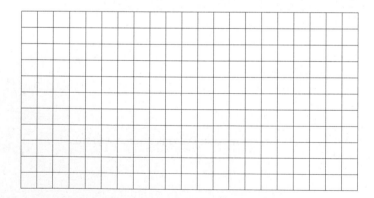

f. How long will it take for iodine-131 to decrease to half its original value? Explain how you determined your answer.

Activity 3.4

Population Growth

Objectives

1. Determine the annual growth or decay rate of an exponential function represented by a table of values or an equation.

2. Graph an exponential function having equation $y = a(1 + r)^x$.

1. According to U.S. Census Bureau estimates, the city of Charlotte, North Carolina, had a population of approximately 687,000 in 2008.

   a. Assuming that the population increases at a constant percent rate of 3.0%, determine the population of Charlotte (in thousands) in 2009.

   b. Determine the population of Charlotte (in thousands) in 2010.

   c. Divide the population in 2009 by the population in 2008 and record this ratio.

   d. Divide the population in 2010 by the population in 2009 and record this ratio.

   e. What do you notice about the ratios in parts c and d? What do these ratios represent?

Linear functions represent quantities that change at a constant average rate (slope). Exponential functions represent quantities that change at a constant percent rate.

Example 1   *Population growth, sales and advertising trends, compound interest, spread of disease, and concentration of a drug in the blood are examples of quantities that increase or decrease at a constant percent rate.*

2. a. Let t represent the number of years since 2008 ($t = 0$ corresponds to 2008). Use the results from Problem 1 to complete the following table.

t, Years (Since 2008)	0	1	2	3	4	5
P, Population (in Thousands)						

Once you know the growth factor, b, and the initial value, a, you can write the exponential equation. In this situation, the initial value is the population in thousands, in 2008 ($t = 0$), and the growth factor is $b = 1.030$.

   b. Write the exponential equation, $P = a \cdot b^t$, for the population of Charlotte.

3. a. Write the growth rate, $r = 3.0\%$, as a decimal.

   b. Add 1 to the decimal form of the growth rate, r.

The growth factor, b, is determined from the growth rate, r, by writing r in decimal form and adding 1:

$$b = 1 + r.$$

Example 2 *Determine the growth factor, b, for a growth rate of r = 8%.*

SOLUTION

$$r = 8\% = 0.08, b = 1 + r = 1 + 0.08 = 1.08$$

c. Solve the equation for the growth factor, $b = 1 + r$, for r.

The growth rate, r, is determined from the growth factor, b, by subtracting 1 from b and writing the result in percent form.

Example 3 *Determine the growth rate, r, for a growth factor of b = 1.054.*

SOLUTION

$$r = b - 1 = 1.054 - 1 = 0.054 = 5.4\%$$

The growth rate is 5.4%.

4. a. Complete the following table.

t	CALCULATION FOR POPULATION (IN THOUSANDS)	EXPONENTIAL FORM	$P(t)$, POPULATION IN THOUSANDS
0	687	$687(1.030)^0$	687
1	(687)1.030	$687(1.030)^1$	
2	(687)(1.030)(1.030)		
3			

b. Use the pattern in the table in part a to help you write the equation for $P(t)$, the population of Charlotte (in thousands), using t, the number of years since 2008, as the input value. How does your result compare to the equation obtained in Problem 2?

Example 4 **a.** *Determine the growth factor and the growth rate of the function defined by $f(x) = 250(1.7)^x$.*

SOLUTION

The growth factor $1 + r$ is the base 1.7. To determine the growth rate, solve the equation $1 + r = 1.7$ for r.

$$r = 0.7, \text{ or } 70\%$$

b. *If the growth rate of a function is 5%, determine the growth factor.*

SOLUTION

If $r = 5\%$ or 0.05, the growth factor is $1 + r = 1 + 0.05 = 1.05$.

5. a. In the Charlotte population function $P(t) = 687(1.030)^t$, determine the growth factor.

b. Determine the growth rate. Express your answer as a percent.

6. a. Using the function defined by $P(t) = 687(1.030)^t$, determine the population of Charlotte in 2015. That is, determine $P(t)$ when $t = 7$.

b. Graph the function with your graphing calculator. Use the window Xmin $= 0$, Xmax $= 100$, Ymin $= 0$, and Ymax $= 15000$. Your graph should resemble the following.

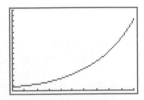

c. Determine $P(0)$. What is the graphical and practical meaning of $P(0)$?

7. a. Use the model to predict Charlotte's population in the year 2018.

b. Verify your prediction on the graph.

8. a. Use the graph to estimate when Charlotte's population will reach 1,000,000, assuming it continues to grow at the same rate. Remember, $P(t)$ is the number of thousands of people.

b. Evaluate $P(32)$ and describe what it means.

9. Use the model to estimate the population of Charlotte in 2010 and in 2030. In which prediction would you be more confident? Why?

10. a. Assuming the growth rate remains constant, how long will it take for the population of Charlotte to double its 2008 population?

b. Explain how you reached your conclusion in part a.

Wastewater Treatment Facility

You are working at a wastewater treatment facility. You are presently treating water contam
nated with 18 micrograms (μg) of pollutant per liter. Your process is designed to remo
20% of the pollutant during each treatment. Your goal is to reduce the pollutant to less th.
3 micrograms per liter.

11. a. What percent of pollutant present at the start of a treatment remains at the end of th
treatment?

b. The concentration of pollutants is 18 micrograms per liter at the start of the first
treatment. Use the result of part a to determine the concentration of pollutant at the
end of the first treatment.

c. Complete the following table. Round the results to the nearest tenth.

n, Number of Treatments	0	1	2	3	4	5
C(n), Concentration of Pollutant, in μg/l, at the End of the nth Treatment	18	14.4				

d. Write an equation for the concentration, $C(n)$, of the pollutant as a function of the
number of treatments, n.

The equation $C(n) = 18(0.80)^n$ has the general form $C = C_0(1 - r)^n$, where r is the
decay rate, $(1 - r)$ is the **decay factor** or the base of the exponential function, n is the
number of treatments, and C_0 is the initial value, the concentration when $n = 0$.

Example 5   **a. Determine the decay factor and the decay rate of the function
defined by $h(x) = 123(0.43)^x$.**

SOLUTION

The decay factor $1 - r$ is the base, 0.43. To determine the decay rate, solve the equatio
$1 - r = 0.43$ for r.

$$r = 0.57, \text{ or } 57\%$$

b. If the decay rate of a function is 5%, determine the decay factor.

SOLUTION

If $r = 5\%$, or 0.05, the decay factor is $1 - r = 1 - 0.05 = 0.95$.

12. a. If the decay rate is 2.5%, what is the decay factor?

b. If the decay factor is 0.76, what is the decay rate?

13. a. Use the function defined by $C(n) = 18(0.8)^n$ to predict the concentration of
contaminants at the wastewater treatment facility after seven treatments.

b. Sketch a graph of the concentration function on your graphing calculator. Use the table in Problem 11c to set a window. Does the graph look like you expected it would? Explain.

c. What is the vertical intercept? What is the practical meaning of the intercept in this situation?

d. Reset the window of your graphing calculator to Xmin $= -5$, Xmax $= 15$, Ymin $= -10$, and Ymax $= 50$. Does the graph have a horizontal asymptote? Explain what this means in this situation.

14. Use the table or trace feature of your graphing calculator to estimate the number of treatments necessary to bring the concentration of pollutant below 3 micrograms per liter.

SUMMARY: ACTIVITY 3.4

1. **Exponential functions** are used to describe phenomena that grow or decay by a constant percent rate per unit time.

2. If r represents the **growth rate**, the exponential function that models the quantity, P, can be written as

$$P(t) = P_0(1 + r)^t,$$

where P_0 is the initial amount, t represents the amount of elapsed time, and $1 + r$ is the growth factor.

3. If r represents the **decay rate**, the exponential function that models the amount remaining can be written as

$$P(t) = P_0(1 - r)^t,$$

where $1 - r$ is the decay factor.

EXERCISES: ACTIVITY 3.4

1. Determine the growth and decay factors and growth and decay rates in the following tables.

GROWTH FACTOR	GROWTH RATE
1.02	
	2.9%
2.23	
	34%
1.0002	

DECAY FACTOR	DECAY RATE
0.77	
	68%
0.953	
	19.7%
0.9948	

2. In 2006 the U.S. Census Bureau estimated the populations of Helena, Montana, as 27,885 and Butte, Montana, as 30,752. Since the 1990 census, Helena's population has been increasing at approximately 1.13% per year. Butte's population has been decreasing at approximately 1.0% per year. Assume the growth and decay rates stay constant.

   a. Let $P(t)$ represent the population t years after 2006. Determine the exponential functions that model the populations of both cities.

   b. Use your models to predict the populations of both cities in the year 2010.

   c. Estimate the number of years for the population of Helena, Montana, to double.

   d. Using the table and/or graphs for these functions, predict when the populations will be equal.

3. You have just taken over as the city manager of a small city. The personnel expenses were $8,500,000 in 2010. Over the previous 5 years, the personnel expenses have increased at a rate of 3.2% annually.

   a. Assuming that this rate continues, write an equation describing personnel costs, $C(t)$, in millions of dollars, where $t = 0$ corresponds to 2010.

   b. Sketch a graph of this function up to the year 2025 ($t = 15$).

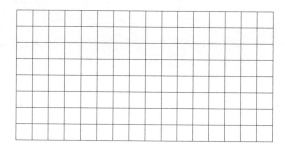

   c. What are your projected personnel costs in the year 2015?

   d. What is the vertical intercept? What is the practical meaning of the intercept in this situation?

   e. In what year will the personnel expenses be double the 2010 personnel expenses?

4. According to the U.S. Census Bureau, the population of the United States from 1930 to 2000 can be modeled by $P(t) = 120.6 \cdot 1.0125^t$, where t represents the number of years since 1930.

   a. Sketch a graph of the U.S. population model from 1930 to 2000.

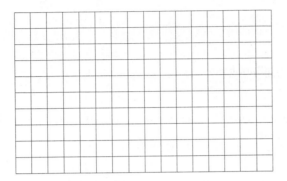

   b. Determine the annual growth rate and the growth factor from the equation.

   c. Use the population equation to determine the population (in millions) of the United States in 2000. How does your answer compare to the actual population of 281.4 million?

5. You have recently purchased a new car for $20,000 by arranging financing for the next 5 years. You are curious to know what your new car will be worth when the loan is completely paid off.

   a. Assuming that the value depreciates at a constant rate of 15%, write an equation that represents the value, $V(t)$, of the car t years from now.

   b. What is the decay rate in this situation?

   c. What is the decay factor in this situation?

   d. Use the equation from part a to estimate the value of your car 5 years from now.

   e. Use the trace and table features of your graphing calculator to check your results in part d.

   f. Use the trace or table features of your graphing calculator to determine when your car will be worth $10,000.

6. Suppose the inflation rate is 7% per year and remains the same for the next 7 years.

   a. Complete the following table for the cost of a pair of sneakers that costs $45 now. Round to the nearest cent:

t, Years From Now	0	1	2	3	4	5	6	7
c(t), Cost of Sneakers ($)	45							

   b. Determine the growth factor for a 7% inflation rate.

   c. If the yearly inflation rate remains at 7%, what exponential function would you use to determine the cost of $45 sneakers after t years?

   d. Use the equation in part c to determine the cost of the sneakers in 10 years. What assumption are you making regarding the inflation rate?

Congratulations, you have inherited $20,000! Your grandparents suggest that you use half of the inheritance to start a retirement fund. Your grandfather claims that an investment of $10,000 could grow to over half a million dollars by the time of retirement. You are intrigued by this statement and decide to investigate if this could happen.

Activity 3.5

Time Is Money

Objective

- Apply the compound interest and continuous compounding formulas to a given situation.

1. a. Suppose the $10,000 is deposited in a bank at a 6.5% annual interest. What is the interest earned after 1 year?

b. Suppose you left the money in the account for 10 years. Can the total amount of interest on your investment be calculated by multiplying your answer in part a by 10? What assumption are you making if you said yes?

Recall that for **simple interest** on an investment, the interest earned during the first period does not earn interest for the rest of the life of the investment. Therefore, at 6.5% simple annual interest, the $10,000 would earn a total interest of $10,000(0.065)(10) = \$6500$.

The interest paid on savings accounts in most banks is **compound interest**. The interest earned for each period is added to the previous principal before the next interest calculation is made. Simply stated, interest earns interest.

For example, if you deposit $10,000 in the bank at 6.5%, the balance after 1 year is

$$10,000 + 0.065(10,000) = 10,000 + 650 = 10,650.$$

The interest, $650, earned during the year becomes part of the new balance. At the end of the second year, your balance is

$$10,650 + 0.065(10,650) = 10,650 + 692.25 = 11,342.25.$$

Note that you made interest on the original deposit, plus interest on the first year's interest. In this situation, we say that interest is compounded. Usually, the compounding occurs at fixed intervals (typically at the end of every year, quarter, month, or day). In this example, interest is compounded annually.

If interest is compounded, then the current balance is given by the formula

$$A = P\left(1 + \frac{r}{n}\right)^{nt},$$

where A is the current balance, or compound amount in the account,

P is the principal (the original amount deposited),

r is the annual interest rate (in decimal form),

n is the number of times per year that interest is compounded, and

t is the time in years the money has been invested.

The given formula for the compound amount A is called the **compound interest formula**.

Example 1   *You invest $100 at 4% compounded quarterly. How much money do you have after 5 years?*

SOLUTION

The principal is $100, so $P = 100$. The annual interest rate is 4%, so $r = 0.04$. Interest is compounded quarterly, that is, 4 times per year, so $n = 4$. The money is invested for 5 years, so $t = 5$. Substituting the values for the P, r, n, and t in the compound interest formula, you have

$$A = 100\left(1 + \frac{0.04}{4}\right)^{4 \cdot 5} = \$122.02.$$

2. a. Suppose you deposit $10,000 in an account that has a 6.5% annual interest rate (usually referred to as APR, for annual percentage rate), and whose interest is compounded annually ($n = 1$). Substitute the appropriate values for P, n, and r into the compound interest formula to get the balance, A, as a function of time, t.

b. Use the compound interest formula from part a to determine your balance, A, at the end of the first year ($t = 1$).

c. What will be the amount of interest earned in the first year?

d. Use the compound interest formula developed in part a to complete the following table.

t, Year	0	1	2	3	4
A, Balance	10,000.00	10,650.00			

e. The compound interest formula in part a defines A as an exponential function of t. Identify the base.

f. Is the base a growth or decay factor? Explain.

3. a. Suppose you deposit the $10,000 into an account that has the same interest rate (APR) of 6.5%, with compounding quarterly ($n = 4$) rather than annually ($n = 1$). Write a new formula for your balance, A, as a function of time.

b. What would be your balance after the first year?

c. Use the table feature of your calculator to determine the balance at the end of each year for 10 years, and record the values in the table in Problem 4 under $n = 4$ (compounded quarterly).

d. Write the function $A = 10{,}000 \cdot (1.01625)^{4t}$ in the form $A = ab^t$

e. What is the base of this exponential function from problem 3d?

4. Now deposit your $10,000 into a 6.5% APR account with *monthly* compounding ($n = 12$) and then in an account with *daily* compounding ($n = 365$). Use your graphing calculator and the appropriate formula to complete the following table.

COMPARISON OF $10,000 PRINCIPAL IN 6.5% APR ACCOUNTS WITH VARYING COMPOUNDING PERIODS			
t	*n* = 4	*n* = 12	*n* = 365
0			
5			
10			
15			
20			
25			
30			
35			
40			

5. In Problem 4, you calculated the balance on a deposit of $10,000 at an annual interest rate of 6.5% that was compounded at different intervals. After 40 years, which account has the higher balance? Does this seem reasonable? Explain.

Continuous Compounding

You could extend this problem so that interest is compounded every hour or every minute or even every second. However, compounding more frequently than every hour does not increase the balance very much.

To discover why this happens, take a closer look at the exponential functions from Problems 1–3.

$$n = 1 \qquad A = 10,000\left(1 + \tfrac{0.065}{1}\right)^{1 \cdot t} = 10,000\left[\underline{(1 + 0.065)^1}\right]^t$$

$$n = 4 \qquad A = 10,000\left(1 + \tfrac{0.065}{4}\right)^{4 \cdot t} = 10,000\left[\underline{\left(1 + \tfrac{0.065}{4}\right)^4}\right]^t$$

$$n = 12 \qquad A = 10,000\left(1 + \tfrac{0.065}{12}\right)^{12 \cdot t} = 10,000\left[\underline{\left(1 + \tfrac{0.065}{12}\right)^{12}}\right]^t$$

$$n = 365 \qquad A = 10,000\left(1 + \tfrac{0.065}{365}\right)^{365 \cdot t} = 10,000\left[\underline{\left(1 + \tfrac{0.065}{365}\right)^{365}}\right]^t$$

Can you discover a pattern in the form of the underlined expressions?

Each formula can be expressed as $A = 10,000b^t$, where $b = \left(1 + \tfrac{0.065}{n}\right)^n$ for $n = 1, 4, 12,$ and 365. The number b is called the **growth** factor, and n is the number of compounding periods per year.

Example 2

If $n = 4$ in the formula $b = \left(1 + \tfrac{0.065}{n}\right)^n$, then
$b = \left(1 + \tfrac{0.065}{4}\right)^4 = 1.06660.$ *The number **1.06660** is the growth factor.*

6. Determine the value of b in the following table, where $b = \left(1 + \frac{0.065}{n}\right)^n$. Round to fiv
decimal places.

n, Number of Compounding Periods	1	4	12	365
b, Growth Factor				

The growth rate is the percentage by which the balance grows by in 1 year. It is called th
effective yield, r_e. Notice that as the number of compounding periods increases, the effectiv
yield increases. This means that with the same annual interest rate (APR), your investmen
will earn more with more compounding periods.

Procedure

To calculate the effective yield

1. Determine the growth factor $b = \left(1 + \frac{r}{n}\right)^n$.

2. Subtract 1 from b and write the result as a decimal.

$$r_e = b - 1 = \left(1 + \frac{r}{n}\right)^n - 1$$

Example 3   *Determine the effective yield for an APR of 4.5% compounded monthly.*

SOLUTION

$r = 4.5\% = 0.045, n = 12$

$$r_e = b - 1 = \left(1 + \frac{r}{n}\right)^n - 1 = \left(1 + \frac{0.045}{12}\right)^{12} - 1 \approx 1.04594 - 1 = 0.04594$$

$r_e = 4.594\%$

---

7. a. If interest is compounded hourly, then $n = 365 \cdot 24 = 8760$. Compute the growth
factor, b, for compounding hourly, using an APR of 6.5%.

b. Determine the effective yield associated with each of the growth factors in the table
from Problem 6.

n	1	4	12	365
Growth Factor, b	1.065	1.0666	1.06697	1.06715
Effective Yield r_e				

c. Write a sentence comparing the growth factor b for compounding hourly, $n = 8760$,
to that for daily compounding, $n = 365$.

If the compounding periods become shorter and shorter (compounding every hour, every minute, every second), n gets larger and larger. If you consider the period to be so short that it's essentially an instant in time, you have what is called **continuous compounding**. Some banks use this method for compounding interest.

The compound interest formula $A = P\left(1 + \frac{r}{n}\right)^{nt}$ is no longer used when interest is compounded continuously. The following develops a formula for continuous compounding.

Step 1. Rewrite the given formula as indicated using properties of exponents.

$$A = P\left(1 + \frac{r}{n}\right)^{nt} = P\left[\left(1 + \frac{r}{n}\right)^{n/r}\right]^{rt}, \text{ since } \frac{n}{r} \cdot rt = nt$$

Step 2. Let $\frac{n}{r} = x$. It follows that $\frac{r}{n} = \frac{1}{x}$. Note that as n gets very large, the value of x also gets very large.

Step 3. Substituting x for $\frac{n}{r}$ and $\frac{1}{x}$ for $\frac{r}{n}$, in the rewritten formula in step 1, you have

$$A = P\left[\left(1 + \frac{r}{n}\right)^{n/r}\right]^{rt} = P\left[\left(1 + \frac{1}{x}\right)^{x}\right]^{rt}.$$

8. a. Now take a closer look at the expression $\left(1 + \frac{1}{x}\right)^{x}$. Enter $\left(1 + \frac{1}{x}\right)^{x}$ into your calculator as a function of x. Display a table that starts at 0 and is incremented by 100. The results are displayed below.

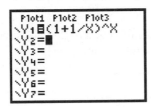  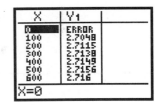

b. In the table of values, why is there an error at $x = 0$?

c. Determine the output for $x = 4000$, 5000, and 10,500. What happens to the output, $\left(1 + \frac{1}{x}\right)^{x}$, as the input, x, gets very large?

The letter e is used to represent the number that $\left(1 + \frac{1}{x}\right)^{x}$ approaches as x gets very large. This notation was devised by mathematician Leonhard Euler (pronounced oiler) (1707–1783). Euler used the letter e to denote this number. The number is irrational and its decimal representation never ends and never repeats.

9. The number e is a very important number in mathematics. Find it on your calculator, and write its decimal approximation below. How does this approximation compare to the result in Problem 8a?

You are now ready to complete the compound interest formula for continuous compounding.

Substituting e for $\left(1 + \frac{1}{x}\right)^{x}$ in $A = P\left[\left(1 + \frac{1}{x}\right)^{x}\right]^{rt}$, you obtain the following continuous compounding formula:

$$A = Pe^{rt},$$

where A is the current amount, or balance, in the account;

   P is the principal;

   r is the annual interest rate (annual percentage rate in decimal form);

   t is the time in years that your money has been invested; and

   e is the base of the continuously compounded exponential function.

Example 4   *You invest $100 at a rate of 4% compounded continuously. How much money will you have after 5 years?*

SOLUTION

The principal is $100, so $P = 100$. The annual interest rate is 4%, so $r = 0.04$. The money is invested for 5 years, so $t = 5$. Because interest is compounded continuously, you use the formula for continuous compounding as follows.

$$A = 100e^{0.04 \cdot 5} \approx \$122.14$$

---

10. **a.** Calculate the balance of your $10,000 investment in 10 years with an annual interest rate of 6.5% compounded continuously.

The formula used for the preceding result was $A = 10,000e^{0.065t}$. Comparing $A = 10,000e^{0.065}$ with $A = 10,000b^t$ shows that the growth factor is $b = e^{0.065}$.

   b. Determine the growth factor in this situation.

   c. What is the effective yield of an annual interest rate of 6.5% compounded continuously?

11. **a.** Historically, investments in the stock market have yielded an average rate of 11.7% per year. Suppose you invest $10,000 in an account at an 11% annual interest rate that compounds continuously. Use the formula $A = Pe^{rt}$ to determine the balance after 35 years.

   b. What is the balance after 40 years?

   c. Your grandfather claimed that $10,000 could grow to more than half a million dollars by retirement time (40 years). Is your grandfather correct in his claim?

SUMMARY: ACTIVITY 3.5

1. The formula for **compounding interest** is $A = P\left(1 + \frac{r}{n}\right)^{nt}$.

2. The formula for **continuous compounding** is $A = Pe^{rt}$.

3. If the number of **compounding periods** is large, $A = P\left(1 + \frac{r}{n}\right)^{nt}$ is approximated by $A = Pe^{rt}$.

EXERCISES: ACTIVITY 3.5

1. You inherit $25,000 and deposit it into an account that earns 4.5% annual interest compounded quarterly.

   a. Write an equation that gives the amount of money in the account after t years.

   b. How much money will be in the account after 10 years?

   c. You want to have approximately $65,000 in the bank when your first child begins college. Use your graphing calculator to determine in how many years you will reach this goal.

   d. If the interest were to be compounded continuously at 4.5%, how much money would be in the account after 10 years?

   Use your graphing calculator to determine in how many years you would reach your goal of $65,000, if the interest is compounded continuously.

   f. Should you look for an investment account that will be compounded continuously?

2. You deposit $2000 in an account that earns 5% annual interest compounded monthly.

   a. What will be your balance after 2 years?

   b. Estimate how long it would take for your investment to double.

   c. Identify the annual growth rate and the growth factor.

3. Your friend deposits $1900 in an account that earns 6% compounded continuously.

   a. What will be her balance after 2 years?

   b. Estimate how long it will take for your friend's investment to double.

4. You are 25 years old and begin to work for a large company that offers you two different retirement options.

   Option 1.   You will be paid a lump sum of $20,000 for each year you work for the company.

   Option 2.   The company will deposit $10,000 into an account that will pay you 12% annual interest compounded monthly. When you retire, the money will be given to you.

   Let A represent the amount of money you will have for retirement after t years.

   a. Write an equation that represents option 1.

   b. Write an equation that represents option 2.

   c. Use your graphing calculator to sketch a graph of the two options on the same axis.

   d. If you plan to retire at age 65, which would be the better plan? Explain.

   e. If you decide to retire at age 55, which would be the better plan? Explain.

   f. Use your graphing calculator to determine at what age it would not make a difference which plan you choose.

5. The compound interest formula that gives the balance in an account with a principal of $1500 that earns interest at the rate of 4.8% compounded monthly is $A = 1500\left(1 + \dfrac{0.048}{12}\right)^{12t}$. Compare this formula to the exponential equation $A = 1500 \cdot b^t$.

   a. What takes the place of b in the compound interest formula?

   b. What is the value of the growth factor b?

   c. What is the effective yield?

6. The compound interest formula that gives the balance in an account with a principal of $1500 that earns interest at the rate of 4.8% compounded continuously is $A = 1500 \cdot e^{0.048t}$. Compare this formula to the exponential equation $A = 1500 \cdot b^t$.

   a. What takes the place of b in the compound interest formula?

   b. What is the value of the growth factor b?

   c. What is the effective yield?

Objectives

1. Discover the relationship between the equations of exponential functions defined by $y = ab^t$ and the equations of continuous growth and decay exponential functions defined by $y = ae^{kt}$.

2. Solve problems involving continuous growth and decay models.

3. Graph base e exponential functions.

The U.S. Census Bureau estimated that the U.S. population in July 2007 was 301,290,332. The U.S. population in July 2008 was estimated to be 304,059,724.

1. Assuming exponential growth, the U.S. population y can be modeled by the equation $y = a \cdot b^t$, where t is the number of years since July 2007. Therefore, $t = 0$ corresponds to July 2007.

   a. What is the initial value, a?

   b. Determine the annual growth factor, b, for the U.S. population.

   c. What is the annual growth rate?

   d. Write the equation for the U.S. population as a function of t.

The U.S. population did not remain constant at 301,290,332 from July 2007 to June 2008 and then jump to 304,059,724 on July 1, 2008. The population grew continuously throughout the year. The exponential function used to model continuous growth is the same function used to model continuous compounding for an investment.

Recall from Activity 3.5 that the formula for continuous compounding is $A = Pe^{rt}$, where A (output) is the amount of the investment, P is the initial principal, r is the compounding rate, t (input) is time, and e is the constant irrational number. When this function is used more generally, it is written as

$$y = ae^{kt},$$

   where A has been replaced by y, the output;

   P replaced by a, the initial value; and

   r replaced by k, the continuous growth rate.

Now, the exponential growth model for the U.S. population determined in Problem 1d, written in the form $y = ab^t$, can be rewritten equivalently in the continuous growth form, $y = ae^{kt}$. Since y represents the same output value in each case, $ab^t = ae^{kt}$. Since a represents the same initial value in each model, it follows that $b^t = e^{kt}$.

   2. a. Notice that e^{kt} can be written as $(e^k)^t$. How are b and e^k related?

   b. Set the value of b determined in Problem 1 ($b = 1.01$) equal to e^k and solve for k, the continuous growth rate. Solve the equation graphically by entering $Y_1 = e^x$ and $Y_2 = 1.01$. Use the window Xmin = 0, Xmax = 0.02, Ymin = 1, and Ymax = 1.02.

   c. Rewrite the U.S. population function $y = 301{,}290{,}332(1.01)^t$ in the form $y = a \cdot e^{kt}$.

Notice that an annual growth rate of $r = 0.01 = 1\%$ is equivalent to a continuous growth rate of $k = 0.00995 = 0.995\%$.

The exponential model used to describe continuous growth at a constant percent rate is $y = ae^{kt}$, where k is the constant continuous growth rate, a is the amount present initially (when $t = 0$), and e is the constant irrational number approximately equal to 2.718.

Example 1

a. Rewrite the equation $y = 42(1.23)^t$ into a continuous growth equation of the form $y = ae^{kt}$.

SOLUTION

$$1.23 = e^k, k = 0.207, y = 42e^{0.207t}$$

b. What is the continuous growth rate?

SOLUTION

$$0.207 = 20.7\%$$

c. What is the initial amount present (when $t = 0$)?

SOLUTION

$$42(1.23)^0 = 42$$

Now, consider a situation which involves continuous decay at a constant percentage rate Tylenol (acetaminophen) is metabolized in your body and eliminated at the rate of 24% per hour. You take two Tylenol tablets (1000 milligrams) at 12 noon.

3. Assume that the amount of Tylenol in your body can be modeled by an exponential function $Q = ab^t$, where t is the number of hours from 12 noon.

   a. What is the initial value, a, in this situation?

   b. Determine the decay factor, b, for the amount of Tylenol in your body.

   c. Write an exponential equation for the amount of Tylenol in your body as a function of t.

Of course, the amount of Tylenol in your body does not decrease suddenly by 24% at the end of each hour; it is metabolized and eliminated continuously. The equation $y = ae^{kt}$ can also be used to model a quantity that decreases at a continuous rate.

Recall that in Problem 2a, you compared $y = ab^t$ to $y = ae^{kt}$ and established that $b = e^k$.

4. **a.** The value of b for the Tylenol equation is 0.76. Set $b = 0.76$ equal to e^k and solve for k graphically as in Problem 2b. Use the window Xmin $= -1$, Xmax $= 0$, Ymin $= 0$, and Ymax $= 1$.

   b. Write the equation for the amount of Tylenol in your body, $Q = 1000(0.76)^t$, in the form $Q = ae^{kt}$.

Notice that the value of k in Problem 4 is negative. Whenever $0 < b < 1$, then b is a decay factor and the value of k will be negative. A decreasing exponential function written in the form $y = ae^{kt}$ will have $k < 0$, and $|k|$ is the continuous rate of decrease.

For exponential decrease (decay) at a continuous constant percent rate, the model $y = ae^{kt}$ is used, where $k < 0$, $|k|$ is the constant continuous decrease (decay) rate, a is the amount present initially, when $t = 0$, and e is the constant irrational number approximately equal to 2.718.

Example 2

a. Rewrite the decay equation $y = 12.5(0.83)^t$ in the form $y = ae^{kt}$.

SOLUTION

$$0.83 = e^k, k = -0.186, y = 12.5e^{-0.186t}$$

b. What is the continuous percentage rate of decay?

SOLUTION

$$0.186 = 18.6\%$$

c. What is the initial amount present when $t = 0$?

SOLUTION

$$12.5$$

Graphs of Exponential Functions Having Base e

5. Consider the exponential function defined by $y = e^x$.

    a. What is the domain of the function?

    b. Is this function increasing or decreasing? Explain.

    c. Complete the following table. If necessary, round the y-values to the nearest two decimal places.

x	−3	−2	−1	0	1	2	3	4	5
$y = e^x$									

    d. Sketch a graph of $y = e^x$. Verify using a graphing calculator.

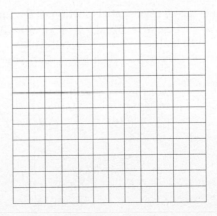

  e. What is the horizontal asymptote?

  f. What is the range of the function?

  g. What are the intercepts of the graph?

  h. Is the function continuous over its domain?

6. A vertical shift of the graph of $y = e^x$ changes the horizontal asymptote.

  a. What is the equation of the horizontal asymptote of the graph of $y = e^x + 2$?

  b. What is the equation of the horizontal asymptote of the graph of $y = e^x - 2$?

7. Identify the given exponential function as increasing or decreasing. In each case give the initial value and rate of increase or decrease.

  a. $P = 2500e^{0.04t}$

  b. $Q = 400(0.86)^t$

  c. $A = 75(1.032)^t$

  d. $R = 12e^{-0.12t}$

SUMMARY: ACTIVITY 3.6

1. When a quantity increases or decreases continuously at a constant percent rate, the amount present at time t can be modeled by $y = ae^{kt}$, where a is the initial quantity at $t = 0$. If the quantity is increasing, then $k > 0$ and k is the continuous rate of increase. If the quantity is decreasing, then $k < 0$ and $|k|$ is the continuous rate of decrease.

2. The graph of an increasing exponential function of the form $y = ab^x$ or $y = ae^{kt}$, where $a > 0$ and $b > 1$ or $k > 0$, will be shaped like

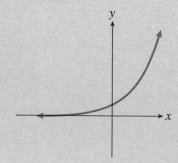

3. The graph of a decreasing exponential function of the form
$y = ab^x$ or $y = ae^{kt}$, where $a > 0$ and $0 < b < 1$, or $k < 0$, will be shaped like

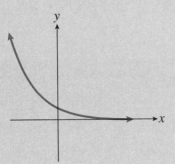

Notation

The function $y = ae^{kt}$ may be written in other forms, such as $y = y_0e^{kt}$, where the initial value is y_0 (y sub zero) or $Q = Q_0e^{kt}$, where the initial value is Q_0 (Q sub zero).

EXERCISES: ACTIVITY 3.6

1. In 2007, Charlotte, North Carolina was ranked nineteenth in population of all U.S. cities.
   The population of Charlotte has been increasing steadily, and the exponential function defined
   by $f(x) = 537.6(1.03)^x$ models the population from 2000 to 2007, where x represents the number of years since 2000 and $f(x)$ represents the total population in thousands.
   (*Source*: U.S. Census Bureau)

   a. Is this an increasing or decreasing exponential function? Explain.

   b. Determine the annual growth or decreasing (decay) rate from the model.

   c. According to the model, what is the initial value? What does the initial value mean in this situation?

   d. The equation for continuous growth is $y = ae^{kt}$. Set the value of b in your model equal to e^k, and use your graphing calculator to determine the value for k graphically. This will be the continuous growth or decay rate.

   e. Rewrite the population function in the form of $y = ae^{kt}$.

   f. Use the function in part e and predict the population of Charlotte in 2015.

2. The table shows the smoking prevalence among U.S. adults (18 years and over) as a percent of the population.

Where There's Smoke

SMOKING PREVALENCE AMONG U.S. ADULTS							
YEAR	1970	1980	1990	2000	2002	2004	2007
Years Since 1970	0	10	20	30	32	34	37
Percent of Population	37.4	33.2	25.5	23.3	22.5	20.9	20.8

Source: The U.S. Centers for Disease Control & Prevention.

This data can be modeled by an exponential function $p(t) = 37.6(0.984)^t$, where t equals the number of years since 1970 and p is the smoking prevalence among adults as a percent of the U.S. population.

a. Is this an increasing or decreasing exponential function? Explain.

b. Sketch a graph of the function using the data in the table. Does the graph reinforce your answer in part a?

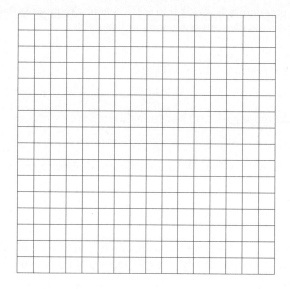

c. Determine the growth or decreasing (decay) rate from the model.

d. According to the model, what is the initial value?

e. The equation for the continuous growth or decay is $y = ae^{kt}$. Set the value of b in your model equal to e^k, and use your graphing calculator to determine the value for k graphically. This will be the continuous growth or decay rate.

f. Rewrite the smoking function in the form of $y = ae^{kt}$.

g. Use the function from part f and determine the percent of adults in the U.S. population that will be smoking in 2013.

3. Identify the given exponential function as increasing or decreasing. In each case give the initial value and rate of increase or decrease.

a. $R(t) = 33(1.097)^t$

b. $f(x) = 97.8e^{-0.23x}$

c. $S = 3250(0.73)^t$

d. $B = 0.987e^{0.076t}$

4. Sketch a graph of each of the following and verify using your graphing calculator.

a. $y = 20e^{0.08x}$ **b.** $f(x) = 10e^{-0.3x}$

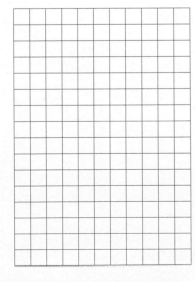
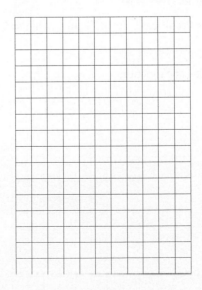

c. Compare the two graphs. Include shape, direction, and initial values.

5. Strontium 90 is a radioactive material that decays according to the function defined by $y = y_0 e^{-0.0244t}$, where y_0 is the amount present initially and t is time in years.

a. If there are 20 grams of strontium 90 present today, how much will be present in 20 years?

b. Use the graph of the function to approximate how long it will take for 20 grams to decay to 10 grams, 10 grams to decay to 5 grams, and 5 grams to decay to 2.5 grams. The length of time is called the *half-life*. In general, a half-life is the time required for half of a radioactive substance to decay.

c. Identify the annual decay rate and the decay factor.

6. When drugs are administered into the bloodstream, the amount present decreases continuously at a constant rate. The amount of a certain drug in the bloodstream is modeled by the function $y = y_0 e^{-0.35t}$, where y_0 is the amount of the drug injected (in milligrams) and t is time (in hours).

a. Suppose that 10 milligrams are injected at 10:30 A.M. How much of the drug is still in the bloodstream at 2:00 P.M.?

b. If another dose needs to be administered when there is 1 milligram of the drug present in the bloodstream, approximately when should the next dose be given (to the nearest quarter hour)?

7. The amount of credit-card spending from Thanksgiving to Christmas has increased continuously by 14% per year since 1987. The amount, A, in billions of dollars of credit-card spending during the holiday period in a given year can be modeled by

$$A = f(x) = 36.2e^{0.14x},$$

where x represents the number of years since 1987.

a. How much was spent using credit cards from Thanksgiving to Christmas in 1996?

b. Sketch a graph of the credit-card function.

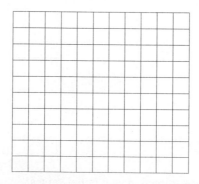

c. What is the vertical intercept of the graph? What is the practical meaning of the intercept in this situation?

d. Determine, graphically and numerically, the year when credit-card spending reached 75 billion dollars.

e. What is the doubling time?

8. *E. coli* bacteria are capable of very rapid growth, doubling in number approximately every 49.5 minutes. The number, N, of *E. coli* bacteria per milliliter after x minutes can be modeled by the equation

$$N = 500,000e^{0.014x}.$$

a. What is the initial number of bacteria per milliliter?

b. How many *E. coli* bacteria would you expect after 99 minutes? (*Hint:* There will be two doublings.) Verify your estimate using the equation.

c. Use a graphing or numerical approach to determine the elapsed time when there would be 20,000,000 *E. coli* bacteria per milliliter.

Activity 3.7

Bird Flu

Objectives

1. Determine the regression equation of an exponential function that best fits the given data.

2. Make predictions using an exponential regression equation.

3. Determine whether a linear or exponential model best fits the data.

In 2005, the Avian Flu, also known as Bird Flu, received international attention. Although there were very few documented cases of the Avian Flu infecting humans worldwide, world health organizations, including the Centers for Disease Control in Atlanta, expressed concer that a mutant strain of the Bird Flu virus that could infect humans via human-to-human con tact would develop and produce a worldwide pandemic.

The infection rate (the number of people that any single infected person will infect) and the incubation period (the time between exposure and the development of symptoms) of this fl cannot be known precisely but they can be approximated by studying the infection rates and incubation periods of existing strains of the virus.

A very conservative infection rate would be 1.5 and a reasonable incubation period would be about 15 days or roughly half of a month. This means that the first infected person could be expected to infect 1.5 people in about 15 days. After 15 days, that person cannot infect anyone else. This assumes that the spread of the virus is not checked by inoculation or vaccination.

So the total number of infected people 0.5 month after the first person was infected would be 2.5, the sum of the original infected person and the 1.5 newly infected people. During the sec ond half month, the 1.5 newly infected people would infect 1.5 × 1.5 = 2.25 new people This means you have 2.25 people to add to the 2.5 people previously infected, or approxi mately 5 people infected with Bird Flu at the end of the first month.

1. The following table represents the total number of people who could be infected with a mutant strain of Bird Flu over a period of 5 months. Complete the table. Round each value to the nearest whole person.

The Spread of Bird Flu

Months Since the First Person Was Infected	0	0.5	1	1.5	2	2.5	3	3.5	4	4.5	5
Number of Newly Infected		1.5	2.25	3.375	5						
Total Number of People Infected	1	2.5	5	8	13						

2. Let t represent the number of months since the first person was infected and T represent the total number of people infected with the Bird Flu virus. Create a scatterplot of the given data.

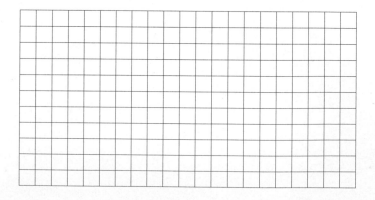

3. Does the scatterplot indicate a linear relationship between t and T? Explain.

4. a. Use your calculator to model the data with an exponential function. Use option 0:ExpReg in the STAT CALC menu to determine an exponential function that best fits the given data. Record the regression equation of the exponential model below. Round a and b to the nearest 0.001. Your screen should appears as follows:

 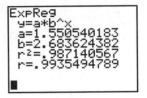

b. Sketch a graph of the exponential model using your calculator and add it to the scatterplot in Problem 2.

c. What is the practical domain of this function?

d. What is the y-intercept of the graph? How does it compare to the actual initial value $(t = 0)$ from the table?

The procedure used to develop the equation of an exponential function that best fits a set of data is quite different from the least-squares method used to develop a linear regression equation. Therefore, a correlation coefficient r cannot be calculated for exponential regression equations. However, some statisticians use a value called the coefficient of multiple determination, denoted by R^2, to measure the goodness of fit of an exponential model.

The values of R^2 vary from 0 to 1. The closer to 1, the better the likelihood an exponential regression equation fits the data. The value of R^2 is calculated using the graphing calculator and appears on the same screen as the exponential regression equation.

5. What is the value of R^2 for the exponential function determined in Problem 4? Does the value confirm your answer in Problem 3?

6. a. Use the exponential model to determine the total number of infected people 1 year after the initial infection $(t = 12)$ provided the virus is unchecked. Round your result to the nearest whole person.

b. Use the exponential function to write an equation that can be used to determine when the virus will first infect 2,000,000 people.

c. Solve the equation in part b using a graphing approach. Use the intersect feature of your calculator; the screen containing the solution should resemble the following.

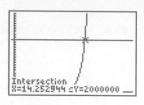

Intersection
X=14.252944 Y=2000000

d. Interpret the meaning of your solution in part c.

Increasing Exponential Model

In recent years there has been a great deal of discussion about global warming and the possibility of developing alternative energy sources that are not carbon based. One of the fastest growing sources has been wind energy. The following table shows the rapid increase in installed capacity of wind power in the United States from 2000 to 2008. The wind power capacity is measured in megawatts, MW.

Wind Power

Year	2001	2002	2003	2004	2005	2006	2007	2008
Installed Capacity, MW	4261	4685	6374	6740	9149	11,575	16,596	25,176

7. Let t represent the number of years since 2001 ($t = 0$ corresponds to 2001, $t = 1$ corresponds to 2002, etc.). Let P represent the installed wind power capacity (in megawatts) at time t. Sketch a scatterplot of the given data on your graphing calculator. Your scatterplot should appear as follows.

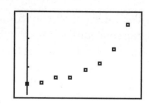

8. a. Use your graphing calculator to determine the regression equation of an exponential function that best fits the given data.

b. Sketch a graph of the exponential function using your graphing calculator.

c. What is the practical domain of this exponential function?

d. What is the vertical intercept of the graph? How does it compare to the actual initial value ($t = 0$) from the table?

9. a. What is the base of the exponential function? Is it a growth or decay factor? How do you know?

b. What is the annual growth rate?

10. a. Use the exponential model to determine the installed wind capacity in 2015 ($t = 14$).

b. Use the exponential model to write an equation to determine the year in which there will be an installed wind capacity of 50,000 megawatts.

c. Solve the equation in part b using a graphing approach. Use the intersect feature of your graphing calculator; the screen containing the solution should appear as follows. How confident are you in this prediction?

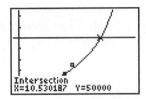

```
Intersection
X=10.530187  Y=50000
```

11. What is the doubling time for your exponential model? That is, approximately how many years will it take for a given installed wind capacity to double?

Decreasing Exponential Model

Students in U.S. public schools have had much greater access to computers in recent years. The following table shows the number of students per computer in a large school district in selected years.

Year	1994	1995	1996	1998	2000	2003	2006	2010
Number of Students per Computer	125	75	50	32	22	16	10	5.7

12. a. Use your graphing calculator to determine the regression equation of an exponential function that models the given data. Let your input, t, represent the number of years since 1994.

b. Sketch a graph of the exponential model using your graphing calculator.

c. What is the base of the exponential model? Is the base a growth or decay factor? How do you know?

d. What is the annual decay rate?

e. Does the graph have a horizontal asymptote? What is the practical meaning of this asymptote in the context of the situation?

EXERCISES: ACTIVITY 3.7

1. The total amount of money spent on health care in the United States is increasing at an alarming rate. The following table gives the total national health care expenditures in billions of dollars in selected years from 1980 through 2006.

Billions for Health

Year	1980	1990	1995	2000	2004	2006
Total Spent (billions of dollars)	253	714	1016	1354	1852	2106

Source: National Center for Health Statistics.

a. Would the data in the preceding table be better modeled by a linear model, $y = mx + b$ or an exponential model $y = ab^x$? Explain.

b. Sketch a scatterplot of this data.

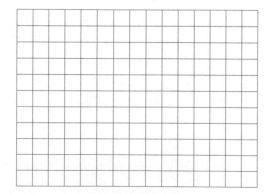

c. Does the graph reinforce your conclusion in part a? Explain.

d. Use your graphing calculator to determine the exponential regression equation that best fits the health care data in the table. Let your input, t, represent the number of years since 1980.

e. Using the regression equation from part d, determine the predicted total health care expenditures for the year 2005.

f. According to the exponential model, what is the growth factor for the total health care costs per year?

g. What is the growth rate?

h. According to the exponential model, in what year did the total heath care costs first exceed $1.5 trillion?

i. What is the doubling time for your exponential model?

2. a. Consider the following data set for the variables x and y.

x	5	8	11	15	20
y	70.2	50.7	35.1	22.6	9.5

Plot these points on the following grid.

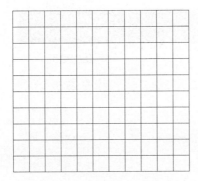

b. Use your graphing calculator to determine both a linear regression and an exponential regression model of the data. Record the equations for these models here.

c. Use the graph of each equation in part b to determine which model appears to fit the data better? Explain.

d. Use the better model to determine y when $x = 13$ and y when $x = 25$.

e. For the exponential model, what is the decay factor?

f. What does it mean that the decay factor is between 0 and 1?

g. What is the half-life for the exponential model?

3. Use the graph of $y = 5(2)^x$ as a model, and summarize the properties of the exponential function defined by $y = ab^x$, where $a > 0$.

a. What is the domain?

b. What is the range?

c. For what value of x is $y = ab^x$ positive?

d. For what value of x is $y = ab^x$ negative?

e. What is the vertical intercept of the graph of $y = ab^x$?

4. The number of transistors that can be placed on a single chip has grown significantly between 1971 and 2004. The following table gives the number of transistors (in millions) that can be placed on a specific chip in a given year.

YEAR	x, NUMBER OF YEARS SINCE 1970	CHIP	TRANSISTORS (IN MILLIONS)
1971	1	4004	0.0023
1986	16	386DX	0.275
1989	19	486DX	1.2
1993	23	Pentium	3.3
1995	25	P6	5.5
1997	27	Pentium II	7.5
1999	29	Pentium III	9.5
2000	30	Pentium IV	42
2004	34	Pentium IV Prescott	125

Source: Intel

a. Use a graphing calculator to determine an exponential regression model of the data. Let the independent variable x represent the number of years since 1970.

b. Assuming the rate of growth continues, approximate the number of transistors that can be placed on a chip in 2012.

Collecting and Analyzing Data

5. You need to find some real-world data that appears to be increasing exponentially. Newspapers, magazines, scientific journals, almanacs, and the Internet are good resources. Once the data has been obtained, model the data by an exponential function defined by $y = ab^x$. Explain why an exponential function would best represent the data. Be sure to describe the meaning of a and b in the exponential model in terms of the situation. Make a prediction about the dependent variable y for a specific value of the independent variable x. Describe the reliability of this prediction.

Cluster 1   What Have I Learned?

1. Consider a linear function defined by $g(x) = mx + b$, $m \neq 0$, and an exponential function defined by $f(x) = ab^x$. Explain how you can determine from the equation whether the function is increasing or decreasing.

2. Suppose you have an exponential function of the form $f(x) = ab^x$, where $a > 0$ and $b > 0$ and $b \neq 1$. By inspecting the graph of f, can you determine if $b > 1$ or if $0 < b < 1$? Explain.

3. You are given a function defined by a table, and the input values are in increments of 1. By looking at the table, can you determine whether or not the function can be approximated by an exponential model? Explain.

4. Explain the difference between growth rate and growth factor.

5. An exponential function $y = ab^x$ passes through the point $(0, 2.6)$. What can you conclude about the values of a and b?

6. You have just received a substantial tax refund of P dollars. You decide to invest the money in a CD for 2 years. You have narrowed your choices to two banks. Bank A will give you 6.75% interest compounded quarterly. Bank B offers you 6.50% compounded continuously. Where do you deposit your money? Explain.

7. Explain why the base in an exponential function cannot equal 1.

Cluster 1   How Can I Practice?

1. You are planning to purchase a new car and have your eye on a specific model. You know that new car prices are projected to increase at a rate of 4% per year for the next few years.

   a. Write an equation that represents the projected cost, C, of your dream car t years in the future, given that it costs $17,000 today.

   b. Identify the growth rate and the growth factor.

   c. Use your equation in part a to project the cost of your car 3 years from now.

   d. Use your graphing calculator to approximate how long it will take for your dream car to cost $30,000, if the price continues to increase at 4% per year.

2. Without using your graphing calculator, match the graph with its equation.

   a. $g(x) = 2.5(0.47)^x$ **b.** $h(x) = 1.5(1.47)^x$

   i. 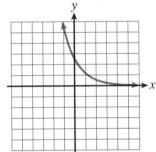   **ii.**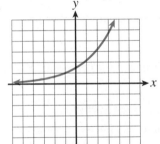

3. Explain the reasons for your choices in Problem 2.

4. Complete the following tables representing exponential functions. Round calculations to two decimal places whenever necessary.

   a.

x	0	1	2	3	4
y	2.00	5.10			

   b.

x	0	1	2	3	4
y	3.50	2.10			

c.

x	0	1	2	3	4
y	$\frac{1}{6}$	6			

5. Write the equation of the exponential function that represents the data in each table in Problem 4.

a.   **b.**   **c.**

6. Without graphing, classify each of the following functions as increasing or decreasing, and determine $f(0)$. (Use your graphing calculator to verify.)

a. $f(x) = 1.3(0.75)^x$ **b.** $f(x) = 0.6(1.03)^x$ **c.** $f(x) = 3\left(\frac{1}{5}\right)^x$

7. a. Given the following table, do you believe that the table can be approximately modeled by an exponential function?

x	0	1	2	3	4	5	6
y	2	5	12.5	31.3	78.1	195.3	488.3

b. If you answered yes to part a, what is the constant ratio of successive output values?

c. Determine an exponential equation that models this data.

8. Your starting salary for a new job is $22,000 per year. You are offered two options for salary increases:

Plan 1: an annual increase of $1000 per year or

Plan 2: an annual percentage increase of 4% of your salary

Your salary is a function of the number of years of employment at your job.

a. Write an equation to determine the salary, S, after x years on the job using plan 1; using plan 2.

b. Complete the following table using the equations from part a.

x	0	1	3	5	10	15
S, PLAN 1						
S, PLAN 2						

c. Which plan would you choose? Explain.

9. The number of victims of a flu epidemic is increasing at a continuous rate of 7.5% per week.

a. If 2000 people are currently infected, write an exponential model of the form
$N = f(t) = N_0 e^{rt}$, where

N is the number of victims in thousands,

N_0 is the initial number infected in thousands,

r is the weekly percent rate expressed as a decimal, and

t is the number of weeks.

b. Use the exponential model to predict the number of people infected after 8 weeks.

c. Sketch the graph of the flu function using your graphing calculator.

d. Use a graphing approach to predict when the number of victims of the flu will triple.

10. a. Complete the following tables.

x	−3	−2	−1	0	1	2	3
$h(x) = 4^x$							

x	−3	−2	−1	0	1	2	3
$g(x) = \left(\frac{1}{4}\right)^x$							

b. Sketch graphs of h and g on the following grid.

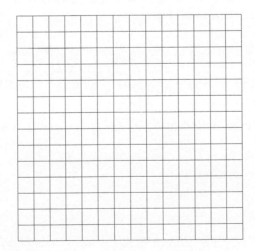

c. Use the tables and graphs in part a and b to complete this table.

FUNCTION	BASE, b	GROWTH OR DECAY FACTOR	x-INTERCEPT	y-INTERCEPT	HORIZONTAL ASYMPTOTE	INCREASING OR DECREASING
$h(x) = 4^x$						
$g(x) = \left(\frac{1}{4}\right)^x$						

11. You are a college freshman and have a credit card. You immediately purchase a stereo system for $415. Your credit limit is $500. Let's assume that you make no payments and purchase nothing more and there are no other fees. The monthly interest rate is 1.18%.

   a. What is your initial credit-card balance?

   b. What is the growth rate of your credit-card balance?

   c. What is the monthly growth factor of your credit-card balance?

   d. Write an exponential function to determine how much you will owe (represented by $f(x)$) after x months with no more purchases or payments.

   e. Use your graphing calculator to graph this function. What is the vertical intercept?

   f. What is the practical meaning of this intercept in this situation?

   g. How much will you owe after 10 months? Use the table feature on your graphing calculator to determine the solution.

   h. When you reach your credit limit of $500, the bank will expect a payment. How long do you have before you will have to start paying the money back? Use the trace feature on your grapher to approximate the solution.

12. You are working part-time for a computer company while going to college. The following table shows the hourly wage, $w(t)$, in dollars, that you earn as a function of time, t. Time is measured in years since the beginning of 2010 when you started working.

Time, t, Years, Since 2010	0	1	2	3	4	5
Hourly Wage, $w(t)$, ($)	12.50	12.75	13.01	13.27	13.53	13.81

   a. Calculate the ratios of the outputs to determine if the data in the table is exponential. Round each ratio to the nearest hundredth.

   b. What is the growth factor?

   c. Write an exponential equation that models the data in the table.

   d. What percent raise did you receive each year?

   e. If you continue to work for this company, what can you expect your hourly wage to be in 2020?

f. For approximately how many years will you have to work for the company in order for your hourly wage to double? (Assume you will receive the same percentage increase each year.)

13. You deposited $10,000 in an account that pays 12% annual interest compounded monthly.

a. Write an equation to determine the amount, A, you will have in t years.

b. How much will you have in 5 years?

c. Use your graphing calculator to determine in how many years your investment will double.

d. Write an equation to determine the amount, A, you will have in t years if the interest is compounded continuously.

e. Use the equation in part d to determine how much you will have in 5 years. Compare your answer to your answer in part b.

14. The number of farms in the United States has declined from 1950 to 2007, as the data in the following table shows. The data is estimated from the National Agricultural Statistics Service, U.S. Department of Agriculture.

Year	1950	1960	1970	1980	1990	2000	2007
Number of Farms, In Millions, N	5.8	4	3	2.5	2.2	2.2	2.1

a. Make a scatterplot of this data.

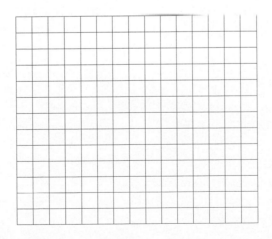

b. Does the scatterplot show that the data would be better modeled by a linear model or by an exponential model? Explain.

c. Use your graphing calculator to determine the exponential regression equation that best fits the U.S. farm data. Let x represent the number of years since 1950.

d. Use the regression equation to predict the total number of farms in the United States in 2015.

e. According to your exponential model, what is the decay factor for the total number of farms in the United States?

f. What is the decay rate?

g. Explain the meaning of the decay rate determined in part f in this situation.

h. Use your graphing calculator to determine the halving time for your exponential model.

Cluster 2   Logarithmic Functions

Activity 3.8

The Diameter of Spheres

Objectives

1. Define *logarithm*.

2. Write an exponential statement in logarithmic form.

3. Write a logarithmic statement in exponential form.

4. Determine log and ln values using a calculator.

Spheres are all around you (pardon the pun). You play sports with spheres like baseballs, basketballs, and golf balls. You live on a sphere. Earth is a big ball in space, as are the other planets, the Sun, and the Moon. All spheres have properties in common. For example, the formula for the volume, V, of any sphere is $V = \frac{4}{3}\pi r^3$, and the formula for the surface area, S, of any sphere is $S = 4\pi r^2$, where r represents the radius of the sphere.

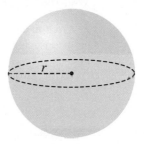

However, not all spheres are the same size. The following table gives the diameter, d, of some spheres you all know. Recall that the diameter, d, of a sphere is twice the radius, r.

SPHERE	DIAMETER, d, IN METERS
Golf ball	0.043
Baseball	0.075
Basketball	0.239
Moon	3,476,000
Earth	12,756,000
Jupiter	142,984,000

If you want to determine either the volume or surface area of any of the spheres in the preceding table, the diameter of the given sphere would be the input value and would be referenced on the horizontal axis. But how would you scale this axis?

1. **a.** Plot the values in the first three rows of the table. Scale the axis starting at 0 and incrementing by 0.02 meter.

    b. Can you plot the values in the last three rows of the table on the same axis? Explain.

2. **a.** Plot the values in the last three rows of the table on a different axis. Scale the axis starting at 0 and incrementing by 10,000,000 meters.

b. Can you plot the values in the first three rows of the table on the axis in part a? Explain.

Logarithmic Scale

3. There is a way to scale the axis so that you can plot all the values in the table on th same axis.

   a. Starting with the leftmost tick mark, give the first tick mark a value of 0.01 meter. Write 0.01 as a power of 10 as follows: $0.01 = \dfrac{1}{100} = \dfrac{1}{10^2} = 10^{-2}$ meters. Give th next tick mark a value of 0.1, written as 10^{-1} meters. Continue in this way by giving each consecutive tick mark a value that is one power of 10 greater than the preceding tick mark.

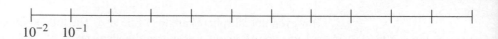

10^{-2} 10^{-1}

   b. Complete the following table by writing all of the diameters from the preceding table in scientific notation.

SPHERE	DIAMETER, d, IN METERS	d, IN SCIENTIFIC NOTATION
Golf ball	0.043	
Baseball	0.075	
Basketball	0.239	
Moon	3,476,000	
Earth	12,756,000	
Jupiter	142,984,000	

   c. To plot the diameter of a golf ball, notice that 0.043 meter is between $10^{-2} = 0.01$ meter and $10^{-1} = 0.1$ meter. Now using the axis in part a, plot 0.043 meter between the tick mark labeled 10^{-2} and 10^{-1} meters, closer to the tick mark labeled 10^{-2} meters.

   d. To plot the diameter of Earth, notice that 12,756,000 meters is between $10^7 = 10,000,000$ meters and $10^8 = 100,000,000$ meters. Now plot 12,756,000 meters between the tick marks labeled 10^7 and 10^8 meters, closer to the tick mark labeled 10^7.

   e. Plot the remaining data in the same way by first determining between which two powers of 10 the number lies.

The scale you used to plot the diameter values is a *logarithmic* or *log scale*. The tick marks on a logarithmic scale are usually labeled with just the exponent of the powers of 10.

4. a. Rewrite the axis from Problem 3a by labeling the tick marks with just the exponents of the powers of 10.

b. The axis looks like a standard axis with tick marks labeled -2, -1, 0, 1, and so on. However, it is quite different. Describe the difference between this log scale and a standard axis labeled in the same way. Focus on the values between consecutive tick marks.

Definition

The exponents used to label the tick marks of the preceding axis are **logarithms** or simply **logs**. Since these are exponents of powers of 10, the exponents are logs **base** 10, known as **common logarithms** or common logs.

Example 1

a. *The common logarithm of 10^3 is the exponent to which 10 must be raised to obtain a result of 10^3. Therefore, the common log of 10^3 is 3.*

b. *The common log of 10^{-2} is -2.*

c. *The common log of $100 = 10^2$ is 2.*

5. Determine the common log of each of the following.

a. 10^{-1} **b.** 10^4 **c.** 1000

d. 100,000   **e.** 0.0001

Logarithmic Notation

Remember that **a logarithm is an exponent.** The common log of x is an exponent, y, to which the base, 10, must be raised to get result x. That is, in the equation $10^y = x$, y is the logarithm. Using log notation, $\log_{10} x = y$. Therefore, $\log_{10} 10,000 = \log_{10} 10^4 = 4$.

Example 2

x, THE NUMBER	y, THE EXPONENT (LOGARITHM) TO WHICH THE BASE, 10, MUST BE RAISED TO GET x	LOG NOTATION $\log_{10} x = y$
10^3	3	$\log_{10} 10^3 = 3$
10^{-2}	-2	$\log_{10} 10^{-2} = -2$
100	2	$\log_{10} 100 = 2$

When using logs base 10, the notation $\log_{10}$ is shortened by dropping the 10. Therefore,

$$\log_{10} 10^3 = \log 10^3 = 3; \log_{10} 100 = \log 100 = 2$$

6. Determine each of the following. Compare your result with those from Problem 5.

a. $\log 10^{-1}$ **b.** $\log 10^4$ **c.** $\log 1000$

d. $\log 100{,}000$ **e.** $\log 0.0001$

Bases for Logarithms

The logarithmic scale for the diameter of spheres situation was labeled with the exponents of powers of 10. Using 10 as the base for logarithms is common since the number 10 is the base of our number system. However, other numbers could be used as the base for logs. For example, you could use exponents of powers of 5 or exponents of powers of 2.

> **Example 3** *Base-5 logarithms: The log base 5 of a number, x, is the exponent to which the base, 5, must be raised to obtain x. For example,*
>
> **a.** $\log_5 5^4 = 4$ or, in words, log base 5 of 5 to the fourth power equals 4
> **b.** $\log_5 125 = \log_5 5^3 = 3$
> **c.** $\log_5 \frac{1}{25} = \log_5 5^{-2} = -2$
>
> *Base-2 logarithms: The log base 2 of a number, x, is the exponent to which 2 must be raised to obtain x. For example,*
>
> **a.** $\log_2 2^5 = 5$ or log base 2 of 2 to the fifth power equals 5
> **b.** $\log_2 16 = \log_2 2^4 = 4$
> **c.** $\log_2 \frac{1}{8} = \log_2 2^{-3} = -3$

In general, a statement in logarithmic form is $\log_b x = y$, where b is the base of the logarithm, x is a power of b, and y is the exponent. The base b for a logarithm can be any positive number except 1.

7. Determine each of the following.

a. $\log_4 64$ **b.** $\log_2 \frac{1}{16}$ **c.** $\log_3 9$ **d.** $\log_3 \frac{1}{27}$

The examples and problems so far in this activity demonstrate the following property of logarithms.

Property of Logarithms

In general, $\log_b b^n = n$, where $b > 0$ and $b \neq 1$.

8. Determine each of the following:

a. $\log 1$ **b.** $\log_5 1$ **c.** $\log_{\frac{1}{2}} 1$

d. $\log 10$ **e.** $\log_5 5$ **f.** $\log_{1/2} \left(\frac{1}{2}\right)$

9. a. Referring to Problem 8a–c, write a general rule for $\log_b 1$.

   b. Referring to Problem 8d–f, write a general rule for $\log_b b$.

> **Property of Logarithms**
>
> In general, $\log_b 1 = 0$ and $\log_b b = 1$, where $b > 0, b \neq 1$.

Natural Logarithms

Because the base of a log can be any positive number except 1, the base can be the number e. Many applications involve the use of log base e. Log base e is called the **natural** log and has the following special notation:

$$\log_e x \text{ is written as } \ln x, \text{ read simply as el-n-x.}$$

Example 4   **a.** $\ln e^2 = \log_e e^2 = 2$ **b.** $\ln \frac{1}{e^4} = \ln e^{-4} = -4$

10. Evaluate the following.

a. $\ln e^7$ **b.** $\ln\left(\frac{1}{e^3}\right)$ **c.** $\ln 1$

d. $\ln e$ **e.** $\ln \sqrt{e}$

Logarithmic and Exponential Forms

Because logarithms are exponents, logarithmic statements can be written as exponential statements, and exponential statements can be written as logarithmic statements.

For example, in the statement $3 = \log_5 125$, the base is 5, the exponent (logarithm) is 3, and the result is 125. This relationship can also be written as the equation $5^3 = 125$.

> In general, the logarithmic equation $y = \log_b x$ is equivalent to the exponential equation $b^y = x$.

Example 5   *Rewrite the exponential equation $e^{0.5} = x$ as an equivalent logarithmic equation.*

SOLUTION

In the equation $e^{0.5} = x$, the base is e, the result is x, and the exponent (logarithm) is 0.5. Therefore, the equivalent logarithmic equation is $0.5 = \log_e x$, or $0.5 = \ln x$.

11. Rewrite each exponential equation as a logarithmic equation and each log equation a an exponential equation.

a. $3 = \log_2 8$

b. $\ln e^3 = 3$

c. $\log_2 \frac{1}{16} = -4$

d. $6^3 = 216$

e. $e^1 = e$

f. $3^{-2} = \frac{1}{9}$

Logarithms and the Calculator

The numbers whose logarithms you have been working with have been exact powers of th base. However, in many situations, you have to evaluate a logarithm where the number is no an exact power of the base. For example, what is log 20 or ln 15? Fortunately, the commo log (base 10) and the natural log (base e) are functions on your calculator.

12. Use your calculator to evaluate the following.

a. $\log 20$

b. $\ln 15$

c. $\ln \frac{1}{2}$

d. $\log 0.02$

e. Use your calculator to check your answers to Problems 6 and 10.

13. a. Use your calculator to complete the following table.

SPHERE	DIAMETER, d, IN METERS	d, IN SCIENTIFIC NOTATION	$\log(d)$
Golf ball	0.043	4.3×10^{-2}	
Baseball	0.075	7.5×10^{-2}	
Basketball	0.239	2.39×10^{-1}	
Moon	3,476,000	3.476×10^6	
Earth	12,756,000	1.2756×10^7	
Jupiter	142,984,000	1.4298×10^8	

b. Plot the values from the log column in the preceding table on the following axis.

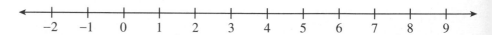

c. Compare the preceding plot with the plot on the log-scaled axis in Problem 3a and comment.

SUMMARY: ACTIVITY 3.8

1. The notation for logarithms is $\log_b x = y$, where b is the base of the log, x is the resulting power of b, and y is the exponent. The base b can be any positive number except 1; x can be any positive number. The range of y values includes all real numbers.

2. The notation for **common logarithm** or base-10 logarithms is $\log_{10} x = \log x$.

3. The notation for the **natural logarithm** or base e logarithm is $\log_e x = \ln x$.

4. The **logarithmic equation** $y = \log_b x$ is equivalent to the **exponential equation** $b^y = x$.

5. If $b > 0$ and $b \ne 1$,

   a. $\log_b 1 = 0$

   b. $\log_b b = 1$

   c. $\log_b b^n = n$

EXERCISES: ACTIVITY 3.8

1. Use the definition of logarithm to determine the exact value of each of the following.

   a. $\log_2 32$

   b. $\log_3 27$

   c. $\log 0.1$

   d. $\log_2 \left(\frac{1}{64} \right)$

   e. $\log_5 1$

   f. $\log_{1/2} \left(\frac{1}{4} \right)$

   g. $\log_7 \sqrt{7}$
   (*Hint:* $\sqrt{7} = 7^{1/2}$)

   h. $\log_{100} 10$

   i. $\log 1$

   j. $\log_2 1$

   k. $\ln e^5$

   l. $\ln \left(\frac{1}{e^2} \right)$

   m. $\ln 1$

2. Evaluate each common logarithm without the use of a calculator.

   a. $\log \left(\dfrac{1}{1000} \right) = $ _____

   b. $\log \left(\dfrac{1}{100} \right) = $ _____

   c. $\log \left(\dfrac{1}{10} \right) = $ _____

   d. $\log 1 = $ _____

   e. $\log 10 = $ _____

   f. $\log 100 = $ _____

   g. $\log 1000 = $ _____

   h. $\log \sqrt{10} - $ _____

3. Rewrite the following equations in logarithmic form.

   a. $3^2 = 9$

   b. $\sqrt{121} = 11$ (*Hint:* First rewrite $\sqrt{121}$ in exponential form.)

   c. $4^t = 27$

   d. $b^3 = 19$

4. Rewrite the following equations in exponential form.

   a. $\log_3 81 = 4$

   b. $\frac{1}{2} = \log_{100} 10$

   c. $\log_9 N = 12$

   d. $y = \log_7 x$

   e. $\ln \sqrt{e} = \frac{1}{2}$

   f. $\ln \left(\frac{1}{e^2}\right) = -2$

5. Estimate between what two integers the solutions for the following equations fall. Then solve each equation exactly by changing it to log form. Use your calculator to approximate your answer to three decimal places.

   a. $10^x = 3.25$

   b. $10^x = 590$

   c. $10^x = 0.0000045$

Activity 3.9

Walking Speed of Pedestrians

Objectives

1. Determine the inverse of the exponential function.

2. Identify the properties of the graph of a logarithmic function.

3. Graph the natural logarithmic function.

On a recent visit to Boston, you notice that people seem rushed as they move about the city. Upon returning to college, you mention this observation to your psychology instructor. The instructor refers you to a psychology study that investigates the relationship between the average walking speed of pedestrians and the population of the city. The study cites statistics presented graphically as follows.

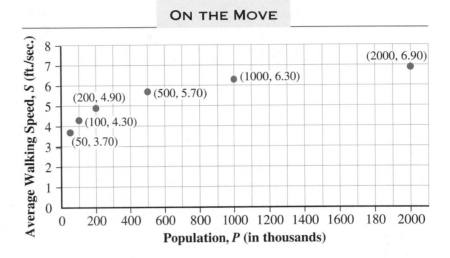

ON THE MOVE

1. a. Does the data appear to be linear? Explain.

b. Does the data appear to be exponential? Explain.

This data is actually logarithmic. Situations that can be modeled by logarithmic functions will be the focus of this and the following activity.

Introduction to the Logarithmic Function

The logarithmic function base b is defined by $y = \log_b x$, where

i. b represents the base of the $\log (b > 0, b \neq 1)$,

ii. x is the input and represents a power of the base b (x is also called the argument), and y is the output and is the exponent needed on the base b to obtain x.

2. a. Evaluate $\log_{10} (-100)$ using your calculator. What do you observe? Does it seem reasonable? Explain.

b. Is it possible to determine $\log (0)$? Explain.

c. What is the domain for the function defined by $y = \log x$?

d. What is the range? Remember, the output y is an exponent.

3. The exponential function defined by $f(x) = 10^x$ has a special relationship with the corresponding logarithmic function defined by $g(x) = \log_{10} x = \log x$.

a. Complete the following tables for $f(x) = 10^x$ and $g(x) = \log x$.

x	$f(x) = 10^x$
−2	
−1	
0	
1	
2	

x	$g(x) = \log x$
0.01	
0.1	
1	
10	
100	

b. Compare the input and output values for functions f and g.

c. Sketch the graphs of Y1 $= 10^x$ and Y2 $= \log_{10} x$ using your graphing calculator. Use the window Xmin $= -4$, Xmax $= 4$, Ymin $= -3$, and Ymax $= 3$. Your screen should appear as follows.

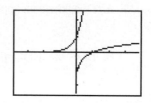

d. Graph $y = x$ on the same coordinate axes as functions f and g. Describe in a sentence or two the symmetry you observe in the graphs of f and g.

Recall the concept of an inverse function from Chapter 2. The inverse function interchanges the domain and range of the original function. Also, the graph of an inverse function is the reflection of the original function about the line $y = x$. Therefore, the results in Problem 3 demonstrate that $f(x) = 10^x$ and $g(x) = \log x$ are inverse functions.

You can determine the equation of the inverse function by solving the defining equation for the input (x-value) and then interchanging the input (x-value) and the output (y-value).

Example 1   *Determine the equation of the inverse of the function defined by* $y = 5^x$.

SOLUTION

Step 1. Solve the equation for x by writing the statement in logarithmic notation. $x = \log_5 y$

Step 2. Interchange the x and y variables. $y = \log_5 x$

4. Use the algebraic approach demonstrated in Example 1 to verify that $y = \log x$ is the inverse of $y = 10^x$.

Problems 2, 3, and 4 illustrate the following properties of the common logarithmic function.

> **Properties of the Common Logarithmic Function defined by $f(x) = \log x$**
>
> **1.** The domain of f is the set of all positive real numbers $(x > 0)$.
>
> **2.** The range of f is all real numbers.
>
> **3.** f is the inverse of the function defined by $g(x) = 10^x$.

The Graph of the Natural Logarithmic Function

5. a. Using your calculator, complete the following table. Round your answers to three decimal places.

x	0.1	0.5	1	5	10	20	50
$y = \ln x$							

b. Sketch a graph of $y = \ln x$.

c. Verify your graph in part b using your graphing calculator. Using the window Xmin $= -1$, Xmax $= 4$, Ymin $= -2.5$, and Ymax $= 2.5$, your screen should appear as follows.

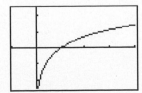

d. What are the domain and range of the function defined by $y = \ln x$?

e. Determine the intercepts of the graph.

f. Does the graph of $y = \ln x$ have a horizontal asymptote? Explain.

g. Complete the following table using your calculator. Round your answers to the nearest tenth.

x	1	0.5	0.25	0.1	0.01	0.001
$y = \ln x$						

h. As the input values take on values closer and closer to 0, what happens to the corresponding output values?

The y-axis (the line $x = 0$) is a vertical asymptote of the graph of $y = \ln x$.

--- **Definition** ---

A **vertical asymptote** is a vertical line, $x = a$, that the graph of a function becomes very close to but never touches. As the input values get closer and closer to $x = a$, the output values get larger and larger in magnitude. That is, the output values become very large positive or very large negative values.

Example 2   *The vertical asymptote of the graphs of $y = \log x$ and $y = \ln x$ is the vertical line $x = 0$ (the y-axis).*

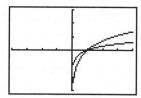

6. a. Graph $y = e^x$, $y = \ln x$, and $y = x$ on the same set of coordinate axes using the window Xmin $= -7.5$, Xmax $= 7.5$, Ymin $= -5$, and Ymax $= 5$. Describe the symmetry that you observe.

b. Use an algebraic approach to determine the inverse of the exponential function defined by $y = e^x$.

7. A horizontal shift of the graph of $y = \ln x$ changes the vertical asymptote.

a. What is the equation of the vertical asymptote of the graph of $y = \ln(x + 3)$?

b. What is the equation of the vertical asymptote of the graph of $y = \ln(x - 2)$?

SUMMARY: ACTIVITY 3.9

1. Properties of the logarithmic function defined by $y = \log_b x$, where $b > 1$.

a. The domain of f is $x > 0$.

b. The range of f is all real numbers.

c. f is the inverse of the function defined by $g(x) = b^x$.

2. The graph of a logarithmic function defined by $y = \log_b x$, where $b > 1$.

a. is increasing for all $x > 0$

b. has an x-intercept of $(1, 0)$

c. has no y-intercept

d. has a vertical asymptote of $x = 0$, the y-axis

e. resembles the following graph:

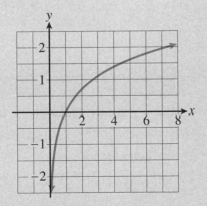

3. The **common logarithmic function** is defined by

$$y = \log x = \log_{10} x.$$

4. The **natural logarithmic function** is defined by

$$y = \ln x = \log_e x.$$

5. Since the base of the natural log function is $e > 1$, the graph of the natural logarithmic function $y = \ln x$

a. is increasing for all $x > 0$

b. has an x-intercept of $(1, 0)$

c. has a vertical asymptote of $x = 0$, the y-axis

6. The logarithmic function defined by $y = \log_b x$, where $b > 1$, is continuous over its domain, $x > 0$.

7. You can determine the equation of the inverse of the function by solving the equation of the function for x, and then interchanging the input (x-values) and the output (y-values) in the new equation.

EXERCISES: ACTIVITY 3.9

1. Using the graph of $y = \log x$ as a check, summarize the following properties of the common logarithmic function.

   a. What is the domain?

   b. What is the range?

   c. For what values of x is $\log x$ positive?

   d. For what values of x is $\log x$ negative?

   e. For what values of x does $\log x = 0$?

   f. For what values of x does $\log x = 1$?

2. a. Complete the following table using your calculator. Round your answers to the nearest tenth.

x	0.001	0.01	0.1	0.25	0.5	1
$y = \log x$						

   b. As the positive input values take on values closer and closer to 0, what happens to the corresponding output values?

   c. Determine the vertical asymptote of the graph of $y = \log x$.

3. The exponential function defined by $y = 2^x$ has an inverse. Determine the equation of the inverse function. Write your answer in logarithmic form.

4. Using the graph of $y = \ln x$ as a check, summarize the following properties of the natural logarithmic function.

   a. What is the domain?

   b. What is the range?

Exercise numbers appearing in color are answered in the Selected Answers appendix.

c. For what values of x is $\ln x$ positive?

e. For what values of x is $\ln x$ negative?

e. For what values of x does $\ln x = 0$?

f. For what values of x does $\ln x = 1$?

5. The life expectancy for a piece of equipment is the number, n, of years for the equipment to depreciate to a known salvage value, V. The life expectancy, n, is given by the formula

$$n = \frac{\log V - \log C}{\log (1 - r)},$$

where C is the initial cost of the piece of equipment and r is the annual rate of depreciation expressed as a decimal. If a backhoe costs $45,000 and has a salvage value of $2500, what is the life expectancy if the annual rate of depreciation is 40%?

6. The first 2-year college, Joliet Junior College in Chicago, was founded in 1901. The number of 2-year colleges grew rapidly, especially during the 1960s. Since then, the growth has lessened. The number y of 2-year colleges in the United States can be modeled by

$$y = 513 + 175.6 \ln x, \quad x \geq 1,$$

where x represents the number of years since 1960 (that is, $x = 1$ to 1961, etc.).

a. Determine the number of 2-year colleges in the year 2000 ($x = 40$).

b. Approximate the number of 2-year colleges in the year 2010.

Activity 3.10

Walking Speed of Pedestrians, continued

Objectives

1. Compare the average rate of change of increasing logarithmic, linear, and exponential functions.

2. Determine the regression equation of a natural logarithmic function having equation $y = a + b \ln x$ that best fits a set of data.

In Activity 3.9 you looked at a psychology study that investigated the relationship between the average walking speed of pedestrians and the population of the city. Graphically the data was presented as follows.

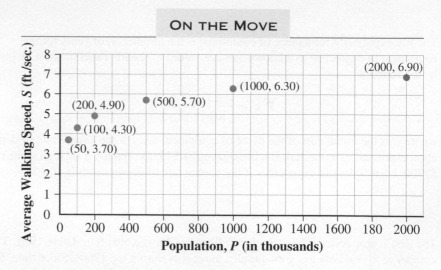

ON THE MOVE

1. a. Does the data appear to be logarithmic? Explain.

b. Use the data in the graph to complete the following table.

Population, P (in thousands)	50	100	200	500	1000	200
Average Walking Speed, S						

The natural logarithmic function can be used to model a variety of scientific and natural phenomena. The natural logarithmic function is so prevalent that on most graphing calculators it has its own built-in regression finder.

2. a. Use your graphing calculator and the table in Problem 1b to produce a scatterplot of the average walking speed data.

b. Use the regression feature of your calculator to produce a natural logarithmic curve that approximates the data in the table. Use option 9 from the STAT CALC menu.

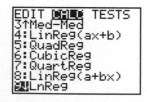

The LnReg option will generate a regression equation of the form $y = a + b \ln x$. Round a and b to the nearest thousandth, and record the function below.

c. Enter the function from part b into your graphing calculator. Verify visually that this function is a good model for your data.

d. What is the practical domain of this function?

e. Use the function from part b to predict the average walking speed in Boston, population 589,121. (*Note:* P is in thousands (589.121 thousands).)

f. Use the model to predict the average walking speed in New York City, population 8,008,278.

3. a. If the average walking speed in a certain city is 5.2 feet per second, write an equation that can be used to estimate the population P of the city.

b. Solve the equation using a graphical approach.

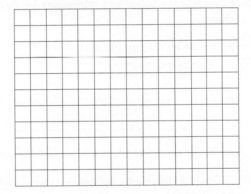

Comparing the Average Rate of Change of Logarithmic, Linear, and Exponential Functions

4. a. Complete the following table using the function defined by
$S = 0.303 + 0.868 \ln P$.

P, Population (thousands)	10	20	150	250
S, Average Walking Speed (ft./sec.)				

b. Determine the average rate of change of S as the population increases from

i. 10 to 20 thousand

ii. 20 to 150 thousand

iii. 150 to 250 thousand

c. What can you say in general about the average rate of change in the walking speed as the population increases?

You should have discovered that the average rate of change in this situation is always positive. This means that the walking speed increases as the population increases. Nevertheless, in general, the increase gets smaller as the population increases. This is characteristic of logarithmic functions.

> As the input of a logarithmic function with $b > 1$ increases, the output increases at a slower rate (the graph becomes less steep).

5. Complete the following statements by describing the rate at which the output values change.

a. For an increasing linear function, as the input variable increases, the output

_____

b. For an increasing exponential function, as the input increases, the output

_____

c. For an increasing logarithmic function, as the input increases, the output

_____

6. Consider the graphs of

i. $f(x) = e^x$ **ii.** $h(x) = x$ **iii.** $g(x) = \ln x$

using the window Xmin $= -7.5$, Xmax $= 7.5$, Ymin $= -5$, and Ymax $= 5$.

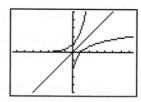

a. Which of the functions are increasing?

b. Which of the functions are decreasing?

c. As the input values get larger, which of the functions grows fastest?

d. As the input values get larger, which of the functions grows most slowly?

e. Do any of these functions have a horizontal asymptote?

f. Do any of these functions have a vertical asymptote?

g. Compare the domains of these functions.

h. Compare the ranges of these functions.

Problem 6 illustrates some of the relationships between $f(x) = b^x$, where $b > 1$; $g(x) = \log_b x$, where $b > 1$; and $y = mx + b$, where $m > 0$.

Application

7. You are working on the development of an "elastic" ball for the IBF Toy Company. The question you are investigating is, "If the ball is launched straight up, how far has it traveled vertically when it hits the ground for the tenth time?"

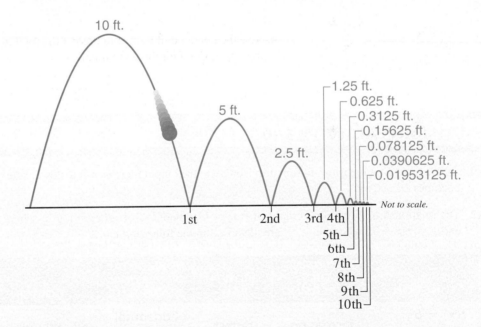

Your launcher will project the ball 10 feet into the air. This means it will travel 20 feet (10 feet up and 10 down) before it hits the ground the first time. Assuming that the ball returns to 50% of its previous height, it will rebound 5 feet and travel 10 feet before it hits the ground again. The following table summarizes this situation.

N, Times the Ball Hits the Ground	1	2	3	4	5	6
Distance Traveled Since Last Time (ft.)	20	10	5	2.5	1.25	0.625
T, Total Distance Traveled (ft.)	20	30	35	37.5	38.75	39.375

a. Using the window Xmin = 0, Xmax = 7, Ymin = 0, and Ymax = 45, a plot of N versus T should resemble the following.

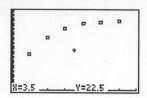

b. Do the table and scatterplot indicate that the data is linear, exponential, or logarithmic

c. Use your graphing calculator to produce linear, exponential, and natural log regression equations for the given data.

d. Graph each equation and visually determine which of the regression models best fits the data.

e. Use the equation of best fit to predict the total distance traveled by the ball when it hits the ground for the tenth time.

SUMMARY: ACTIVITY 3.10

1. As the input of a logarithmic function increases, the output increases at a slower rate (the graph becomes less steep).

2. The relationships among the graphs of $f(x) = b^x$, where $b > 1$: $g(x) = \log_b x$, where $b > 1$: and $y = mx + b$, where $m > 0$, are identified in the following table.

FUNCTION	INCREASING OR DECREASING	GROWTH RATE	HORIZONTAL OR VERTICAL ASYMPTOTE	DOMAIN	RANGE
$f(x) = b^x$, $b > 1$	increasing	fastest	horizontal asymptote	all real numbers	$y > 0$
$g(x) = \log_b x$, $b > 1$	increasing	slowest	vertical asymptote	$x > 0$	all real numbers
$y = mx + b$, $m > 0$	increasing	constant	none	all real numbers	all real numbers

EXERCISES: ACTIVITY 3.10

1. *Chlamydia trachomatis* infections are the most commonly reported notifiable disease in the United States. These are among the most prevalent of all sexually transmitted diseases. The following data from the Centers for Disease Control indicates the reported rates, R, in rates per 100,000 population from 1985 to 2006. Let t represent the number of years since 1985.

Years, t, Since 1985	1	3	5	9	13	15	17	19	21
Reported Rates: U.S. (Rate per 100,000 Population), R	40	90	160	200	250	260	288	317	344

a. Plot the points on the following grid.

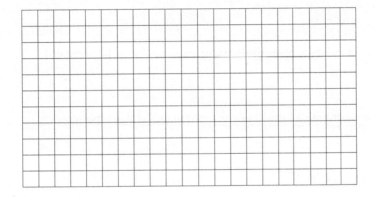

b. Does the scatterplot indicate that the data is logarithmic? Explain.

c. Determine the natural log regression equation. Record the regression equation below and add a sketch of the regression curve to the scatterplot in part a.

d. Does this appear to be a good fit? Explain.

e. Use your model to predict the reported rate of *Chlamydia trachomatis* infections per 100,000 population in 2015.

2. a. Consider a data set for the variables x and y.

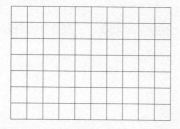

x	1	4	7	10	13
$f(x)$	3.0	4.5	5.0	5.2	5.8

Plot these points on the following grid.

b. Does the scatterplot indicate that the data is more likely linear, exponential, or logarithmic? Explain.

c. Use your graphing calculator to determine a logarithmic regression model that represents this data.

d. Use your model to determine $f(11)$ and $f(20)$.

3. The barometric pressure, P, in inches of mercury at a distance x miles from the eye of a moderate hurricane can be modeled by

$$P = f(x) = 0.48 \ln(x + 1) + 27.$$

a. Determine $f(0)$. What is the practical meaning of the value in this situation?

b. Sketch a graph of this function.

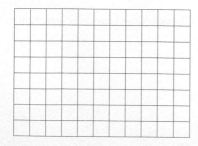

c. Describe how air pressure changes as you move away from the eye of the hurricane.

4. The following data was collected during an experiment in science class. The table shows the yield y of a substance (in milligrams) after x minutes of a chemical reaction.

Minutes, x	1	2	3	4	5	6	7	8
Yield, y (in milligrams)	1.5	7.4	10.2	13.4	15.8	16.3	18.2	18.3

a. Produce a scatterplot of the chemical reaction data.

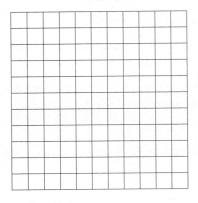

b. Use the regression feature of a calculator to obtain a natural logarithmic model and a linear model that approximates the data in the table. Graph each equation on the scatterplot in part a.

c. Determine which model best fits the data.

5. The formula $R = 80.4 - 11 \ln x$ is used to approximate the minimum required ventilation rate, R, as a function of the air space per child in a public school classroom. The rate R is measured in cubic feet per minute, and x is measured in cubic feet.

a. Sketch a graph of the rate function for $100 \leq x \leq 1500$.

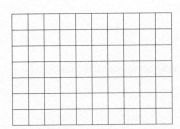

b. Determine the required ventilation rate if the air space per child is 300 cubic feet.

6. You have recently accepted a job working in the coroner's office of a large city. Because of the large numbers of homicides, it has been very difficult for the coroners to complete all of their work. Your job is, in part, to assist them in the paperwork. On one particular day, you are working on a case in which you are attempting to establish the time of death.

The coroner tells you that to establish the time of death, he uses the formula

$$t = 4 \ln \frac{98.6 - T_s}{T_b - T_s},$$

where t is the number of hours the victim has been dead,

   T_b represents the temperature of the body when discovered, and

   T_s represents the temperature of his surroundings.

The coroner also tells you that the thermostat was set at 68°F in the apartment in which the body was found and that the victim's body temperature was 78°F.

a. Using the preceding formula, determine the number of hours the victim has been deceased. Use your calculator to approximate your answer to one decimal place.

b. If the body was discovered at 10:07 P.M., what do you estimate for the time of death?

7. The following formula can be used to determine the time t it takes for an investment to double or triple in value:

$$t = \frac{\ln m}{n \ln \left(1 + \dfrac{r}{n}\right)},$$

where m represents the number of times the investment is to grow in
       value ($m = 2$ is double, $m = 3$ is triple),
   r is the annual interest rate expressed as a decimal,
   n is the number of corresponding compounding periods per year,

a. How many years will it take an investment to double if you are receiving an annual rate of 5.5% compounded quarterly ($n = 4$)?

b. How many years will it take the investment in part a to triple in value?

c. Suppose the interest on the investment in part a was compounded monthly ($n = 12$). How long will it take the value to double?

Collecting and Analyzing Data

8. The cost of a gallon of gas has increased dramatically over the past few years. In the spring of 2008, it was speculated that the price of a gallon of regular unleaded gasoline would soon exceed $4.00 per gallon.

Go to the Web site www.fueleconomy.gov to obtain the average price for 1 gallon of regular unleaded gasoline in the United States in each year from 1986 through 2007.

a. Beginning with 1986, make a scatterplot of the data collected. Let the independent variable represent the average price (in cents) of 1 gallon of regular unleaded gasoline and the dependent variable represent the number of years since 1986. Describe any patterns you observe.

b. Determine a logarithmic regression equation for the data.

c. How well does the logarithmic model in part b fit the data? Explain.

d. Predict the year in which the average price of 1 gallon of regular unleaded gasoline will reach $4.00 per gallon.

e. How reliable is this prediction? Explain.

Activity 3.11

The Elastic Ball

Objectives

1. Apply the log of a product property.

2. Apply the log of a quotient property.

3. Apply the log of a power property.

4. Discover the change of base formula.

You are continuing your work on the development of the elastic ball. You are still investigating the question, "If the ball is launched straight up, how far has it traveled vertically when it hits the ground for the tenth time?" However, your supervisor tells you that you cannot count the initial launch distance. You must calculate only the rebound distance.

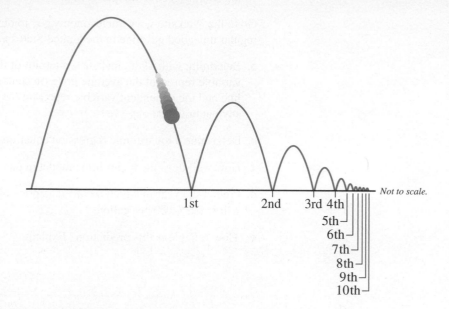

Using some physical properties, timers, and your calculator, you collect the following data.

N, Number of Times the Ball Hits the Ground	1	2	3	4	5	6
T, Total Rebound Distance (ft.)	0	9.0	13.5	16.3	18.7	21.0

1. Does the data seem reasonable? Explain.

2. Use your graphing calculator to construct a scatterplot of the data with N as the input and T as the output. Using a window of Xmin = 0, Xmax = 7, Ymin = 0, and Ymax = 25, your graph should resemble the following.

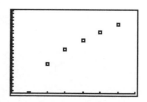

3. Do you believe the data can be modeled by a logarithmic function? Explain.

4. This data can be modeled by $T = 26.75 \log N$. Use your graphing calculator to verify visually that this is a reasonable model for the given data.

5. a. Using the log model, complete the following table. Round values to the nearest hundredth.

N	2	5	10
T = 26.75 log N			

b. How are the T-values for $N = 2$ and $N = 5$ related to the T-value for $N = 10$?

c. Using the results from part b, how could you determine the total rebound distance after 10 bounces?

The results from Problem 5 can be written as follows.

$$\underbrace{26.75} = \underbrace{8.05} + \underbrace{18.70}$$
$$26.75 \log 10 = 26.75 \log 2 + 26.75 \log 5$$
$$26.75 \log (2 \cdot 5) = 26.75 \log 2 + 26.75 \log 5$$

Dividing both sides by 26.75, you have

$$\log (2 \cdot 5) = \log 2 + \log 5.$$

This result illustrates an important property of logarithms.

Property of the Logarithm of a Product

If $A > 0, B > 0$, then $\log_b (A \cdot B) = \log_b A + \log_b B$, where $b > 0, b \neq 1$. Expressed verbally, this property states that the logarithm of a product is the sum of the individual logarithms.

Example 1   **a.** $\log_2 32 = \log_2 (4 \cdot 8) = \log_2 4 + \log_2 8 = 2 + 3 = 5$

b. $\log (5st) = \log 5 + \log s + \log t$

c. $\ln (xy) = \ln x + \ln y$

6. Use the property of the logarithm of a product to write the following as the sum of two or more logarithms.

a. $\log_b (7 \cdot 13)$ **b.** $\log_3 (xyz)$

c. $\log 15$ **d.** $\ln (3xy)$

7. Write the following as the logarithm of a single expression.

a. $\ln a + \ln b + \ln c$ **b.** $\log_4 3 + \log_4 9$

Logarithm of a Quotient

Consider the following table from Problem 5.

N	2	5	10
$T = 26.75 \log N$	8.05	18.70	26.75

This table also indicates that the rebound distance after this ball has hit the floor twic (8.05 feet) is the total rebound distance when the ball has hit the ground 10 times (26.75 fee minus the total rebound distance when the ball has hit the ground 5 times (18.70 feet).

This can be written as

$$\underbrace{8.05}_{26.75 \log 2} = \underbrace{26.75}_{26.75 \log 10} - \underbrace{18.70}_{26.75 \log 5}$$

$$\log 2 = \log 10 - \log 5.$$

Substituting $\log\left(\frac{10}{5}\right)$ for $\log 2$, you have

$$\log\left(\frac{10}{5}\right) = \log 10 - \log 5.$$

This suggests another important property of logarithms. The property is demonstrated furthe in Problem 8.

8. a. Complete the following table. Round your answers to the nearest thousandth.

x	Y1 = log $\left(\frac{x}{4}\right)$	Y2 = log x − log 4
1		
5		
10		
23		

b. Is the expression $\log\left(\frac{x}{4}\right)$ equivalent to $\log x - \log 4$? Explain.

c. Sketch the graph of $y = \log\left(\frac{x}{4}\right)$ and $y = \log x - \log 4$ using your graphing calcula-tor. What do the graphs suggest about the relationship between $\log\left(\frac{x}{4}\right)$ and $\log x - \log 4$?

Property of the Logarithm of a Quotient

If $A > 0, B > 0$, then $\log_b\left(\frac{A}{B}\right) = \log_b A - \log_b B$, where $b > 0, b \neq 1$. Expressed ver-bally, this property states that the logarithm of a quotient is the difference of the loga-rithm of the numerator and the logarithm of the denominator.

Example 2

a. $\log_3\left(\frac{81}{27}\right) = \log_3 81 - \log_3 27 = 4 - 3 = 1$

Note that $\log_3\left(\frac{81}{27}\right) = \log_3 3 = 1$.

b. $\log\left(\frac{2x}{y}\right) = \log 2x - \log y = \log 2 + \log x - \log y$

c. $\ln\left(\frac{x^2}{5}\right) = \ln x^2 - \ln 5$

9. Use the properties of logarithms to write the following as the sum or difference of logarithms.

    a. $\log_6 \frac{17}{3}$

    b. $\ln \frac{x}{23}$

    c. $\log_3 \frac{2x}{y}$

    d. $\log \frac{3}{2z}$

10. Write the following expressions as the logarithm of a single expression.

    a. $\log x - \log 4 + \log z$

    b. $\log x - (\log 4 + \log z)$

11. a. Use your graphing calculator to sketch the graphs of $y = \log x + \log 4$ and $y = \log (x + 4)$.

    b. How do these graphs compare?

    c. What do the graphs suggest about the relationship between $\log (A + B)$ and $\log A + \log B$?

Logarithm of a Power

Before calculators, logarithms were used to help in computing products and quotients of numbers. More importantly, logarithms were used to compute powers such as 734.21^3 and $\sqrt{0.0761} = (0.0761)^{1/2}$. In such a case, the first step was to take the logarithm of the power and rewrite the resulting expression. To determine how to rewrite $\log 734.21^3$, you can investigate the expression $\log x^3$.

12. a. Complete the following table. Round to the nearest thousandth.

x	Y1 = log x³	Y2 = 3 log x
2		
7		
15		

    b. Sketch the graphs of $y = \log x^3$ and $y = 3 \log x$ using your graphing calculator.

    c. What do the results of part a and part b demonstrate about the relationship between $\log x^3$ and $3 \log x$?

The results in Problem 12 illustrate another property of logarithms.

Property of the Logarithm of a Power

If $A > 0$ and p is any real number, then $\log_b A^p = p \cdot \log_b A$, where $b > 0, b \neq 1$. In words, the property states that the logarithm of a power is equivalent to the exponent times the logarithm of the base.

Example 3

a. $\log_3 9^2 = 2 \log_3 9 = 2 \cdot 2 = 4$ **b.** $\log_5 x^4 = 4 \log_5 x$

c. $\ln (xy)^7 = 7 \ln (xy)$ **d.** $\ln x^{1/4} = \frac{1}{4} \ln x$

e. $\log \sqrt{63} = \log 63^{1/2} = \frac{1}{2} \log 63$

13. Use the properties of logarithms to write the given logarithms as the sum or difference of two or more logarithms, or as the product of a real number and a logarithm. All variables represent positive numbers.

a. $\log_3 x^{1/2}$ **b.** $\log_5 x^3$

c. $\ln t^2$ **d.** $\log \sqrt[3]{50}$ (*Hint:* $\sqrt[3]{50} = 50^{1/3}$)

e. $\log_5 \frac{x^2 y^3}{z}$ **f.** $\log_3 \frac{3x^2}{y^3}$

14. Write each of the following as the logarithm of a single expression with coefficient 1.

a. $2 \log_3 5 + 3 \log_3 2$ **b.** $\frac{1}{2} \log x^4 - \frac{1}{2} \log y^5$

c. $3 \log_b 10 - 4 \log_b 5 + 2 \log_b 3$ **d.** $3 \ln 4 - (4 \ln 5 + 2 \ln 3)$

Using the properties of logarithms to solve exponential equations algebraically will be investigated in the next activity.

Change of Base Formula

Because the TI-83/T1-84 Plus has only the log base 10 (log) and the log base e (ln) keys, you cannot graph a logarithmic function such as $y = \log_2 x$ directly. Consider the following argument to rewrite the expression $\log_2 x$ as an equivalent expression using log base 10.

By definition of logs, $y = \log_2 x$ is the same as $x = 2^y$. Taking the log base 10 of both sides of the second equation, $x = 2^y$, you have

$$\log x = \log 2^y.$$

Using the property of the log of a power, $\log x = y \log 2$. Solving for y, you have

$$y = \frac{\log x}{\log 2}.$$

Therefore, the equation $y = \log_2 x$ is equivalent to $y = \frac{\log x}{\log 2}$.

15. To graph $y = \log_2 x$, enter $\log(X)/\log(2)$ for Y1 in your TI-83/TI-84 Plus calculator. Your graph should resemble the following.

 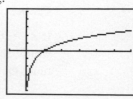

16. a. Write $y = \log_6 x$ as an equivalent equation using base 10.

b. Use the result from part a to graph $y = \log_6 x$.

c. What is the domain of the function?

d. What is the x-intercept of the graph?

The formula you used in Problems 15 and 16 for graphing log functions of different bases is a special case of the formula

$$\log_b x = \frac{\log_a x}{\log_a b}, \text{ where } a > 0, a \neq 1, b > 0, \text{ and } b \neq 1$$

This is often called the **change of base formula**.

The change of base formula is used to change from base b to base a. Because most calculators have log base 10 (log) and log base e (ln) keys, you usually convert to one of those bases. For those bases,

$$\log_b x = \frac{\log x}{\log b} \quad \text{or} \quad \log_b x = \frac{\ln x}{\ln b}.$$

Example 4   *Change the equation $y = \log_5 x$ to an equivalent equation in base* **10** *and/or base e.*

$$y = \log_5 x = \frac{\log x}{\log 5} \quad \text{or} \quad y = \log_5 x = \frac{\ln x}{\ln 5}$$

17. Use each of the change of base formulas to determine $\log_4 1024$.

a. Using base 10:   **b.** Using base e:

c. How do the results in parts a and b compare?

SUMMARY: ACTIVITY 3.11

Properties of the Logarithmic Function

If $A > 0, B > 0, b > 0$, and $b \neq 1$, then

1. $\log_b (A \cdot B) = \log_b A + \log_b B$

2. $\log_b \left(\dfrac{A}{B}\right) = \log_b A - \log_b B$

3. $\log_b (x + y) \neq \log_b x + \log_b y$

4. $\log_b A^p = p \log_b A$

5. You can use the calculator to change logarithms in base b to common or natural logarithms by

$$\log_b x = \frac{\log x}{\log b} \quad \text{or} \quad \log_b x = \frac{\ln x}{\ln b}.$$

EXERCISES: ACTIVITY 3.11

1. Use the properties of logarithms to write the following as a sum or difference of two or more logarithms.

   a. $\log_b (3 \cdot 7)$

   b. $\log_3 (3 \cdot 13)$

   c. $\log_7 \frac{13}{17}$

   d. $\log_3 \frac{xy}{3}$

2. Write the following expressions as the logarithm of a single number.

   a. $\log_3 5 + \log_3 3$

   b. $\log 25 - \log 17$

   c. $\log_5 x - \log_5 5 + \log_5 7$

   d. $\ln (x + 7) - \ln x$

3. a. Sketch the graphs of $y = \log (2x)$ and $y = \log x + \log 2$ on your graphing calculator.

   b. Are you surprised by the results? Explain.

4. a. Sketch the graphs of $y = \log \left(\frac{3}{x}\right)$ and $y = \log x - \log 3$ on your graphing calculator.

   b. Are you surprised by your results? Explain.

   c. If your graphs in part a are not identical, can you modify the second function to make the graphs identical? Explain.

5. You have been hired to handle the local newspaper advertising for a large used car dealership in your community. The owner tells you that your predecessor in this position used the formula

$$N(A) = 7.4 \log A$$

to decide how much to spend on newspaper advertising over a 2-week period. The owner admitted that he didn't know much about the formula except that $N(A)$ represented the number of cars that the owner could expect to sell, and A was the amount of money that was spent on local newspaper advertising. He also indicated that the formula seemed to work well. You can purchase small ads in the local paper for $15 per day, larger ads for $50 per day, and giant ads for $750 per day.

a. How many cars do you expect to sell if you purchase one small ad?

b. To understand the relationship between the amount spent on advertising and the number of cars sold, you set up a table. Complete the following table.

AD COST, A	EXPECTED CAR SALES, $N(A)$
15	
50	
750	

c. How do the expected car sales from one small ad and one larger ad compare to the expected car sales from just one giant ad?

d. Are the results in the table in part b consistent with what you know about the properties of logarithms? Explain.

e. What are you going to advise the owner regarding the purchase of a giant ad?

6. Use the properties of logarithms to write the given logarithms as the sum or difference of two or more logarithms or as the product of a real number and a logarithm. Simplify, if possible. All variables represent positive numbers.

a. $\log_3 3^5$

b. $\log_2 2^x$

c. $\log_b \dfrac{x^3}{y^4}$

d. $\ln \dfrac{\sqrt[3]{x}\sqrt[4]{y}}{z^2}$

e. $\log_3 (2x + y)$

7. Write each of the following as the logarithm of a single expression with coefficient 1.

a. $2 \log_2 7 + \log_2 5$

b. $\frac{1}{4}\log x^3 - \frac{1}{4}\log z^5$

c. $2 \ln 10 - 3 \ln 5 + 4 \ln z$

d. $\log_5 (x + 2) + \log_5 (x + 1) - 2 \log_5 (x + 3)$

8. Given that $\log_a x = 6$ and that $\log_a y = 25$, determine the numeric value of each of the following.

a. $\log_a \sqrt{y}$

b. $\log_a x^3$

c. $3 + \log_a x^2$

d. $\log_a \dfrac{x^2 y}{a}$

9. Use the change of base formula and your calculator to determine a decimal approximation of each of the following to the nearest ten thousandth.

a. $\log_7 5$

b. $\log_6 \sqrt{15}$

c. $\log_{13} 47$

d. $\log_5 \sqrt[3]{31}$

10. The formula

$$P = 95 - 30 \log_2 t$$

gives the percentage, P, of students who could recall the important content of a classroom presentation as a function of time, t, where t is the number of days that have passed since the presentation was given.

a. Sketch a graph of the function.

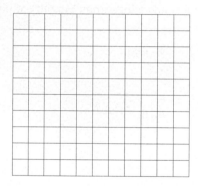

b. After 3 days, what percentage of the students will remember the important content of the presentation?

c. According to the model, after how many days do only half $(P = 50)$ of the students remember the important features of the presentation? Use a graphing approach.

You are a criminal justice major at the local community college. The following statistics appeared in one of your required readings about the inmate population of U.S. federal prisons.

Activity 3.12

Prison Growth

Objective

1. Solve exponential equations both graphically and algebraically.

Year	1979	1986	1990	1994	1998	2000	2003	2007
Total Sentenced Population, P_T (in thousands)	21.5	31.8	47.8	76.2	95.5	112.3	158.0	179.2
Total Sentenced Drug Offenders, P_D (in thousands)	5.5	12.1	25.0	46.7	56.3	63.9	86.9	95.4

You decide to do an analysis of the prison growth situation for a project in your criminology course.

1. Although the years in the table are not evenly spaced, you notice that each of the populations seems to grow rather slowly at first and more quickly later. Do you think the data will be better modeled by linear or exponential functions?

2. Let t represent the number of years since 1975. Use your graphing calculator to produce a scatterplot of the total inmate population, P_T. Your screen should appear as follows.

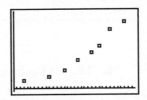

3. Use your graphing calculator to determine the regression equation of an exponential function that best represents the total inmate population, P_T. Remember that the input variable is t, the number of years since 1975. In your regression equation, $P_T = ab^t$, round the value for a to two decimal places and the value for b to three decimal places. Write your model below.

4. Use your graphing calculator to visually check how well the equation in Problem 3 fits the data. Using the window Xmin $= -3$, Xmax $= 40$, Ymin $= 0$, Ymax $= 140$, your graph should resemble the following.

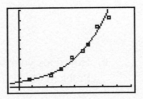

5. Use the exponential regression model from Problem 3 to determine the total federal prison inmate population in the year 2015.

6. a. Using your model from Problem 3, write an equation that can be used to determine the year in which the total federal inmate population, P_T, is 300,000. Remember, the population is given in the model in thousands.

b. Solve this equation using a graphing approach. Your screen should resemble the following. What is the equation of the horizontal line in the graph?

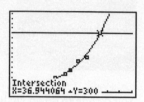

To solve the equation $14.73(1.085)^t = 300$ for t using an algebraic approach, you need to remove t as an exponent. The following problem guides you through this process. As you will discover, logarithms are essential in this algebraic approach.

7. Solve $14.73(1.085)^t = 300$ for t using an algebraic approach.

a. Isolate the exponential factor $(1.085)^t$ on one side of the equation.

b. Take the log (or ln) of each side of the equation in part a.

c. Apply the appropriate property of logarithms on the left side of the equation in order to "remove" t as an exponent.

d. Solve the resulting equation in part c for t.

e. How does your solution in part d compare to the estimate obtained graphically in Problem 6b?

8. You notice that over the years, the number of drug offenders seems to become a bigger percentage of the total population.

a. Determine an exponential model for the number of total sentenced drug offenders, P_D. Let the input variable t represent the number of years since 1975. In the regression equation $P_D = ab^t$, round the value of a to two decimal places and the value of b to three decimal places.

b. Use the exponential model to predict the total sentenced drug offenders in federal prisons in the year 2015.

c. Write an equation that can be used to determine the year in which the total sentenced drug offenders will reach 200,000.

d. Solve the equation in part c using an algebraic approach.

Radioactive Decay

Radioactive substances, such as uranium-235, strontium-90, iodine-131, and carbon-14, decay continuously with time. If P_0 represents the original amount of a radioactive substance, then the amount P present after a time t (usually measured in years) is modeled by

$$P = P_0 e^{-kt},$$

where k represents the rate of continuous decay.

Example 1 *One type of uranium decays at a rate of 0.35% per day. If 40 pounds of this uranium are available today, how much will be available after 90 days?*

SOLUTION

The uranium decays at a constant rate of $0.35\% = 0.0035$ per day. The initial amount, the amount available on the first day, is 40 pounds, so the equation for the amount available after t days is

$$P = 40e^{-0.0035t}.$$

To determine the amount available after 90 days, let $t = 90$. The amount available 90 days from now is

$$P = 40e^{-.0035(90)} \approx 29.2 \text{ lb.}$$

9. Strontium-90 decays continuously at a constant rate of 2.4% per year. Therefore, the equation for the amount P of strontium-90 after t years is

$$P = P_0 e^{-0.024t}.$$

a. If 10 grams of strontium-90 are present initially, determine the number of grams present after 20 years.

b. How long will it take for the given quantity to decay to 2 grams?

c. How long would it take for the given amount of strontium-90 to decay to one-half of its original size (called its half-life)? Round to the nearest whole number.

d. Do you think that the half-life of strontium-90 is 29 years regardless of the initial amount? Answer part c using P_0 as the initial amount. (*Hint:* Find t when $P = \frac{1}{2}P_0$.)

SUMMARY: ACTIVITY 3.12

To solve exponential equations of the form $ab^x = c$, where $a > 0$, $b > 0$, $b \neq 1$, and $c > 0$:

1. Isolate the exponential factor on one side of the equation.

2. Take the log (or ln) of each side of the equation.

3. Apply the property $\log b^x = x \log b$ to remove the variable x as an exponent.

4. Solve the resulting equation for the variable.

EXERCISES: ACTIVITY 3.12

1. The number of arrests for possession of marijuana in New York City has decreased since it reached a peak of 50,000 in 2000. The number of arrests per year can be modeled by

$$N(t) = 49.6(0.91)^t,$$

where $N(t)$ represents the number of arrests in thousands and t represents the number of years since 2000.

a. According to the model, how many arrests were made in the year 2005?

b. According to the model, when will the number of arrests in New York City be down to 20,000 arrests per year?

2. The U.S. Department of Transportation recommended that states adopt a 0.08% blood-alcohol concentration as the legal measure of drunk driving. Medical research has shown that as the concentration of alcohol in the blood increases, the risk of having a car accident increases exponentially. The risk, R, expressed as a percentage, is modeled by

$$R(x) = 6e^{12.77x},$$

where x is the blood-alcohol concentration, expressed as a percent.

a. What is the risk of having a car accident if your blood-alcohol concentration is 0.08% ($x = 0.08$)?

b. What blood-alcohol concentration has a corresponding 25% risk of a car accident?

3. In 1990, the International Panel on Climate Change projected the following future amounts of carbon dioxide (in parts per million or ppm) in the atmosphere.

Year	1990	2000	2075	2175	2275
Amount of Carbon Dioxide (ppm)	353	375	590	1090	2000

a. Use your graphing calculator to create a scatterplot of the data. Let t represent the number of years since 1990 and $A(t)$ represent the amount of carbon dioxide (in ppm) in the atmosphere. Do the carbon dioxide levels appear to be growing exponentially?

b. Use your graphing calculator to determine the regression equation of an exponential model that best fits the data.

c. Use the model in part b to determine in what year the 1990 carbon dioxide level is expected to double.

d. Verify your result in part c graphically.

In Exercises 4–9, solve each equation using an algebraic approach. Verify your answers graphically.

4. $2^x = 14$

5. $3^{2x} = 8$

6. $1000 = 500(1.04)^t$

7. $e^{0.05t} = 2$ (*Hint:* Take the natural log of both sides.)

8. $2^{3x+1} = 100$

9. $e^{-0.3t} = 2$

10. a. Iodine-131 disintegrates at a continuous constant rate of 8.6% per day. Determine its half-life. Use the model

$$P = P_0 e^{-0.086t},$$

where t is measured in days. Round your answer to the nearest whole number.

b. If dairy cows eat hay containing too much iodine-131, their milk will be unsafe to drink. Suppose that hay contains five times the safe level of iodine-131. How many days should the hay be stored before it can be fed to dairy cows?

(*Hint:* Find t when $P = \dfrac{1}{5} P_0$.)

11. a. In 1969 a report written by the National Academy of Sciences (U.S.) estimated that Earth could reasonably support a maximum world population of 10 billion. The world's population was approximately 3.6 billion and growing continuously at 2% per year. If this growth rate remained constant, in what year would the world population reach 10 billion, referred to as Earth's carrying capacity? Use the model

$$P = P_0 e^{kt},$$

where P is the population (in billions), $P_0 = 3.6$, $k = 0.02$, and t is the number of years since 1969.

b. According to your growth model, when would this 1969 population double?

c. The world population in 1995 was approximately 5.7 billion. How does this compare with the population predicted by your growth model in part a?

d. The growth rate in 1995 was 1.5%. Assuming this growth rate remains constant, determine when Earth's carrying capacity will be reached. Use the model $P = P_0 e^{kt}$.

Collecting and Analyzing Data

12. The combined populations of China and India currently represent 38% of the world's population. Go to the United Nations' Population Database Web site http://esa.un.org/unpp to obtain the population of each country for every 5 years from 1960 to 2005.

a. Beginning with 1960, make a scatterplot of the data from each country's population. Let the independent variable represent the number of years since 1960 and the dependent variable represent the population of the country in billions. Describe any patterns you observe.

b. Determine whether a linear, exponential, or logarithmic function best fits each set of data. Explain.

c. Determine a regression equation for each set of population data. Which country has the larger rate of increase in population?

d. Sketch the graph of each regression equation on the appropriate scatterplot in part a.

e. Predict the population of each country in 2010.

f. Determine the year in which the population of each country should reach 1.5 billion.

g. Use the intersect feature of a graphing calculator to estimate the year when the populations of China and India will be equal.

Cluster 2   What Have I Learned?

1. A logarithm is an exponent. Explain how this fact relates to the following properties of logarithms.

   a. $\log_b (x \cdot y) = \log_b x + \log_b y$

   b. $\log_b \frac{x}{y} = \log_b x - \log_b y$

   c. $\log_b x^n = n \cdot \log_b x$

2. You have $20,000 to invest. Your broker tells you that the value of shares of mutual fund A has been growing exponentially for the past 2 years and that shares of mutual fund B have been growing logarithmically over the same period. If you make your decision based solely on the past performances of the funds, in which fund would you choose to invest? Explain.

3. Study the following graphs, which show various types of functions you have encountered in this course.

a.
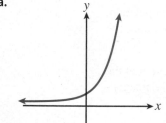

b.
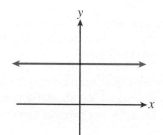

c.
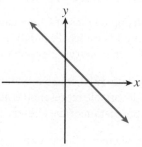

d.
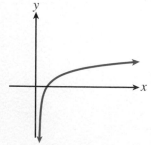

e.
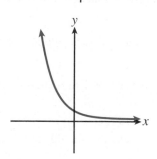

f.
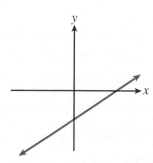

Complete the following table with respect to the preceding graphs.

DESCRIPTION	GRAPH LETTER	GENERAL EQUATION
Constant function		
Linearly decreasing function		
Logarithmically increasing function		
Exponentially decreasing function		
Exponentially increasing function		
Linearly increasing function		

4. The graph of $y = \log_b x$ will never be located in the second or third quadrants. Explain.

5. What function would you enter into Y1 on your graphing calculator to graph the function $y = \log_4 x$?

6. What values of x cannot be inputs in the function $y = \log_b (3x - 2)$?

7. What is the relationship between the functions $y = \log x$ and $y = 10^x$? How are the graphs related?

Cluster 2 How Can I Practice?

1. Write each equation in logarithmic form.

   a. $4^2 = 16$

   b. $0.0001 = 10^{-4}$

   c. $3^{-4} = \frac{1}{81}$

2. Write each equation in exponential form.

   a. $\log_2 32 = 5$

   b. $\log_5 1 = 0$

   c. $\log_{10} 0.001 = -3$

   d. $\ln e = 1$

3. Solve each equation for the unknown variable.

   a. $\log_4 x = -3$

   b. $\log_b 32 = 5$

   c. $\log_5 125 = y$

4. a. Complete the table of values for the function $f(x) = \log_4 x$.

x	0.25	0.5	1	4	16	64
f(x)						

   b. Sketch a graph of the function f.

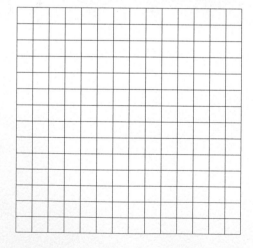

   c. Use your graphing calculator to check your result in parts a and b.

   d. Determine the x-intercept.

e. What is the domain of the function?

f. What is the range?

g. Is the function increasing or decreasing?

h. Does the graph have a vertical or horizontal asymptote?

i. Use your graphing calculator to determine $f(32)$.

j. Use your graphing calculator to determine x when $f(x) = 3.25$.

5. Write each of the following as a sum, difference, or multiple of logarithms. Assume that x, y, and z represent positive numbers.

a. $\log_b \frac{xy^2}{z}$

b. $\log_3 \frac{\sqrt{x^3 y}}{z}$

c. $\log_5 \left(x\sqrt{x^2 + 4}\right)$

d. $\log_4 \sqrt[3]{\frac{xy^2}{z^2}}$

6. Rewrite the following as the logarithm of a single quantity.

a. $\log x + \frac{1}{3} \log y - \frac{1}{2} \log z$

b. $3 \log_3 (x + 3) + 2 \log_3 z$

c. $\frac{1}{3} \log_3 x - \frac{2}{3} \log_3 y - \frac{4}{3} \log_3 z$

7. Use the change of base formula and your calculator to approximate the following.

a. $\log_5 17$

b. $\log_{13} \sqrt[3]{41}$

8. Solve each of the following using an algebraic approach.

a. $25 + 3 \ln x = 10$

b. $1.5 \log_4 (x - 1) = 7$

9. Solve the following algebraically. Check your solutions using graphs or tables.

    a. $3^x = 17$ **b.** $42 = 3e^{1.7x}$

10. Data collected from over 100 countries in the year 2000 showed that the relationship between per capita health care expenditures, H, in dollars and average life expectancy, E, could be modeled by the formula $E = 0.035 + 9.669 \ln(H)$, where $0 < H \le 4500$.

    a. Sketch a graph of the health care/life expectancy model using the domain $0 < H \le 4500$.

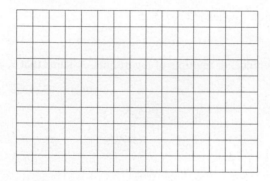

    b. Use this model to predict the average life expectancy in a country whose per capita health care expenditure is $1500 per year.

    c. Use the model and your graphing calculator to predict per capita health care expenditures in a country whose average life expectancy is 77 years.

The bracketed numbers following each concept indicate the activity in which the concept is discussed.

CONCEPT/SKILL	DESCRIPTION	EXAMPLE
Exponential functions [3.1]	The exponential functions can be defined by $y = b^x, b > 0, b \neq 1$.	$y = 3^x$
Growth factor of an exponential function [3.1]	If $b > 1$, the function $y = b^x$ is increasing and b is called the growth factor.	The exponential function $y = 3^x$ has a growth factor of 3.
Vertical intercept of an exponential function [3.1]	The vertical intercept (y-intercept) of an exponential function $y = b^x$ is $(0, 1)$.	The graph of $y = 2^x$ passes through the point $(0, 1)$.
Horizontal asymptote of an exponential function [3.1]	The line $y = 0$ is a horizontal asymptote of an exponential function $y = b^x$.	As x gets smaller, the output values of $y = 3^x$ approach 0.
Decay factor of an exponential function [3.2]	If $0 < b < 1$, the function $y = b^x$ is decreasing and b is called the decay factor.	The exponential function $y = \left(\frac{1}{2}\right)^x$ has a decay factor of $\frac{1}{2}$.
Doubling time [3.3]	The doubling time of an increasing exponential function is the time it takes for an output to double. The doubling time is set by the growth factor and remains the same for all output values.	Example 2, Activity 3.3; see pages 279–280
Half-life [3.3]	The half-life of a decreasing exponential function is the time it takes for an output to decay by one-half. The half-life is determined by the decay factor and remains the same for all output values.	Example 3, Activity 3.3; see page 281
Growth model [3.4]	If r represents the annual percentage growth rate, the exponential function that models the quantity P can be written as $P(t) = P_0(1 + r)^t$, where P_0 is the initial amount, t represents the number of elapsed years, and $1 + r$ is the growth factor.	Example 4, Activity 3.4; see page 290
Decay model [3.4]	If r represents the annual percent that decays, the exponential function that models the amount remaining can be written as $P(t) = P_0(1 - r)^t$, where $1 - r$ is the decay factor.	Example 5, Activity 3.4; see page 292
Compound interest [3.5]	The formula for compounding interest is $A = P\left(1 + \frac{r}{n}\right)^{nt}$.	Example 1, Activity 3.5; see page 297

CONCEPT/SKILL	DESCRIPTION	EXAMPLE
Continuous compounding [3.5]	The formula for continuous compounding is $A = Pe^{rt}$.	Example 4, Activity 3.5; see page 302
Continuous growth at a constant percentage rate [3.5], [3.6]	Whenever growth is continuous at a constant percentage rate, the exponential model used is $y = y_0 e^{rt}$.	Problem 10, Activity 3.5; see page 302
Continuous decay at a constant percentage rate [3.6]	Whenever decay is continuous at a constant rate, the model used is $y = y_0 e^{-rt}$.	Example 2, Activity 3.6; see page 307
Logarithm [3.8]	In the equation $y = b^x$, where $b > 0$ and $b \neq 1$, x is called a logarithm or log.	For the equation $3^4 = 81$, 4 is the logarithm of 81, base 3.
Notation for logarithms [3.8]	The notation for logarithms is $\log_b x = y$, where b is the base of the log, x (a positive number) is the power of b, and y is the exponent.	In the equation $\log_2 16 = 4$, 2 is the base, 4 is the log or exponent, and 16 is the power of 2.
Common logarithm [3.8]	A common logarithm is a base-10 logarithm. The notation is $\log_{10} x = \log x$.	$1000 = 10^3$. The common logarithm of 10^3 is 3; i.e., $\log 1000 = 3$.
Natural logarithm [3.8]	A natural logarithm is a base-e logarithm. The notation is $\log_e x = \ln x$.	$\log_e e^3 = \ln e^3 = 3$
Logarithmic equation [3.8]	The logarithmic equation $y = \log_b x$ is equivalent to the exponential equation $b^y = x$.	The equations $6 = \log_4 x$ and $x = 4^6$ are equivalent.
Basic properties of logarithms [3.8]	If $b > 0$ and $b \neq 1$, $\log_b 1 = 0$, $\log_b b = 1$, and $\log_b b^n = n$.	$\log_4 1 = 0$, $\log_7 7 = 1$, $\log_6 6^4 = 4$
Logarithmic function [3.9]	If $b > 0$ and $b \neq 1$, the logarithmic function is defined by $y = \log_b x$.	$y = \log_4 x$
Graph of the logarithmic function [3.9]	The graph of $y = \log_b x$, where $b > 1$ is increasing for all $x > 0$, has an x-intercept of $(1, 0)$ and has a vertical asymptote of $x = 0$, the y-axis.	
Comparison of the graphs of $f(x) = b^x$, where $b > 1$, and $g(x) = \log_b (x)$, where $b > 1$ [3.10]	Both graphs increase. The exponential function increases faster as x increases; the log function increases slower as x increases. The domain of the exponential function is the range of the log, which is all real numbers; the range of the exponential function is the domain of the log, which is the interval $(x > 0)$.	Problem 6, Activity 3.10; see pages 346–347

CONCEPT/SKILL	DESCRIPTION	EXAMPLE
If $A > 0, B > 0, b > 0$, and $b \neq 1$, then $\log_b (A \cdot B) = \log_b A + \log_b B$ [3.11]	The logarithm of a product is the sum of the logarithms.	$\log_2 (4 \cdot 8) = \log_2 (4) + \log_2 (8)$ $= 2 + 3 = 5$
If $A > 0, B > 0, b > 0$, and $b \neq 1$, then $\log_b \left(\frac{A}{B}\right) = \log_b A - \log_b B$ [3.11]	The logarithm of a quotient is the difference of the logarithms.	$\log_3 \left(\frac{81}{27}\right) = \log_3 81 - \log_3 27$ $= 4 - 3 = 1$
If $A > 0, B > 0, b > 0$, and $b \neq 1$, then $\log_b (A + B) \neq \log_b (A) + \log_b (B)$ [3.11]	The logarithm of a sum is not the sum of the logarithms.	$\log 2 + \log 3 =$ $0.3010 + 0.4771 = 0.7781$ $\log (2 + 3) = \log 5 = 0.6990$
If $A > 0$, p is a real number, $b > 0$, and $b \neq 1$, then $\log_b A^p = p \log_b A$ [3.11]	The logarithm of a power of A is the exponent times the logarithm of A.	$\log_5 x^4 = 4 \log_5 x$ $\log_3 \sqrt{x} = \frac{1}{2} \log_3 x$
Change of base formula [3.11]	The logarithm of any positive number x to any base can be found using the formula $\log_b x = \frac{\log x}{\log b} \text{ or } \log_b x = \frac{\ln x}{\ln b}.$	$\log_2 (2.5) = \frac{\log (2.5)}{\log 2} = 1.3219$

1. Due to inflation your tuition will increase 6% each year.

    a. If tuition is $300 per credit now, determine how much it will be in 5 years; in 10 years.

    b. Calculate the average rate of change in tuition over the next 5 years.

    c. Calculate the average rate of change in tuition over the next 10 years.

    d. If the inflation stays at 6%, approximately when will the tuition double?

2. a. Determine some of the output values for the function $f(x) = 8^x$ by completing the following table.

x	-1	$-\frac{1}{3}$	0	1	$\frac{4}{3}$	2	3
$f(x) = 8^x$							

    b. Sketch the graph of the function f.

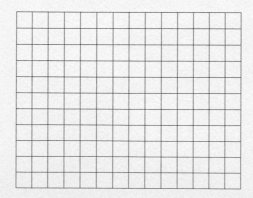

Answers to all Gateway exercises are included in the Selected Answers appendix.

378

c. Is this function increasing or decreasing? Explain how you know this by looking at the equation of the function.

d. What is the domain?

e. What is the range?

f. What are the x- and y-intercepts?

g. Are there any asymptotes? If yes, write the equations of the asymptotes.

h. Compare the graph of f to the graph of $g(x) = \left(\frac{1}{8}\right)^x$. What are the similarities and the differences?

i. In what way does the graph of $h(x) = 8^x + 5$ differ from that of $f(x) = 8^x$?

j. Write the equation of the function that is the inverse of the function $f(x)$.

3. Complete the table for each exponential function. Use your graphing calculator to check your work.

FUNCTION	BASE, b	GROWTH OR DECAY FACTOR	x-INTERCEPT	y-INTERCEPT	HORIZONTAL ASYMPTOTE	INCREASING OR DECREASING
$h(x) = 6^x$						
$g(x) = \left(\frac{1}{3}\right)^x$						
$p(x) = 5(2.34)^x$						
$q(x) = 3(0.78)^x$						
$r(x) = 2^x - 4$						

4. Use your graphing calculator to help you determine the domain and range for each function.

Function	$f(x) = 0.8^x$	$h(x) = 6^x + 2$	$t(x) = 3^x - 5$	$q(x) = \log_4 x$	$r(x) = \ln(x - 3)$
Domain					
Range					

5. a. Given the following table, determine whether the given data can be approximately modeled by an exponential function. If it can, what is the growth or decay factor?

x	0	1	2	3	4
y	10	15.5	24	36	55.5

b. Determine an exponential equation that models this data.

6. Complete the following tables, which represent exponential functions. Round calculations to two decimal places whenever necessary.

a.

x	0	1	2	3	4
y	3.00	6.12			

b.

x	0	1	2	3	4
y	4.50	3.15			

c.

x	0	1	2	3	4
y	$\frac{1}{4}$	4			

d. Write the equation of the exponential function that represents the data in each table in parts a, b, and c.

i.   **ii.**   **iii.**

7. a. Your salary has increased at the rate of 1.5% annually for the past 5 years, and your boss projects this will remain unchanged for the next 5 years. You were making $15,000 annually in 2010. Complete the following table.

2010	2011	2012	2013	2014	2015

b. Write the exponential growth function that models your annual salary during this period of time. Let x represent the number of years since 2010.

c. If your increase in salary continues at this rate, how much will you make in 2018? Is this realistic?

d. You would like to double your salary. How many years will you have to work before your salary will be twice the salary you made in 2010?

8. a. You just inherited $5000. You can invest the money at a rate of 6.5% compounded continuously. In 8 years, your oldest child will be going to college. How much will be in the bank for her education? Use the equation $A = A_0 e^{rt}$.

b. You actually need to have $12,000 for your child's first year of college. For how many years would you have to leave the money in the bank to have the $12,000?

9. Determine the value of each of the following without using your calculator.

    a. $25^{3/2}$

    b. $81^{3/4}$

    c. $64^{-5/6}$

    d. $\sqrt[3]{125^2}$

    e. $\log_3 \frac{1}{9}$

    f. $\log_5 625$

    g. $\log 0.001$

    h. $\ln e^2$

0. Write each equation in logarithmic form.

    a. $6^2 = 36$

    b. $0.000001 = 10^{-6}$

    c. $2^{-5} = \frac{1}{32}$

11. Write each equation in exponential form.

    a. $\log_3 81 = 4$

    b. $\log_7 1 = 0$

    c. $\log_{10} 0.0001 = -4$

    d. $\ln e = 1$

    e. $\log_q y = b$

12. Solve each equation for the unknown variable.

    a. $\log_5 x = -3$

    b. $\log_b 256 = 4$

    c. $\log_2 64 = y$

    d. $\log_4 x = \dfrac{3}{2}$

13. a. Complete the table of values for the function $f(x) = \log_5 x$.

x	0.008	0.04	0.2	1	5	25
f(x)						

    b. Sketch a graph of the function.

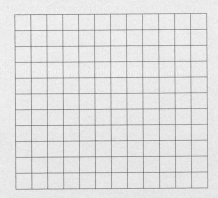

c. Use your graphing calculator to check your result in parts a and b.

d. Determine the x-intercept.

e. What is the domain of the function?

f. What is the range?

g. Does the graph have a vertical or horizontal asymptote?

h. Use your graphing calculator to determine $f(23)$.

i. Use your graphing calculator to determine x when $f(x) = 2.46$.

14. Use the change of base formula and your calculator to approximate the following.

    a. $\log_7 21$ **b.** $\log_{15} \frac{8}{9}$

15. Write each of the following as a sum, difference, or multiple of logarithms. Assume that x, y, and z are all greater than 0.

    a. $\log_2 \dfrac{x^3 y}{z^{1/2}}$ **b.** $\log \sqrt[3]{\dfrac{x^4 y^3}{z}}$

16. Rewrite the following as the logarithm of a single quantity.

    a. $\log x + \frac{1}{4} \log y - 3 \log z$ **b.** $\frac{1}{3}\left(\log x - 2 \log y - \log z\right)$

17. Solve the following algebraically.

    a. $3^{3+x} = 7$ **b.** $\log_2 (4x + 9) = 4$

    c. $50 + 6 \ln x = 85$

8. a. Sketch the graph of the function using the data from the given table.

x	0.1	0.5	1	2	4	16
f(x)	−1.66	−0.5	0	.5	1	2

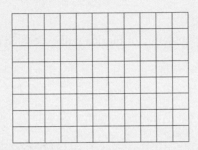

b. Use the table and the graphing feature of your calculator to verify that the equation that defines function f is $f(x) = 0.5 \log_2 x$.

c. Use the function to determine the value of $f(54)$.

d. If $f(x) = 2.319$, determine the value of x.

e. Use your graphing calculator to verify that the function $g(x) = 4^x$ is the inverse of f.

19. The populations of New York State and Florida (in millions) can be modeled by the following:

New York State	$P_N = 19.43e^{0.0031t}$
Florida	$P_F = 18.20e^{0.0126t}$

where t represents the number of years since 2007.

a. Determine the population of New York and Florida in 2007 ($t = 0$).

b. Sketch a graph of each function on the same coordinate axes.

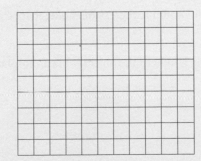

c. Determine graphically the year when the population of Florida was equal to the population of New York State.

d. Determine algebraically the year when the population of Florida will exceed 25 million.

e. Will the population of Florida ever exceed the population of New York? Explain. Assume that population growth given by both populations will continue. Solve algebraically.

20. Atmospheric pressure decreases with increasing altitude. The following data is collected during a science experiment to investigate the relationship between the height of an object above ground level and the pressure exerted on the object.

Atmospheric Pressure, x (in millimeters of mercury)	760	740	725	700	650	630	600	580	550
Height, y (in kilometers)	0	0.184	0.328	0.565	1.079	1.291	1.634	1.862	2.235

a. Use a graphing calculator to determine a logarithmic regression model that best fits the data.

b. Use the log equation in part a to predict the height of the object if the atmospheric pressure is 500 millimeters of mercury.

Quadratic and Higher-Order Polynomial Functions

Introduction to Quadratic Functions

Activity 4.1

Baseball and the Willis Tower

Objectives

1. Identify functions of the form $f(x) = ax^2 + bx + c$ as quadratic functions.

2. Explore the role of c as it relates to the graph of $f(x) = ax^2 + bx + c$.

3. Explore the role of a as it relates to the graph of $f(x) = ax^2 + bx + c$.

4. Explore the role of b as it relates to the graph of $f(x) = ax^2 + bx + c$.

Note: $a \neq 0$ in Objectives 1–4.

Imagine yourself standing on the roof of the 1450-foot-high Willis Tower (formerly called the Sears Tower) in Chicago. When you release and drop a baseball from the roof of the tower, *the ball's height above the ground, H* (in feet), can be described as a function of the time, *t* (in seconds), since it was dropped. This height function is defined by

$$H(t) = -16t^2 + 1450.$$

1. Sketch a diagram illustrating the Willis Tower and the path of the baseball as it falls to the ground.

2. **a.** Complete the following table.

TIME, t (sec.)	$H(t) = -16t^2 + 1450$
0	1450
1	
2	
3	
4	
5	
6	
7	
8	
9	
10	

385

b. How far does the baseball fall during the first second?

c. How far does it fall during the interval from 1 to 3 seconds?

3. Using the height function, $H(t) = -16t^2 + 1450$, determine the average rate of change of H with respect to t over the given interval. Remember:

$$\text{average rate of change} = \frac{\text{change in output}}{\text{change in input}}.$$

a. $0 \leq t \leq 1$ **b.** $1 \leq t \leq 3$

c. Based on the results of parts a and b, do you believe that $H(t) = -16t^2 + 1450$ is a linear function? Explain.

4. a. What is the value of H when the baseball strikes the ground? Use the table in Problem 2a to estimate the time when the ball is at ground level.

b. What is the practical domain of the height function?

c. Determine the practical range of the height function.

d. On the following grid, plot the points in Problem 2a that satisfy part b (practical domain) and sketch a curve representing the height function.

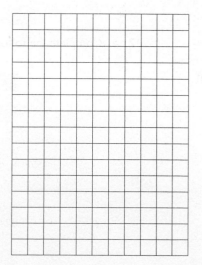

e. Is the graph of the height function in part d the actual path of the object (see Problem 1)? Explain.

Some interesting properties of the function defined by $H = -16t^2 + 1450$ arise when you ignore the falling object context. Replace H with y and t with x and consider the general function defined by $y = -16x^2 + 1450$.

5. a. Graph the function defined by $y = -16x^2 + 1450$, setting the window parameters at Xmin $= -10$ and Xmax $= 10$ for the input and Ymin $= -50$ and Ymax $= 1500$ for the output. Your graph should appear as follows.

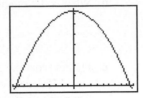

b. Describe the important features of the graph of $y = -16x^2 + 1450$. Discuss the shape, symmetry, and intercepts.

Quadratic Functions

The graph of the function defined by $y = -16x^2 + 1450$ is a parabola. The graph of a **parabola** is a $\cup$-shaped figure that opens upward, $\cup$, or downward, $\cap$. Parabolas are graphs of a special category of functions called quadratic functions.

Definition

Any function defined by an equation of the form $y = ax^2 + bx + c$ or $f(x) = ax^2 + bx + c$, where a, b, and c represent real numbers and $a \neq 0$, is called a **quadratic function**. The output variable y is defined by an expression having three terms: the **quadratic term**, ax^2, the **linear term**, bx, and the **constant term**, c. The numerical factors of the quadratic and linear terms, a and b, are called the **coefficients** of the terms.

Example 1   *$H(t) = -16t^2 + 1450$ defines a quadratic function. The quadratic term is $-16t^2$. The linear term is $0t$, although it is not written as part of the expression defining $H(t)$. The constant term is 1450. The numbers -16 and 0 are the coefficients of the quadratic and linear terms, respectively. Therefore, $a = -16$, $b = 0$, and $c = 1450$.*

6. For each of the following quadratic functions, identify the value of a, b, and c.

QUADRATIC FUNCTION	a	b	c
$y = 3x^2$			
$y = -2x^2 + 3$			
$y = x^2 + 2x - 1$			
$y = -x^2 + 4x$			

The Constant Term c: A Closer Look

Consider once again the height function $H(t) = -16t^2 + 1450$ from the beginning of the activity.

   7. a. What is the vertical intercept of the graph? Explain how you obtained the results.

   b. What is the practical meaning of the vertical intercept in this situation?

   c. Predict what the graph of $H(t) = -16t^2 + 1450$ would look like if the constant term 1450 were changed to 800. That is, the baseball is dropped from a height of 800 feet rather than 1450 feet. Verify your prediction by graphing $H(t) = -16t^2 + 800$. What does the constant term tell you about the graph of the parabola?

The constant term c of a quadratic function $f(x) = ax^2 + bx + c$ *always* indicates the vertical intercept of the parabola. The vertical intercept of any quadratic function is $(0, c)$ since $f(0) = a(0)^2 + b(0) + c = c$.

8. Graph the parabolas defined by the following quadratic equations. Note the similarities and differences among the graphs, especially the vertical intercepts. Be careful in your choice of a window.

   a. $f(x) = 1.5x^2$ **b.** $g(x) = 1.5x^2 + 7$

   c. $q(x) = 1.5x^2 + 4$ **d.** $s(x) = 1.5x^2 - 4$

The Effects of the Coefficient a on the Graph of $y = ax^2 + bx + c$

9. a. Graph the quadratic function defined by $y_1 = 16x^2 + 1450$ on the same screen as $y_2 = -16x^2 + 1450$. Use the window settings Xmin $= -10$, Xmax $= 10$, Ymin $= -50$, and Ymax $= 3000$.

   b. What effect does the sign of the coefficient of x^2 appear to have on the graph of the parabola?

10. Graph the functions $y_3 = -16t^2 + 100$, $y_4 = -6t^2 + 100$, $y_5 = -40t^2 + 100$ in the same window. What effect does the magnitude of the coefficients of x^2 (namely, $|-16| = 16$, $|-6| = 6$, and $|-40| = 40$) appear to have on the graph of that particular parabola? Use window settings Xmin $= -15$, Xmax $= 15$, Ymin $= -200$, and Ymax $= 200$.

The results from Problems 9 and 10 regarding the effects of the coefficient a can be summarized as follows.

> The graph of a quadratic function defined by $f(x) = ax^2 + bx + c$ is called a parabola.
>
> - If $a > 0$, the parabola opens upward.
> - If $a < 0$, the parabola opens downward.
> - The magnitude of a affects the width of the parabola. The larger the absolute value of a, the narrower the parabola.

11. a. Is the graph of $h(x) = 0.3x^2$ wider or narrower than the graph of $f(x) - x^2$?

b. How do the output values of h and the output values of f compare for the same input value?

c. Is the graph of $g(x) = 3x^2$ wider or narrower than the graph of $f(x) = x^2$?

d. How do the output values of g and f compare for the same input value?

e. Describe the effect of the magnitude of the coefficient a on the width of the graph of the parabola.

f. Describe the effect of the magnitude of the coefficient of a on the output value.

The Effects of the Coefficient b on the Turning Point

Assume for the time being that you are back on the roof of the 1450-foot Willis Tower. Instead of merely releasing the ball, suppose you *throw it down* with an initial velocity of 40 feet per second. Then the function describing its height above ground as a function of time is modeled by

$$H_{down}(t) = -16t^2 - 40t + 1450.$$

If you tossed the ball straight up with an initial velocity of 40 feet per second, then the function describing its height above ground as a function of time is modeled by

$$H_{up}(t) = -16t^2 + 40t + 1450.$$

12. Predict what features of the graphs of H_{down} and H_{up} have in common with

$$H(t) = -16t^2 + 1450.$$

13. a. Graph the three functions $H(t)$, $H_{down}(t)$, and $H_{up}(t)$ using the same window settin given in Problem 5a.

b. What effect do the $-40t$ and $40t$ terms seem to have upon the turning point of the graphs?

If $b = 0$, the turning point of the parabola is located on the vertical axis. If $b \neq 0$, the turning point will not be on the vertical axis.

14. For each of the following quadratic functions, identify the value of b and then, witho graphing, determine whether or not the turning point is on the y-axis. Verify your concl sion by graphing the given function. Set the window of your calculator to Xmin $= -$ Xmax $= 8$, Ymin $= -20$, and Ymax $= 20$.

a. $y = x^2$

b. $y = x^2 - 4x$

c. $y = x^2 + 4$

d. $y = x^2 + x$

e. $y = x^2 - 3$

15. Match each function with its corresponding graph below, and then verify using your graphing calculator.

a. $f(x) = x^2 + 4x + 4$ **b.** $g(x) = 0.2x^2 + 4$ **c.** $h(x) = -x^2 + 3x$

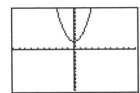 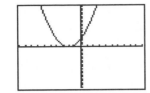 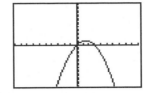

SUMMARY: ACTIVITY 4.1

1. The equation of a **quadratic function** with x as the input variable and y as the output variable has the standard form

$$y = ax^2 + bx + c,$$

where a, b, and c are real numbers and $a \neq 0$.

2. The graph of a quadratic function is called a **parabola**.

3. For the quadratic function defined by $f(x) = ax^2 + bx + c$:

- If $a > 0$, the parabola opens upward.

- If $a < 0$, the parabola opens downward.

   The magnitude of a affects the width of the parabola. The larger the absolute value of a, the narrower the parabola.

4. If $b = 0$, the turning point of the parabola is located on the vertical axis. If $b \neq 0$, the turning point will not be on the vertical axis.

5. The constant term, c, of a quadratic function $f(x) = ax^2 + bx + c$ always indicates the vertical intercept of the parabola. The vertical intercept of any quadratic function is $(0, c)$.

EXERCISES: ACTIVITY 4.1

1. a. Complete the following table for $f(x) = x^2$.

x	3	2	−1	0	1	2	3
$f(x) = x^2$							

b. Use the results of part a to sketch a graph $y = x^2$. Verify using a graphing calculator.

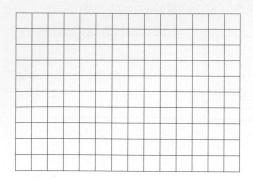

c. What is the coefficient of the term x^2?

d. From the graph, determine the domain and range of the function.

e. Create a table similar to the one in Exercise 1a to show the output for $g(x) = -x^2$.

x	-3	-2	-1	0	1	2	3
$g(x) = -x^2$							

f. Sketch the graph of $g(x) = -x^2$ on the same coordinate axis in part a. Verify using a graphing calculator.

g. What is the coefficient of the term $-x^2$?

h. How can the graph of $y = -x^2$ be obtained from the graph of $y = x^2$?

2. In each of the following functions defined by an equation of the form $y = ax^2 + bx + c$, identify the value of a, b, and c.

a. $y = -2x^2$

b. $y = \frac{2}{5}x^2 + 3$

c. $y = -x^2 + 5x$

d. $y = 5x^2 + 2x - 1$

3. Predict what the graph of each of the following quadratic functions will look like. Use your graphing calculator to verify your prediction.

a. $f(x) = 3x^2 + 5$

b. $g(x) = -2x^2 + 1$

c. $h(x) = 0.5x^2 - 3$

4. Graph the following pairs of functions, and describe any similarities as well as any differences that you observe in the graphs.

a. $f(x) = 3x^2, g(x) = -3x^2$ b. $h(x) = \frac{1}{2}x^2, f(x) = 2x^2$

c. $g(x) = 5x^2, h(x) = 5x^2 + 2$ d. $f(x) = 4x^2 - 3, g(x) = 4x^2 + 3$

e. $f(x) = 6x^2 + 1, h(x) = -6x^2 - 1$

5. Use your graphing calculator to graph the two functions $y_1 = 3x^2$ and $y_2 = 3x^2 + 2x - 2$.

a. What is the vertical intercept of the graph of each function?

b. Compare the two graphs to determine the effect of the linear term $2x$ and the constant term -2 on the graph of $y_1 = 3x^2$.

For Exercises 6–10, determine

a. whether the parabola opens upward or downward and

b. the vertical intercept.

6. $f(x) = -5x^2 + 2x - 4$ 7. $g(t) = \frac{1}{2}t^2 + t$ 8. $h(v) = 2v^2 + v + 3$

a.   a.   a.

b.   b.   b.

9. $r(t) = 3t^2 + 10$ 10. $f(x) = -x^2 + 6x - 7$

a.   a.

b.   b.

11. Does the graph of $y = -2x^2 + 3x - 4$ have any horizontal intercepts? Explain.

12. a. Is the graph of $y = \frac{3}{5}x^2$ wider or narrower than the graph of $y = x^2$?

b. For the same input value, which graph would have a larger output value?

13. Put the following in order from narrowest to widest.

a. $y = 0.5x^2$ **b.** $y = 8x^2$ **c.** $y = -2.3x^2$

. Determine the vertex
or turning point of a
parabola.

. Identify the vertex as a
maximum or minimum.

. Determine the axis of
symmetry of a parabola.

. Identify the domain
and range.

. Determine the y-intercept
of a parabola.

. Determine the
x-intercept(s) of a parabola
using technology.

. Interpret the practical
meaning of the vertex
and intercepts in a
given problem.

Parabolas are good models for a variety of situations that you encounter in everyday life.
Examples include the path of a golf ball after it is struck, the arch (cable system) of a bridge,
the path of a baseball thrown from the outfield to home plate, the stream of water from a
drinking fountain, and the path of a cliff diver.

Consider the 2008 men's Olympic shot put event, which was won by Poland's Tomasz
Majewski with a throw of 70.57 feet. The path of his winning throw can be approximately
modeled by the quadratic function defined by

$$y = -0.015375x^2 + x + 6,$$

where x is the horizontal distance in feet from the point of the throw and y is the vertical
height in feet of the shot above the ground.

1. a. After inspecting the equation for the path of the winning throw, which way do you
expect the parabola to open? Explain.

b. What is the vertical intercept of the graph of the parabola? What practical meaning
does this intercept have in this situation?

2. Use your graphing calculator to produce a plot of the path of the winning throw. Be sure
to adjust your window settings so that all of the important features of the parabola
(including x-intercepts) appear on the screen. Your graph should resemble the following:

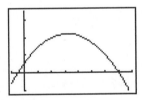

3. a. Use the graph to estimate the practical domain of the function.

b. What does the practical domain mean in the shot put situation?

c. Use the graph to estimate the practical range of the function.

d. What does the practical range mean in the shot put situation?

4. a. Use the table feature of your graphing calculator to complete the following table.

x	10	20	30	40	50
$H(x)$					

b. Use the table to estimate the horizontal distance from the point of release when the shot put reaches its maximum height above the ground.

Vertex of a Parabola

An important feature of the *graph* of any quadratic function defined by $f(x) = ax^2 + bx +$ is its **turning point**, also called the **vertex**. The turning point of a parabola that opens down ward or upward is the point at which the parabola changes direction from increasing decreasing or decreasing to increasing.

5. Use the results from Problems 3c and 4b to approximate the coordinates of the vertex the shot put function H.

6. The vertex is often very important in a situation. What is the significance of the coord nates of the turning point in this problem?

The coordinates of the vertex of a parabola having equation $y = ax^2 + bx + c$ can be dete mined from the values of a and b in the equation.

Definition

The **vertex** is the turning point of a parabola. The vertex of a parabola with the equation $y = f(x) = ax^2 + bx + c$ has coordinates

$$\left(-\frac{b}{2a}, f\left(-\frac{b}{2a}\right)\right),$$

where a is the coefficient of the x^2 term and b is the coefficient of the x term.

Note that the y-coordinate (output) of the vertex is determined by substituting the x-coordina of the vertex into the equation of the parabola and evaluating the resulting expression.

Example 1   *Determine the vertex of the parabola defined by the equation* $y = -3x^2 + 12x + 5.$

SOLUTION

Step 1. Determine the x-coordinate of the vertex by substituting the values of a and b into the formula $x = \dfrac{-b}{2a}$.

Because $a = -3$ and $b = 12$, you have

$$x = \frac{-(12)}{2(-3)} = \frac{-12}{-6} = 2.$$

Step 2. The y-value of the vertex is the corresponding output value for $x = 2$. Substituting 2 for x in the equation, you have

$$y = -3(2)^2 + 12(2) + 5 = 17.$$

Therefore, the vertex is $(2, 17)$.

Because the parabola in Example 1 opens downward ($a = -3 < 0$), the vertex is the high point (maximum) of the parabola, as demonstrated by the following graph of the parabola defined by $y = -3x^2 + 12x + 5$.

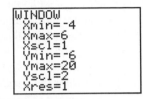 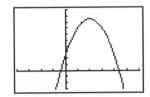

7. Use the method demonstrated in Example 1 to determine the vertex of the parabola defined by $y = -0.015375x^2 + x + 6$ (the shot put function).

You can also determine the vertex of a parabola by selecting the maximum (or minimum) option in the CALC menu of your graphing calculator. Follow the prompts to obtain the coordinates of the maximum point (vertex) of $y = -3x^2 + 12x + 5$.

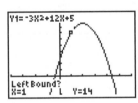 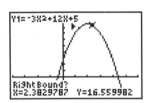

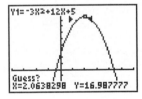 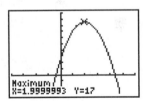

Appendix

For further help with the TI-83/84 Plus, see Appendix C.

8. a. Use your graphing calculator to determine the vertex of the parabola having equation $y = -0.015375x^2 + x + 6$ (the shot put function).

b. How do the coordinates you determined using your graphing calculator compare with your results in Problem 7?

c. What is the practical meaning of the coordinates of the vertex in this situation?

Axis of Symmetry of a Parabola

> **Definition**
>
> The **axis of symmetry** of a parabola is a vertical line that divides the parabola into two symmetrical parts that are mirror images in the line.

Example 2 *Consider the parabola having equation $y = -3x^2 + 12x + 5$ from Example 1. The axis of symmetry of the parabola is $x = 2$. Note that the line of symmetry passes through the vertex of the parabola.*

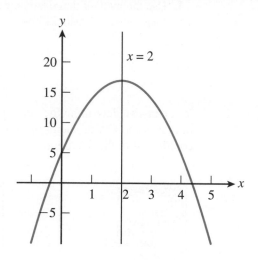

Because the vertex (turning point) of a parabola lies on the axis of symmetry, the equation of the axis of symmetry is

$$x = \frac{-b}{2a}.$$

9. What is the axis of symmetry of the shot put function, H?

Intercepts of the Graph of a Parabola

The y-intercept (vertical intercept) of the graph of the parabola defined by $y = -3x^2 + 12x + 5$ (see Example 1) can be determined directly from the equation. If $x = 0$, then

$$y = -3(0)^2 + 12(0) + 5 = 5$$

and the y-intercept is $(0, 5)$.

In general, the y-intercept of the parabola defined by $y = ax^2 + bx + c$ is $(0, c)$.

Because the vertex $(2, 17)$ of the parabola having equation $y = -3x^2 + 12x + 5$ is a point above the x-axis (the y-coordinate is positive) and the parabola opens downward, the parabola must intersect the horizontal axis in two places. This is verified by the following graph.

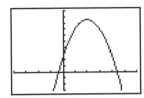

The x-intercepts of $y = -3x^2 + 12x + 5$ can be determined using the zero option in the CALC menu of your graphing calculator. Follow the prompts to obtain one x-intercept at a time. The screens should appear as follows.

 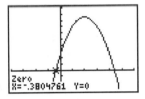 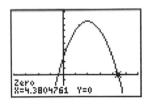

For further help with the TI-83/84 Plus, see Appendix C.

10. a. Use the zero option of your graphing calculator to determine the x-intercept(s) for the shot put function having equation $y = -0.015375x^2 + x + 6$. The rightmost intercept appears in the following screen:

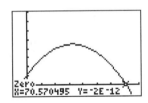

b. Is either x-intercept determined in part a significant to the problem situation? Explain.

11. a. Use your result from Problem 10 to determine the practical domain of the shot put function. How does this compare with your answer in Problem 3a?

b. Sketch the path of the winning throw of the shot put. Be sure to label all key points, including the vertex and intercepts.

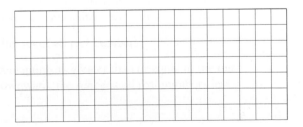

EXERCISES: ACTIVITY 4.2

For Exercises 1–8, determine the following characteristics of each quadratic function:

a. *The direction in which the graph opens*

b. *The axis of symmetry*

c. *The turning point (vertex); determine if maximum or minimum*

d. *The y-intercept*

1. $f(x) = x^2 - 3$

  a.

  b.

  c.

  d.

2. $g(x) = x^2 + 2x - 8$

  a.

  b.

  c.

  d.

3. $y = x^2 + 4x - 3$

  a.

  b.

  c.

  d.

4. $f(x) = 3x^2 - 2x$

  a.

  b.

  c.

  d.

5. $h(x) = x^2 + 3x + 4$

  a.

  b.

  c.

  d.

6. $g(x) = -x^2 + 7x - 6$

  a.

  b.

  c.

  d.

7. $y = 2x^2 - x - 3$

  a.

  b.

  c.

  d.

8. $f(x) = x^2 + x + 3$

  a.

  b.

  c.

  d.

For Exercises 9–16, use your graphing calculator to sketch the graphs of the functions, and then determine each of the following:

a. *The coordinates of the x-intercepts for each function, if they exist*

b. *The domain and range for each function*

c. *The horizontal interval over which each function is increasing*

d. *The horizontal interval over which each function is decreasing*

9. $g(x) = -x^2 + 7x - 6$

  a.

  b.

  c.

  d.

10. $h(x) = 3x^2 + 6x + 4$

  a.

  b.

  c.

  d.

Exercise numbers appearing in color are answered in the Selected Answers appendix.

11. $y = x^2 - 12$

   a.

   b.

   c.

   d.

12. $f(x) = x^2 + 4x - 5$

   a.

   b.

   c.

   d.

13. $g(x) = -x^2 + 2x + 3$

   a.

   b.

   c.

   d.

14. $h(x) = x^2 + 2x - 8$

   a.

   b.

   c.

   d.

15. $y = -5x^2 + 6x - 1$

   a.

   b.

   c.

   d.

16. $f(x) = 3x^2 - 2x + 1$

   a.

   b.

   c.

   d.

17. You shoot an arrow vertically into the air from a height of 5 feet with an initial velocity of 96 feet per second. The height, h, in feet above the ground, at any time, t (in seconds), is modeled by

$$h(t) = 5 + 96t - 16t^2.$$

   a. Determine the maximum height the arrow will attain.

   b. Approximately when will the arrow reach the ground?

   c. What is the significance of the vertical intercept?

   d. What are the practical domain and practical range in this situation?

   e. Use your graphing calculator to determine the horizontal intercepts. Determine the practical meaning of these intercepts in this situation.

18. As part of a recreational waterfront grant, the city council plans to enclose a rectangular area along the waterfront of Lake Erie and create a park and swimming area. The budget calls for the purchase of 3000 feet of fencing. (*Note:* There is no fencing along the lake.)

   a. Draw a picture of the planned recreational area. Let x represent the length of one of the two equal sides that are perpendicular to the water.

b. Write an expression that represents the width (side opposite the water) in terms of x. (*Note:* You have 3000 feet of fencing.)

c. Write an equation that expresses the area, $A(x)$, of this rectangular site as a quadratic function of x.

d. Determine the value of x for which $A(x)$ is a maximum.

e. What is the maximum area that can be enclosed?

f. What are the dimensions of the maximum enclosed area?

g. Use your graphing calculator to graph the area function. What point on the graph represents the maximum area?

h. What is the vertical intercept? Does this point have any practical meaning in this situation?

i. From the graph, determine the horizontal intercepts. Do they have any practical meaning in this situation? Explain.

19. The average cost to produce pewter oil lamps is given by

$$\overline{C}(x) = 2x^2 - 120x + 2000,$$

where x represents the number of oil lamps produced and $\overline{C}(x)$ is the average cost of producing x oil lamps.

a. Use your graphing calculator to graph the average cost function and determine the coordinates of the turning point.

b. Determine the vertex algebraically.

c. How do your answers in parts a and b compare?

d. Is the vertex a minimum or maximum point?

e. What is the practical meaning of the vertex in this situation?

f. What is the vertical intercept? What is the practical meaning of this intercept?

20. You are manufacturing ceramic lawn ornaments. After several months, your accountant tells you that your profit, $P(n)$, can be modeled by

$$P(n) = -0.002n^2 + 5.5n - 1200,$$

where n is the number of ornaments sold each month.

a. Use your graphing calculator to produce a graph of this function. Use the table feature set at TblStart $= 0$ and ΔTbl $= 500$ to help you set your window. Include the x-intercepts and the vertex.

b. Determine the x-intercepts of the graph of the profit function.

c. Determine the practical domain of the profit function.

d. Determine the practical range of the profit function.

e. How many ornaments must be sold to maximize the profit?

f. Write the equation that must be solved to determine the number of ornaments that must be sold to produce a profit of $2300.

g. Solve the equation in part f graphically.

Activity 4.3

Per Capita Personal Income

Objectives

1. Solve quadratic equations numerically.

2. Solve quadratic equations graphically.

3. Solve quadratic inequalities graphically.

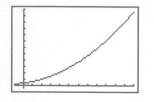

According to statistics from the U.S. Department of Commerce, the per capita personal income (or the average annual income) of each resident of the United States from 1960 to 2005 can be modeled by the equation

$$P(t) = 12.6779t^2 + 177.3069t + 1,836.8354,$$

where t equals the number of years since 1960.

1. What is the practical domain for the model represented by the function P?

2. Let $t = 0$ correspond to the year 1960. Use your graphing calculator to complete the following table of values for t, the number of years since 1960, and $P(t)$, the per capita income. Round your output to the nearest dollar.

Year	1960	1970	1980	1990	1995	2000	2005
t	0	10	20	30	35	40	45
$P(t)$							

3. Sketch a graph of the function using your graphing calculator and using the window Xmin $= -5$, Xmax $= 60$, Ymin $= -2000$, Ymax $= 60,000$. The graph appears to the left.

4. Estimate the per capita personal income in the year 1989 ($t = 29$).

5. You want to determine in which year the per capita personal income reached $20,500. In other words, for what value of t does $P(t) = 20,500$? Write an equation to determine the value of t when $P(t) = 20,500$.

The equation in Problem 5 is a quadratic equation because it involves a polynomial expression of degree 2.

> The standard form of a quadratic equation is $ax^2 + bx + c = 0$, $a \neq 0$.

Example 1   *Solve the quadratic equation $x^2 + 3x - 1 = 9$ numerically (using tables of data).*

SOLUTION

Create a table in which x is the input and $y = x^2 + 3x - 1$ is the output. The solution is the x-value corresponding to a y-value of 9. Using the graphing calculator, the solution is $x = 2$.

```
Plot1 Plot2 Plot3
\Y1◻X²+3X-1
\Y2◻9■
\Y3=
\Y4=
\Y5=
\Y6=
\Y7=
```

X	Y1	Y2
1	3	9
2	9	9
3	17	9
4	27	9
5	39	9
6	53	9
7	69	9
X=1		

A second solution is $x = -5$; try it yourself.

6. Determine the solution to $20{,}500 = 12.6779t^2 + 177.3069t + 1{,}836.8354$ numerically using a table of appropriate data points (see Problem 2). What is your approximation using this approach?

Solving Quadratic Equations Graphically

A second method of solving the quadratic equation in Problem 5 is a graphical approach using your graphing calculator. Recall from Chapter 1 that you can solve the equation $12.6779t^2 + 177.3069t + 1836.8354 = 20{,}500$ by solving the following system of equations graphically:

$$y_1 = 12.6779t^2 + 177.3069t + 1{,}836.8354$$

$$y_2 = 20{,}500.$$

The expression for y_1 gives the per capita personal income in any given year. The value y_2 is the specific per capita personal income in which you are interested. The solution to the equation is the x-value for which $y_1 = y_2$. To do this, determine the point of intersection of these two graphs. Using the intersect option under the CALC menu, the graph should appear as follows.

 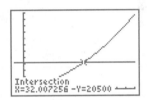

Another graphical method for solving the problem is to rearrange the quadratic equation

$$20{,}500 = 12.6779t^2 + 177.3069t + 1{,}836.8354$$

so the left-hand side is equal to zero. Subtracting 20,500 from each side, you have

$$0 = 12.6779t^2 + 177.3069t - 18{,}663.1646 \tag{1}$$

If you let $y = 12.6779t^2 + 177.3069t - 18{,}663.1646$, then the solution to equation 1 is the t-value for which $y = 0$, if it exists. This is the t-value of the t-intercept of the graph, also called the zero of the function.

7. a. Use your graphing calculator to sketch a graph of

$$y = 12.6779t^2 + 177.3069t - 18{,}663.1646.$$

The screen should appear as follows.

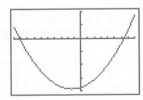

b. Use the zero option of the CALC menu to determine the t-intercepts of the new function defined by $y = 12.6779t^2 + 177.3069t - 18{,}663.1646$.

c. Using the results from part b, determine whether the solutions to the equation are the solutions to the equation $20{,}500 = 12.6779t^2 + 177.3069t + 1{,}836.8354$? Are both of the values relevant to our problem? Explain.

8. Describe two different ways to solve the equation $2x^2 - 4x + 3 = 2$ using a graphin▸ approach. Solve the equation using each graphing method. How do your answers compare▸

Solving Quadratic Inequalities Graphically

You are interested in determining in which years the per capita personal income was mor▸ than \$15,000. To answer this problem you need to solve the inequality

$$12.6779t^2 + 177.3069t + 1836.8354 > 15{,}000,$$

where t equals the number of years since 1960.

The following example demonstrates a procedure for solving an inequality similar to th▸ preceding one.

Example 2   *Solve the inequality $2x^2 - 4x + 3 > 7$ using a graphing approach.*

SOLUTION

Form the following system of equations

$$Y_1 = 2x^2 - 4x + 3$$
$$Y_2 = 7$$

and graph each equation. The screen should resemble the following.

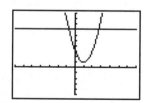

Use the intersection option on the CALC menu to determine where $Y_1 = Y_2$.

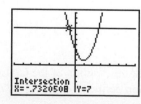
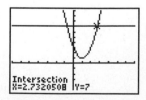

The solutions to the inequality are the values of x where $Y_1 > Y_2$; that is, the x-values o▸ points where the graph of Y_1 is above the graph of Y_2. Therefore, the solutions are $x < -0.732$ or $x > 2.732$.

If the problem had been the reverse inequality $2x^2 - 4x + 3 < 7$, then the solution would be x-values of points where the graph of Y_1 is below the graph of Y_2. The solution would be $-0.732 < x < 2.732$.

9. **a.** Solve the inequality $12.6779t^2 + 177.3069t + 1,836.8354 > 15,000$ using a graphing approach. Be careful. The practical domain is $t \geq 0$.

b. In which years was the per capita personal income more than \$15,000?

c. There are negative values of t for which $Y1 > 15,000$. Determine them graphically.

d. Explain why the values determined in part c are not relevant to the original problem situation.

10. Solve the following quadratic inequalities using a graphing approach.

a. $x^2 - x - 6 < 0$

b. $x^2 - x - 6 > 0$

SUMMARY: ACTIVITY 4.3

1. A **quadratic equation** is an equation involving polynomial expressions of degree 2. The standard form of a quadratic equation is $ax^2 + bx + c = 0, a \neq 0$.

2. To solve $f(x) = c$ **numerically**, construct a table, and determine the x-values that produce c as an output.

3. To solve $f(x) = c$ **graphically**,

   a. graph $y = f(x)$, graph $y = c$, and determine the x-values of the points of intersection.

   b. or graph $y = f(x) - c$, and determine the x-intercepts.

4. To solve $f(x) > c$ graphically, graph $y = f(x)$, graph $y = c$, and determine all x-values for which the graph of f is above the graph of $y = c$.

5. To solve $f(x) < c$ graphically, graph $y = f(x)$, graph $y = c$, and determine all x-values for which the graph of f is below the graph of $y = c$.

EXERCISES: ACTIVITY 4.3

In Exercises 1–4, solve the quadratic equation numerically (using tables of x- and y-values). Verify your solutions graphically.

1. $-4x = -x^2 + 12$

2. $x^2 + 9x + 18 = 0$

3. $2x^2 = 8x + 90$

4. $x^2 - x - 3 = 0$

In Exercises 5–8, solve the quadratic equation graphically using at least two different approaches. When necessary, give your solutions to the nearest hundredth.

5. $x^2 + 12x + 11 = 0$

6. $2x^2 - 3 = 2x$

7. $16x^2 - 400 = 0$

8. $4x^2 + 12x = -4$

In Exercises 9–12, solve the equation by using either a numerical or a graphical approach.

9. $x^2 + 2x - 3 = 0$

10. $x^2 + 11x + 24 = 0$

11. $x^2 - 2x - 8 = x + 20$

12. $x^2 - 10x + 6 = 5x - 50$

In Exercises 13–14, solve the given inequality using a graphing approach.

13. a. $x^2 - 4x - 1 < 11$

b. $x^2 - 4x - 1 > 11$

14. a. $2x^2 + 5x - 3 < 0$

b. $2x^2 + 5x - 3 \geq 0$

15. The stopping distance, d (in feet), for a car moving at a velocity (speed) v miles per hour is modeled by the equation

$$d(v) = 0.04v^2 + 1.1v.$$

a. What is the stopping distance for a velocity of 55 miles per hour?

b. What is the speed of the car if it takes 200 feet to stop?

16. An international rule for determining the number, n, of board feet (usable finished lumber) in a 16-foot log is modeled by the equation

$$n(d) = 0.22d^2 - 0.71d,$$

where d is the diameter of the log in inches.

a. How many board feet can be obtained from a 16-foot log with a 14-inch diameter?

b. Sketch a graph of this function. What is the practical domain of this function?

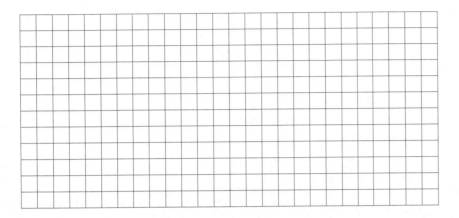

c. Use the graph to approximate the horizontal intercept(s). What is the practical meaning in this situation?

d. What is the diameter of a 16-foot log that has 200 board feet?

e. What inequality would you solve to determine the diameter when the board feet are at most 200?

f. Solve the inequality by using the graph of the function.

Activity 4.4

Sir Isaac Newton

Objectives

1. Factor expressions by removing the greatest common factor.

2. Factor trinomials using trial and error.

3. Use the Zero-Product Property to solve equations.

4. Solve quadratic equations by factoring.

Sir Isaac Newton XIV, a descendant of the famous physicist and mathematician, takes you the top of a building to demonstrate a physics property discovered by his famous ancesto He throws your math book straight up into the air. The book's distance, s, above the groun as a function of time, t, is modeled by

$$s = -16t^2 + 16t + 32.$$

1. When the book strikes the ground, what is the value of s?

2. Write the equation that you must solve to determine when the book strikes the ground.

The quadratic equation in Problem 2 can be solved by using a numerical or a graphical ap proach. However, an algebraic technique is efficient in this case and will give an exact answe The algorithm is based on the algebraic principle known as the **Zero-Product Property**.

> **Zero Product Property**
>
> If a and b are any numbers and $a \cdot b = 0$, then either a or b, or both, must be equal to zero.

Example 1 *Solve the equation $x(x + 5) = 0$.*

SOLUTION

The two factors in this equation are x and $x + 5$. The Zero-Product Property says one o these factors must equal zero. That is,

$$x = 0 \quad \text{or} \quad x + 5 = 0.$$

The first equation tells you that $x = 0$ is a solution. To determine a second solution, solve $x + 5 = 0$.

$$\begin{array}{r} x + 5 = 0 \\ -5 \quad -5 \\ \hline x = -5 \end{array}$$

There are two solutions, $x = 0$ and $x = -5$.

Your graphing calculator verifies the solutions as follows.

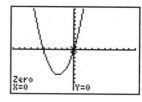 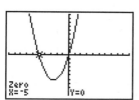

3. Solve each of the following equations using the Zero-Product Property.

a. $3x(x - 2) = 0$ **b.** $(2x - 3)(x + 2) = 0$

c. $(x + 2)(x + 3) = 0$

For the Zero-Product Property to be applied, one side of the equation must be zero. Therefore, at first glance, the Zero-Product Property can be used to solve the quadratic equation $3x^2 - 6x = 0$. However, a second condition must be satisfied. The nonzero side of the equation must be written as a product.

The process of writing an expression such as $3x^2 - 6x$ as a product is called factoring.

Definition

Rewriting an expression as a product is called **factoring**.

Factoring Common Factors

A **common factor** is a number or an expression that is a factor of each term of the entire expression. Whenever you wish to factor a polynomial, look first for a common factor.

Procedure

Removing a Common Factor from a Polynomial: First, identify the common factor, and then apply the distributive property in reverse.

Example 2 *Given the binomial $3x + 6$, 3 is a common factor because 3 is a factor of both terms $3x$ and 6. Applying the distributive property in reverse, you write*

$$3x + 6 \text{ as } 3(x + 2).$$

You may always check the factored binomial by multiplying:

$$3(x + 2) = 3(x) + 3(2) = 3x + 6.$$

When you look for a common factor, determine the largest or **greatest common factor** (or GCF). You can see that 3 is a common factor of $6x + 24$ because 3 is a factor of both 6 and 24. However, there is a larger common factor, 6. Therefore,

$$6x + 24 = 6(x + 4).$$

Example 3 *Given $6x^2 + 14x - 30$, you can see that 2 is a common factor. Is 2 the greatest common factor? Yes, because no larger number is a factor of every term.*

If you divide each term by 2, you obtain $3x^2 + 7x - 15$. The expression $6x^2 + 14x - 30$ can now be written in factored form as $2(3x^2 + 7x - 15)$. Check the factored trinomial by multiplying.

Example 4 *Factor $4x^3 - 8x^2 + 28x$.*

SOLUTION

Four is a factor of each term, but x is as well. Therefore, the greatest common factor is $4x$. You remove the GCF by dividing each term by $4x$. This leads to the factored form $4x(x^2 - 2x + 7)$.

You can check your factoring by applying the distributive property.

4. Factor the following polynomials by removing the greatest common factor.

a. $9a^6 + 18a^2$

b. $21xy^3 + 7xy$

c. $3x^2 - 21x + 33$

d. $4x^3 - 16x^2 - 24x$

Factoring Trinomials

With patience, you can factor trinomials of the form $ax^2 + bx + c$ by trial and error, using the FOIL method in reverse.

Procedure

Factoring Trinomials by Trial and Error

1. Remove the greatest common factor, GCF.

2. To factor the resulting trinomial into the product of two binomials, try combinations of factors for the first and last terms in two binomials.

3. Check the outer and inner products to match the middle term of the original trinomial.

   a. If the constant term, c, is positive, both of its factors are positive or both are negative.

   b. If the constant term is negative, one factor is positive and one is negative.

4. If the check fails, repeat steps 2 and 3.

Example 5   *Factor $6x^2 - 7x - 3$.*

SOLUTION

Step 1. There is no common factor, so go to step 2.

Step 2. You could factor the first term, $6x^2$, as $6x(x)$ or as $2x(3x)$. The last term, -3, has factors $3(-1)$ or $-3(1)$. Try $(2x + 1)(3x - 3)$.

Step 3. The outer product is $-6x$. The inner product is $3x$. The sum is $-3x$, not $7x$. The check fails.

Step 4. Try $(2x - 3)(3x + 1)$. The outer product is $2x$. The inner product is $-9x$. The sum is $-7x$. It checks.

5. Factor the following trinomials.

a. $x^2 - 7x + 12$

b. $x^2 - 8x - 9$

c. $x^2 + 14x + 49$

d. $25 + 10w + w^2$

Solving Quadratic Equations by Factoring

The following example demonstrates the procedure for solving quadratic equations written in standard form, $ax^2 + bx + c = 0$, by factoring.

Example 6   *Solve the equation $3x^2 - 2 = -x$ by factoring.*

Step 1.   Rewrite the equation in the form $ax^2 + bx + c = 0$ (called standard form).

$$3x^2 - 2 = -x$$
$$\underline{+x \qquad\qquad +x}$$
$$3x^2 + x - 2 = 0$$

Step 2.   Factor the expression on the nonzero side of the equation.

$$(x + 1)(3x - 2) = 0$$

Step 3.   Use the Zero-Product Property to set each factor equal to zero, and then solve each equation.

$$(x + 1)(3x - 2) = 0$$

$x + 1 = 0$	$3x - 2 = 0$
$x = -1$	$3x = 2$
	$x = \frac{2}{3}$

Therefore, the solutions are $x = -1$ and $x = \frac{2}{3}$.

These solutions can be verified graphically as follows.

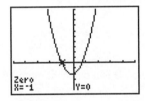
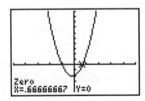

6. a.   Returning to the math book problem from the beginning of this activity, solve the equation from Problem 2 by factoring.

b.   Are both solutions to the equation ($t = 2$ and $t = -1$) also solutions to the question, "At what time does the book strike the ground"? Explain.

7. a.   You want to know at what time the book is 32 feet above the ground. Write a quadratic equation that represents this situation.

b. Solve the quadratic equation in part a by factoring.

8. Solve each of the following quadratic equations by factoring.

a. $2x^2 - x - 6 = 0$ **b.** $3x^2 - 6x = 0$ **c.** $x^2 + 4x = -x - 6$

9. Determine the zeros of the function defined by $f(x) = 2x^2 - 3x + 1$.

SUMMARY: ACTIVITY 4.4

1. To remove a **common factor** from a polynomial, first

  a. identify the common factor, and then

  b. apply the distributive property in reverse.

2. The **Zero-Product Property** says that if $ab = 0$ is a true statement, then either $a = 0$ or $b = 0$.

3. To factor trinomials of the form $ax^2 + bx + c$ by **trial and error**,

  a. remove the greatest common factor.

  b. try combinations of factors for the first and last terms in two binomials.

  c. check the outer and inner products to match the middle term of the original trinomial.

   • If the constant term, c, is positive, both factors of c are positive or both are negative.

   • If the constant term is negative, one factor is positive and one is negative.

  d. If the check fails, repeat steps 3b and 3c.

4. To solve equations by **factoring**,

  a. use the addition principle to remove all terms from one side of the equation; this results in a polynomial being set equal to zero.

  b. combine like terms, and then factor the nonzero side of the equation.

  c. use the Zero-Product Property to set each factor containing a variable equal to zero, and then solve the equations.

  d. check your solutions in the original equation.

EXERCISES: ACTIVITY 4.4

In Exercises 1–4, factor the polynomials by removing the GCF (greatest common factor).

1. $12x^5 - 18x^8$

2. $14x^6y^3 - 6x^2y^4$

3. $2x^3 - 14x^2 + 26x$

4. $5x^3 - 20x^2 - 35x$

In Exercises 5–13, completely factor the polynomials. Remember to look first for the GCF.

5. $x^2 + x - 6$

6. $p^2 - 16p + 48$

7. $x^2 + 7xy + 10y^2$

8. $x^2 - 4x - 32$

9. $12 + 8x + x^2$

10. $2x^2 + 7x - 15$

11. $3x^2 + 19x - 14$

12. $8x^4 - 47x^3 - 6x^2$

13. $20b^4 - 65b^3 - 60b^2$

In Exercises 14–21, solve each quadratic equation by factoring.

14. $x^2 - 5x + 6 = 0$

15. $x^2 + 2x - 3 = 0$

16. $x^2 - x = 6$

17. $x^2 - 5x = 14$

18. $3x^2 + 11x - 4 = 0$

19. $3x^2 - 12x = 0$

20. $x^2 - 7x = 18$

21. $3x(x - 6) - 5(x - 6) = 0$

Exercise numbers appearing in color are answered in the Selected Answers appendix.

22. Your neighbors have just finished installing a new swimming pool at their home. The pool measures 15 feet by 20 feet. They would like to plant a strip of grass of uniform width around three sides of the pool, the two short sides and one of the longer sides.

   a. Sketch a diagram of the pool and the strip of lawn, using x to represent the width of the uniform strip.

   b. Write an equation for the area, A, in terms of x that represents the lawn area around the pool.

   c. Your neighbors have enough seed for 168 square feet of lawn. Write an equation that relates the quantity of seed to the area of the uniform strip of lawn.

   d. Solve the equation in part c to determine the width of the uniform strip that can be seeded.

In turbulent economic times, investors often buy gold. The following table gives the price of gold per ounce, in dollars, each year from 2000 to 2008.

YEAR	PRICE PER OUNCE (dollars)
2000	290
2001	275
2002	285
2003	330
2004	400
2005	420
2006	510
2007	605
2008	880

1. a. Use your graphing calculator to sketch a scatterplot of the given data. Let the input, t, represent the number of years since 2000.

   b. This data can be modeled by the function

$$P = 12.78t^2 - 37.42t + 303.85,$$

   where P represents the price of gold per ounce. Graph this function on the same co-ordinate axis as the scatterplot. Is this function a good model for the data? Explain.

   c. You want to determine when the price of gold was $450 per ounce. That is, you want to determine the value of t that yields $P = 450$. Write the equation that must be solved.

The Quadratic Formula

The quadratic equation in Problem 1c can be solved by using a numerical or a graphical ap-proach. However, the algebraic technique of factoring cannot be applied.

The technique of solving quadratic equations by factoring is very limited. In most real-world applications involving quadratic equations, the quadratic is not factorable. In those cases, you can use a formula to solve the quadratic equation.

Beginning with the standard quadratic function defined by $y = ax^2 + bx + c, a \neq 0$, set $y = 0$ to obtain the equation

$$0 = ax^2 + bx + c.$$

The quadratic equation $ax^2 + bx + c = 0$ has two solutions,

$$x_1 = \frac{-b + \sqrt{b^2 - 4ac}}{2a} \quad \text{and} \quad x_2 = \frac{-b - \sqrt{b^2 - 4ac}}{2a}.$$

These solutions are often written as a single expression,

$$x = \frac{-b \pm \sqrt{b^2 - 4ac}}{2a}.$$

This formula is known as the **quadratic formula**.

Appendix

For the details showing the equation $ax^2 + bx + c = 0$ can be solved using the quadratic formula, see Appendix A. The section is called "Derivation of the Quadratic Formula."

The following example demonstrates the procedure for using the quadratic formula to solve an equation of the form $ax^2 + bx + c = 0$.

Example 1   *Solve $x(3x + 4) = 5$ using the quadratic formula.*

SOLUTION

Step 1. Write the equation in standard form, $ax^2 + bx + c = 0$.

$$x(3x + 4) = 5 \qquad \text{Apply the distributive property on the left side.}$$
$$3x^2 + 4x = 5$$
$$\underline{ -5 \quad -5} \qquad \text{Subtract 5 from both sides.}$$
$$3x^2 + 4x - 5 = 0$$

Step 2. Identify the coefficients a, b, and the constant term, c.

$$a = 3, b = 4, c = -5$$

Step 3. Substitute the values a, b, and c into the quadratic formula, and simplify.

$$x = \frac{-b \pm \sqrt{b^2 - 4ac}}{2a}$$

$$x = \frac{-4 \pm \sqrt{4^2 - 4(3)(-5)}}{2(3)}$$

$$x = \frac{-4 \pm \sqrt{16 - (-60)}}{6} = \frac{-4 \pm \sqrt{76}}{6}$$

$$x \approx \frac{-4 \pm 8.7178}{6} = -2.1196 \quad \text{or} \quad 0.7863$$

```
(-4+√(4²-4*3*-5)
)/(2*3)
          .7862996478
(-4-√(4²-4*3*-5)
)/(2*3)
         -2.119632981
```

Step 4. Check your solutions. The following graphs verify the solutions.

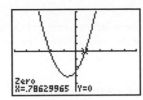 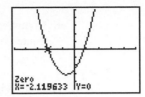

2. a. Solve the equation from Problem 1c, $12.78t^2 - 37.42t + 303.85 = 450$, using the quadratic formula. Check your solution graphically.

b. Use the results from part a to predict the years in which the price of gold will be (was) $450 per ounce.

3. a. Sketch a graph of $y = 2x^2 + 9x - 5$.

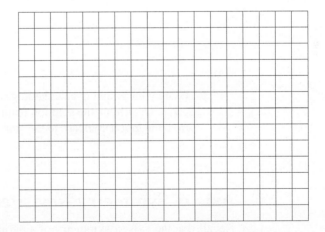

b. Write the equation that you would need to solve to determine the x-intercepts of the graph.

c. Solve the equation from part b using the quadratic formula.

d. Determine the x-intercepts of the graph using your graphing calculator.

e. Compare the solution using the quadratic formula (part c) with the x-intercepts determined from the graph.

4. The following data from the National Health and Nutrition Examination Survey indi◆ cates that the number of American adults who are overweight or obese is increasing.

Years Since 1960, t	1	12	18	31	39	44
Percentage of Americans Who Are Overweight or Obese, $P(t)$	45	47	47	56	64.5	66.3

This data can be modeled by the equation $P(t) = 0.0101t^2 + 0.0530t + 44.537$.

a. Use your graphing calculator to create a scatterplot of the data and a graph of the model P.

b. Does the model appear to be a good fit for the data? Explain.

c. Using the quadratic formula, determine the year when the model predicts that the percentage of overweight or obese Americans will first exceed 75%.

d. Using your graphing calculator, graph Y2 $= 75$ on the same axis as the function in part a. Determine the solution from the graph. How does the solution determined from the graph compare to the solution determined in part c?

Axis of Symmetry Revisited

Recall that the axis of symmetry of a parabola is defined by $f(x) = ax^2 + bx + c, a \neq 0$ is given by the formula $x = -\dfrac{b}{2a}$.

If you write the quadratic formula in a slightly different form, you obtain

$$x = -\frac{b}{2a} \pm \frac{\sqrt{b^2 - 4ac}}{2a} \quad \text{or} \quad x_1 = \frac{-b}{2a} + \frac{\sqrt{b^2 - 4ac}}{2a}, \quad x_2 = \frac{-b}{2a} - \frac{\sqrt{b^2 - 4ac}}{2a}.$$

The next problem uses the rewritten form of the quadratic formula to help identify a relationship between the x-intercepts and the axis of symmetry of the graph of f.

5. Consider the function $f(x) = 2x^2 + 9x - 5$.

   a. Determine the equation of the axis of symmetry of the graph of f.

   b. What is the value of $\dfrac{\sqrt{b^2 - 4ac}}{2a}$ for function f?

   c. Sketch a graph of the function f, and label the axis of symmetry. Show where the value computed in part b is located graphically. What are the x-intercepts of the graph?

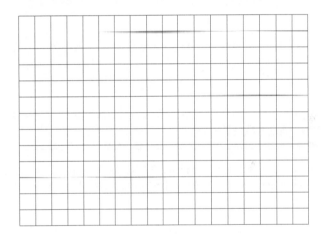

   d. What is the relationship between the axis of symmetry and the x-intercepts of a parabola?

6. For each of the following quadratic functions, determine the x-intercepts and the axis of symmetry of the graph. Solve the appropriate equation using the quadratic formula. Round your answers to the nearest hundredth.

   a. $f(x) = 2x^2 - 6x - 3$ **b.** $h(x) = x^2 - 8x + 16$

SUMMARY: ACTIVITY 4.5

1. To solve a quadratic equation of the form $ax^2 + bx + c = 0, a \neq 0$, using the **quadratic formula** $x = \dfrac{-b \pm \sqrt{b^2 - 4ac}}{2a}$,

   a. rewrite the quadratic equation (if necessary) so that one side is equal to zero.

   b. identify the coefficients a and b, and the constant term, c.

   c. substitute these values into the formula, and simplify.

   d. check your solutions graphically.

2. For a parabola with x-intercepts, the axis of symmetry is always midway between its x-intercepts.

3. The distance from the axis of symmetry to either x-intercept is $\dfrac{\sqrt{b^2 - 4ac}}{2a}$.

EXERCISES: ACTIVITY 4.5

1. The height of a bridge arch located in the Thousand Islands is modeled by the function $f(x) = -0.04x^2 + 28$, where x is the distance, in feet, from the center of the arch and $f(x)$ is the height of the arch.

   a. Sketch a picture of this arch on a grid using the vertical axis as the center of the arch.

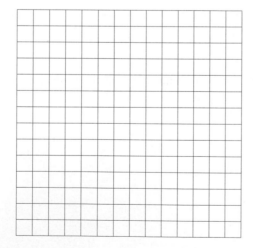

   b. Determine the vertical intercept. What is the practical meaning of this intercept in this situation?

c. Determine the x-intercepts algebraically using the quadratic formula.

d. Graph the function on your graphing calculator and check the accuracy of the intercepts you determined in part c.

e. If the arch straddles the river exactly, how wide is the river?

f. A sailboat is approaching the bridge. The top of the mast is 30 feet above the water. Will the boat clear the bridge? Explain.

g. You want to install a flagpole on the bridge at an arch height of 20 feet. Write the equation that you must solve to determine how far to the right or left of center the arch height is 20 feet.

h. Solve the equation in part g using the quadratic formula. Use your graphing calculator to check your result.

In Exercises 2–8, identify the values of a, b, and c, and then solve the equations using the quadratic formula. Round your answers to the nearest hundredth. Verify your solutions graphically.

2. $x^2 + 6x - 3 = 0$

3. $4x^2 + 4x + 1 = 0$

4. $x^2 + 5x = 13$

5. $2x^2 - 6x + 3 = 0$

6. $2x^2 - 3x = 5$

7. $(2x - 1)(x + 2) = 1$

8. $(x + 2)^2 + x^2 = 44$

In Exercises 9–11, determine the x-intercept(s) of the graph algebraically. Then check your results graphically.

9. $y = 3x^2 + 6x$ **10.** $y = x^2 - x - 6$

11. $f(x) = 2x^2 - x - 5$

12. The number n (in millions) of cell-phone subscribers in the United States from 2000 to 2006 is given in the following table.

Year	2000	2001	2002	2003	2004	2005	2006
Number of Subscribers (millions)	109.5	128.4	140.8	156.7	182.1	207.9	233.0

This data can be approximated by the quadratic model

$$n(t) = 1.393t^2 + 12.029t + 111.293,$$

where $t = 0$ corresponds to the year 2000.

a. Use your graphing calculator to sketch a graph of the function.

b. Use the graph to estimate the year in which there will be 300 million cell-phone subscribers.

c. Use the quadratic formula to answer part b. How does your answer compare to the estimate you obtained using a graphical approach?

d. How confident are you in your prediction? Explain.

13. The quadratic function defined by the equation

$$d = 2r^2 - 16r + 34$$

gives the density of smoke, d, in millions of particles per cubic foot for a certain type of diesel engine. The input variable, r, represents the speed of the engine in hundreds of revolutions per minute.

a. Determine the density of smoke when $r = 3.5$ (350 revolutions per minute).

b. Determine the number of revolutions per minute for minimum smoke. What is the minimum output?

c. If the density of smoke is determined to be 100 million particles per cubic foot, determine the speed of the engine.

Activity 4.6

Heat Index

Objectives

1. Determine quadratic regression models using a graphing calculator.

2. Solve problems using quadratic regression models.

On very hot and humid summer days, it is common for the National Weather Service to issue warnings due to a very high heat index. The heat index is a measurement that combines air temperature and relative humidity to determine a relative temperature. The heat index is the temperature you perceive (how it feels) on a hot and humid day. Heat index is similar to windchill, the temperature you perceive on a cold and windy day. Heat indices are not usually calculated until the air temperature reaches 80 degrees Fahrenheit.

The following table gives the heat index for various temperatures when the relative humidity is 90%. Both air temperature and the heat index (perceived temperature) are measured in degrees Fahrenheit.

Temperature (degrees Fahrenheit), t	80	82	84	86	88	90	9
Heat Index (°F), h	86	91	98	105	113	122	13

Note: Relative humidity is 90%.

1. Sketch a scatterplot of the data. Let the air temperature, t, represent the input variable and the heat index, h, represent the output variable. Does the data appear to be quadratic? Explain.

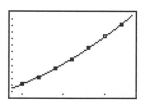

2. Use the regression feature of your graphing calculator to determine and plot a quadratic function that best fits this data. Your graph should appear as follows.

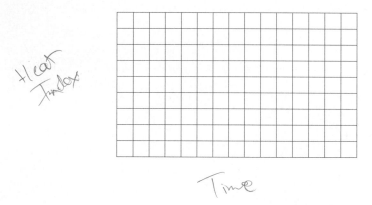

3. **a.** How does the plot of the quadratic regression equation compare with your scatterplot of the data?

   b. Do you believe the quadratic regression model is a good model for the heat index when the relative humidity is 90%?

4. What is the practical domain of this function?

5. a. When the humidity is 90%, use the quadratic regression equation to estimate the heat index for each of the following temperatures.

   i. 45 degrees   **ii.** 85 degrees   **iii.** 100 degrees

   b. Which, if any, of these estimates do you think is most reliable? Explain.

6. a. If the relative humidity is 90%, estimate the temperatures for which the heat index is equal to 100° using each of the following methods.

   i. The given table (numerical)

   ii. The graph of the quadratic regression equation (graphical)

   b. Use the quadratic formula (algebraic) to estimate the temperature for which the heat index is 100 degrees when the relative humidity is 90%.

   c. How do the results in parts a and b compare?

7. a. In Problem 6b, the stated relative humidity was 90%, and you wrote an equation to determine the temperature for which the heat index (perceived temperature) was equal to 100 degrees. Write an equation to determine the temperature for which the air temperature and the heat index are equal.

   b. Solve the equation in part a graphically to estimate the temperatures for which the air temperature and the heat index are equal.

8. a. The National Weather Service does not calculate heat index unless the air temperature is at least 80 degrees Fahrenheit and the relative humidity is at least 40%. Based on your results in Problems 5, 6, and 7, does this policy seem reasonable? Explain.

b. According to this policy, will the heat index reported by the National Weather Service ever be equal to the actual air temperature if the relative humidity is 90%? Explain.

9. The following data from the National Health and Nutrition Examination Survey appeared in Problem 4 of Activity 4.5. The data indicates that the number of American adults who are overweight or obese is increasing.

Years Since 1960, t	1	12	18	31	39	44	46
Percentage of Americans Who Are Overweight or Obese, P(t)	45	47	47	56	64.5	66.3	66.9

Use the regression feature of your graphing calculator to verify that this data can be modeled by the equation $P(t) = 0.0101t^2 + 0.0530t + 44.537$.

SUMMARY: ACTIVITY 4.6

1. Parabolic data can be modeled by a **quadratic regression equation**.

EXERCISES: ACTIVITY 4.6

1. During one game, the Buffalo Bills punter was called upon to punt the ball 8 times. On one of these punts, the punter struck the ball at his own 30-yard line. The height, h, of the ball above the field in feet as a function of time, t, in seconds can be partially modeled by the following table.

t	0	0.6	1.2	1.8	2.4	3.0
h(t)	2.50	28.56	43.10	46.12	37.12	17.60

a. Sketch a scatterplot of the data using your graphing calculator.

b. Use your graphing calculator to obtain a quadratic regression function for this data. Round the values of a, b, and c to four decimal places.

c. Graph the equation from part b on the same coordinate axes as the data points. Does the curve appear to be a good fit for the data? Explain.

d. In this model, what is the practical domain of the quadratic regression function?

e. Estimate the practical range of this model.

f. How long after the ball was struck did the ball reach 35 feet above the field? Explain.

g. How many results did you obtain for part f? Do you think you have all of the solutions? Explain.

2. Use the following data set to perform the tasks in parts a–e.

x	0	3	6	9	12
y	5	28	86	180	310

a. Determine an appropriate scale, and plot these points.

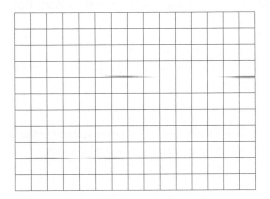

b. Use your graphing calculator to determine the quadratic regression equation for this data set.

c. Graph the regression equation using your graphing calculator.

d. Use the table feature of your graphing calculator to compare the predicted outputs with the outputs given in the table.

e. Predict the output for $x = 7$ and for $x = 15$.

3. The following table shows the stopping distance for a car at various speeds on dry pavement.

Speed (mph)	25	35	45	55	65	75
Distance (ft.)	65	108	167	245	340	450

a. Use your graphing calculator to determine a quadratic regression equation that represents this data.

b. Use the regression equation to predict the stopping distance at 90 mph.

c. What speed would produce a stopping distance of 280 feet? (Round to the nearest tenth.) Explain how you arrived at your conclusion.

4. Downturns in the high-tech field during the early 2000s have caused a similar downturn in computer science graduates. The following data from the Computer Research Association approximates the number of bachelor's degrees granted since 1998.

Years Since 1998, t	0	1	2	3	4	5	6	7	8	9
Bachelors Degrees Granted (thousands), N	7.5	8.3	10.3	11.4	12.3	13.1	14.1	11.9	10.5	8.0

a. Create a scatterplot of the data using your graphing calculator.

b. Use your graphing calculator to obtain a quadratic regression function for these data. Round the values of a, b, and c to four decimal places.

c. Graph the equation from part b on the same coordinate axes as the data points. Does the curve appear to be a good fit for the data? Explain.

d. What does the regression equation predict bachelor's degrees in computer science in the year 2015? Does this seem reasonable? Use the graph to determine the practical domain.

e. Use the model to determine the year that the number of bachelor's degrees in computer science will drop below 5000.

5. The shape of the main support cable in a suspension bridge is a parabola.

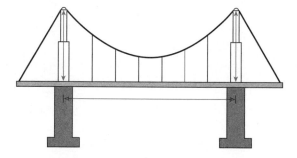

Use resources from the library or the Internet to determine specific dimensions, such as the distance between the main support columns, for one of the following bridges: Golden Gate Bridge, George Washington Bridge, or Verrazano Bridge.

Use the dimensions to develop points that lie on the graph of the main support cable of the bridge you selected. Draw a sketch of the support cable on graph paper. The orientation of the coordinate system, especially the origin, is very important. You will need at least three points that lie on the graph of the support cable in your drawing.

Use the points to determine a quadratic regression equation to represent the main support cable of the bridge. Use the model to determine the minimum distance from the cable to the highway.

Be prepared to give a presentation of your findings.

Activity 4.7

Complex Numbers

Objectives

1. Identify the imaginary unit $i = \sqrt{-1}$.

2. Identify a complex number.

3. Determine the value of the discriminant $b^2 - 4ac$.

4. Determine the types of solutions to a quadratic equation.

5. Solve a quadratic equation in the complex number system.

Recall that the solutions to the quadratic equation $ax^2 + bx + c = 0$ correspond to the x-intercepts of the parabola having equation $y = ax^2 + bx + c$.

Do all parabolas possess x-intercepts? Consider the graph of $y = 2x^2 + x + 5$. If you graph the function in the window Xmin $= -5$, Xmax $= 5$, Ymin $= -3$, and Ymax $= 15$, the graph will resemble the following.

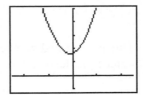

1. **a.** Based on what you know about parabolas, will the graph of $y = 2x^2 + x + 5$ have any x-intercepts? Explain.

   b. What can you say about the solutions to $2x^2 + x + 5 = 0$?

Your response to Problem 1b should have been, "There are no real solutions." This is consistent with the graph. Because there are no x-intercepts for the graph of $y = 2x^2 + x + 5$, there are no real-valued solutions to the equation $2x^2 + x + 5 = 0$. Would you have discovered this if you had tried to solve the equation $2x^2 + x + 5 = 0$ algebraically using the quadratic formula? Problem 2 addresses this question.

2. Use the quadratic formula to solve $2x^2 + x + 5 = 0$. Where does the solution process break down? Explain.

Complex Numbers

Problem 2 illustrates that the breakdown with the quadratic formula occurs when you are asked to evaluate a radical with a negative radicand. If you try to evaluate $\sqrt{-39}$ using your TI-83/TI-84 Plus, the following screens will appear.

This tells you there is no real number that is the square root of -39. Now change the MODE on the calculator from *real* to $a + bi$.

Now try to evaluate $\sqrt{-39}$ again. This time your calculator returns a value.

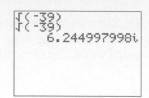

This is not a real number. Such a number is a **complex number** (an extension of the real numbers). The distinguishing characteristic of the complex numbers is the **imaginary unit** $i = \sqrt{-1}$.

The quadratic formula solution to $2x^2 + x + 5 = 0$ uses the values $a = 2, b = 1,$ and $c = 5$. The solution is

$$x = \frac{-1 \pm \sqrt{1^2 - 4(2)(5)}}{2(2)} = \frac{-1 \pm \sqrt{1 - 40}}{4} = -\frac{1}{4} \pm \frac{\sqrt{-39}}{4}.$$

The problem is that you cannot evaluate $\sqrt{-39}$ in the real number system because any real number multiplied by itself is nonnegative. Therefore, you need to introduce the imaginary unit, $i = \sqrt{-1}$, and interpret $\sqrt{-39}$ as

$$\sqrt{39(-1)} = \sqrt{39}\sqrt{-1} = \sqrt{39} \cdot i = i\sqrt{39}.$$

This approach can be used to rewrite any radical expression with a negative radicand.

Example 1

    a. $\sqrt{-16} = \sqrt{16}\sqrt{-1} = 4i$

    b. $\sqrt{-25} = \sqrt{25}\sqrt{-1} = 5i$

    c. $\sqrt{-53} = \sqrt{53}\sqrt{-1} = \sqrt{53}i$ or $i\sqrt{53}$

3. Rewrite each of the following in the form bi, where $i = \sqrt{-1}$.

    a. $\sqrt{-26}$ **b.** $\sqrt{-5}$ **c.** $\sqrt{-64}$

    d. $\sqrt{3 \cdot (-7)}$ **e.** $\sqrt{-18}$ **f.** $\sqrt{-27}$

    g. $\sqrt{-\frac{3}{4}}$ **h.** $\sqrt{-\frac{15}{27}}$

Definition

Numbers of the form bi, where b is a real number and $i = \sqrt{-1}$, are called **pure imaginary numbers**. Numbers of the form $a \pm bi$, where a and b are real and $i = \sqrt{-1}$, are called **complex numbers**. The term a is called the *real part*. The term bi is called the *imaginary part*. Imaginary numbers are complex numbers of the form $0 + bi$. Real numbers are complex numbers of the form $a + 0i$.

Example 2

a. *The numbers* $-3i, \frac{2}{3}i,$ *and 7.4i are pure imaginary numbers.*

b. *The numbers* $-4 + 3i, \frac{1}{2} - \frac{2}{3}i, 4i, -2i,$ *and 5 $-$ 6i are complex numbers.*

Note that the set of real numbers is contained within the set of complex numbers. A real number a may be thought of as the complex number $a + 0i$.

In the sixteenth century, complex numbers were first used as solutions to polynomial equations. The notation $\sqrt{-1}$ was used during this time. Such numbers were called imaginary because their existence was not clearly understood. In 1777, Leonhard Euler introduced the notation i and wrote complex numbers in the form $a + bi$. Caspar Wessel in 1797 and Carl Friedrich Gauss in 1799 used the geometric interpretation of complex numbers as points in a plane. This made such numbers more concrete and less mysterious. Finally, in 1833, Sir William Hamilton showed that if the number i is defined to have the property

$$i^2 = -1,$$

then the set of real numbers can be extended to include numbers like $\sqrt{-1}$.

Today, complex numbers are used in a variety of applications, including chaos theory (fractals) and engineering.

Operations with Complex Numbers

The operations of addition, subtraction, and multiplication of complex numbers are demonstrated in the following example.

Example 3

a. *To add complex numbers, add the real parts and the imaginary parts.*
$$(3 + 2i) + (5 - 7i) = 8 - 5i$$

b. *To subtract complex numbers, add the opposite.*
$$(2 - 2i) - (-6 + i) = 2 - 2i + 6 - i = 8 - 3i$$

c. *To multiply complex numbers, multiply each term of the first by each term of the second, and simplify.*

$$(3 - 5i)(-1 + 8i) = -3 + 24i + 5i - 40i^2$$
$$= -3 + 29i + 40$$
$$= 37 + 29i$$

Remember that $i \cdot i = i^2 = -1$.

The TI-83/TI-84 Plus is capable of operations with complex numbers. You first need to change the mode of the calculator from Real to $a + bi$. Note that the i key is 2nd ▢ period.

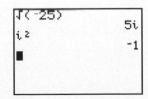
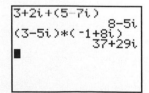

4. Perform the following operations with complex numbers. Use your graphing calculato to check your results.

   a. $(3 + 4i) + (-5 + 6i)$

   b. $(5 - 7i) - (-2 + 5i)$

   c. $5i(2 - 4i)$

   d. $(3 - 2i)(4 + 5i)$

Discriminant

In the complex number system, every quadratic equation has at least one solution. In th quadratic formula

$$x = \frac{-b \pm \sqrt{b^2 - 4ac}}{2a}$$

the expression $b^2 - 4ac$ is called the **discriminant** because its value determines the numbe and type of solutions of a quadratic equation $ax^2 + bx + c = 0$. There are three possibl cases, depending on whether the value of the discriminant is positive, zero, or negative Problems 5, 6, and 7 investigate this relationship.

5. For each of the following quadratic functions, determine the sign of the discriminant Then sketch a graph using your graphing calculator, and determine the number o x-intercepts.

   a. $y = 2x^2 - 7x - 4$

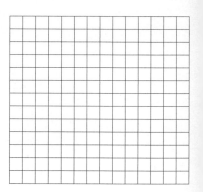

   b. $y = 3x^2 + x + 1$

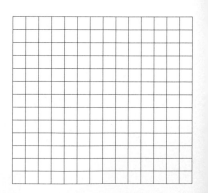

c. $y = x^2 + 2x + 1$

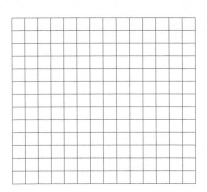

Recall that the real solutions to the equation $ax^2 + bx + c = 0$ and the x-intercepts of the graph of $y = ax^2 + bx + c$ are the same. Since the graph of $y = 2x^2 + x + 5$ (see Problem 1) has no x-intercept, it has no real solutions.

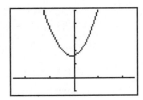

But the quadratic equation $2x^2 + x + 5 = 0$ does have exactly two solutions in the complex number system. The two solutions must be complex and not real. Similarly, if the graph of a quadratic function has two x-intercepts, the solutions to the equation $ax^2 + bx + c = 0$ must be two real numbers.

6. If the graph of $y = ax^2 + bx + c$ has exactly one x-intercept, what are the number and type (real or complex) of solutions to the equation $ax^2 + bx + c = 0$?

7. Return to Problem 5. Use the value of the discriminant $b^2 - 4ac$ and the number of x-intercepts in parts a, b, and c to complete the following table.

SOLUTIONS TO $ax^2 + bx + c = 0$ IN THE COMPLEX NUMBER SYSTEM	
$b^2 - 4ac$	NUMBER AND TYPE OF SOLUTIONS
Positive	
Zero	
Negative	

8. a. Evaluate the discriminant for each of the following equations, and indicate the number and type of solutions to the equations.

   i. $2x^2 - 7x - 4 = 0$ **ii.** $3x^2 + x + 1 = 0$

   iii. $x^2 - 2x + 1 = 0$ **iv.** $3x^2 + 2x = -1$

b. Determine the solutions to each of the equations in part a in order to verify your results from part a.

   i.   **ii.**

   iii.   **iv.**

SUMMARY: ACTIVITY 4.7

1. The **imaginary unit** denoted by i is the number $\sqrt{-1}$. The number i is defined to have the property $i^2 = -1$.

2. Any number that can be written in the form $a + bi$, where a and b are real numbers and i is the imaginary unit, is called a **complex number**. The term a is called the **real part**. The term bi is called the **imaginary part**.

3. In the quadratic formula

$$x = \frac{-b \pm \sqrt{b^2 - 4ac}}{2a}$$

the expression $b^2 - 4ac$ is called the **discriminant**. Its value determines the number and type of solutions of a quadratic equation $ax^2 + bx + c = 0$.

4. Solutions of the quadratic equation $ax^2 + bx + c = 0$ in the complex number system are summarized in the following table.

$b^2 - 4ac$	NUMBER AND TYPE OF SOLUTIONS
Positive	2 real solutions
Zero	1 real solution
Negative	2 complex solutions

EXERCISES: ACTIVITY 4.7

In Exercises 1–8, write each of the following in the form bi, where $i = \sqrt{-1}$.

1. $\sqrt{-25}$ **2.** $\sqrt{-20}$ **3.** $\sqrt{-36}$

4. $\sqrt{-10}$ **5.** $\sqrt{-48}$ **6.** $\sqrt{-80}$

7. $\sqrt{-\frac{9}{16}}$ **8.** $\sqrt{\frac{-20}{75}}$

Exercises 9–13, perform the operations, and express your answer in the form $a + bi$. Use your graphing calculator to verify the results.

9. $(2 + 8i) + (-7 + 2i)$

10. $(5 - 3i) - (2 - 6i)$

11. $5i + (3 - 7i)$

12. $3i(-2 + 4i)$

13. $(4 - 3i)(1 + 2i)$

14. Complex numbers are used in electronics to describe the current in an electric circuit. In an alternating current, the resistance, R, in ohms, is the measure of how much the circuit resists (or impedes) the flow of current through it. The resistance, R, is related to the voltage, V, and current, I, by Ohm's Law:

$$V = IR$$

a. If $I = 0.3 + 2i$ amperes and $R = 0.5 - 3i$ ohms, determine the voltage, V.

b. If $I = 2 - 3i$ amperes and $R = 3 + 5i$ ohms, determine the voltage, V.

In Exercises 15–18, solve the quadratic equations in the complex number system using the quadratic formula. Verify your real solutions graphically. Verify that no real solutions mean no x-intercepts.

15. $3x^2 - 2x + 7 = 0$

16. $x^2 + x = 3$

17. $2x^2 + 5x = 7$

18. $0.5x^2 - x + 3 = 0$

In Exercises 19–24, determine the number and type of solutions of each equation by examining the discriminant.

19. $2x^2 + 3x - 5 = 0$

20. $6x^2 + x + 5 = 0$

21. $4x^2 - 4x + 1 = 0$

22. $9x^2 + 6x + 1 = 0$

23. $12x^2 = 4x - 3$

24. $3x^2 = 5x + 7$

Cluster 1   What Have I Learned?

1. **a.** In order for the graph of the equation $y = ax^2 + bx + c$ to be a parabola, the value of the coefficient of x^2 cannot be zero. Explain.

   b. What is the vertex of a parabola having an equation of the form $y = ax^2$?

   c. Describe the relationship between the vertex and the vertical intercept of the graph of $y = ax^2 + c$.

2. Determine if the vertex is a minimum point or a maximum point of $y = ax^2 + bx + c$ in each of the following situations.

   a. $a < 0$ **b.** $a > 0$

3. **a.** What are the possibilities for the number of vertical intercepts of a quadratic function?

   b. What are the possibilities for the number of horizontal intercepts of a parabola?

4. What is the relationship between the vertex and the x-intercept of the graph of $y = x^2 - 4x + 4$?

5. **a.** The vertex of a parabola is $(3, 1)$. Using this information, complete the following table.

x	1	2	3	4	5
y	5	2			

   b. If the vertex of a parabola is $(2, 4)$, complete the following table.

x	−2	0	2	4	6
y	0	3			

6. **a.** Given the following graph, explain why choices i, ii, and iii do not fit the curve.

   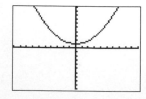

   i. $f(x) = ax^2 + bx$ with $a > 0, b < 0$

ii. $g(x) = ax^2 + c$ with $a < 0, c > 0$

iii. $h(x) = ax^2 + bx + c$ with $a < 0, b > 0, c < 0$

b. What restrictions on a, b, and c are necessary to fit $y = ax^2 + bx + c$ to the given graph?

7. Review the steps in the following solution. Is the solution correct? Explain why or why not.

$$x^2 - 3x - 4 = 6$$
$$(x - 4)(x + 1) = 6$$

$x - 4 = 6$	$x + 1 = 6$
$x = 10$	$x = 5$

8. Describe how you would determine the solutions to $ax^2 + bx + c > 5$ graphically.

9. Which of the following statements are true? In each case, justify your decision.

a. $3 + 2i$ is a pure imaginary number.

b. $\sqrt{-7}$ is a complex number.

c. 0 is a complex number.

10. a. Describe the relationship between the x-intercepts (if they exist) of the graph of $y = ax^2 + bx + c$ and the solutions to the equation $ax^2 + bx + c = 0$.

b. Describe the relationship between the x-intercepts (if they exist) of the graph of $y = ax^2 + bx + c$ and the discriminant $b^2 - 4ac$.

11. Consider the quadratic equation $ax^2 + bx + c = 0$. If the quadratic expression $ax^2 + bx + c$ is factorable, what can you say about the sign of the discriminant, $b^2 - 4ac$? Is it positive, negative, or zero? Explain.

12. For what values of c are the solutions to $2x^2 - 5x + c = 0$ imaginary?

13. For what values of k does $x^2 - kx + k = 0$ have only one solution? (*Hint:* Examine the discriminant.)

Cluster 1 | How Can I Practice?

1. Complete the following table.

EQUATION OF THE FORM $y = ax^2 + bx + c$	VALUE OF a	VALUE OF b	VALUE OF c
$y = 5x^2$			
$y = \frac{1}{3}x^2 + 3x - 1$			
$y = -2x^2 + x$			

For Exercises 2–7, determine the following characteristics for each graph:

a. *The direction in which the parabola opens*

b. *The equation of the axis of symmetry*

c. *The vertex; determine if maximum or minimum*

d. *The y-intercept*

2. $y = -2x^2 + 4$

3. $y = \frac{2}{3}x^2$

4. $f(x) = -3x^2 + 6x + 7$

5. $f(x) = 4x^2 - 4x$

6. $y = x^2 + 6x + 9$

7. $y = x^2 - x + 1$

For Exercises 8–11, use your graphing calculator to sketch the graph of each quadratic function, and then determine the following for each function:

a. *The coordinates of the x-intercepts (if they exist)*

b. *The domain and range*

c. *The horizontal interval over which the function is increasing*

d. *The horizontal interval over which the function is decreasing*

Answers to all How Can I Practice exercises are included in the Selected Answers appendix.

8. $y = -x^2 + 4$

9. $y = x^2 - 5x + 6$

10. $y = -3x^2 - 6x + 8$

11. $y = 0.22x^2 - 0.71x + 2$

12. Use your graphing calculator to approximate the vertex of the graph of the parabola defined by the equation $y = -2x^2 + 3x + 25$.

13. Completely factor the following polynomials (if possible).

   a. $9a^5 - 27a^2$

   b. $24x^3 - 6x^2$

   c. $4x^3 - 16x^2 - 20x$

   d. $5x^2 - 16x + 6$

   e. $x^2 - 5x - 24$

   f. $y^2 + 10y + 25$

14. Determine one solution of the following quadratic equations numerically. That is, construct a table of (x, y) ordered pairs, where $y = f(x)$, and estimate the value of x (input) that results in the required y-value (output).

   a. $5x^2 = 7$

   b. $x^2 - 7x + 10 = 5$

   c. $3x^2 - 5x = 2$

a.

x	1	1.1	1.2	1.3	1.4	1.5
y						

b.

x	0.5	0.6	0.7	0.8	0.9	1
y						

c.

x	0	1	2	3	4	5
y						

15. Solve each of the equations from Exercise 14 using the quadratic formula. When necessary, round your solutions to the nearest tenth. Check your solutions by graphing.

16. Solve each of the following equations by factoring.

a. $4x^2 - 8x = 0$

b. $x^2 - 6 = 7x + 12$

c. $2x(x - 4) = -6$

d. $x^2 - 8x + 16 = 0$

e. $x^2 - 2x - 24 = 0$

f. $y^2 - 2y - 35 = -20$

g. $a^2 + 2a + 1 = 3a + 7$

h. $4x^2 + 4x - 3 = -3x - 1$

17. Solve the following inequalities using a graphing approach.

a. $x^2 + 6x - 16 < 0$

b. $x^2 + 6x - 16 > 0$

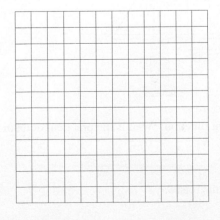

18. A fastball is hit straight up over home plate. The ball's height, h (in feet), from the ground is modeled by

$$h(t) = -16t^2 + 80t + 5,$$

where t is measured in seconds.

a. What is the maximum height of the ball above the ground?

b. Write an equation to determine how long it will take for the ball to reach the ground. Solve the equation using the quadratic formula. Check your solution by graphing.

c. Write the equation you would need to determine when the ball is 101 feet above the ground.

d. Solve the equation you determined in part c algebraically to determine the time it will take for the ball to reach a height of 101 feet. Verify your results graphically.

19. A suspension bridge (shown in the accompanying figure) is 100 meters long. The bridge is supported by cables attached to the tops of towers 35 meters high at each end of the bridge. The cables hang from the towers approximately in the shape of a parabola. The height, $h(t)$ (in meters), of the cables above the surface of the roadway is modeled by

$$h(x) = 0.01x^2 - x + 35,$$

where x is the horizontal distance measured from the point where the tower and roadway meet.

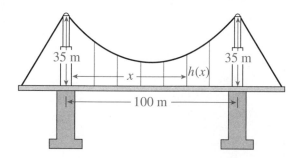

a. Use your graphing calculator to examine the height function. What is the practical domain of this function?

b. What is the minimum distance of the cables from the roadway?

0. Use the following data set to perform the tasks in parts a–f.

x	0	1	3	5	7	8
y	10	4	−18	−54	−107	−145

a. Determine an appropriate scale, and plot these points.

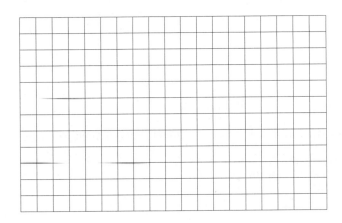

b. Use your graphing calculator to determine the quadratic regression equation for this data set.

c. Graph the regression equation on the same coordinate axes as the data points.

d. Compare the predicted outputs with the given outputs in the table.

e. What is the predicted output for $x = 4$ and for $x = 9$?

f. For what value of x is $y = -40$? Use the quadratic formula.

Write each of the following in the form bi, where $i = \sqrt{-1}$.

21. $\sqrt{-49}$ **22.** $\sqrt{-45}$ **23.** $\sqrt{-121}$

24. $\sqrt{-15}$ **25.** $\sqrt{-112}$ **26.** $\sqrt{-125}$

27. $\sqrt{-\frac{16}{25}}$ **28.** $\sqrt{\frac{-24}{42}}$

Perform the following operations, and express your answer in the form $a + bi$. Use your graphing calculator to verify the results.

29. $(7 + 5i) + (-3 - 2i)$

30. $(3 - 3i) - (8 - 9i)$

31. $3i + (6 - 7i)$

32. $-3i(8 - 4i)$

33. $(3 - 4i)(-1 + 2i)$

34. In parts i–iv, perform the following tasks.

    a. Identify the values of a, b, and c in $ax^2 + bx + c = 0$.

    b. Determine the type of solution by examining the sign of the discriminant.

    c. Solve the given equation using the quadratic formula. If necessary, round your solutions to the nearest hundredth.

    d. Check your solutions by graphing as well as by substitution.

      i. $3x^2 - x = 7$

      ii. $x^2 - 4x + 10 = 0$

      iii. $2x^2 - 3x = 2x + 3$

      iv. $3x(3x - 2) + 1 = 0$

35. Each of the following graphs represents a quadratic function. For each graph, determine if the discriminant is positive, negative, or zero. Explain your decision.

    a. 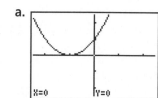   **b.** 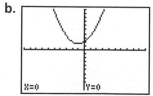   **c.**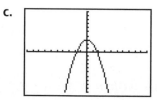

    i. graph a

    ii. graph b

    iii. graph c

Cluster 2 Curve Fitting and Higher-Order Polynomial Functions

Activity 4.8

The Power of Power Functions

Objectives

- Identify a direct variation function.

- Determine the constant of variation.

- Identify the properties of graphs of power functions defined by $y = kx^n$, where n is a positive integer, $k \neq 0$.

You are traveling in a hot air balloon when suddenly your binoculars drop from the edge of the balloon's basket. At that moment, the balloon is maintaining a constant height of 500 feet. The distance of the binoculars from the edge of the basket is modeled by

$$s = 16t^2.$$

The following table gives the distance, s (in feet), from the drop point at various times, t (in seconds).

t, Time (seconds)	0	1	2	3	4
s, Distance (feet)	0	16	64	144	256

As the input values (units of time) increase, the corresponding output values (units of distance) increase. Let us look more closely at how this increase takes place.

Because $s = 16t^2$, you can say that the output, s, varies directly as the square of the input, t. Therefore, as t doubles in value from 1 to 2 or from 2 to 4, the corresponding output values become 4 times as large: increasing from 16 to 64 or 64 to 256.

1. a. As t triples from 1 to 3, the corresponding s-values become _____ times as large.

b. In general, if y varies directly as the square of x, then when x becomes n times as large, the corresponding y-values become _____ times as large.

The volume, V, of a sphere is given by $V = \frac{4}{3}\pi r^3$. In this situation, you can say that the output, V, varies directly as the cube of the radius, r.

2. a. Complete the following table. Leave your answers for V in terms of π.

r	1	2	3	4	8
V					

Note that as r-values double from 2 to 4, the corresponding V-values increase from $\frac{32}{3}\pi$ to $\frac{256}{3}\pi$.

b. As r doubles from 2 to 4 or from 4 to 8, the corresponding V-values become _____ times as large.

c. In general, if y varies directly as the cube of x, then when x becomes n times as large, the corresponding y-values become _____ times as large.

d. Sketch a graph of the volume function. What is the practical domain of this function?

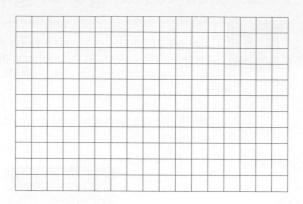

Definition

The equation

$$y = kx^n,$$

where $k \neq 0$ and n is a positive integer, defines a **direct variation** function in which y varies directly as x^n. The constant, k, is called the **constant of variation**.

Example 1 *The constant of variation, k, in the free-falling object situation defined by $s = 16t^2$ is 16.*

3. What is the constant of variation for the direct variation function defined by $V = \frac{4}{3}\pi r^3$?

In the falling binocular situation, you are given the direct variation equation. Suppose you only know that the distance, s, varies directly as the square of t and one data pair. Are you able to determine the direct variation equation? Example 2 demonstrates the process.

Example 2 *Let s vary directly as the square of t. If $s = 64$ when $t = 2$, determine the direct variation equation.*

SOLUTION

Because s varies directly as the square of t, you have

$$s = kt^2,$$

where k is the constant of variation. Substituting 64 for s and 2 for t, you have

$$64 = k(2)^2 \quad \text{or} \quad 64 = 4k \quad \text{or} \quad k = 16.$$

Therefore, the direct variation equation is

$$s = 16t^2.$$

4. For each table, determine the pattern and complete the table. Then write a direct variation equation for each table.

a. y varies directly as x.

x	1	2	4	8	12
y		12			

b. y varies directly as x^3.

x	1	2	3
y		32	

c. y varies directly as x.

x	1	2	3	4	5
y			3		

5. The length, L (in feet), of skid distance left by a car varies directly as the square of the initial velocity, v (in miles per hour), of the car.

a. Write a general equation for L as a function of v. Let k represent the constant of variation.

b. Suppose a car traveling at 40 miles per hour leaves skid distance of 60 feet. Use this information to determine the value of k.

c. Use the function to determine the length of the skid distance left by the car traveling at 60 miles per hour.

Power Functions

The direct variation functions that have equations of the form $y = kx^n$, where n is a positive integer and $k \neq 0$, are also called **power functions**. The graphs of this family of functions are very interesting and are useful in problem solving.

6. Sketch a graph of each of the following power functions. Use your graphing calculator to verify the graph.

a. $y = x$

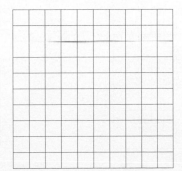

b. $y = x^2$

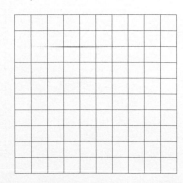

c. $y = x^3$

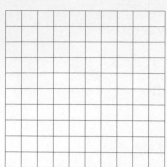

d. $y = x^4$

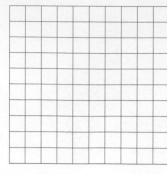

e. $y = x^5$

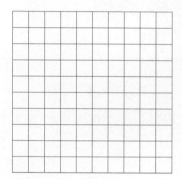

f. $y = x^6$

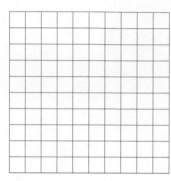

7. Each graph in Problem 6 has an equation of the form $y = x^n$, where n is a positive integer.

    a. What is the basic shape of the graph if

        i. n is even?   **ii.** n is odd?

    b. If n is even, what happens to the graph as n gets larger in value?

    c. If n is odd, is the function increasing or decreasing?

8. Use the patterns from Problem 7 in combination with graphing techniques you have learned previously to sketch a graph of each of the following without using a graphing calculator.

    a. $y = x^2 + 1$ **b.** $y = -2x^4$

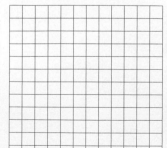

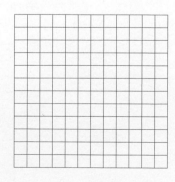

c. $y = 3x^8 + 1$

d. $y = -2x^5$

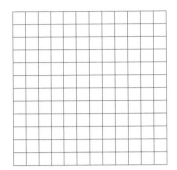

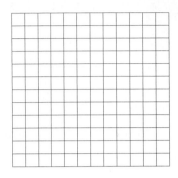

e. $y = x^{10}$

f. $y = 5x^3 + 2$

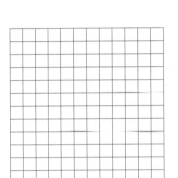

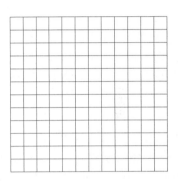

SUMMARY: ACTIVITY 4.8

1. The equation $y = kx^n$, where $k \neq 0$ and n is a positive integer, defines a **direct variation function**. The constant, k, is called the **constant of variation**.

2. The direct variation functions that have equations of the form $y = kx^n$, where n is a positive integer, are also called **power functions**.

   a. Power functions in which n is even resemble parabolas. As n increases in value, the graph flattens near the vertex.

   b. Power functions in which n is odd resemble the graph of $y = kx^3$. If k is positive, the graph is increasing. If k is negative, the graph is decreasing.

EXERCISES: ACTIVITY 4.8

1. For each table, determine the pattern and complete the table. Then write a direct variation equation for each table.

   a. y varies directly as x.

x	$\frac{1}{4}$	1	4	8
y		8		

b. y varies directly as x^3.

x	$\frac{1}{2}$	1	3	6
y		1		

2. The area, A, of a circle is given by the function $A = \pi r^2$, where r is the radius of the circle.

   a. Does the area vary directly as the radius? Explain.

   b. What is the constant of variation k?

3. Assume that y varies directly as the square of x, and that when $x = 2$, $y = 12$. Determine y when $x = 8$.

4. The distance, d, that you drive at a constant speed varies directly as the time, t, that you drive. If you can drive 150 miles in 3 hours, how far can you drive in 6 hours?

5. The number of meters, d, that a skydiver falls before her parachute opens varies directly as the square of the time, t, that she is in the air. A skydiver falls 20 meters in 2 seconds. How far will she fall in 2.5 seconds?

In Exercises 6–10, sketch a graph of the given power function. Verify your graphs using your graphing calculator.

6. $y = -3x^2$ **7.** $y = x^4 + 1$ **8.** $y = -2x^5$

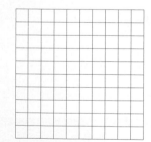   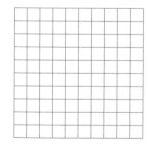   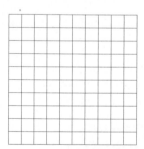

9. $f(x) = x^6$ **10.** $g(x) = 3x^3 - 3$

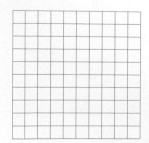   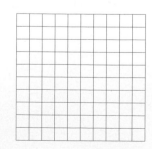

11. Determine the x-interval over which the function $f(x) = \frac{1}{2}x^4$ is increasing.

12. Does the function $g(x) = -\frac{1}{2}x^6$ have a maximum or a minimum point? Explain.

13. For $x > 1$, is the graph of $y = x^2$ rising faster or slower than the graph of $y = x^3$? Explain.

14. Is the graph of $y = \frac{3}{2}x^4$ wider or narrower than the graph of $y = x^4$?

15. How are the graphs of $y = -2x^3$ and $y = 2x^3 + 1$ different? How are the graphs similar?

16. a. For $x > 0$, is the graph of $y = x^2$ rising faster or slower than the graph of $y = 2^x$? Explain.

   b. For $x > 0$, is the graph of $y = x^5$ rising faster or slower than the graph of $y = 2^x$? Explain.

Activity 4.9

Volume of a Storage Tank

Objectives

1. Identify equations that define polynomial functions.

2. Determine the degree of a polynomial function.

3. Determine the intercepts of the graph of a polynomial function.

4. Identify the properties of the graphs of polynomial functions.

The volume, V (in cubic feet), of a partially cylindrical storage tank of liquid fertilizer is re resented by the formula

$$V = 4.2r^3 + 37.7r^2,$$

where r is the radius (in feet) of the cylindrical part of the tank.

1. Determine the volume of the tank if its radius is 3 feet.

Polynomial Functions

Recall from Chapter 2 that **polynomial functions** are defined by equations of the form

$$\text{output} = \text{polynomial expression involving the input.}$$

If x and y represent the input and output, respectively, then y must equal sums and difference of terms of the form ax^n, where n is a positive integer. The largest exponent on the input var able, n, is called the **degree** of the function. The following example gives several types c polynomial functions.

Example 1   *Examples of polynomial functions are listed in the following table.*

POLYNOMIAL FUNCTION	DEGREE OF THE POLYNOMIAL	NAME
$y = 3x - 2$	1	linear
$y = 2x^2 + 3x - 4$	2	quadratic
$y = 3x^3 - x - 4$	3	cubic
$y = 0.2x^4 - 2x^2 + 7x - 1$	4	quartic
$y = -2x^5 + 3x^4 + 2x - 6$	5	quintic

Note that the cubic function defined by $y = 3x^3 - x - 4$ can be written as $y = 3x^3 + 0x^2 - x - 4$.

Since the expression $4.2r^3 + 37.7r^2$ is a polynomial, the volume function defined by $V = 4.2r^3 + 37.7r^2$ is a polynomial function. Because the largest exponent on the input vari able r is 3, this function is a **third-degree polynomial function** or a **cubic function**.

Polynomial Functions of Degree 3 or Greater

You have already studied polynomial functions of degree 1 (linear) and degree 2 (quadratic). The remainder of this activity explores some of the properties and shapes of the graphs of polynomial functions having degree 3 or greater.

2. a. What is the domain of the cubic function defined by $y = 2x^3 - 8x^2 - 10x$?

b. Determine the y-intercept of the graph of the cubic function in part a.

c. Use your graphing calculator to sketch a graph. Use the window Xmin $= -5$, Xmax $= 8$, Ymin $= -50$, Ymax $= 10$, Yscl $= 5$. The graph should appear as follows.

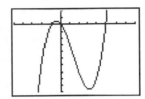

d. Write an equation to determine the x-intercepts of the graph.

e. Solve the equation in part d using a graphing approach. Use the zero option in the CALC menu of your graphing calculator.

Can the equation $2x^3 - 8x^2 - 10x = 0$ be solved using an algebraic approach? Yes! The solution process is demonstrated in the following example.

Example 2   *Solve the equation $2x^3 - 8x^2 - 10x = 0$ using an algebraic approach.*

SOLUTION

Step 1. The equation is already written in the form *polynomial expression* $= 0$.

Step 2. Completely factor the cubic expression on the left side of the equation.

$$2x^3 - 8x^2 - 10x = 0 \quad \text{Factor out the GCF.}$$
$$2x(x^2 - 4x - 5) = 0 \quad \text{Factor the trinomial.}$$
$$2x(x + 1)(x - 5) = 0$$

Step 3. Apply the Zero-Product Property, and set each factor equal to zero. Solve each resulting equation.

$$2x(x + 1)(x - 5) = 0$$

$2x = 0$	$x + 1 = 0$	$x - 5 = 0$
$x = 0$	$x = -1$	$x = 5$

3. Using an algebraic approach (factoring), determine the x-intercepts of each of the following polynomial functions. Verify your results using a graphing approach.

a. $y = 2x^3 + 5x^2 - 12x$

b. $f(x) = x^2(x^2 - 5) + 4$

c. $g(x) = 2x^5 - 18x^3$

4. a. Returning to the cubic function defined by $y = 2x^3 - 8x^2 - 10x$, the graph shows a high point (maximum point) in quadrant II and a low point (minimum point) in quadrant IV. Note that the maximum and minimum points occur at turning points of the graph. Use the maximum and minimum options in the CALC menu of your graphing calculator to approximate the coordinates of each of these points. Your screens should appear as follows.

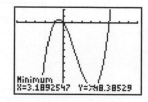 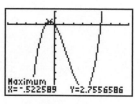

b. Using the results from part a, determine the interval along the x-axis where the function is

i. increasing.

ii. decreasing.

5. Using your graphing calculator, plot the following third-degree polynomials. Be careful of your choice of windows.

a. $f(x) = x^3$

b. $i(x) = 3x^3 - x - 4$

c. $g(x) = 0.2x^3 - 2x + 7$

d. $j(x) = -5x^3 + 1$

e. $h(x) = -0.6x^3 + 2x^2 - 1$

f. Use the graphs of cubic functions in parts a–e to write a few sentences comparing and contrasting the graph of the general quadratic equation, $y = ax^2 + bx + c, a \neq 0$, and the general cubic equation, $y = ax^3 + bx^2 + cx + d, a \neq 0$. Include comments on turning points and general trends, such as increasing and decreasing intervals.

6. Using your graphing calculator, plot the following fourth-degree polynomials. Be careful of your choice of windows.

a. $f(x) = x^4$

b. $i(x) = 3x^4 - x - 4$

c. $g(x) = 0.2x^4 - 2x^2 + 7x - 1$

d. $j(x) = -5x^4 + 1$

e. $h(x) = -0.6x^4 + 2x^3 - x + 1$

f. Use the graphs of quartic functions in parts a–e to write a few sentences comparing and contrasting the graph of the general quadratic equation, $y = ax^2 + bx + c, a \neq 0$, and the general quartic equation, $y = ax^4 + bx^3 + cx^2 + dx + e, a \neq 0$. Include comments on turning points and general trends, such as increasing and decreasing intervals.

SUMMARY: ACTIVITY 4.9

1. **Polynomial functions** are defined by equations of the form

$$\text{output} = \text{polynomial expression involving the input.}$$

2. The largest exponent, n, on the input variable is called the **degree** of the function.

3. Polynomial functions are continuous with the domain of all real numbers.

4. **Polynomial equations** of the form

$$\text{polynomial expression} = 0$$

can be solved graphically by locating the x-intercepts of the graph of the function defined by

$$y = \text{polynomial expression}$$

or algebraically using factoring (if possible) and the Zero-Product Property.

EXERCISES: ACTIVITY 4.9

In Exercises 1–3, determine the x-intercept(s) of the graphs of each polynomial function using an algebraic approach (factoring). Verify your answer using your graphing calculator.

1. $f(x) = x^3 + 3x^2 + 2x$

2. $g(x) = 2x^2(x^2 - 4)$

3. $h(x) = x^4 - 13x^2 + 36$

4. Determine the vertical intercept of each of the functions in Exercises 1–3.

5. Consider the function defined by $f(x) = x^4 - 6x^3 + 8x^2 + 1$.

a. Use your graphing calculator to sketch a graph of the function.

b. Determine the domain of the function, f.

c. Determine the range of f.

d. Use your graphing calculator to determine the x-intercepts of the function, f.

e. Use your graphing calculator to determine any maximum and/or minimum points.

6. Describe any symmetry of the graph of $y = x^4 - 4x^2 - 2$.

7. As the value of the input variable x increases without bound (say 10 to 100 to 1000 and so on), do the output values decrease without bound for the function $y = x^3 + 3x^2 - x - 4$? Use a graph of the function to help answer the question.

8. Is the graph of $y = -x^3 - x + 3$ increasing or decreasing?

9. Consider the following graph of $y = f(x)$.

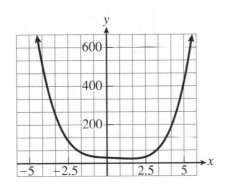

a. As x decreases without bound, the corresponding y-values _____.

b. Is the function f increasing or decreasing for $-2 < x < 2$?

c. How many turning points does the curve have?

10. Using your graphing calculator, sketch a graph of $y = (x - 2)^4$. What is the relationship between the minimum point of the graph and its horizontal intercept?

Activity 4.10

Recycling

Objective

1. Determine the regression equation of a polynomial function that best fits the data.

The state of Washington has been collecting recycling data since 1986. The following table shows the total tonnage (in thousands of tons) of recycled materials in the state for selected years since 1986.

Year	1986	1990	1994	1998	2002	2006
Total Recycled Materials (thousands of tons)	450	1889	2493	2165	2513	4021

1. Let x represent the number of years since 1986. Let y represent the total tonnage of recycled materials. Plot these points on your graphing calculator. The scatterplot should resemble the following.

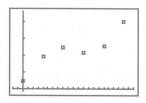

2. a. Using your graphing calculator, determine the regression equations of the first-, second-, and third-degree curves of best fit.

b. Use your calculator to fit each of your models from part a to your scatterplot. Your graphs should resemble the following.

Linear model

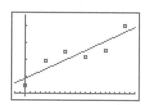

Quadratic model

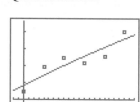

Cubic model

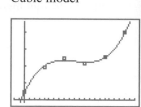

c. Which of these curves best represents the data? Explain.

3. a. What is the practical domain of the introductory recycling situation?

b. Would you consider this a discrete situation (consisting of separate, isolated points) or a continuous situation? Explain.

c. What is the practical range of this problem?

4. a. In 2000 the state of Washington recycled 2463 thousand tons of material. Is this result consistent with the curve you chose as the best model for the given data? Explain.

b. In 2007 the state of Washington recycled 4001 thousand tons of material. Is this result consistent with the curve you chose as the best to describe the given data? Explain.

c. Include the two data points from parts a and b on the lists in your graphing calculator. Then recalculate your curve of best fit. Describe the changes.

d. Using the function generated in part c, predict the total tonnage of material the state of Washington will recycle in the year 2020. Is this a realistic prediction? Explain.

SUMMARY: ACTIVITY 4.10

1. The graphing calculator can be used to model data with cubic and quartic polynomial functions as well as linear and quadratic polynomial functions.

EXERCISES: ACTIVITY 4.10

1. According to the Research and Innovative Technology Administration, the following table summarizes the average number of gallons of fuel consumed per vehicle in the United States from 1960 to 2007.

Year	1960	1970	1980	1990	2000	2003	2006	2007
Average gal of fuel consumed per vehicle, g	784	830	712	677	720	718	697	692

a. Sketch a scatterplot of the data. Let t represent the number of years since 1960. Does the data appear to be linear? Explain.

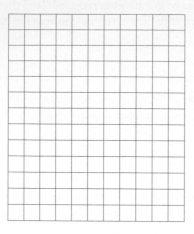

b. Use your graphing calculator to create a STATPLOT; then determine and plot the quadratic, cubic, and quartic regression equations for this data.

2. a. Using the results from Exercise 1b, select the equation that best models the average number of gallons of fuel consumed per vehicle, g, as a function of the number of years since 1960, t.

b. What is the practical domain of this function?

c. What is the practical range of this function? Explain.

d. Use the equation in part a to estimate the average number of gallons of fuel consumed per vehicle in 1955, in 1985, and in 2015. In which of these estimates do you have most confidence? Explain.

3. The following table gives the annual consumption of cigarettes (in billions) in the United States for specific years.

WHERE THERE'S SMOKE								
Year	2000	2001	2002	2003	2004	2005	2006	2007
Cigarette Consumption in Billions	430.0	425.0	415.0	400.0	388.0	376.0	372.0	360.0

a. Let $t = 0$ correspond to the year 2000. Sketch a scatterplot of the data.

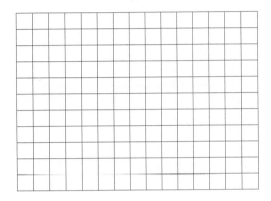

b. Determine a linear model (equation) for the data.

c. Determine a quadratic model (equation) for the data.

d. Which model best represents the data? Explain.

e. Use each model to predict the consumption of cigarettes in the year 2015.

f. How confident are you in your predictions? Explain.

Cluster 2   What Have I Learned?

1. In a hurricane, the wind pressure varies directly as the square of the wind velocity (speed). If the wind speed doubles in value, what change in the wind pressure do you experience?

2. Is the graph of $y = 3x^4$ narrower or wider than the graph of $y = x^2$? Explain.

3. The graph of any cubic (third-degree polynomial) function must have one of the four following general shapes.

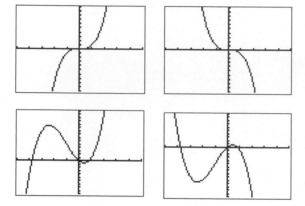

   a. Complete the following table, which gives the maximum number of turning points for a given family of polynomial functions.

DEGREE OF POLYNOMIAL FUNCTION	MAXIMUM NUMBER OF TURNING POINTS
1 (linear)	
2 (quadratic)	
3 (cubic)	
4 (quartic)	

   b. If n represents the degree, then write an expression that represents the maximum number of turning points.

4. a. Sketch a graph of $y = x^4 - 4x^2$. Describe any symmetry that you observe.

   b. Do all graphs of quartic (fourth-degree) functions have symmetry? Explain.

5. a. Does the graph of any cubic function have a horizontal intercept? Can the graph have more than one horizontal intercept? Explain.

b. Does the graph of any cubic function have at least one vertical intercept? Explain.

6. Given the following graph, determine whether any of the three functions in parts a–c fit the curve. Explain.

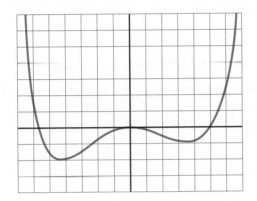

a. $f(x) = ax^4 + bx^3 + cx^2 + dx + e,$ $a > 0, e = 0$

b. $g(x) = ax^3 + bx^2 + cx + d,$ $a > 0, d = 0$

c. $h(x) = ax^4 + bx^3 + cx^2 + dx + e,$ $a < 0, e = 0$

Cluster 2   How Can I Practice?

1. y varies directly as x^2. When $x = 3$, $y = 45$. Determine y when $x = 6$.

2. Have you ever noticed that, during a thunderstorm, you see lightning before you hear the thunder? This is true because light travels faster than sound. If d represents the distance (in feet) of the lightning from the observer, then d varies directly as the time, t (in seconds), it takes to hear the thunder. The relationship is modeled by

$$d = 1080t.$$

   a. As the time t doubles (say from 3 to 6), the corresponding d-values _____.

   b. What is the value of k, the constant of variation, in this situation? What significance does k have in this problem?

3. The velocity, v, of a falling object varies directly as the time, t, of the fall. After 3 seconds, the velocity of the object is 60 feet per second. What will be its velocity after 4 seconds?

4. Sketch a graph of each of the following pairs of functions. Describe the differences and the similarities in the graphs.

   a. $y = 3x^2, y = 3x^2 + 5$

   b. $y = 5x^4, y = -5x^4$

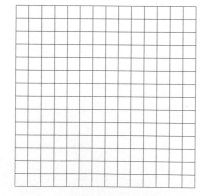

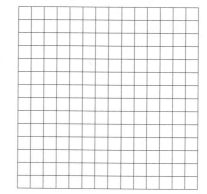

c. $y = 2x^3 + 1, y = 2x^3 - 4$

d. $y = 4x^2, y = 4(x - 1)^2$

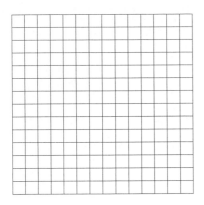

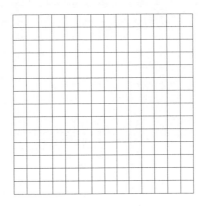

5. Using your graphing calculator, graph each of the following polynomial functions. For each graph,

   i. determine the vertical intercepts.

   ii. approximate the horizontal intercepts (if they exist).

   iii. determine the coordinates of any turning points.

   a. $y = x^3 + 2x^2 - 8x$

   b. $y = -x^4 + 2x + 3$

6. The average miles per gallon (mpg) for U.S. cars has increased steadily over the past several decades. The following table gives the average mpg for selected years.

AVERAGE MILES PER GALLON FOR PASSENGER CARS IN THE UNITED STATES FROM 1960 TO 2006	
YEAR	AVERAGE mpg
1960	14.3
1970	13.5
1980	16.0
1990	20.3
1995	21.1
2000	21.9
2002	22.0
2004	22.5
2006	22.4

a. Draw a scatterplot of the data points. Let x = number of years since 1960.

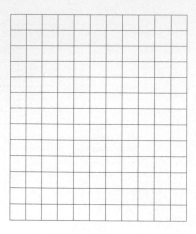

b. Determine the equations of the cubic and quartic functions that best fits the data.

c. Use the regression equations to predict the average mpg in the year 2015.

d. In which model, if either, do you have the most confidence? Explain.

7. Your bathtub is partially filled. You finish filling the tub and settle in for a nice hot bath. The drain plug is broken, and water is slowly leaking out. The amount of water (in gallons) in the bathtub is given by

$$W(t) = 10 + 7t^2 - t^3 \quad (t \geq 0),$$

where t is time in minutes and $W(t)$ represents the amount of water in gallons.

a. With the aid of your graphing calculator, sketch the graph of $W(t)$. Don't go beyond $t = 10$. Why?

b. How much water was in the bathtub at the time you began to fill it?

c. Determine the maximum amount of water in the tub from the graph. Explain your result.

d. Use the zero features of your graphing calculator to determine when the tub will be completely empty, to the nearest 0.01 minute.

The bracketed numbers following each concept indicate the activity in which the concept is discussed.

CONCEPT/SKILL	DESCRIPTION	EXAMPLE
Quadratic function [4.1]	The quadratic function with the input variable x has the standard form $y = ax^2 + bx + c$, where a, b, and c represent real numbers and $a \neq 0$.	$y = 2x^2 - 3x - 2$
Graph of a quadratic function (a parabola) [4.1]	For the quadratic function defined by $f(x) = ax^2 + bx + c$, if $a > 0$ the parabola opens upward; if $a < 0$ the parabola opens downward.	The graph of $y = 2x^2 - 3x - 2$ is a parabola that opens upward.
Vertical intercept of the graph of a quadratic function [4.1]	The constant term c of a quadratic function $f(x) = ax^2 + bx + c$ always indicates the vertical intercept of the parabola. The vertical intercept of any quadratic function is $(0, c)$.	The vertical intercept of the graph of $y = 2x^2 - 3x - 2$ is $(0, -2)$.
Axis of symmetry [4.2]	The axis of symmetry of a parabola is a vertical line that separates the parabola into two mirror images. The equation of the vertical axis of symmetry is given by $x = \frac{-b}{2a}$.	The axis of symmetry of the parabola defined by $y = 2x^2 - 3x - 2$ is $x = \frac{3}{4}$.
Vertex (turning point) [4.2]	The vertex of a parabola defined by $f(x) = ax^2 + bx + c$ is the point where the graph changes direction. It is given by $\left(\frac{-b}{2a}, f\left(-\frac{b}{2a}\right)\right)$.	The vertex of the parabola defined by $y = 2x^2 - 3x - 2$ is $\left(\frac{3}{4}, -\frac{25}{8}\right)$.
x-intercept(s) [4.2]	An x-intercept is the point or points (if any) where the graph crosses the x-axis (that is, where its y-coordinate is zero).	The x-intercepts of the parabola defined by $y = 2x^2 - 3x - 2$ are $(2, 0)$ and $(-0.5, 0)$.
Domain of quadratic functions [4.2]	The domain of any quadratic function is all real numbers.	The domain of $y = 2x^2 - 3x - 2$ is all real numbers.
Range of quadratic functions [4.2]	If the parabola opens upward, the range is [output value of turning point, ∞). If the parabola opens downward, the range is $(-\infty$, output value of turning point].	The range of the parabola defined by $y = 2x^2 - 3x - 2$ is $y \geq -\frac{25}{8}$.
Solving $f(x) = c$ graphically [4.3]	Graph $y = f(x)$, graph $y = c$, and determine the x-values of the points of intersection. Or graph $y = f(x) - c$ and determine the x-intercepts.	Problem 7, Activity 4.3.

CONCEPT/SKILL	DESCRIPTION	EXAMPLE
Solving $f(x) > c$ graphically [4.3]	Graph $y = f(x)$, graph $y = c$, and determine all x-values for which the graph of f is above the graph of $y = c$. Or graph $y = f(x) - c$ and determine all x-values for which the graph of $f(x) - c$ is above the x-axis.	Example 2, Activity 4.3
Solving $f(x) < c$ graphically [4.3]	Graph $y = f(x)$, graph $y = c$, and determine all x-values for which the graph of f is below the graph of $y = c$. Or graph $y = f(x) - c$ and determine all x-values for which the graph of $f(x) - c$ is below the x-axis.	Example 2, Activity 4.3
Greatest common factor (or GCF) [4.4]	The GCF is the largest factor common to all terms in an expression.	The GCF of $3x^4 - 6x^3 + 18x^2$ is $3x^2$.
Zero-Product Property [4.4]	If a and b are any numbers and $a \cdot b = 0$, then either a or b, or both, must be equal to zero.	Example 1, Activity 4.4
Factoring trinomials by trial and error [4.4]	To factor trinomials by trial and error, 1. remove the GCF. 2. try combinations of factors for the first and last terms in two binomials. 3. check the outer and inner products to match middle term of the original trinomial. 4. if the check fails, repeat steps 2 and 3.	Example 5, Activity 4.4
Solving quadratic equations by factoring [4.4]	To solve a quadratic equation by factoring, 1. use the addition principle to remove all terms from one side of the equation. This results in a quadratic polynomial being set equal to zero. 2. combine like terms, and then factor the nonzero side of the equation. 3. use the Zero-Product Property to set each factor containing a variable equal to zero, and then solve the equations. 4. check your solutions in the original equation.	Example 6, Activity 4.4
Quadratic formula [4.5]	$$x = \dfrac{-b \pm \sqrt{b^2 - 4ac}}{2a}$$	Example 1, Activity 4.5

CONCEPT/SKILL	DESCRIPTION	EXAMPLE
Solving a quadratic equation of the form $ax^2 + bx + c = 0, a \neq 0$, using the quadratic formula [4.5]	To solve a quadratic equation of the form $ax^2 + bx + c = 0, a \neq 0$, using the quadratic formula $$x = \frac{-b \pm \sqrt{b^2 - 4ac}}{2a},$$ 1. rewrite the quadratic equation with one side zero. 2. identify the coefficients a and b and the constant term c. 3. substitute these values into the formula, and simplify. 4. check your solutions.	Example 1, Activity 4.5
Imaginary unit [4.7]	The imaginary unit is the number $\sqrt{-1}$. The notation for the imaginary unit is i.	
Complex number [4.7]	Any number that can be written in the form $a + bi$, where a and b are real numbers and i is the imaginary unit, is called a complex number.	$2 + 6i$
Discriminant [4.7]	In the quadratic formula $$x = \frac{-b \pm \sqrt{b^2 - 4ac}}{2a}$$ the expression $b^2 - 4ac$ is called the discriminant. Its value determines the number and type of solutions of a quadratic equation $ax^2 + bx + c = 0$.	For the quadratic equation $y = 2x^2 - 7x - 4$, the discriminant is $49 - 4(2)(-4) = 81$. The equation has two real solutions.
Direct variation function [4.8]	The equation $y = kx^n$, where $k \neq 0$ and n is a positive integer, defines a direct variation function in which y varies directly as x^n.	$y = 4x^3$
Constant of variation [4.8]	In the direct variation equation $y = kx^n$, the constant, k, is called the constant of variation.	In $y = 4x^3$, the constant of variation is 4.
Power functions [4.8]	The direct variation function having an equation of the form $y = kx^n$, where n is a positive integer, is also called a power function.	$y = 4x^3$ is a third-degree power function.

CONCEPT/SKILL	DESCRIPTION	EXAMPLE
Polynomial functions [4.9]	Polynomial functions are defined by equations of the form $$\text{output} = \frac{\text{polynomial expression}}{\text{involving the input.}}$$	$y = 5x^4 + 7x^2 - 3x + 1$
Degree of a polynomial [4.9]	The largest exponent on the input variable n is called the degree of the function.	$y = 5x^4 + 7x^2 - 3x + 1$ is a fourth-degree polynomial function.

In Exercises 1–8, determine the following characteristics of each quadratic function by inspecting its equation.

a. *the direction in which the graph opens*

b. *the equation of the axis of symmetry*

c. *the vertex*

d. *the y-intercept*

1. $f(x) = x^2 + 2$

2. $F(x) = -3x^2$

3. $g(x) = -3x^2 + 4$

4. $f(x) = 2x^2 - x$

5. $h(x) = x^2 + 5x + 6$

6. $F(x) = x^2 - 3x + 4$

7. $f(x) = x^2 - 2x + 1$

8. $g(x) = -x^2 + 5x - 6$

In Exercises 9–15, sketch the graph of each quadratic function using your graphing calculator. Then determine each of the following using the graph.

a. *the coordinates of the x-intercepts, if they exist*

b. *the domain and the range of the function*

c. *the horizontal interval in which the function is increasing*

d. *the horizontal interval in which the function is decreasing*

9. $g(x) = x^2 + 4x + 3$

10. $f(x) = x^2 + 2x - 3$

Answers to all Gateway exercises are included in the Selected Answers appendix.

11. $F(x) = x^2 - 3x + 1$

12. $h(x) = 2x^2 + 8x + 5$

13. $F(x) = -2x^2 + 8$

14. $f(x) = -3x^2 + 4x - 1$

15. $g(x) = 4x^2 + 5$

In Exercises 16–19, solve the quadratic equation numerically (using tables). Verify your solutions graphically.

16. $x^2 + 4x + 4 = 0$

17. $x^2 - 5x + 6 = 0$

18. $3x^2 = 18x + 10$

19. $-x^2 = 3x - 10$

In Exercises 20–21, solve the equation using two different approaches. Round your answer to the nearest tenth when necessary.

20. $8x^2 = 10$

21. $5x^2 + 25x = -5$

22. Completely factor the following polynomials.

a. $9a^5 - 27a^2$

b. $24x^3 - 6x^2$

c. $4x^3 - 16x^2 - 20x$

d. $5x^2 - 16x + 6$

e. $x^2 - 5x - 24$

f. $t^2 + 10t + 25$

In Exercises 23–27, solve each equation by factoring. Verify your answer graphically or by substitution of the solutions in the equations.

23. $x^2 - 9 = 0$

24. $-x^2 + 36 = 0$

25. $x^2 - 7x + 12 = 0$

26. $x^2 - 6x = 27$

27. $x^2 = -x$

In Exercises 28–32, write each of the equations in the form $ax^2 + bx + c = 0$. Then identify a, b, and c, and solve the equation using the quadratic formula. Verify your solutions by substitution.

28. $x^2 + 5x + 3 = 0$

29. $2x^2 - x = -3$

30. $x^2 = 81$

31. $3x^2 + 5x = 12$

32. $2x^2 = 3x + 5$

33. For the quadratic function $f(x) = 2x^2 - 8x + 3$, determine the x-intercepts of the graph, if they exist. First, approximate the intercepts using your graphing calculator. Second, solve the equation using the quadratic formula. Approximate your answers to the nearest hundredth.

34. Write each of the following using the imaginary unit, i.

   a. $\sqrt{-49}$

   b. $\sqrt{-48}$

   c. $\sqrt{-9}$

   d. $\sqrt{-23}$

   e. $\sqrt{-\frac{5}{9}}$

   f. $\sqrt{-\frac{17}{16}}$

35. Perform the following operations with complex numbers. Use your graphing calculator to check your results.

   a. $(2 + 7i) + (-7 + 10i)$

   b. $(4 - 9i) - (-1 + 7i)$

   c. $4i(3 - 8i)$

   d. $(4 - i)(6 + 3i)$

In Exercises 36–39, determine the type of solution to each of the equations considering only its discriminant.

36. $2x^2 - 3x + 1 = 0$

37. $4x^2 + 16x = 0$

38. $x^2 - 9 = 0$

39. $3x^2 + 2x + 2 = 0$

40. Solve the equation in Problem 39 in the complex number system using the quadratic formula. Verify your solution graphically.

41. Solve the following inequalities using a graphing approach.

a. $x^2 - x - 6 < 0$ **b.** $x^2 - x - 6 > 0$

42. a. Suppose y varies directly as x. When $x = 3, y = 12$. Determine y when $x = 5$.

b. Suppose y varies directly as x^2. When $x = 4, y = 8$. Determine y when $x = 8$.

c. Suppose y varies directly as x^3. When $x = 1, y = 5$. Determine y when $x = 2$.

In Exercises 43–47, graph the function using your graphing calculator. Then answer the following questions, referring to the graphing calculator.

a. *Determine the x-intercepts of the function, if it has any.*

b. *Determine the domain and range of the function.*

c. *Determine the values of x for which the function is increasing and the values of x for which the function is decreasing.*

43. $y = x^3 - 8$ **44.** $y = -2x^3 - 2$

45. $y = x^4 - 8$ **46.** $y = x^4 + 2x$

47. $y = x^4 + 5$

48. The height, h (in feet), of a golf ball is a function of the time, t (in seconds), it has been in flight. A golfer strikes a golf ball with an initial upward velocity of 80 feet per second. The flight path of the ball is a parabola. The approximate height of the ball above the ground is modeled by

$$h(t) = -16t^2 + 80t.$$

a. Sketch a graph of the function. What is the practical domain in this situation?

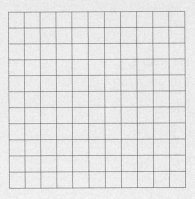

b. Determine the vertex of the parabola. What is the practical meaning of this point?

c. What is the vertical intercept, and what is its practical meaning in this situation?

d. Determine the horizontal intercepts. What is the significance of these intercepts?

e. What assumption are you making in this situation about the elevation of the spot where the ball is struck and the point where the ball lands?

49. To use the regression feature of your calculator to determine the equation of a parabola, you need three distinct points. The stream of water flowing out of a water fountain is in the shape of a parabola. Suppose you let the origin of a coordinate system correspond to the point where the water begins to flow out of the nozzle (see figure).

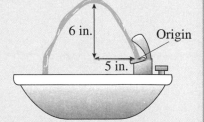

The maximum height of the water stream occurs approximately 5 inches measured horizontally from the nozzle. The maximum height of the stream of water is measured to be approximately 6 inches.

a. What is the vertex of the parabola?

b. You already have two points that lie on the parabola. What are they? Use symmetry to obtain a third point.

c. Using these three points and the regression feature of your graphing calculator, determine the equation of the stream of water.

50. A fastball is hit straight up over home plate. The ball's height, h (in feet), from the ground is modeled by

$$h(t) = -16t^2 + 80t + 5,$$

where t is measured in seconds.

a. What is the maximum height of the ball above the ground?

b. How long will it take for the ball to reach the ground?

51. Safe automobile spacing, S (in feet), is modeled by

$$S(v) = 0.03125v^2 + v + 18,$$

where v is average velocity in feet per second.

a. Suppose a car is traveling at 44 feet per second. To be safe, how far should it be from the car in front of it?

b. If the car is following 50 feet behind a van, what is a safe speed for the car to be traveling? How fast is this in miles per hour (60 miles per hour $\approx$ 88 feet per second)?

Rational and Radical Functions

Cluster 1   Rational Functions

Activity 5.1

peed Limits

Objectives

1. Determine the domain and range of a function defined by $y = \dfrac{k}{x}$, where k is a nonzero real number.

2. Determine the vertical and horizontal asymptotes of the graph of $y = \dfrac{k}{x}$.

3. Sketch a graph of functions of the form $y = \dfrac{k}{x}$.

4. Determine the properties of graphs having equation $y = \dfrac{k}{x}$.

The speed limit on the New York State Thruway is 65 miles per hour.

1. If you maintain an average speed of 65 miles per hour, how long will it take you to make a 200-mile trip on the Thruway? Recall that $distance = rate \cdot time$, $time = \dfrac{distance}{rate}$.

2. Complete the following table, in which the input variable r represents the average speed in miles per hour and the output variable t represents the time in hours to complete a 200-mile trip.

r (mph)	20	30	40	50	60	70	80
t (hr.)							

3. Write an equation that defines travel time, t, as a function of the average speed, r.

4. As the average speed, r, increases, what happens to the travel time, t? What does this mean in practical terms?

5. During a winter storm, a combination of drifting snow and icy conditions reduces your average speed to almost a standstill. Complete the following table for a 200-mile trip on the New York State Thruway.

r (mph)	10	7	5	3	2	1
t (hr.)						

6. As the average speed r gets closer to zero, what happens to the travel time t? Explain what this means in practical terms.

7. Can zero be used as an input value? Explain.

8. a. What is the practical domain of the function given in Problem 3?

b. Sketch a graph of this function using the table values in Problem 2 and 5.

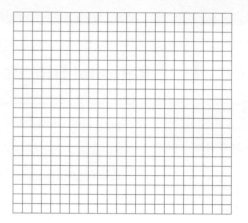

9. a. What are the horizontal and vertical intercepts of the graph?

b. Describe the relationship between the horizontal axis $(t = 0)$ and the graph of the function as the values of r get very large.

> In this situation, the horizontal line $t = 0$ is called a **horizontal asymptote**. Recall that a horizontal asymptote is a horizontal line that a graph approaches as the input values of r get very large in a positive direction (or get very large in a negative direction).

c. Describe the relationship between the vertical axis $(r = 0)$ and the graph of the function as the values of r get close to zero.

In this situation, the vertical line $r = 0$ is called a **vertical asymptote**.

> In general, a **vertical asymptote** is a vertical line, $x = a$, that the graph of a function defined by $y = f(x)$ becomes very close to but never touches. As the x-values get closer and closer to $x = a$, the y-values get larger and larger in magnitude. That is, the y-values become very large positive or very large negative values.

10. Using the graph in Problem 8, approximate your average speed if the 200-mile trip takes 5 hours.

Functions Defined by $y = \frac{k}{x}$, Where k is a Nonzero Constant

The function rule $t = \frac{200}{r}$ gives the relationship between the average speed, r, the time, t, and the given value for distance (200). This function belongs to a family of functions having a general equation of the form $f(x) = \frac{k}{x}$, where k represents some nonzero constant.

Example 1 *Examples of this type of function are $f(x) = \dfrac{1}{x}$, $g(x) = \dfrac{5}{x}$, and $h(x) = \dfrac{10}{x}$.*

11. a. What is the domain of functions f, g, and h defined in Example 1?

b. Complete the following table.

x	−20	−10	−5	−1	−0.5	−0.1	0	0.1	0.5	1	5	10	20
f(x)													
g(x)													
h(x)													

c. Sketch graphs of f, g, and h on the same coordinate system. Verify using your graphing calculator with window Xmin = −5, Xmax = 5, Ymin = −25, and Ymax = 25.

12. Using the table and graphs in Problem 11, answer each of the following questions.

    a. What happens to the y-values as the x-values increase in magnitude infinitely (without bound) in both the positive and negative directions?

    b. What is the horizontal asymptote for each graph?

    c. What happens to the y-values as the positive x-values get closer to zero?

    d. What happens to the y-values as the negative x-values get closer to zero?

    e. What is the vertical asymptote for each graph?

13. **a.** Do the graphs of f, g, or h in Problem 11 have x- or y-intercepts?

    b. Do the functions f, g, and h have a maximum function value or a minimum function value? Explain.

    c. Are any of the functions in Problem 11 continuous? Explain.

14. **a.** Complete the following table, where $Q(x) = \dfrac{-1}{x}$.

x	− 10	− 5	− 1	− 0.5	− 0.1	0	0.1	0.5	1	5	10
Q(x)											

    b. Sketch a graph of Q on the first grid below. Verify using your graphing calculator with window Xmin = − 4, Xmax = 4, Ymin = − 4, and Ymax = 4.

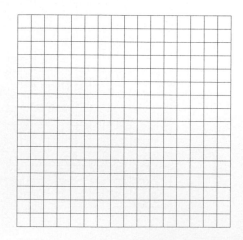
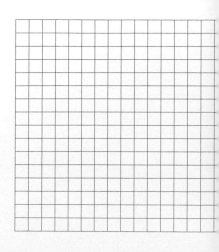

c. Sketch the graph of $f(x) = \dfrac{1}{x}$ on the second grid in part b.

d. Describe how the graph of $Q(x) = \dfrac{-1}{x}$ can be obtained from the graph of $f(x) = \dfrac{1}{x}$.

SUMMARY: ACTIVITY 5.1

Functions defined by $f(x) = \dfrac{k}{x}$, where k represents some nonzero constant, have the following properties.

1. The domain and the range consist of all real numbers except zero.

2. If $k > 0$, the graph of f has the following general shape.

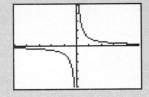

3. If $k < 0$, the graph of f has the following general shape.

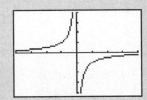

4. The vertical line $x = 0$ is the vertical asymptote.

5. The horizontal line $y = 0$ is the horizontal asymptote.

6. The graph does not intersect either axis (there are no intercepts).

7. There is no maximum or minimum y-value.

8. The function is not continuous at $x = 0$.

EXERCISES: ACTIVITY 5.1

1. You are a member of a group of distance runners who compete in races ranging in length from 5 to 25 kilometers. In these races, each runner who finishes is told his or her time. Given the time and the length of the race, you can calculate your average running speed.

   a. If you finish a 20-kilometer race in 1 hour 15 minutes, what is your average speed?

   b. Complete the following table for a 20-kilometer race.

t (hr.)	1.00	1.25	1.50	1.75	2.00	2.25	2.50
s (km/hr.)							

   c. Write an equation that expresses the average speed, s, as a function of time t in a 20-kilometer race.

   d.   i. What is the domain of the function?

      ii. What is the practical domain?

      iii. Sketch a graph of the function using the practical domain.

   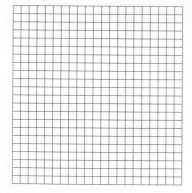

   e. As the running time, t, gets longer, what happens to the average speed, s?

   f. As the running time, t, is reduced (gets closer to zero), what happens to the average speed, s?

2. a. Sketch the graphs of the following pair of functions on separate coordinate systems. Use labels or colors to differentiate the graphs. Verify your sketches using your graphing calculator.

$$f(x) = \frac{5}{x}, \qquad g(x) = \frac{-5}{x}$$

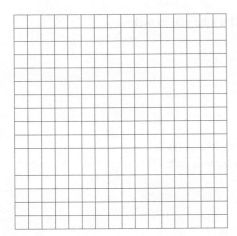 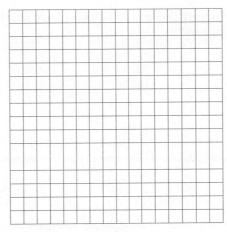

b. Describe how the graph of g can be obtained from the graph of function f.

3. A commercial refrigerator has an initial cost, C, and a scrap value, V. If the life of the refrigerator is N years, then the amount, D, that can be depreciated each year is modeled by the formula

$$D = \frac{C - V}{N}.$$

a. If the initial cost is $1400 and the scrap value is $200, write an equation for D as a function of N.

b. Complete the following table using the equation in part a.

N	1	2	3	6	12	24
D						

c. Sketch a graph of the function.

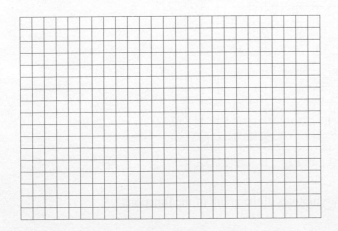

d. If the refrigerator is well constructed, it should have a long, useful life. Will an increase in the useful life of the refrigerator increase or decrease the amount, D, that can be depreciated each year? Explain.

4. The speed limit on Route 66 in Arizona is 75 miles per hour.

   a. If you maintain an average speed of 75 miles per hour, how long will it take you to make a 350-mile trip on Route 66?

   b. Write an equation that defines t as a function of r, in which r represents the average speed in miles per hour and t represents the time in hours to complete the 350-mile trip.

   c. Complete the following table for the equation you determined in part b.

Input, r, mph	25	35	45	55	65	75	85
Output, $t = f(r)$, hr.							

   d. As your average speed increases, what happens to the time it takes to complete the trip?

   e. As your average speed for the trip gets closer to zero, what happens to the time it takes to complete the 350-mile trip?

   f. What is the practical domain?

   g. Sketch a graph.

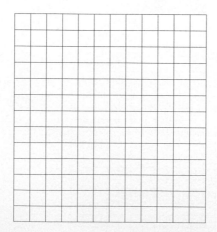

   h. What are the vertical and horizontal asymptotes of the graph of the function?

Loudness of a Sound

Objectives

- Graph a function defined by an equation of the form $y = \dfrac{k}{x^n}$, where n is any positive integer and k is a nonzero real number, $x \neq 0$.

- Describe the properties of graphs having equation $y = \dfrac{k}{x^n}$, $x \neq 0$.

- Determine the constant of proportionality (also called the constant of variation).

The loudness (or intensity) of any sound is a function of the listener's distance from the source of the sound. In general, the relationship between the intensity, I, and the distance, d, can be modeled by an equation of the form

$$I = \frac{k}{d^2},$$

where I is measured in decibels, d is measured in feet, and k is a constant determined by the source of the sound and the nature of the surroundings.

1. The intensity, I, of a human voice can be given by the formula $I = \dfrac{1500}{d^2}$. Complete the following table.

d (ft.)	0.1	0.5	1	2	5	10	20	30
I (dB)								

2. a. What is the practical domain of the function?

b. Sketch a graph that shows the relationship between intensity of sound and distance from the source of the sound. Use the table in Problem 1 to help determine an appropriate scale.

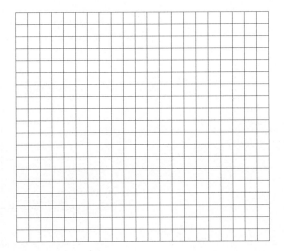

3. As you move closer to the person speaking, what happens to the intensity of the sound?

4. As you move away from the person speaking, what happens to the intensity of the sound?

Functions Defined by $y = \frac{k}{x^2}$, Where k is a Nonzero Constant

The function defined by $I = \frac{1500}{d^2}$ belongs to a family of functions having an equation of the form $y = \frac{k}{x^2}$, where k represents some nonzero constant.

Example 1 *Examples of functions defined by equations of the form $y = \frac{k}{x^2}$ are*
$$f(x) = \frac{1}{x^2} \text{ and } g(x) = \frac{10}{x^2}.$$

5. a. What is the domain of functions f and g defined in Example 1?

b. Complete the following table.

x	−20	−10	−5	−1	−0.5	−0.1	0	0.1	0.5	1	5	10	20
f(x)													
g(x)													

c. Sketch the graphs of f and g on the same coordinate system. Verify your sketch using your graphing calculator.

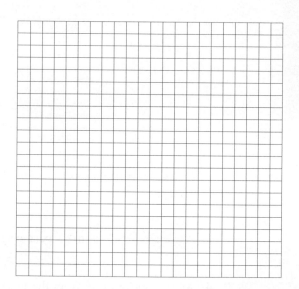

d. Explain why no part of each graph appears below the x-axis.

e. What happens to the y-values as the x-values increase infinitely in the positive direction or decrease infinitely in the negative direction?

f. What is the horizontal asymptote for each graph?

g. What happens to the *y*-values as the positive *x*-values get closer to zero?

h. What happens to the *y*-values as the negative *x*-values get closer to zero?

i. What is the vertical asymptote for each graph?

j. Do the functions have a maximum function value or a minimum function value?

k. For a given *x*-value, how is the output of *g* related to the output of *f*? Describe this relationship graphically.

l. Are the functions *f* and *g* continuous?

6. Describe how the graphs of $y = \dfrac{1}{x}$ and $y = \dfrac{1}{x^2}$ are similar and how they are different.

7. a. Complete the following table for $g(x) = \dfrac{10}{x^2}$ and $h(x) = \dfrac{-10}{x^2}$.

x	−20	−10	−5	−1	−0.5	−0.1	0	0.1	0.5	1	5	10	20
$g(x) = \frac{10}{x^2}$													
$h(x) = \frac{-10}{x^2}$													

b. Sketch graphs of functions *g* and *h* on separate coordinate systems. Verify your sketches using your graphing calculator.

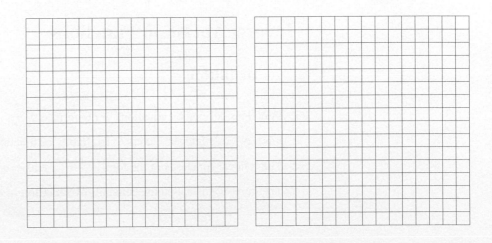

    c. Describe how to obtain the graph of function h using the graph of function g.

8. Sketch a graph of $h(x) = \dfrac{1}{x^3}$. Is the graph similar to the graph of $f(x) = \dfrac{1}{x}$ or $g(x) = \dfrac{1}{x^2}$?

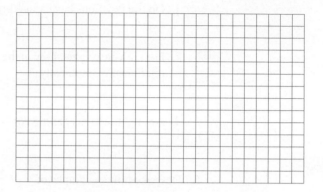

9. Sketch a graph $R(x) = \dfrac{1}{x^4}$. Is the graph similar to the graph of $f(x) = \dfrac{1}{x}$ or $g(x) = \dfrac{1}{x^2}$?

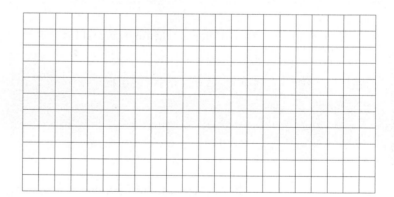

Inverse Variation Functions

Definition

Functions defined by equations of the form $y = \dfrac{k}{x^n}$, where k is a nonzero constant and n is a positive integer, belong to the family of functions called **rational functions**.

The rational functions of the form $y = \dfrac{k}{x^n}$ are also called **inverse variation functions**.

The number k is called the **constant of variation** or **constant of proportionality**.

Example 2

*For the inverse variation function given by $y = \dfrac{4}{x}$, y varies inversely as x, or y is inversely proportional to x. **4** is the constant of variation. The following table demonstrates that as x doubles in value, the corresponding y-values are reduced by half.*

x	2	4	8	16
$y = \dfrac{4}{x}$	2	1	$\dfrac{1}{2}$	$\dfrac{1}{4}$

10. For the function defined by $I = \dfrac{1500}{d^2}$ (see Problem 1), answer the following questions.

a. I varies inversely as what quantity?

b. What is the constant of proportionality?

c. If d is doubled, what is the effect on I?

In many applications, the constant of proportionality, k, is unknown. The procedure for determining k for an inverse variation is demonstrated in Example 3.

Example 3

y varies inversely as the square of x and y = 5 when x = 2. Determine the equation for y.

SOLUTION

Step 1. Write the equation relating x to y using k for the constant of proportionality.

$$y = \frac{k}{x^2}$$

Step 2. Substitute the given values of x and y ($y = 5, x = 2$) into the equation from step 1.

$$5 = \frac{k}{2^2} \quad \text{or} \quad 5 = \frac{k}{4}$$

Step 3. Solve the equation in step 2 for k.

$$5 \cdot 4 = \frac{k}{4} \cdot 4$$
$$20 = k$$

Step 4. Rewrite the equation from step 1 by substituting the value of k determined in step 3.

$$y = \frac{20}{x^2}$$

11. In this activity, the relationship between the intensity, I, of a human voice and the distance, d, from the individual was given by $I = \frac{1500}{d^2}$, where I is measured in decibels, d is measured in feet, and 1500 is the constant of proportionality. The constant of proportionality depends on the source of the sound and the nature of the surroundings. If the source of the sound changes, the value of the constant of proportionality will also change.

a. The intensity of the sound made by a heavy truck 60 feet away is 90 decibels. Determine the constant of proportionality.

b. Write a formula for the intensity, I, of the sound made by the truck when it is d feet away.

c. Use the formula from part b to determine the intensity of the sound made by the truck when it is 100 feet away.

SUMMARY: ACTIVITY 5.2

- Functions defined by equations of the form $f(x) = \dfrac{k}{x^n}$, where k is a nonzero real number, have the following properties.

1. The domain consists of all real numbers except zero.

2. The graph of f has the following general shape.

a. Where $k > 0$, and n is an even integer

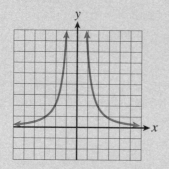

b. Where $k > 0$, and n is an odd integer

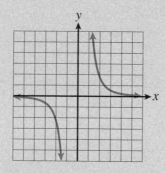

c. Where $k < 0$, and n is an even integer

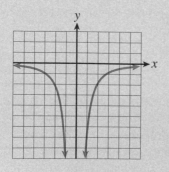

d. Where $k < 0$, and n is an odd integer

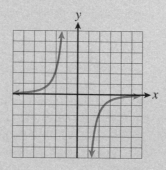

3. The vertical asymptote is the vertical line $x = 0$.

4. The horizontal asymptote is the horizontal line $y = 0$.

5. There are no vertical or horizontal intercepts.

6. There is no maximum or minimum function value.

7. The function f is not continuous at $x = 0$.

- Functions defined by $y = \dfrac{k}{x^n}$ are called **inverse variation functions** in which

  1. y is said to vary inversely as the nth power of x

  2. k is called the constant of variation or constant of proportionality

EXERCISES: ACTIVITY 5.2

1. Doctors sometimes use a patient's body-mass index to determine whether or not the patient should lose weight. The model for the body-mass index, B, is the formula

$$B = \frac{705w}{h^2},$$

where w is the weight in pounds and h is height in inches.

a. What is your body-mass index?

b. Suppose your friend weighs 170 pounds. Substitute this value into the body-mass index formula to obtain an equation for B in terms of height.

c. What is the practical domain of the body-mass index function in part b?

d. Complete the following table using the formula for the body-mass index of a 170-pound person.

h, Height in Inches	60	64	68	72	76	80
B Body-Mass Index						

e. Sketch a graph of the function defined by $B = \dfrac{119,850}{h^2}$. Use the data values in part d to help determine an appropriately scaled axis.

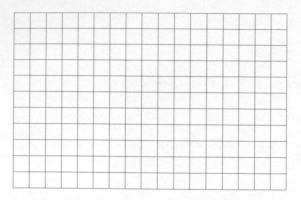

f. What happens to the body-mass index as height increases? Does this make sense in the context of the situation? Explain why or why not.

g. It is recommended that a person's body-mass index be between 19 and 25. Use the graph and the trace key on your calculator to approximate the values of h for which $19 < B < 25$.

2. Sketch a graph of the functions $f(x) = \dfrac{3}{x^2}$ on the first grid and $g(x) = \dfrac{-3}{x^2}$ on the second grid. Describe how the graph of g is related to the graph of f.

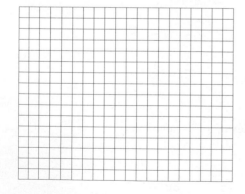
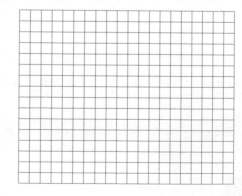

3. Match the following functions with the accompanying graphs.

i. $f(x) = \dfrac{10}{x^4}$

ii. $g(x) = \dfrac{100}{x^5}$

iii. $h(x) = \dfrac{-10}{x^3}$

iv. $F(x) = \dfrac{-1}{x^2}$

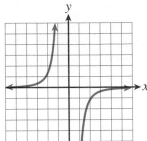

a.

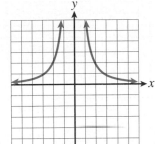

b.

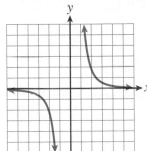

c.

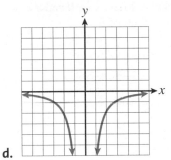

d.

i.

ii.

iii.

iv.

4. Describe how the graphs of $y = \dfrac{1}{x^2}$ and $y = \dfrac{1}{x^3}$ are similar and how they are different.

5. Consider the family of functions of the form $f(x) = \dfrac{k}{x^n}$, where k is a nonzero constant and n is a positive integer.

  a. What is the domain of f?

b. Use several different values of k and n, where $k > 0$ and n is an odd positive integer, to determine the general shape of the graph of f.

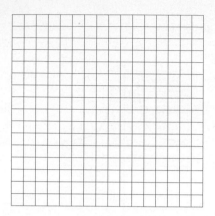

c. Use several different values of k and n, where $k > 0$ and n is an even positive integer, to determine the general shape of the graph of f.

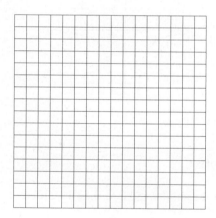

6. How will the general shapes of the graphs in Exercise 5 change if $k < 0$?

7. Complete the following tables of ordered pairs for the given inverse variations. Determine the constant of variation and write the equation that is represented by the table.

a. y varies inversely as x.

x	y
$\frac{1}{2}$	
1	2
2	
6	

b. y varies inversely as x^3.

x	y
$\frac{1}{2}$	
1	
2	1
6	

8. If y varies inversely as the cube of x, determine the constant of proportionality if $y = 16$ when $x = 2$.

9. The amount of current, I (in amps), in a circuit varies inversely as the resistance, R (in ohms). A circuit containing a resistance of 10 ohms has a current of 12 amperes. Determine the current in a circuit containing a resistance of 15 ohms.

10. The intensity, I, of light varies inversely as the square of the distance, d, between the source of light and the object being illuminated. A light meter reads 0.25 unit at a distance of 2 meters from a light source. What will the meter read at a distance of 3 meters from the source?

11. You are investigating the relationship between the volume, V, and pressure, P, of a gas. In a laboratory, you conduct the following experiment: While holding the temperature of a gas constant, you vary the pressure and measure the corresponding volume. The data that you collect appears in the following table.

P (psi)	20	30	40	50	60	70	80
V (ft.³)	82	54	41	32	27	23	20

a. Sketch a graph of the data.

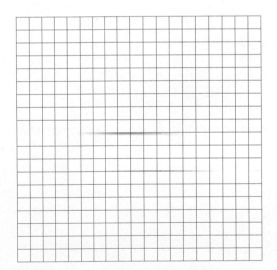

b. One possible model for the data is that V varies inversely as the square of P. Does the data fit the model $V = \dfrac{k}{P^2}$? Explain.

c. Another possible model for the data is that V varies inversely as P. Does $V = \dfrac{k}{P}$ model the data? Explain.

d. Predict the volume of the gas if the pressure is 65 pounds per square inch, using the graph in part a.

e. Verify your answer in part d using the model from part c.

Activity 5.3

Percent Markup

Objectives

1. Determine the domain of a rational function defined by an equation of the form $y = \dfrac{k}{g(x)}$, where k is a nonzero constant and $g(x)$ is a first-degree polynomial.

2. Identify the vertical and horizontal asymptotes of $y = \dfrac{k}{g(x)}$.

3. Sketch graphs of rational functions defined by $y = \dfrac{k}{g(x)}$.

You are a buyer for a national chain of retail stores. You purchase merchandise at a wholesale cost. The merchandise is then sold at a retail price (called the selling price). The retailer's markup is the difference between the selling price (what the consumer pays) and the wholesale cost.

1. **a.** You acquire a line of sports jackets at a wholesale cost of $80 per jacket. If the jackets sell for $120 each at the retail level, what is the amount of the markup?

   b. The markup is what percent of the selling price? (This percent is called the percent markup of the selling price.)

2. The relationship among the selling price, S, the wholesale cost, C, and the percent markup, P, of the selling price (expressed as a decimal) is modeled by

$$S = \frac{C}{1 - P}.$$

   If the wholesale cost of a sports jacket is $80, write an equation for S in terms of P.

3. **a.** Complete the following table for $S = \dfrac{80}{1 - P}$.

P (% markup)	0	0.01	0.05	0.10	0.25	0.50	0.75	0.95
S (selling price) ($)								

   b. As the values of P approach 1, what happens to the values of S? What does this mean in practical terms?

4. **a.** Can the percent markup of the selling price be 100% (i.e., can $P = 1$)? Explain.

   b. What is the practical domain of this function?

5. Sketch a graph of the function defined by $S = \dfrac{80}{1 - P}$. Use the table of data pairs in Problem 3 to help determine an appropriate scale. Verify your sketch using your graphing calculator

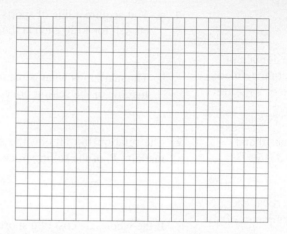

Graphs of $f(x) = \dfrac{k}{g(x)}$

Because S and P represent real-world quantities, the practical domain limits our investigation of the function defined by $S = \dfrac{80}{1 - P}$.

6. Consider the general function f defined by $f(x) = \dfrac{80}{1 - x}$.

   a. What is the domain of function f?

   b. Complete the following table.

x	− 10	− 5	0	0.50	0.75	0.90	1	1.10	1.25	1.50	2	5	10
f(x)													

   c. Sketch a graph of the function f. Use the table in part b to determine an appropriate scale.

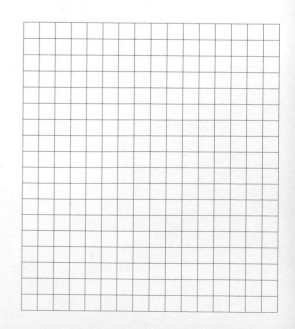

d. Does the graph of *f* have a horizontal asymptote? Explain why or why not. If you answer yes, what is the equation of the horizontal asymptote?

e. Does the graph of *f* have a vertical asymptote? Explain why or why not. If you answer yes, what is the equation of the vertical asymptote?

f. For what value of *x* is $f(x)$ maximum?

g. Does the graph have any intercepts?

7. a. Sketch a graph of $f(x) = \dfrac{80}{1 - x}$ using your graphing calculator. The screens should appear as follows.

b. The vertical asymptote $(x = 1)$ appears to be part of the graph. What do you think has happened?

The calculator setting in Problem 7 was for connected mode. In this mode, it will connect all the points in plots. If the window is such that the calculator does not try to plot a point at $x = 1$, then it may connect a point for an *x*-value slightly less than 1 to one slightly greater than 1. This creates the appearance of an asymptote. To avoid this, change the mode to dot mode, as follows:

 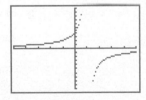

Note: Some TI-83/TI-84 Plus calculators will not display a vertical asymptote even in connected mode. It depends on the operating system in the calculator, and there is no way to tell simply by looking at the calculator.

8. Consider the function defined by $g(x) = \dfrac{80}{x + 1}$.

a. What is the domain of *g*?

b. Construct a table of data points for g.

x	−4	−3	−2	−1	0	1	2
g(x)							

c. Sketch a graph of g using an appropriate scale.

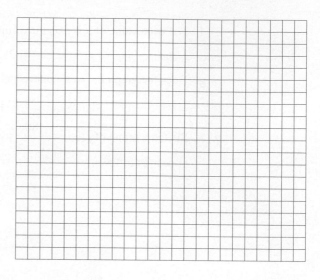

d. Determine the equation of the vertical asymptote. As a graphing aid, if the vertical asymptote is not the y-axis (x = 0), the asymptote is drawn as a dotted vertical line. If you have not done so, draw the vertical asymptote in the graph in part c.

e. Determine the equation of the horizontal asymptote.

f. Does the graph of g have any intercepts?

g. Sketch a graph of $g(x) = \dfrac{80}{1 + x}$ using your graphing calculator. Your screens should appear as follows.

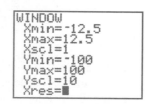
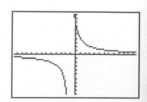

9. How are the graphs of $f(x) = \dfrac{80}{1 - x}$ and $g(x) = \dfrac{80}{x + 1}$ similar? How are they different?

Rational Functions

A function Q, defined by an equation of the form

$$Q(x) = \frac{k}{g(x)},$$

where k is a nonzero constant and $g(x)$ is a polynomial with degree ≥ 1 and $g(x) \neq 0$, belongs to the family of functions known as **rational functions**. The inverse variation function in Activity 5.2 is a special case of a rational function, where $g(x) = x^n$. The only values at which the function is not defined are any values for which the denominator is zero.

If $g(a) = 0$, then $x = a$ is a vertical asymptote of the graph of $Q = \frac{k}{g(x)}$.

The horizontal asymptote of the graph of function Q is the x-axis $(y = 0)$.

Example 1 *Determine the domain and the vertical and horizontal asymptotes for each of the following rational functions.*

a. $f(x) = \dfrac{3}{x}$ **b.** $g(x) = \dfrac{10}{x - 4}$ **c.** $h(x) = \dfrac{-5}{2x + 6}$

SOLUTION

FUNCTION	DOMAIN	VERTICAL ASYMPTOTE	HORIZONTAL ASYMPTOTE
a. $f(x) = \dfrac{3}{x}$	all real numbers except $x = 0$	$x = 0$	$y = 0$
b. $g(x) = \dfrac{10}{x - 4}$	all real numbers except $x = 4$	$x = 4$	$y = 0$
c. $h(x) = \dfrac{-5}{2x + 6}$	all real numbers except $x = -3$	$x = -3$	$y = 0$

10. Consider the function defined by $f(x) = \dfrac{5}{3x - 6}$.

a. Determine the domain of f.

b. Complete the following table.

x	-10	-5	0	1.5	1.9	2	2.1	2.5	3	8	13
$f(x)$											

    c. Determine the vertical and horizontal asymptotes of the graph of f.

    d. Sketch a graph of f.

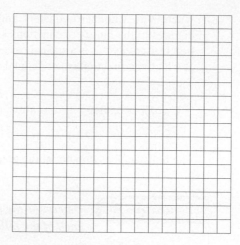

    e. Verify the graph using your graphing calculator.

11. Without using your graphing calculator, match the following functions with the accompanying graphs. Use your graphing calculator to verify your matches.

    i. $f(x) = \dfrac{5}{x}$ **ii.** $g(x) = \dfrac{5}{x^2}$ **iii.** $h(x) = \dfrac{-5}{x}$

    iv. $F(x) = \dfrac{5}{x + 2}$ **v.** $G(x) = \dfrac{5}{2x - 4}$

a. 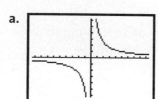   **b.** 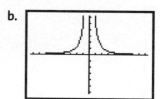   **c.**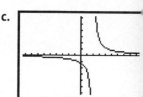

d. 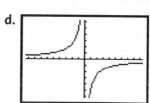   **e.**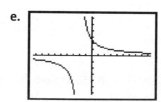

SUMMARY: ACTIVITY 5.3

1. A function Q, defined by an equation of the form

$$Q(x) = \frac{k}{g(x)},$$

where k is any nonzero constant, $g(x)$ is a polynomial with degree ≥ 1 and $g(x) \neq 0$, belongs to a family of functions known as **rational functions**.
Examples include $f(x) = \dfrac{10}{x^2}$, $g(x) = \dfrac{10}{x-4}$, and $h(x) = \dfrac{-5}{2x+6}$.

2. The domain of Q is the set of all real numbers except those values of the input x such that $g(x) = 0$.

3. If $g(a) = 0$, then $x = a$ is a vertical asymptote of $Q(x) = \dfrac{k}{g(x)}$, $k \neq 0$.

4. The horizontal asymptote is the x-axis $(y = 0)$.

EXERCISES: ACTIVITY 5.3

1. To obtain an estimate of the required volume, V, of timber that must be harvested for a logging company to break even, use the model

$$V = \frac{Y + L}{P - S - F - T},$$

where V is the required annual logging volume (in cubic meters),

Y is the yard cost (in dollars),

L is the loading cost (in dollars),

P is the selling price (in dollars per cubic meter),

S is the skidding cost (in dollars per cubic meter),

F is the felling cost (in dollars per cubic meter), and

T is the transportation cost (in dollars per cubic meter).

a. Suppose a logging company estimates that the yard cost will be $25,000, the loading cost will be $55,000, the skidding cost will be $1.50 per cubic meter, the felling cost will be $0.40 per cubic meter, and the transportation cost will be $0.60 per cubic meter. Write a rule for V as a function of P.

b. Complete the following table using the equation in part a.

P ($)	2.50	3.00	5.00	10.00	25.00
V (m³)					

c. As the selling price per cubic meter increases, what happens to the corresponding required logging volume, V?

d. Determine the value of V when $P = 2$. What is the practical meaning of the negative value of V?

e. What is the practical domain of this function?

f. Sketch a graph of the function. Use the table in part b to determine an appropriate scale.

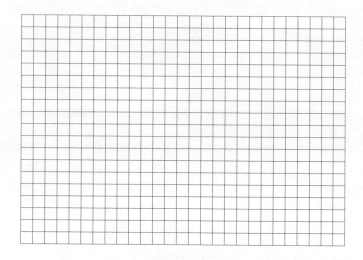

2. Two functions are defined by $f(x) = \dfrac{10}{x - 5}$ and $g(x) = \dfrac{10}{5 - x}$.

   a. Describe how you can determine the vertical asymptote without graphing.

   b. Determine the vertical and horizontal asymptote for the graph of each function.

   c. Verify your answers by graphing each function on your graphing calculator.

3. Without graphing, determine the domain of each of the following functions. Then determine the equation of the vertical asymptote for each function. Verify your answers using your graphing calculator.

a. $f(x) = \dfrac{6}{x - 7}$

b. $g(x) = \dfrac{20}{25 - x}$

c. $h(x) = \dfrac{3}{2x - 10}$

d. $F(x) = \dfrac{13}{0.5x - 7}$

e. $G(x) = \dfrac{-4}{2x + 5}$

4. Give examples of two different rational functions that have a vertical asymptote at $x = 10$.

5. As the input value of a rational function gets closer to a vertical asymptote, the output becomes larger in magnitude, approaching either positive or negative infinity. Consider the functions $f(x) = \dfrac{10}{x - 5}$ and $g(x) = \dfrac{10}{5 - x}$.

a. Determine the equations of the vertical asymptotes for functions f and g.

b. Describe what happens to the output value when x is near the vertical asymptote but to the right of it.

c. Describe what happens to the output value when x is near the vertical asymptote but to the left of it.

Activity 5.4

Blood-Alcohol Levels

Objectives

1. Solve an equation involving a rational expression using an algebraic approach.

2. Solve an equation involving a rational expression using a graphing approach.

3. Determine horizontal asymptotes of the graph of $y = \dfrac{f(x)}{g(x)}$, where $f(x)$ and $g(x)$ are first-degree polynomials.

As of 2009, every state in the United States has adopted a 0.08% blood-alcohol concentration as the legal measure of drunk driving, with the exception of three states. The legal measure of drunk driving in Colorado, Delaware, and Minnesota is a blood-alcohol concentration of 0.10%. If you assume that a regular 12-ounce beer is 5% alcohol by volume and that the normal bloodstream contains 5 liters (or 169 ounces) of fluid, your maximum blood-alcohol concentration, B, can be approximately modeled by the function having the equation

$$B = \frac{600n}{w(169 + 0.6n)},$$

where n is the number of beers consumed in 1 hour and w is your body weight in pounds.

1. **a.** Replace w with your body weight. Write an equation for B in terms of n.

   b. Complete the following table using your equation from part a.

NUMBER OF BEERS CONSUMED IN ONE HOUR, n	BLOOD-ALCOHOL CONCENTRATION, B
1	
2	
3	
4	
5	
6	
7	
8	
9	
10	

2. According to this model, how many beers can you consume in 1 hour without exceeding the recommended legal measure of drunk driving in your state?

3. **a.** A football player friend of yours weighs 232 pounds. Rewrite the equation for B in terms of n. What is his maximum blood-alcohol level if he drinks four beers in 1 hour?

   b. Complete the following table using your equation from part a.

Number of Beers Consumed in One Hour, n	1	2	3	4	5	6	7	8	9	10
Blood-Alcohol Concentration, B										

c. What is the practical domain of the blood-alcohol function of part a?

d. Does the weight of a person have any impact on the practical domain? Explain.

e. What is the vertical intercept? Does this seem reasonable within the context of the problem?

f. Sketch the graph of your blood-alcohol function over the practical domain identified in part c. Use the indicated scale.

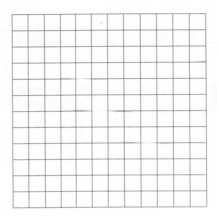

Note that the graph of the blood-alcohol function in Problem 3f looks like a line when drawn in the practical domain. The graph of the general function defined by

$$f(x) = \frac{600x}{232(169 + 0.6x)}$$

appears as follows.

 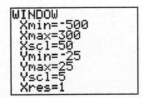 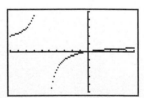

Solving Equations Involving a Rational Expression

4. a. Your 232-pound football player friend is given a breathalyzer test. The result is a blood-alcohol concentration of 0.05%. Using the blood-alcohol concentration function, write an equation that can be solved to determine the number of beers your friend consumed in the previous hour.

b. Solve the equation in part a graphically on the following grid.

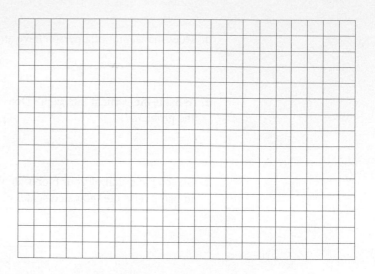

c. Use your graphing calculator to check the answer in part b. Your screen(s) should appear as follows.

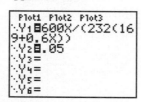 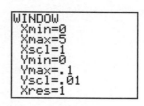 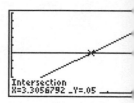

The equation in Problem 4 can be solved using an algebraic approach. If a variable appears in the denominator of a fraction, a general approach is to multiply both sides of the equation by the denominator and solve the resulting equation. Example 1 demonstrates this approach as well as two other methods.

Example 1   *Solve the equation* $\dfrac{16}{x+3} = 2.$

SOLUTION

Method 1.   General Case

To solve the equation $\dfrac{16}{x+3} = 2$, first multiply both sides of the equation by the denominator $x + 3$, as follows.

$$(x+3) \cdot \frac{16}{x+3} = 2(x+3)$$

$$16 = 2x + 6$$

Solving for x, you have

$$10 = 2x$$

$$5 = x.$$

Solutions to an equation involving rational functions should always be checked:

$$\frac{16}{5+3} = \frac{16}{8} = 2.$$

Therefore, 5 is a solution.

Method 2. Cross Multiplication

You can also solve the equation $\dfrac{16}{x+3} = 2$ by applying the following property.

If two ratios $\frac{a}{b}$ and $\frac{c}{d}$ represent the same value, then $\frac{a}{b} = \frac{c}{d}$ is equivalent to $ad = bc$.

This process is called cross multiplication. Therefore,

$$\dfrac{16}{x+3} = \dfrac{2}{1} \qquad \text{Cross multiply.}$$

$$2(x+3) = 16 \cdot 1$$

$$2x + 6 = 16$$

$$\underline{-6 \quad -6}$$

$$2x = 10$$

$$x = 5$$

Method 3. Graphical Approach

You can verify that 5 is a solution to the given equation by graphing $y_1 = \dfrac{16}{x+3}$ and $y_2 = 2$ and determining the x-value of the point of intersection of the two graphs.

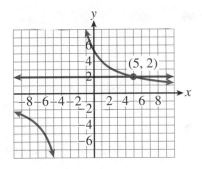

5. Solve each of the following equations using an algebraic approach. Verify your answer graphically.

   a. $\dfrac{45}{x} = 9$

   b. $\dfrac{23}{x+2} = 15$

   c. $\dfrac{13}{x} = \dfrac{2}{5}$

   d. $\dfrac{16}{x^2} = \dfrac{1}{4}$

6. **a.** Solve the equation in Problem 4a using an algebraic approach.

b. How does your solution compare with the result in Problem 4c using a graphical approach?

Horizontal Asymptotes

The graphs you have studied so far in this chapter have at least one feature in common. The horizontal asymptote is the horizontal axis. As the input values increase infinitely in the positive direction or decrease infinitely in the negative direction, the output values have always approached zero.

7. Consider the function defined by $y = \dfrac{2x}{x + 5}$. This equation is in the form $y = \dfrac{f(x)}{g(x)}$ where $f(x) = 2x$ and $g(x) = x + 5$.

a. What is the domain of this function?

b. What is the vertical asymptote?

c. Complete the following table.

x	−100	−50	−10	−5.5	−5.1	−5	−4.9	−4.5	0	50	100
y											

d. What appears to be happening to the y-values as the x-values increase infinitely to the right or decrease infinitely to the left?

e. What is the horizontal asymptote?

f. Sketch a graph of the function; include horizontal and vertical asymptotes as dotted lines.

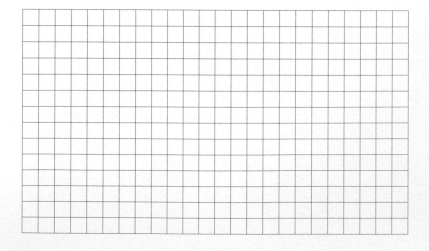

g. Verify the graph in part f using your graphing calculator. As a graphing aid, also sketch the graph of $y = 2$.

8. a. In the blood-alcohol function $B = \dfrac{600n}{232\,(169 + 0.6n)}$, as the positive values of n increase in value, what happens to the corresponding values of B?

b. Extend the window of your graphing calculator until you can see the graph leveling off (becoming horizontal) for large values of n. Estimate the equation of the horizontal asymptote of the graph of the blood-alcohol function.

c. Does this asymptote have any practical significance for this application?

9. When comparing smokers and nonsmokers between the ages of 55 and 64, a recent study determined that smokers in this age group had an incidence ratio of 10 for death due to lung cancer. An incidence ratio of 10 means that smokers in this age group are 10 times more likely than nonsmokers to die of lung cancer.

For a given incidence ratio, x, the percent, P, in decimal form of deaths due to a certain illness caused by smoking can be modeled by

$$P = \frac{x - 1}{x}.$$

a. Determine what percent, P, of the deaths due to lung cancer is caused by smoking, x, in the age group between 55 and 64.

b. Complete the following table.

x	1	2	5	10	30
p					

c. As the incidence ratio, x, increases in value, what happens to the corresponding values of the percent P?

d. Determine the horizontal asymptote. Does this make sense in this situation? Explain.

SUMMARY: ACTIVITY 5.4

1. To solve an equation of the form $\dfrac{f(x)}{g(x)} = \dfrac{a}{b}$, where $g(x) \neq 0$ and $b \neq 0$, using an algebraic approach,

  i. Method 1. Multiply both sides of the equation by the product $b \cdot g(x)$, and solve the resulting equation for x.

  ii. Method 2. Cross multiply to obtain $b \cdot f(x) = a \cdot g(x)$, and solve the resulting equation for x.

2. If the y-values of a function R get closer to a number a as the x-values increase infinitely in the positive direction or decrease infinitely in the negative direction, then the graph of the function R has a horizontal asymptote. The equation of the horizontal asymptote is $y = a$. For example, the horizontal asymptote for $R(x) = \dfrac{3x + 1}{x - 2}$ is $y = 3$. Note that the vertical asymptote is $x = 2$.

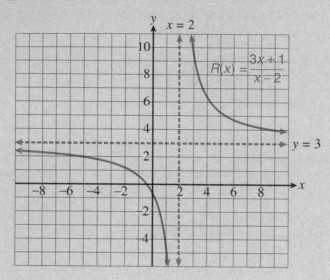

EXERCISES: ACTIVITY 5.4

1. You are on the 5-year reunion committee for your high school. The committee selects a restaurant that charges $500 to rent a large room to accommodate your group and $50 per person for dinner and a 1-hour open bar. Other expenses include $600 for a DJ; $500 for printing invitations, a program, and name tags; and $400 for decorations.

a. What are the total fixed costs?

b. The total cost of the event is a function of the number, n, of people who will attend. Write an expression that represents the total cost in terms of n.

c. Your committee decides to divide the total cost evenly among the people attending. Let m represent the cost per person if n people attend. Write an equation for m in terms of n.

d. What is the cost per person if 100 people attend?

e. What is the practical domain of the cost per person function?

f. Complete the following table.

n, Number of People Attending	50	100	150	200	250
m, Mean Cost Per Person ($)					

g. Does the graph of the cost function have a horizontal asymptote? Does it make sense in this situation? Explain.

h. Sketch a graph of the cost function over its practical domain.

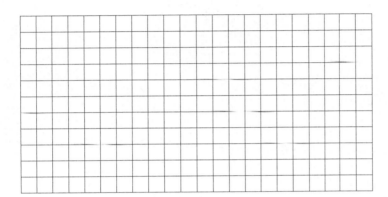

2. For each rational function:

i. Determine the domain.

ii. Determine the vertical asymptotes.

iii. Graph the function using your graphing calculator.

iv. Determine the horizontal asymptote by inspecting the graph of your function.

a. $y = \dfrac{4x}{x + 2}$

   i.

   ii.

   iii.

b. $y = \dfrac{1 - x}{x + 1}$

   i.

   ii.

   iii.

   iv.

   iv.

c. $y = \dfrac{3x}{x - 4}$

    i.

    ii.

    iii.

    iv.

d. $y = 12 - \dfrac{6x}{1 - 2x}$

    i.

    ii.

    iii.

    iv.

3. Solve the following equations algebraically and check your results by graphing.

a. $\dfrac{3x}{2x - 1} = 3$

b. $\dfrac{x + 1}{5x - 3} = 2$

c. $\dfrac{-7x}{2.8 + x} = 3.1$

d. $\dfrac{x + 10}{3x + 2} = 3$

4. In a 20-kilometer race, a runner's average rate (in kilometers per hour) can be expressed as a function of time (in hours) by the equation $r = \dfrac{20}{t}$.

a. Determine your time to complete the race if you average 16 kilometers per hour.

b. What is your time if you average 18 kilometers per hour?

5. The intensity of the human voice varies inversely to the square of the distance from the source. This is given by the formula from Activity 5.2, Loudness of a Sound: $I = \dfrac{1500}{d^2}$, where I is decibels and d is distance in feet.

a. Determine the distance from the source when the intensity of the sound is 15 decibels.

b. What is the distance from the source when the intensity is 8000 decibels?

6. As a fund-raising project, the international club at your college decides to publish and sell a calendar. The cost of photographs and typesetting is $450. It costs $3 to print and assemble each calendar.

a. What is the total cost of printing 200 calendars?

b. What is the average cost per calendar of printing the 200 calendars?

c. Write an expression for the total cost of printing n calendars.

d. Let A represent the average cost per calendar. Write an equation that gives A as a function of n.

e. Complete the following table using the equation in part d.

n (number)	50	75	100	500	750	1000
A (average cost) ($)						

f. As the input n increases, what happens to the output A?

g. What is the horizontal asymptote of this function?

h. Verify your answer in part g graphically.

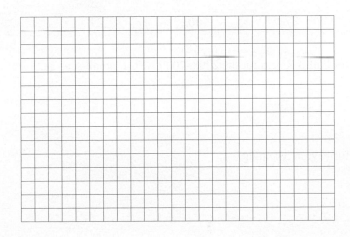

i. Interpret what the horizontal asymptote means in the context of the problem.

j. Suppose you want the average cost to be less than $3.20. Model this problem with an inequality, solve it algebraically for n, and verify it graphically.

7. The following formula is used by the National Football League (NFL) to calculate quarterback ratings:

$$R = \frac{250C + 12.5Y + 1000T - 1250I + 6.25A}{3A},$$

where

$R = $ quarterback rating
$A = $ passes attempted
$C = $ passes completed
$Y = $ passing yardage
$T = $ touchdown passes
$I = $ number of interceptions.

In the 2008–2009 regular season, Philip Rivers, quarterback for the San Diego Chargers, and Peyton Manning, for the Indianapolis Colts, had the following player statistics:

PLAYER	PASSES ATTEMPTED	PASSES COMPLETED	PASSING YARDAGE	NUMBER OF TOUCHDOWN PASSES	NUMBER OF INTERCEPTIONS
Philip Rivers	478	312	4009	34	11
Peyton Manning	555	371	4002	27	12

a. Determine the quarterback rating for Philip Rivers for the 2008–2009 NFL football season.

b. Determine the quarterback rating for Peyton Manning.

c. Suppose an NFL quarterback has the following statistics for the 2009 season: 4000 passing yards, 30 touchdown passes, and 10 interceptions. Write an equation for the quarterback rating R, in terms of the number of passes attempted, A, and the number of passes completed, C.

d. If the quarterback in part c completed 350 passes, how many passes must have been attempted in order to have a quarterback rating of 95?

e. Visit www.nfl.com and select stats to obtain the rating of your favorite quarterback.

8. In a predator-prey model from wildlife biology, the rate, R, at which prey are consumed by one predator is approximated by the function

$$R = \frac{0.623n}{1 + 0.046n},$$

measured in prey per week, where n is the number of prey available per square mile.

a. If the number of prey available per square mile is 30, what is R?

b. Approximately how many prey must be available per square mile for the predator to consume 10 prey per week?

c. Suppose you want to maintain the prey population to ensure that a predator may obtain between 6 and 10 prey per week. One of the necessary equations for this situation was solved in part b above. Write the other equation and solve it.

d. Use the graph of this function to illustrate your solutions to parts b and c.

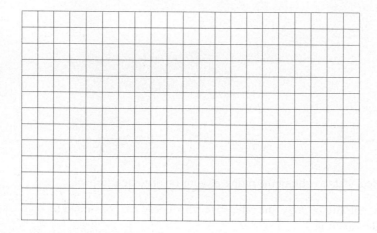

e. How many prey are required for a predator to consume 20 prey per week? Comment on the practical significance and how this shows up on the graph.

Objectives

1. Determine the least common denominator (LCD) of two or more rational expressions.

2. Solve an equation involving rational expressions using an algebraic approach.

3. Solve a formula for a specific variable.

You are an intern at an architecture firm. The company is designing an auditorium that will be annexed to the local high school building. The rate of traffic flow through the exits is an important consideration. The auditorium will have three exit doors. Two exits are single doors of slightly different sizes. The first exit, by itself, can be used to empty the auditorium in 10 minutes. The second exit can be used to empty the room in 8 minutes. The third exit is a double-wide door that, by itself, can be used to empty the auditorium in 5 minutes.

1. If only the first door is open, what fraction of the auditorium can be emptied in 5 minutes? 2 minutes? 1 minute?

2. The rate at which a door can be used to empty the auditorium is the fraction of the job that can be completed in 1 minute. In this case, the units of measurement are auditoriums per minute. Determine the rate of emptying for each of the three exits, and record your answers in the following table.

EXIT	RATE OF EMPTYING
First	
Second	
Third	

Your task is to determine the time, T, it takes to empty the auditorium if all three exit doors are open. The relationship between this time, T, and the individual emptying times is given by the formula

$$\frac{1}{t_1} + \frac{1}{t_2} + \frac{1}{t_3} = \frac{1}{T},$$

where t_1, t_2, and t_3 represent the times that each exit door can be used by itself to empty the auditorium.

Note that three single door rates are added to determine the rate, $\frac{1}{T}$, at which the three doors working together can be used to empty the auditorium.

3. a. Write the equation that can be used to determine the time, T, that it takes for the auditorium to be emptied if all three exits are open.

   b. Solve this equation graphically. Use the window Xmin $= 0$, Xmax $= 10$, Ymin $= 0$, Ymax $= 2$, and Yscl $= 0.5$.

Solution Using an Algebraic Approach

An algebraic approach to solving the equation $\frac{1}{10} + \frac{1}{8} + \frac{1}{5} = \frac{1}{T}$ is to eliminate the fractions from the equation. This can be accomplished by first determining the least common denominator (LCD) of the fractions involved in the equation. The following example demonstrates the procedure for determining the LCD.

Example 1   **a.** *Determine the LCD for* $\dfrac{5}{12}$ *and* $\dfrac{7}{45}$.

SOLUTION

Step 1.   Write the prime factorization of each denominator. Express repeated factors as powers.

$$12 = 2 \cdot 2 \cdot 3 = 2^2 \cdot 3$$
$$45 = 3 \cdot 3 \cdot 5 = 3^2 \cdot 5$$

Step 2.   Identify the different bases (factors) in step 1.

$$2, 3, 5$$

Step 3.   Write the LCD as the product of the highest power of each of the different factors from step 2.

$$\text{LCD} = 2^2 \cdot 3^2 \cdot 5 = 4 \cdot 9 \cdot 5 = 180$$

The smallest number that both 12 and 45 will divide evenly is 180.

b. *Determine the LCD for* $\dfrac{11}{6xy^3}$ *and* $\dfrac{5a}{9x^2y}$.

SOLUTION

Step 1. $6xy^3 = 2 \cdot 3 \cdot x^1 \cdot y^3$

$9x^2y = 3^2 \cdot x^2 \cdot y$

Step 2. $2, 3, x, y$

Step 3. $\text{LCD} = 2 \cdot 3^2 \cdot x^2 \cdot y^3 = 18x^2y^3$

$18x^2y^3$ is evenly divided by both $6xy^3$ and $9x^2y$.

You are now ready to solve the equation $\dfrac{1}{10} + \dfrac{1}{8} + \dfrac{1}{5} = \dfrac{1}{T}$ using an algebraic approach.

4. a. Determine the LCD of the rational expressions in the equation $\dfrac{1}{10} + \dfrac{1}{8} + \dfrac{1}{5} = \dfrac{1}{T}$.

b. Multiply each side of the equation by the LCD, and solve the resulting equation.

c. How does this solution compare to the solution you determined graphically in Problem 3?

5. An auditorium is to be equipped with two ventilation fans. The first fan can exchange the air in the room in 4 hours. The building code requires a complete exchange of air in the room every 3 hours. To model this situation, use the equation

$$\frac{1}{t_1} + \frac{1}{t_2} = \frac{1}{T},$$

where t_1 and t_2 are the exchange times for the fans working alone, and T is the exchange time for the two fans working together.

a. Let x represent the exchange time for the second ventilation fan. Write an equation that can be used to determine x so that the fans working together will satisfy the building code.

b. Solve this equation graphically.

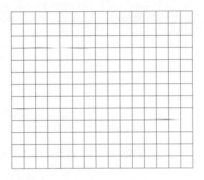

c. Solve the equation in part a algebraically by multiplying both sides by the LCD.

6. Switching to different fans, suppose the first fan can exchange the air in the auditorium twice as fast as the second fan.

a. If x represents the exchange time for the first fan, write an expression that represents the exchange time for the second fan.

b. Working together, the two fans can exchange the air in the room in 4 hours. Using the formula $\frac{1}{t_1} + \frac{1}{t_2} = \frac{1}{T}$, write an equation that can be solved to determine x.

c. Solve this equation algebraically. Verify your solution graphically.

d. Determine the rate for each fan.

Solving a Formula for a Specified Letter

An alternative approach to solving problems involving formulas of the form $\dfrac{1}{t_1} + \dfrac{1}{t_2} = \dfrac{1}{T}$ is to use an equivalent formula that has been solved for the variable T.

Example 2   *Solve* $\dfrac{1}{t_1} + \dfrac{1}{t_2} = \dfrac{1}{T}$ *for T.*

SOLUTION

Step 1. Determine the LCD.

$$\text{LCD} = t_1 t_2 T$$

Step 2. Multiply each side of the equation by the LCD.

$$\frac{t_1 t_2 T}{1} \cdot \left(\frac{1}{t_1} + \frac{1}{t_2} \right) = \frac{1}{T} \cdot \frac{t_1 t_2 T}{1}$$

$$\frac{t_1 t_2 T}{t_1} + \frac{t_1 t_2 T}{t_2} = \frac{t_1 t_2 T}{T}$$

Simplifying, you have $t_2 T + t_1 T = t_1 t_2$.

Step 3. Solve the resulting equation for T.

$$T(t_2 + t_1) = t_1 t_2 \qquad \text{T is the common factor on the left side.}$$

$$T = \frac{t_1 t_2}{t_2 + t_1} \qquad \text{Divide both sides by $t_2 + t_1$.}$$

---

7. a. If t_1 and t_2 are the exchange times for the fans working alone, and T is the exchange time for the two fans working together, determine T if $t_1 = 4$ hours and $t_2 = 12$ hours.
Use the formula $\dfrac{1}{t_1} + \dfrac{1}{t_2} = \dfrac{1}{T}$.

b. Repeat part a using the formula $T = \dfrac{t_1 t_2}{t_2 + t_1}$.

c. Compare the results in parts a and b.

8. The following formula is used in work with lenses and mirrors: $\dfrac{1}{p} + \dfrac{1}{q} = \dfrac{1}{f}$. Solve the formula for f.

SUMMARY: ACTIVITY 5.5

1. To determine an LCD of two or more rational expressions,

Step 1. Write the prime factorization of each denominator. Express repeated factors as powers.

Step 2. Identify the different bases (factors) in step 1.

Step 3. Write the LCD as the product of the highest power of each of the different factors from step 2.

2. To solve an equation involving rational expressions,

Step 1. Determine the LCD of all denominators in the equation.

Step 2. Multiply each side of the equation by the LCD, and simplify the resulting equation.

Step 3. Solve the resulting equation for the desired variable.

3. Many real-life applications involve equations of the form

$$\frac{1}{a} + \frac{1}{b} = \frac{1}{c},$$

where a, b, and $c \neq 0$. For example,

$\dfrac{1}{R_1} + \dfrac{1}{R_2} = \dfrac{1}{R}$ is used for resistance of electrical circuits,

$\dfrac{1}{p} + \dfrac{1}{q} = \dfrac{1}{f}$ is used in work with lenses and mirrors, and

$\dfrac{1}{t_1} + \dfrac{1}{t_2} = \dfrac{1}{T}$ is used to calculate the time it takes to complete a task when two machines or people are working together.

These formulas can also be extended to three or more resistors, lenses, or machines by simply adding additional fractions in each case.

EXERCISES: ACTIVITY 5.5

1. Two pumps are working together to empty a gasoline tank.

   a. The emptying times for pump 1 and pump 2 are 30 minutes and 45 minutes, respectively. Determine the time required to empty the tank if both pumps are working. Use the formula $\dfrac{1}{t_1} + \dfrac{1}{t_2} = \dfrac{1}{T}$, where t_1 and t_2 are the emptying times for pump 1 and pump 2, respectively, and T is the total time required to empty the tank.

   b. Solve the equation $\dfrac{1}{t_1} + \dfrac{1}{t_2} = \dfrac{1}{T}$ for T.

   c. Using the equation developed in part b, determine T if $t_1 = 20$ minutes and $t_2 = 15$ minutes.

   d. If one pump can empty the tank in 40 minutes, how fast must a second pump work for the pumps working together to empty the tank in 10 minutes? Use $\dfrac{1}{t_1} + \dfrac{1}{t_2} = \dfrac{1}{T}$.

   e. Suppose three pumps are working together to empty the tank, with emptying times of 25 minutes, 30 minutes, and 50 minutes. How long will it take to empty the tank if all three pumps are working simultaneously? Use the formula $\dfrac{1}{t_1} + \dfrac{1}{t_2} + \dfrac{1}{t_3} = \dfrac{1}{T}$.

Exercise numbers appearing in color are answered in the Selected Answers appendix.

2. Solve each of the following equations using an algebraic approach. Verify your answers using a graphing approach.

a. $\dfrac{10}{x+1} = 4$

b. $\dfrac{2}{x} + \dfrac{3}{x} = 1$

c. $\dfrac{1}{x} + \dfrac{1}{3x} = \dfrac{1}{5}$

d. $\dfrac{3}{x} + \dfrac{2}{x} = \dfrac{4}{x}$

n Exercises 3–5, use the formula $\dfrac{1}{t_1} + \dfrac{1}{t_2} = \dfrac{1}{T}$.

3. The custodian in the mathematics building can buff the main floor 2 minutes faster than his supervisor. If working together they can buff the floor in 35 minutes, how long does it take the supervisor working alone to buff the main floor of the mathematics building? (*Hint:* Let t represent the supervisor's time, and $t - 2$ represent the custodian's time.)

4. It takes you 4 hours working alone to clean your apartment, and your roommate takes 5 hours and 15 minutes. If you begin at noon and work together, will you complete the cleaning in time to leave for the game at 2:30? How late or early will you be?

5. One of your jobs as a work-study student at your college is sending out mailings (stuffing, sealing, and stamping envelopes). You can work twice as fast as your supervisor. If working together you complete a job in 7 hours, how long would it have taken you to complete the job by yourself? (*Hint:* Let t represent your total time working alone and $2t$ the time of your supervisor working alone.)

6. Solve each of the following formulas for the indicated variable. Express your answer as a single fraction.

   a. $\dfrac{1}{a} + \dfrac{2}{b} = \dfrac{3}{c}$, solve for a.

   b. $\dfrac{1}{x + y} = \dfrac{1}{z}$, solve for x.

   c. $\dfrac{1}{a} + \dfrac{2}{b} = \dfrac{3}{c}$, solve for c.

Activity 5.6

Electrical Circuits

Objectives

1. Multiply and divide rational expressions.

2. Add and subtract rational expressions.

3. Simplify a complex fraction.

In performing some technical work in your new job at a local electronics firm, you need to be familiar with resistors combined in a circuit. The total resistance, R, of two resistors in a parallel circuit is modeled by the formula

$$R = \frac{1}{\frac{1}{R_1} + \frac{1}{R_2}},$$

where R_1 and R_2 are the resistances of the two resistors in the circuit, measured in ohms.

1. Calculate the total resistance for each pair of resistors.

R_1 (OHMS)	R_2 (OHMS)	R (OHMS)
10	10	
10	5	
15	5	
20	10	

If one resistor has to be 10 ohms, then the formula

$$R = \frac{1}{\frac{1}{10} + \frac{1}{R_2}}$$

expresses the total resistance as a function of the second resistor's value.

Now the total resistance, R, is a function of R_2. The right side of the equation is a fraction in which fractions also appear in the denominator.

Definition

A fraction that contains fractions in either its numerator or denominator, or both, is called a **complex fraction**.

Example 1   *The following are examples of complex fractions.*

$$\frac{1}{\frac{1}{10} + \frac{1}{R_2}} \qquad \frac{\frac{50}{x}}{\frac{100}{x^2 + 5x}} \qquad \frac{4 + \frac{1}{x}}{\frac{10}{x^2} - \frac{2}{x}}$$

2. To make the equation $R = \dfrac{1}{\frac{1}{10} + \frac{1}{R_2}}$ less cumbersome to work with, simplify the right side of the equation so that it is written as a single fraction (with only one dividing line). This can be accomplished as follows.

a. Determine the LCD of the fractions $\dfrac{1}{10}$ and $\dfrac{1}{R_2}$

b. Add the fractions $\frac{1}{10}$ and $\frac{1}{R_2}$ by writing each fraction as an equivalent fraction that has the LCD as the denominator.

c. Divide the numerator by the denominator, and simplify.

3. a. Using your graphing calculator, sketch a graph of the function defined by

$$R = \frac{10R_2}{R_2 + 10},$$

where R is the total resistance of two resistors in a parallel circuit, one having resistance 10 ohms and the second having a resistance represented by R_2. Your screens should resemble the following.

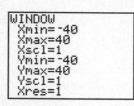
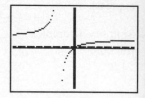

b. What is the domain of this function?

c. What is the practical domain?

d. What is the horizontal asymptote of this function?

e. Interpret the practical meaning of the horizontal asymptote.

4. a. Write an equation that you can use to determine the size of the second resistor you would need to add to the circuit to make a total resistance of 7 ohms.

b. Solve this equation graphically.

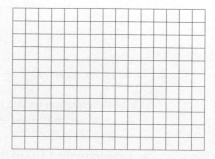

c. Solve the equation in part a algebraically.

5. If resistors are available only in increments of 0.1 ohm, what size would you use to get as close as possible, but still have a total resistance of at least 7 ohms?

Operations with Rational Expressions

Appendix

As you discovered in Problem 2, simplifying complex fractions involves a lot of work with rational expressions. The following examples illustrate how to perform operations with algebraic fractions. Appendix A contains several additional examples and practice exercises involving operations with rational expressions.

Example 2   *Simplify the following complex fractions.*

a. $\dfrac{\frac{50}{x}}{\frac{100}{x^2 + 5x}}$

Both the numerator and the denominator of the complex fraction contain single rational expressions. Therefore, write the complex fraction as a division problem, and divide.

$$\frac{\frac{50}{x}}{\frac{100}{x^2 + 5x}} = \frac{50}{x} \div \frac{100}{x^2 + 5x}$$ Divide using the division rule $\frac{a}{b} \div \frac{c}{d} = \frac{a}{b} \cdot \frac{d}{c}$.

$$= \frac{50}{x} \cdot \frac{x^2 + 5x}{100}$$ Simplify, if possible.

$$= \frac{50}{x} \cdot \frac{x(x + 5)}{2 \cdot 50}$$

$$= \frac{x + 5}{2}$$

b. $\dfrac{4 + \frac{1}{x}}{\frac{10}{x^2} - \frac{2}{x}}$

The rational expressions in the numerator and denominator of the complex fraction can each be combined into a single rational expression.

Step 1. $\dfrac{4}{1} + \dfrac{1}{x} = \dfrac{4}{1} \cdot \dfrac{x}{x} + \dfrac{1}{x} = \dfrac{4x}{x} + \dfrac{1}{x} = \dfrac{4x + 1}{x}$

Step 2. $\dfrac{10}{x^2} - \dfrac{2}{x} = \dfrac{10}{x^2} - \dfrac{2x}{xx} = \dfrac{10}{x^2} - \dfrac{2x}{x^2} = \dfrac{10 - 2x}{x^2}$

Step 3.   Now divide the numerator of the complex fraction by the denominator.

$$\frac{4 + \frac{1}{x}}{\frac{10}{x^2} - \frac{2}{x}} = \frac{4x + 1}{x} \div \frac{10 - 2x}{x^2} = \frac{(4x + 1)}{x} \cdot \frac{x \cdot x}{(10 - 2x)} = \frac{4x^2 + x}{10 - 2x}$$

6. Simplify the following complex fraction:

$$\dfrac{\dfrac{4}{x+3}}{\dfrac{1}{x+2}+\dfrac{3}{x}}$$

SUMMARY: ACTIVITY 5.6

1. To **multiply or divide** rational expressions,

   a. Factor the numerator and denominator of each fraction completely.

   b. Divide out the common factors (cancel).

   c. Multiply remaining factors.

   d. In division, proceed as above after inverting the divisor (the fraction after the division sign).

2. To **add or subtract** rational expressions,

   a. Determine the LCD (least common denominator).

   b. Build each fraction to have the LCD.

   c. Add or subtract numerators.

   d. Place the numerator over the LCD, and simplify if possible.

3. To simplify a complex fraction by simplifying the numerator and denominator,

   a. Express the numerator as a single fraction.

   b. Express the denominator as a single fraction.

   c. Divide the numerator by the denominator.

   d. Simplify, if possible.

EXERCISES: ACTIVITY 5.6

1. An emergency medical services (EMS) vehicle has a single-tone siren with a pitch of approximately 330 hertz (Hz). You are standing on the street corner as the vehicle approaches you at 40 miles per hour. Due to the Doppler effect, the pitch you hear is not 330 hertz. The pitch, h, that you hear is modeled by

$$h = \dfrac{a}{1 - \dfrac{s}{770}},$$

where a is the actual pitch and s is the speed of the source of the sound in miles per hour.

Exercise numbers appearing in color are answered in the Selected Answers appendix.

a. Is the pitch you hear due to the Doppler effect lower or higher than the actual pitch?

b. Simplify the original equation by rewriting the complex fraction on the right side as a single fraction (using only one dividing line).

c. Redo part a using the new equation obtained from part b. How do the results compare?

d. What pitch sound would you hear if the EMS vehicle were traveling at 60 miles per hour?

2. If three resistors having resistances R_1, R_2, and R_3 are connected in parallel, their combined resistance, R, is modeled by the formula

$$R = \frac{1}{\dfrac{1}{R_1} = \dfrac{1}{R_2} = \dfrac{1}{R_3}}.$$

a. Determine R if R_1 is 4 ohms, R_2 is 8 ohms, and R_3 is 12 ohms.

b. Simplify the complex fraction on the right side of the original formula.

c. Redo part a using the new formula from part b. How do the answers compare?

3. Simplify the following complex fractions.

a. $\dfrac{\frac{1}{2} - \frac{2}{x}}{x - 4}$

b. $\dfrac{\frac{1}{x} - \frac{1}{2}}{\frac{1}{x^2} - \frac{1}{4}}$

c. $\dfrac{x + \frac{2x - 6}{x - 1}}{\frac{x}{3} - \frac{3}{x}}$

d. $\dfrac{\frac{x}{2} - 1}{x - \frac{4}{x}}$

4. You decide to buy a new car, but you are concerned about the amount of the monthly payments. The amount, A, of each monthly payment is modeled by the formula

$$A = \dfrac{pi}{1 - \dfrac{1}{(1 + i)^n}},$$

where P represents the principal or the amount borrowed,

   i is the monthly interest rate, and

   n is the number of monthly payments.

a. The car you are looking at costs \$16,000 at 1% monthly interest. If you want to pay for the car over 60 months, how much will your monthly payment be?

b. Simplify the complex fraction on the right side of the original formula.

c. Use the simplified formula to determine the monthly payments. How does your answer compare to the answer in part a?

Cluster 1   What Have I Learned?

1. Make a list of some of the special features of the graphs of rational functions.

2. Describe the connection, if any, between the domain of a rational function and the equation of its vertical asymptote.

3. Describe the algebraic steps required to determine the vertical asymptote(s) of the graph of a rational function.

4. For what values of k will the graph of $y = \dfrac{k}{x}$ be in the second and fourth quadrants?

5. a. Describe the algebraic steps required to solve $10 = \dfrac{35}{1 + 5x}$.

  b. Explain the technique for solving the equation in part a graphically.

6. Explain how you would determine the horizontal asymptote of the rational function $f(x) = \dfrac{6x + 1}{2 - 3x}$.

Cluster 1 How Can I Practice?

1. Describe the relationship between the graphs of $f(x) = \dfrac{1}{x}$ and $g(x) = \dfrac{-1}{x}$.

2. Describe the relationship between the graphs of $f(x) = \dfrac{1}{x}$ and $g(x) = \dfrac{1}{x^3}$.

3. **a.** Suppose you are taking a trip of 145 miles. Assume that you drive the entire distance at a constant speed. Express your time to take this trip as a function of your speed.

   b. What is the practical domain of this function?

   c. Using the equation for the function, determine the domain.

4. Determine the domain of each of the following functions. Then give the equation of the vertical asymptote of each function.

   a. $g(x) = \dfrac{10}{x + 5}$

   b. $f(x) = \dfrac{5}{13 - 2x}$

   c. $g(x) = \dfrac{-3}{5x - 8}$

   d. $h(x) = \dfrac{0.02}{5.7x - 3.2}$

5. The weight of a body above the surface of Earth varies inversely with the square of the distance from the center of Earth. If an object weighs 100 pounds when it is 4000 miles from the center of Earth, how much will it weigh when it is 4500 miles from the center?

6. A manufacturer of lawn mowers uses the function $C(x) = \dfrac{132x + 75{,}250}{x}$ to model the average cost per lawn mower, in dollars, where x is the number of lawn mowers produced.

  a. What is the practical domain of this function?

  b. What is the minimum number of lawn mowers that must be manufactured to bring the average cost per lawn mower down to $199? Solve algebraically and verify graphically with your calculator.

7. The concentration of a drug in the bloodstream, measured in milligrams per liter, can be modeled by the function

$$C = \frac{14t}{3t^2 + 2.5},$$

where t is the number of minutes after injection of the drug.

  a. How long after injection will it take for the concentration to equal 0.05 milligram per liter? Solve algebraically and check graphically.

  b. Use the graph to determine when the drug will be at its highest concentration.

8. Solve each equation algebraically. Verify your answer graphically.

a. $\dfrac{3}{x+1} = 4$

b. $\dfrac{3x}{2x-5} = 10$

c. $\dfrac{4}{x+3} + 12 = 52$

d. $\dfrac{2.4x}{1+0.3x} = 5.8$

9. As an object rises, the effect of Earth's gravitational pull on the object is reduced.

If an object weighs E kilograms at sea level, then the weight, W (also in kilograms), of the object at a distance of h kilometers above sea level is modeled by the function

$$W = \frac{E}{\left(1 + \frac{h}{6400}\right)^2}.$$

a. Suppose you are flying in a commercial jetliner 15 kilometers above sea level. Replace E with your body weight, measured in kilograms (1 kilogram = 2.2 pounds), and calculate your weight at 15 kilometers above sea level.

b. If an astronaut weighs 70 kilograms at sea level, write a function equation that expresses the astronaut's weight as a function of his or her distance above sea level.

c. Complete the following table using the function equation from part b.

h (km)	0	10	100	1000	1500	2000	10,000	20,000
W (kg)								

d. As the height, h, of the space shuttle increases, what happens to the corresponding weight of the astronaut?

e. As the space shuttle reaches its orbiting altitude of 650 kilometers above sea level, what is the weight of the astronaut?

f. What is the practical domain of the weight function?

g. Use your graphing calculator to sketch a graph of the weight function. Use the window Xmin $= 0$, Xmax $= 40000$, Ymin $= 0$, and Ymax $= 80$.

h. At what altitude does the astronaut's weight equal one-half of what it is at sea level?

10. An electrical circuit has three resistors. The total resistance of the circuit, R, is related to the individual resistances R_1, R_2, and R_3 by the equation

$$\frac{1}{R_1} + \frac{1}{R_2} + \frac{1}{R_3} = \frac{1}{R}.$$

a. You know that $R_1 = 4$ ohms, $R_2 = 6$ ohms, and the total resistance of the circuit is 2 ohms. Determine R_3.

b. Solve the equation $\dfrac{1}{R_1} + \dfrac{1}{R_2} + \dfrac{1}{R_3} = \dfrac{1}{R}$ for R.

c. Using the equation developed in part b, determine R if $R_1 = 4$ ohms, $R_2 = 6$ ohms, and $R_3 = 12$ ohms.

11. Solve each of the following equations using an algebraic approach. Verify your answers using a graphing approach.

a. $\dfrac{3}{x-1} = 10$

b. $\dfrac{2}{x} - \dfrac{4}{x} = 2$

c. $\dfrac{1}{x} + \dfrac{1}{4x} = \dfrac{1}{4}$

d. $\dfrac{3}{x} - 4 = \dfrac{2}{x}$

12. The average speed, s, of your round-trip commute from home to campus is modeled by

$$s = \dfrac{2d}{\dfrac{d}{r_1} + \dfrac{d}{r_2}},$$

where d is the one-way distance from home, r_1 is your average morning commute speed, and r_2 is your average afternoon commute speed.

a. Your one-way commute to campus is 15.3 miles. Your average morning commute speed is 45 miles per hour. Your average afternoon commute speed is 40 miles per hour. What is your average speed, s?

b. Simplify the original equation by rewriting the complex fraction on the right side as a single fraction.

c. Redo part a using the new equation obtained in part b. How do the results compare?

13. Simplify the following complex fractions.

a. $\dfrac{4 + \dfrac{2}{x}}{1 - \dfrac{3}{x}}$

b. $\dfrac{\dfrac{x}{5} - \dfrac{5}{x}}{\dfrac{1}{5} + \dfrac{1}{x}}$

c. $\dfrac{\dfrac{1}{x + 2}}{1 + \dfrac{1}{x + 2}}$

Activity 5.7

ky Diving

Objectives

1. Determine the domain of a radical function defined by $y = \sqrt{g(x)}$, where $g(x)$ is a polynomial.

2. Graph functions having equation $y = \sqrt{g(x)}$ and $y = -\sqrt{g(x)}$.

3. Identify the properties of the graph of $y = \sqrt{g(x)}$ and $y = -\sqrt{g(x)}$.

The first recorded successful parachute jump was made from a hot air balloon in 1797. Parachute technology and equipment were significantly refined by the use of parachutes in the military. Modern parachuting, also known as skydiving, has developed into a popular and exciting recreational activity and competitive sport.

Typically, individuals jump out of an aircraft at approximately 14,000 feet and free-fall for a period of time before activating a parachute. When first exiting the aircraft, a skydiver will have a slight feeling of falling. However, during their period of free-fall, skydivers generally do not experience a "falling" sensation because the resistance of the air to their body provides some feeling of weight and direction.

The distance, s (in feet), that a skydiver travels in free-fall is a function of time, t (in seconds). This function is defined by $s = 16t^2$. *Note*: This model neglects air resistance.

1. a. Use the formula $s = 16t^2$ to complete the following table.

t, sec.	0	5	10	15	20	25
s, ft.						

b. Use the table values to sketch a graph of the free-fall distance function.

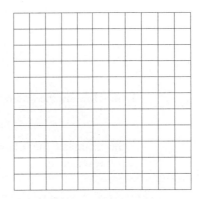

A skydiver needs to know how much time she has in free-fall before having to activate the parachute. The free-fall time will depend on the distance in free-fall.

2. a. Suppose the skydiver jumps from a plane at 12,000 feet and has to activate the parachute at 3000 feet. Determine the free-fall distance.

b. Use the formula $s = 16t^2$ to determine how much time she will have in free-fall.

The process of determining free-fall time for a given free-fall distance would be simplified you had a formula that expresses time t as a function of distance s.

3. a. Solve the equation $s = 16t^2$ for t, where $t > 0$, to determine an equation for the time in free-fall, t, as a function of the distance, s.

b. Use the function from part a to complete the following table.

s, ft.	0	400	1600	3600	6400	10,000
t, sec.						

c. Use the table values to sketch a graph of the function defined by $t = 0.25\sqrt{s}$.

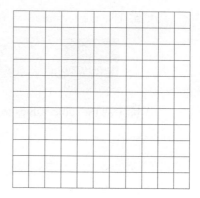

4. a. How are the ordered pairs in the tables in Problems 1a and 3b related?

b. Complete the following sentence. Functions that have their input and output values reversed are _____ functions.

Radical Functions

The investigation of the properties of the function defined by $t = 0.25\sqrt{s}$ is somewhat lim ited because of the restrictions placed on the variables t and s, which represent real-worl quantities.

5. a. Consider the general function defined by $F(x) = 0.25\sqrt{x}$. What is the domain of F

b. Sketch a graph of F.

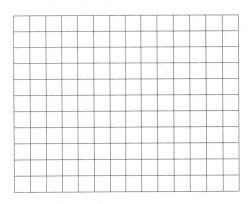

c. What is the range of F?

Recall that the square root of a negative number is not a real number. Therefore, the domain of a function defined by an equation of the form $y = \sqrt{g(x)}$, where $g(x)$ is a polynomial, is the set of all real numbers for x such that $g(x) \geq 0$.

Example 1   *Determine the domain of the function defined by* $f(x) = \sqrt{2x - 10}$.

SOLUTION

You need to determine all values of x such that $2x - 10 \geq 0$. Therefore,

$$2x - 10 \geq 0$$
$$2x \geq 10$$
$$x \geq 5.$$

The domain is all real numbers greater than or equal to 5.

6. Consider the functions defined by the following equations.

$$f(x) = \sqrt{x} \qquad g(x) = \sqrt{x + 2} \qquad h(x) = \sqrt{x - 3}$$

a. Determine the domain of each function.

b. Sketch graphs of the functions f, g, and h on the same coordinate system.

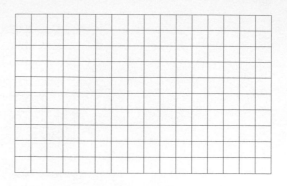

c. Are the functions f, g, and h increasing or decreasing?

7. a. Determine the domain and range of the functions defined by $f(x) = \sqrt{x}$ and $g(x) = -\sqrt{x}$.

b. Sketch graphs of f and g on the same coordinate system.

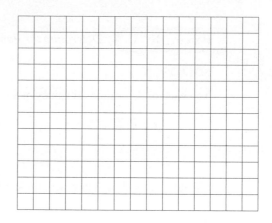

c. How would you obtain the graph of g from the graph of f?

8. a. Determine the domains of the functions defined by $h(x) = \sqrt{x-3}$ and $F(x) = \sqrt{3-x}$.

b. Complete the following tables.

x	3	4	7	12	19	28
h(x)						

x	3	2	−1	−6	−13	−22
F(x)						

c. Sketch the graphs of h and F on the same coordinate system.

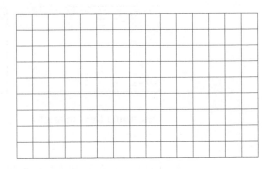

d. Determine whether the functions h and F are increasing or decreasing.

9. Consider the function defined by $f(x) = -\sqrt{3x - 6}$.

a. Determine the domain of f.

b. Determine the x-intercept of f.

c. Complete the following table. If necessary, approximate $f(x)$ to the nearest tenth.

x	2	3	4	5	8	10
f(x)						

d. Sketch a graph of f.

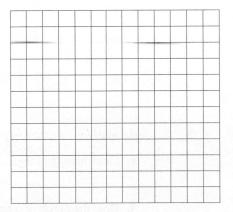

e. Use your graphing calculator to sketch a graph of f. Your screens should appear as follows.

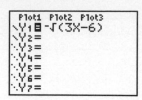 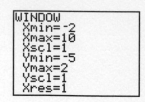 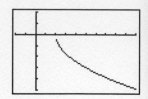

f. Determine the range of f.

g. Does the graph of f have a maximum value? If so, what is it?

10. Consider the function defined by $g(x) = \sqrt{x^2 + 4}$.

a. Determine the domain of g.

b. Complete the following table. If necessary, approximate $g(x)$ to the nearest hundredth.

x	-4	-2	0	2	4	6
$g(x)$						

c. Sketch a graph of g.

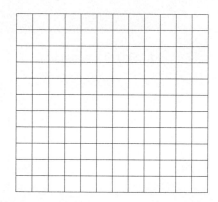

d. Verify your graph in part c using your graphing calculator.

e. Determine the range of g.

f. Does the graph of g have a maximum or minimum value? If so, what is it?

11. a. Complete the following table for $f(x) = \sqrt{x^2} + \sqrt{4}$.

x	-4	-2	0	2	4	6
$f(x)$						

b. How do the outputs in the table in part a compare with the outputs for
$g(x) = \sqrt{x^2 + 4}$ in Problem 13b?

c. Use your graphing calculator to graph $f(x) = \sqrt{x^2} + \sqrt{4}$.

d. How does the graph of $f(x) = \sqrt{x^2} + \sqrt{4}$ compare to the graph of
$g(x) = \sqrt{x^2 + 4}$ in Problem 10c?

e. Is the expression $\sqrt{x^2 + 4}$ equivalent to $\sqrt{x^2} + \sqrt{4}$? Explain.

Problem 11 demonstrates the following important fact about radicals.
$$\sqrt{a + b} \neq \sqrt{a} + \sqrt{b}$$

Recall from Chapter 2 that the expression $\sqrt{x}$ can also be written as $x^{1/2}$. The fractional exponent means that you are taking the positive square root of x. Therefore, the expression $\sqrt{3x - 2}$ can also be written as $(3x - 2)^{1/2}$.

12. Sketch the graph of $y = (3x - 2)^{1/2}$. Use your graphing calculator to verify that this is the same as $y = \sqrt{3x - 2}$. What is the domain?

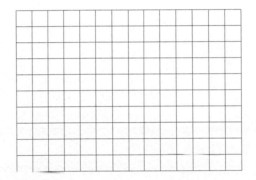

Space and Radicals

13. It is not unreasonable to imagine that someday travel in space will be a common occurrence. According to Einstein's theory of relativity, time would pass more quickly on Earth than it would for someone who is traveling in a spacecraft at a velocity close to the speed of light. As a result, a person on Earth would age more rapidly than a space traveler. The formula

$$A = F\sqrt{1 - \frac{v^2}{c^2}}$$

models the relationship between the aging rate, A, of an astronaut and the aging rate, F, of a person on Earth. The variable v represents the astronaut's velocity in miles per second, and c represents the speed of light (approximately 186,000 miles per second).

a. Suppose you are on a spaceship that is traveling at 80% of the speed of light, that is $v = 0.8c$. What is your aging rate compared to a person on Earth?

b. If you travel at 80% of the speed of light for 1 year (as you perceive it), approximately how much time has passed for a person on Earth?

c. Suppose you are traveling at a velocity very close to the speed of light. Substitute c (speed of light) for v in the formula, and simplify. Interpret your results.

14. Escape velocity is the minimum speed that an object must attain to escape a planet's pull of gravity. Escape velocity, V, is modeled by the formula

$$V = \sqrt{\frac{2Gm}{r}},$$

where G is the universal gravitational constant,

m is the mass of the planet, and

r is the radius of the planet.

If Earth has mass 5.97×10^{24} kilograms and radius 6.37×10^{6} meters, then determin the escape velocity for Earth. Round your answer to the nearest whole number. Us $G = 6.67 \times 10^{-11}$ m^3/kg · sec.^{2}

SUMMARY: ACTIVITY 5.7

Square Root Notation and Terminology

1. $\sqrt[2]{n}$, or simply $\sqrt{n}$, represents the square root of a nonnegative number n. The 2 is called the **index**. In general, when you are working with square roots, the 2 is omitted.

2. The symbol $\sqrt{\ }$ is called the **radical sign**. The expression under the radical is called the **radicand**.

3. $\sqrt{n} \geq 0$, where $n \geq 0$

4. $\sqrt{a \cdot b} = \sqrt{a} \cdot \sqrt{b}$, where $a \geq 0, b \geq 0$

5. $\sqrt{\frac{a}{b}} = \frac{\sqrt{a}}{\sqrt{b}}, b \neq 0$

6. $\sqrt{a + b} \neq \sqrt{a} + \sqrt{b}$

Properties of Radical Functions

1. The function defined by $y = \sqrt{g(x)}$ has domain all real numbers x with $g(x) \geq 0$.

2. The function defined by $y = -\sqrt{g(x)}$ has domain all real numbers x with $g(x) \geq 0$. The graph of $y = -\sqrt{g(x)}$ is the reflection of the graph of $y = \sqrt{g(x)}$ about the x-axis.

EXERCISES: ACTIVITY 5.7

1. Using your calculator, determine the value of each number to the nearest hundredth, if necessary

a. $\sqrt{30}$ b. $6^{1/2}$ c. $\left(\sqrt{13}\right)^4$ d. $\left(9^{1/2}\right)^3$

2. Determine the domain of each function.

a. $f(x) = \sqrt{x - 5}$ b. $g(x) = \sqrt{3x + 2}$

c. $h(x) = \sqrt{6 - 2x}$ d. $R(x) = -\sqrt{2x}$

3. The following table gives the number of undergraduate students receiving federal Stafford loans from the academic years 2000–2001 through 2006–2007.

	YEAR						
	2000–2001	2001–2002	2002–2003	2003–2004	2004–2005	2005–2006	2006–2007
x, Number of Years Since 2000–2001	0	1	2	3	4	5	6
$N(x)$, Number of Undergrads Receiving Stafford Loans (in millions)	4.2	4.5	5.0	5.5	5.9	6.0	6.1

This data can be modeled by the function

$$N(x) = 1.222\sqrt{x + 0.24} + 3.442.$$

a. Determine the N-intercept. What does this intercept represent in this situation?

b. Complete the following table using the given equation. How well does the equation represent the actual data?

x	0	1	2	3	4	5	6
$N(x)$							

c. Sketch a graph of the student loan function represented by $N(x) = 1.222\sqrt{x + 0.24} + 3.442$.

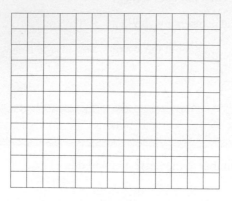

d. Use the model to predict the number of undergrads who will receive Stafford loans in the 2012–2013 school year.

4. Describe how to obtain the graph of the second function from the graph of the first.

a. $g(x) = \sqrt{x}$, $h(x) = \sqrt{x} + 1$

b. $f(x) = \sqrt{x}$, $g(x) = -\sqrt{x}$

5. a. Sketch a graph of $f(x) = x^2$ and $g(x) = \sqrt{x}$ on the same coordinate system. Use the graphs to answer parts b and c.

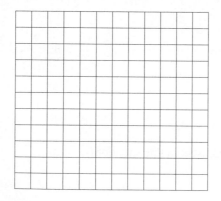

b. Is $x^2 > \sqrt{x}$ for $0 < x < 1$? Explain.

c. Is $x^2 > \sqrt{x}$ for $x > 1$? Explain.

6. Which of the following functions increases more rapidly for $x > 1$: $f(x) = \sqrt{x}$ or $g(x) = \ln(x)$?

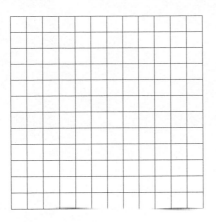

7. For each of the given functions, do the following:

  i. Determine the domain.

  ii. Determine the x- and y-intercepts.

  iii. Sketch a graph.

  a. $h(x) = \sqrt{2x + 3}$

   i.

   ii.

   iii.

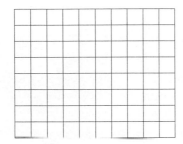

  b. $f(x) = -\sqrt{4x + 8}$

   i.

   ii.

   iii.

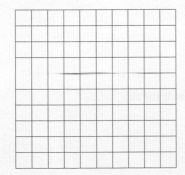

c. $g(x) = \sqrt{5 - x}$

    i.

    ii.

    iii.

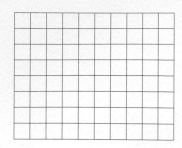

8. The surface area, S, of a cone is modeled by the formula

$$S = \pi r \sqrt{r^2 + h^2},$$

where r is the radius of the base and h is the height.

An umbrella is in the shape of a cone having radius 4 feet and height 2 feet. Determine the amount of material needed to make the umbrella.

9. A shipping carton in the shape of a rectangular box has dimensions 12 inches $\times$ 24 inches $\times$ 17 inches. The diagonal, d, of a rectangular box is modeled by

$$d = \sqrt{w^2 + l^2 + h^2},$$

where w is the width, l is the length, and h is the height.

Will an umbrella measuring 34 inches long fit in the carton?

10. Law enforcement officers investigating a car accident often use the formula $s = \sqrt{30fl}$ to estimate a car's speed, s, in miles per hour based upon the length, l (in feet), of the skid marks. The f in the formula represents the road condition at the time of the accident.

a. On dry pavement the f-value is 0.85. Write a function equation for speed, s, as a function of the length of the skid marks, l, on dry pavement.

b. Estimate the speed of a car if the skid marks on dry pavement are 90 feet long.

c. What is the practical domain for the function in part a?

d. Graph this function using your graphing calculator.

e. Use the graph to determine the length of the skid marks on dry pavement if the car was traveling at 70 miles per hour when the brakes were applied.

Activity 5.8

Falling Objects

Objective

1. Solve an equation involving a radical expression using a graphical and algebraic approach.

If an object is dropped from a tall building, the time, t, in seconds, it takes for the object to strike the ground is modeled by

$$t = \frac{\sqrt{d}}{4} = \frac{1}{4}\sqrt{d},$$

where the input, d, is the distance traveled in feet. The time it takes for the object to hit the ground varies directly to the square root of the distance traveled. The number $\frac{1}{4}$ or 0.25 is the constant of proportionality or constant of variation.

1. **a.** How long will it take an object to fall from the top of the Willis Tower in Chicago, a distance of 1450 feet? Round to the nearest hundredth of a second.

   b. Complete the following table. Round to the nearest hundredth of a second.

d (ft.)	0	100	200	300	500	750	1000
t (sec.)							

   c. Sketch a graph of the given function. Use the table in part b to determine an appropriate scale.

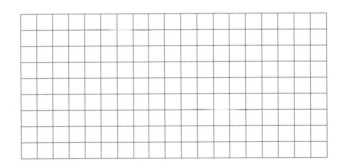

   d. How tall must a building be for an object to take 6 seconds to fall to the ground? Use the graph from part c to approximate your answer.

Solving Equations Involving Radical Expressions

Suppose you are interested in determining the value of d for many different values of t. In such a situation, the process could be simplified if you had a rule for d as a function of t. The equation $t = \frac{\sqrt{d}}{4}$ gives t as a function of d. You need to solve this equation for d.

> **Definition**
>
> An equation in which at least one side contains a radical with a variable in the radicand is called a **radical equation**.

> **Example 1**   *Examples of radical equations include* $t = \dfrac{\sqrt{d}}{4}$, $\sqrt{2x + 1} = 5$, *and*
> $2\sqrt{3x} = \sqrt{5x - 7}$.

Solving an equation algebraically when the variable appears under a radical involves using the following property of equations.

> If a and b are two quantities such that $a = b$, then $a^n = b^n$, where n is a positive integer.

> **Example 2**   *If $t = \sqrt{s}$, then apply the preceding property by squaring both sides of the equation.*
>
> $t^2 = (\sqrt{s})^2$ Rewrite $\sqrt{s}$ as $s^{1/2}$.
>
> $t^2 = (s^{1/2})^2$ Apply the property of exponents $(a^m)^n = a^{mn}$.
>
> $t^2 = s^{1/2 \cdot 2}$ Simplify.
>
> $t^2 = s^1$
>
> Therefore, if $t = \sqrt{s}$, then $t^2 = s$.

2. a. Now solve the equation $t = \dfrac{\sqrt{d}}{4}$ for d by first squaring both sides of the equation.

b. Using the new formula from part a, determine how tall a building must be for an object to take 6 seconds to fall to the ground.

c. How does your answer compare to the result in Problem 1d?

d. You could also answer part b by solving the equation $6 = \dfrac{\sqrt{d}}{4}$. Solve the equation.

How does your answer compare to the result in part b?

The following example demonstrates a general algebraic procedure for solving radical equations.

> **Example 3**   *Solve for x: $2\sqrt{3x} - \sqrt{5x + 7} = 0$.*

SOLUTION

Step 1.   If the equation involves more than one radical term, isolate one radical term on one side of the equation.

$$2\sqrt{3x} - \sqrt{5x + 7} = 0$$
$$\underline{+ \sqrt{5x + 7} + \sqrt{5x + 7}}$$
$$2\sqrt{3x} = \sqrt{5x + 7}$$

Step 2. Square both sides of the equation.

$$(2\sqrt{3x})^2 = (\sqrt{5x + 7})^2$$
$$4 \cdot 3x = 5x + 7$$

Step 3. If a radical remains, repeat steps 1 and 2. Solve the resulting equation.

$$4 \cdot 3x = 5x + 7$$
$$12x = 5x + 7$$
$$7x = 7$$
$$x = 1$$

Step 4. Check all solutions in the original equation.

$$2\sqrt{3(1)} - \sqrt{5(1) + 7} = 2\sqrt{3} - \sqrt{12}$$
$$= 2\sqrt{3} - \sqrt{4 \cdot 3} = 2\sqrt{3} - 2\sqrt{3} = 0$$

You can also check your answer by solving the equation graphically.

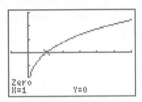

3. Suppose two different objects are dropped: a marble and a large beach ball. Because of air resistance, the beach ball will take longer than the marble to fall the same distance. Assume that the marble falls according to $t = 0.25\sqrt{d}$, as in Problem 1. The time for the beach ball to hit the ground is modeled by $t = k\sqrt{d}$, where the positive constant, k, is determined by experiment.

   The beach ball is dropped from a height of 250 feet, and it takes 4.11 seconds to hit the ground. Determine the constant, k, accurate to the hundredths place. Remember that $(ab)^2 = a^2b^2$.

4. Now suppose the beach ball in Problem 3 is dropped from a height 50 feet lower than the marble. Then $t = 0.26\sqrt{d - 50}$ is the time for the beach ball to drop $d - 50$ feet, where d is the height the marble falls.

   a. Write an equation that can be used to determine from what height the marble must be dropped so the beach ball and marble will hit the ground at the same time.

   b. Solve this equation using an algebraic approach.

   c. Verify your solution in part b using your graphing calculator.

5. Consider the following algebraic solution of the equation $\sqrt{x+3} + 5 = 0$.

$$\sqrt{x+3} + 5 = 0$$
$$\sqrt{x+3} = -5$$
$$(\sqrt{x+3})^2 = (-5)^2$$
$$x + 3 = 25$$
$$x = 22$$

a. It appears that $x = 22$ is a solution to the given equation. Check the solution by substituting 22 for x in the original equation. Does it check?

b. What happened in the solution process to cause an **extraneous solution** (an apparent solution that does not check) to appear?

c. Does the equation $\sqrt{x+3} + 5 = 0$ have a solution? Include a graph to help support your answer.

6. In Exercise 3 on page 551, you were given the following table giving the number of undergraduate students receiving federal Stafford loans from the academic year 2000–2001 through 2006–2007.

	YEAR						
	2000–2001	2001–2002	2002–2003	2003–2004	2004–2005	2005–2006	2006–2007
x, number of years since 2000–2001	0	1	2	3	4	5	6
$N(x)$, number of undergraduates receiving Stafford loans (in millions)	4.2	4.5	5.0	5.5	5.9	6.0	6.1

This date was modeled by the function

$$N(x) = 1.222\sqrt{x + 0.24} + 3.442.$$

Determine the year that this model would predict that the number of undergraduate students receiving Stafford loans would be approximately 8 million.

SUMMARY: ACTIVITY 5.8

1. If a and b are two quantities such that $a = b$, then $a^n = b^n$, where n is a positive integer.

2. To solve an equation involving one radical expression,

   a. Isolate the radical term on one side of the equation.

   b. Square both sides of the equation.

   c. Solve the resulting equation.

   d. Check all solutions in the original equation.

3. To solve an equation involving more than one radical expression,

   a. If the equation involves more than one radical term, isolate one radical term on one side of the equation.

   b. Square both sides of the equation.

   c. If a radical remains, repeat steps a and b. Solve the resulting equation.

   d. Check all solutions in the original equation.

4. When you raise both sides of an equation to an even power, it is possible to introduce **extraneous solutions** into the process. These are values of the variable that appear by the process to be solutions but do not make the original equation true. It is important to check all potential solutions to determine whether or not they are real solutions or extraneous solutions.

EXERCISES: ACTIVITY 5.8

1. Solve the following equations, or determine which does not have a solution. Explain your reasoning, or verify your solution.

   a. $\sqrt{2x + 1} = 3$

   b. $\sqrt{x + 1} + 5 = 1$

   c. $\sqrt{2 - x} = -x$

2. Solve each of the following equations algebraically. Then verify your answers graphically.

a. $\sqrt{x} = 2.5$ **b.** $\sqrt{x} - 3 = 0$ **c.** $\sqrt{2x} = 14$

d. $3\sqrt{x} = 243$ **e.** $4 - 5\sqrt{3x} = 1$ **f.** $\sqrt{x+1} = 9$

3. Solve each of the following equations algebraically and by graphing. Be aware of any extraneous roots.

a. $\sqrt{x+5} = 1$ **b.** $10\sqrt{x+2} = 20$

c. $\sqrt{5-x} = x + 1$

4. Solve algebraically and graphically: $\sqrt{1.4x + 3.2} = \sqrt{3.8x - 1}$.

5. The time, t, in seconds, that it takes for a pendulum to complete one complete period (to swing back and forth one time) is modeled by

$$t = 2\pi\sqrt{\frac{L}{32}},$$

where L is the length of the pendulum, in feet (see diagram).

How long is the pendulum of a clock with a period of 1.95 seconds?

6. In a certain population, there are 28,520 births on a particular day. The number, N, of these people surviving to age x can be modeled by the function $N = 2850\sqrt{100 - x}$.

a. According to this model, how many of the 28,520 babies will survive to age 5?

b. What is the practical domain of this function?

c. When only 5000 of this group are still alive, how old do you expect them to be?

d. After how many years will half of the original population of 28,520 people remain alive?

7. a. A pressure gauge on a bridge indicates a wind pressure, P, of 10 pounds per square foot. What is the velocity, V, of the wind if

$$V = \sqrt{\frac{1000P}{3}},$$

where velocity is measured in miles per hour?

b. What is the wind pressure if the wind is blowing at 70 miles per hour?

8. Artificial gravity can be created in a space station by revolving the station. The number of revolutions required can be modeled by

$$N = \frac{1}{2\pi}\sqrt{\frac{a}{r}},$$

where N is measured in revolutions per second, a is the artificial gravity produced (measured in meters per second squared), and r is the radius of the space station in meters.

a. To produce an artificial acceleration simulating gravity on Earth, a must equal 9.8 meters per second squared. If the space station must revolve at the rate of one revolution every 5 minutes, what must its radius be? Solve both algebraically and graphically. Be careful; N is measured in revolutions per second.

b. Solve the original formula for r.

c. Use the formula in part b to answer part a again. How do the answers compare?

9. The Masteller formula for calculating the adult body surface area, A, is

$$A = \sqrt{\frac{hw}{3131}},$$

where h is the person's height in inches and w is the adult's weight in pounds. A is the surface area in square meters.

a. Determine the body surface area, A, of an adult who is 70 inches tall and weighs 200 pounds.

b. Solve the formula for w.

Activity 5.9

ropane Tank

Objectives

1. Determine the domain of a function defined by an equation of the form $y = \sqrt[n]{g(x)}$, where n is a positive integer and $g(x)$ is a polynomial.

2. Graph $y = \sqrt[n]{g(x)}$.

3. Identify the properties of graphs of $y = \sqrt[n]{g(x)}$.

4. Solve radical equations that contain radical expressions with an index other than 2.

A propane tank is in the shape of a sphere. The radius, r, of the spherical tank is modeled by the formula

$$r = \sqrt[3]{\frac{3V}{4\pi}},$$

where V is the volume of the tank.

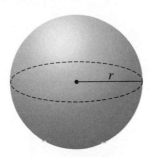

1. a. What is the radius of a propane tank having volume 50 cubic feet (round the answer to the nearest tenth)?

b. Complete the following table.

V (ft.³)	0	5	10	15	20	25	100
r (ft.)							

c. Sketch a graph of the radius function over its practical domain.

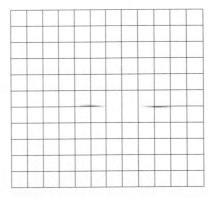

Graphs of $y = \sqrt[n]{g(x)}$, $n = 3, 4, 5$

The investigation of the properties of the radius function defined by $r = \sqrt[3]{\frac{3V}{4\pi}}$ is somewhat limited by the restrictions placed on the variables V and r.

2. a. Consider the cube root function defined by $f(x) = \sqrt[3]{x}$. What is the domain of f?

b. Complete the following table. Round $f(x)$ to nearest hundredth, if necessary.

x	-10	-7	-4	-1	0	1	4	7	10
$f(x)$									

c. Sketch a graph of f. Verify your sketch using your graphing calculator.

d. Is the function f increasing or decreasing?

There are two different ways to enter the cube root function on your calculator.

Method 1: Using fractional exponents, enter the cube root of x as $x \wedge (1/3)$ in the $Y =$ editor.

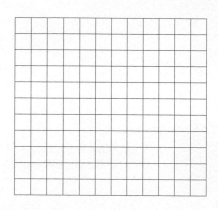

Method 2: Select the $Y =$ editor, highlight the Y_n you want, and then select the MATH menu. Option 4 is the cube root.

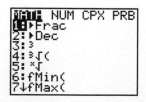

Select option 4 and press (ENTER). Insert the argument x and the right parenthesis to complete the function.

3. Use your graphing calculator to verify that the graphs of $y = x^{1/3}$ and $y = \sqrt[3]{x}$ are identical. Your screens should appear as follows.

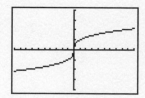

4. Consider functions defined by $f(x) = x^3$ and $g(x) = \sqrt[3]{x}$.

    a. Determine the domain of each function.

    b. Complete the following table.

x	-10	-7	-4	-1	0	1	4	7	10
$f(x)$									
$g(x)$									

    c. Sketch a graph of f and g on the same coordinate system. Verify your sketch using your graphing calculator.

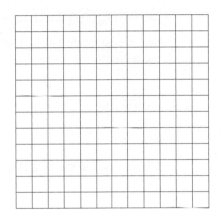

    d. Are f and g inverse functions? Explain.

    e. Determine the composition of f and g. That is, determine $f(g(x))$ and $g(f(x))$.

5. Given the functions f and g defined by

$$f(x) = \sqrt[4]{x} - 1 \quad \text{and} \quad g(x) = \sqrt[5]{x}.$$

    a. Complete the following table.

x	-10	-5	-1	0	1	5	10
$f(x)$							
$g(x)$							

    b. Determine the domains of f and g.

c. Sketch a graph of f and g on the same coordinate axes.

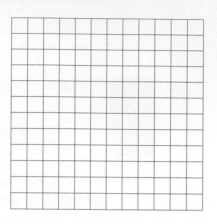

Solving Equations Involving Radical Expressions

Solving an equation such as $\sqrt[3]{x + 1} + 5 = 9$ is similar to solving equations involvin square roots.

Example 1   *Solve $\sqrt[3]{x + 1} + 5 = 9$.*

SOLUTION

Step 1. Isolate the radical term on one side of the equation.

$$\sqrt[3]{x + 1} + 5 = 9$$
$$\sqrt[3]{x + 1} \quad\;\; = 4$$

Step 2. Raise each side of the equation to the power that matches the index of the radical. In this situation, cube each side and simplify.

$$(\sqrt[3]{x + 1})^3 = 4^3$$

Step 3. If a radical remains, repeat steps 1 and 2. Solve the resulting equation.

$$x + 1 = 64$$
$$x = 63$$

Step 4. Check all solutions in the original equation.

$$\sqrt[3]{63 + 1} + 5 = 9$$
$$\sqrt[3]{64} + 5 = 9$$
$$4 + 5 = 9$$
$$9 = 9$$

You can also verify your solutions graphically.

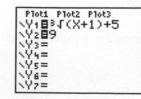
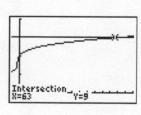

6. a. Returning to the propane tank situation, suppose the amount of space available in all directions for a propane tank is 10 feet. Write an equation to determine the maximum volume of a spherical tank that can fit into the given space.

b. Solve this equation using an algebraic approach.

c. Verify the solution using your graphing calculator. The screen should appear as follows.

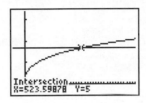

7. Solve the formula $V = I^3$ for I.

8. The basal metabolic rate (BMR) is the number of calories per day a person needs to maintain life. A person's basal metabolic rate is a function of his or her weight and is modeled by

$$B(w) = 70\sqrt[4]{w^3},$$

where $B(w)$ represents the basal metabolic rate measured in calories per day and w is the person's weight in kilograms.

a. Write the expression $70\sqrt[4]{w^3}$ using fractional exponents.

b. If your friend weighs 50 kilograms (approximately 110 pounds), determine her basal metabolic rate. Round your answer to the nearest calorie.

c. Determine your basal metabolic rate. Be sure to convert your weight to kilograms.

d. Suppose a person is on a 2000-calorie-per-day diet. If the number of calories represents the person's basal metabolic rate, write an equation to determine the weight that is associated with this number of calories per day.

e. Solve the equation in part d. To help determine to what power you need to raise eac
side, first consider how you simplify the expression $\left(x^{3/4}\right)^{4/3}$.

f. If the person weighs 210 pounds, is the 2000-calorie diet healthy?

SUMMARY: ACTIVITY 5.9

1. To solve an equation when the variable appears under a radical, use the following two
properties.

    i. If a and b are two quantities such that $a = b$, then $a^n = b^n$, for any positive integer.

    ii. $\left(b^{1/n}\right)^n = b^1$ and $\left(b^{m/n}\right)^{n/m} = b^1$

2. **i.** The domain of $f(x) = \sqrt[n]{x}$, where n is a positive odd integer, is all real numbers.

    ii. The domain of $f(x) = \sqrt[n]{x}$, where n is a positive even integer, is $x \geq 0$.

3. To solve equations involving radicals,

  Step 1. Isolate the radical term on one side of the equation.

  Step 2. Raise each side of the equation to the power that matches the index of the radical.

  Step 3. If a radical remains, repeat steps 1 and 2. Solve the resulting equation.

  Step 4. Check all solutions in the original equation.

EXERCISES: ACTIVITY 5.9

1. If possible, determine the exact value of each of the following.

  a. $\sqrt[3]{64}$ **b.** $\sqrt[4]{16}$ **c.** $(-27)^{1/3}$ **d.** $(625)^{1/4}$

  e. $\sqrt{\frac{1}{36}}$ **f.** $(-81)^{1/4}$ **g.** $(100{,}000)^{1/5}$ **h.** $(-1)^{1/6}$

2. If the volume of a cube is 728 cubic centimeters, what is the length of one edge to the nearest
tenth of a centimeter?

3. If the volume of a cube is decreased from 1450 cubic inches to 1280 cubic inches (and still
remains a cube), by how much has the length of one edge decreased?

Exercise numbers appearing in color are answered in the Selected Answers appendix.

4. The volume of a sphere is 520 cubic meters. What is the diameter of the sphere?

5. What is the domain of each function?

a. $y = \sqrt[3]{x + 6}$

b. $f(x) = \sqrt[4]{x - 3}$

c. $g(x) = \sqrt[5]{2 - x}$

d. $f(x) = (2 - x)^{1/6}$

6. Solve each of the following algebraically and graphically.

a. $\sqrt[3]{x + 4} = 3$

b. $\sqrt[4]{x + 5} = 2$

c. $\sqrt[3]{2x - 3} + 4 = 3$

d. $\sqrt[4]{2 - x} = 5$

7. Solve each of the following algebraically, and verify your results graphically.

a. $x^{2/3} = 16$

b. $2x^{3/4} = 54$

8. a. The diameter d of a sphere is modeled by the formula

$$d = \sqrt[3]{\frac{6V}{\pi}},$$

where V represents the volume of a sphere. Approximate the diameter of a sphere having a volume of 10 cubic inches.

b. Determine the volume of a sphere having diameter 5 feet.

9. The radius, r, of a sphere is modeled by

$$r = \sqrt[3]{\frac{3V}{4\pi}},$$

where V is the volume of the sphere.

a. Determine the radius of a sphere that has a volume equal to 40 cubic centimeters.

b. Determine the volume of a sphere that has a radius equal to 3.5 feet.

c. Solve the formula for V, expressing volume as a function of the radius.

Cluster 2   What Have I Learned?

1. Explain the steps involved in solving the equation $\sqrt{2x + 3} = 5$ using an algebraic approach. Why must you be sure to check your solution?

2. What is the domain for the variable b in each of the following?

a. $\sqrt[n]{b}$, where n is an even positive integer

b. $\sqrt[n]{b}$, where n is an odd positive integer

3. Is it possible for an extraneous solution to appear when the equation $\sqrt[3]{2x + 1} = -3$ is being solved? Explain.

4. Describe two ways to check for extraneous solutions when you are solving an equation by squaring both sides.

5. Determine whether the following statements are true or false. Explain your answer.

a. If two numbers are equal, then their squares are equal.

b. If the squares of two numbers are equal, then the two numbers are equal.

6. a. For a given value of x, which is greater, $\sqrt[5]{x}$ or $x^{1/3}$? Explain how you determine your answer.

b. In part a, did you assume that $x > 0$? Does your answer change if $x < 0$?

Chapter 5

Cluster 2　How Can I Practice?

1. Solve each of the following equations algebraically and check graphically.

   a. $\sqrt{x + 2} = 10$

   b. $(x - 5)^{1/2} = 6$

   c. $\sqrt{2x + 1} - 5 = 0$

   d. $\sqrt[3]{x^2 + 3} = 4$

   e. $\sqrt{x} = \sqrt{x + 2}$

   f. $(2 - x)^{1/3} = -2$

   g. $\sqrt[4]{2x - 5} = 2$

   h. $(2.3x + 1.9)^{1/3} = 1.6$

2. Identify the domain of each of the following functions.

   a. $f(x) = \sqrt{6 - x}$

   b. $g(x) = (2x - 9)^{1/3}$

   c. $h(x) = (x^2 - 4)^{1/4}$

3. If the volume of a cube is 458 cubic inches, what is the length of one edge? Determine the value to the nearest hundredth of an inch.

4. If the volume of a sphere is 620 cubic centimeters, what is its radius?

5. When a stone is dropped to the ground, its velocity is modeled by the function $v = \sqrt{64d}$, where d is the distance the stone has fallen, in feet, and v is its velocity in feet per second. If the stone hits the ground at 100 feet per second, from what height was it dropped?

6. A cardboard box with a square bottom has a height of 10 inches and a volume of 422.5 cubic inches. What are the dimensions of the bottom of the box?

7. Describe the similarities and differences between the graphs of $y = \sqrt{2 - x}$ and $y = \sqrt{x - 2}$.

The bracketed numbers following each concept indicate the activity in which the concept is discussed.

CONCEPT/SKILL	DESCRIPTION	EXAMPLE
Domain and range of $y = \frac{k}{x}$ [5.1]	The domain and range consist of all real numbers except zero.	$y = \frac{3}{x}$
Graph of $y = \frac{k}{x}$ [5.1]	The graph is in the first and third quadrants if $k > 0$ and in the second and fourth if $k < 0$.	$y = \frac{3}{x}$ 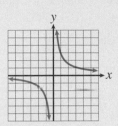
Asymptotes of the graph of $y = \frac{k}{x}$ [5.1]	The y-axis, $x = 0$, is the vertical asymptote. The x-axis, $y = 0$, is the horizontal asymptote.	$y = \frac{3}{x}$ 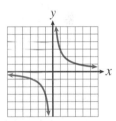
Domain of $y = \frac{k}{x^n}$ [5.2]	The domain consists of all real numbers except zero.	$y = \frac{4}{x^3}$
Graph of $y = \frac{k}{x^n}$ [5.2]	The graphs will vary depending on the values of k and n.	See the summary at the end of Activity 5.2.
Asymptotes of the graph of $y = \frac{k}{x^n}$ [5.2]	The y-axis, $x = 0$, is the vertical asymptote. The x-axis, $y = 0$, is the horizontal asymptote.	$y = \frac{4}{x^3}$ 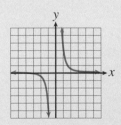
Inverse variation functions [5.2]	Functions defined by $y = \frac{k}{x^n}$ are called inverse variation functions in which y is said to vary inversely as the nth power of x; k is called the constant of variation.	For the function $y = \frac{4}{x^3}$, y varies inversely as the cube of x, and 4 is the constant of variation.
Rational function [5.3]	A function Q, defined by an equation of the form $Q(x) = \frac{k}{g(x)}$, where k is a nonzero constant and $g(x)$ is a polynomial, belongs to the family of functions known as rational functions.	$f(x) = \frac{10}{x - 3}$

CONCEPT/SKILL	DESCRIPTION	EXAMPLE
Domain of a rational function [5.3]	The domain of $y = \frac{k}{g(x)}$ is the set of all real numbers except those values of the input x such that $g(x) = 0$.	The domain of $f(x) = \frac{10}{x-3}$ is all real numbers except 3.
Vertical asymptote of a rational function [5.3]	The vertical asymptote is the vertical line that passes through the x-value for which $g(x) = 0$.	The vertical asymptote of $f(x) = \dfrac{10}{x-3}$ is $x = 3$.
Horizontal asymptote of a rational function $y = \frac{k}{g(x)}$ [5.3]	The horizontal asymptote of $f(x) = \frac{k}{g(x)}$ is the x-axis ($y = 0$).	$f(x) = \dfrac{10}{x-3}$
Rational equations [5.4]	Method 1. To solve an equation of the form $\frac{f(x)}{g(x)} = \frac{a}{b}$, where $g(x) \neq 0$ and $b \neq 0$, multiply both sides of the equation by the product $b \cdot g(x)$, and solve the resulting equation for x.	See Example 1, Activity 5.4, page 512.
Rational equations [5.4]	Method 2. To solve an equation of the form $\frac{f(x)}{g(x)} = \frac{a}{b}$, where $g(x) \neq 0$ and $b \neq 0$, cross multiply to obtain $b \cdot f(x) = a \cdot g(x)$, and solve the resulting equation for x.	See Example 1, Activity 5.4, page 513.
Horizontal asymptotes of rational functions [5.4]	Suppose the output values of a rational function R get closer and closer to a number a as the input values increase infinitely in both the positive and negative directions. Then, the graph of the function R has a horizontal asymptote. The equation of the horizontal asymptote is $y = a$.	The horizontal asymptote of $R(x) = \frac{3x+1}{x-2}$ is $y = 3$.
Determine an LCD of two or more expressions [5.5]	1. Write the prime factorization of each denominator. Express repeated factors as powers. 2. Identify the different bases (factors) in step 1. 3. Write the LCD as the product of the highest power of each of the different factors from step 2.	See Example 1, Activity 5.5, page 524.
Solving an equation involving rational expressions [5.5]	1. Determine the LCD of all denominators in the equation. 2. Multiply each side of the equation by the LCD, and simplify the resulting equation. 3. Solve the resulting equation for the desired variable.	See Problem 4, pages 524.
Simplifying rational expressions [5.6]	1. Factor the numerator and the denominator. 2. Divide the numerator and the denominator by the common factors.	See Appendix A.

CONCEPT/SKILL	DESCRIPTION	EXAMPLE
Multiplying or dividing rational expressions [5.6]	1. Factor the numerator and denominator of each fraction completely. 2. Divide out the common factors (cancel). 3. Multiply remaining factors. 4. In division, proceed as above after inverting the divisor (the fraction after the division sign).	See Appendix A.
Adding or subtracting rational expressions [5.6]	1. Determine the LCD. 2. Build each fraction to have the LCD. 3. Add or subtract numerators. 4. Place the numerator over the LCD, and simplify if necessary.	See Appendix A.
Simplifying a complex fraction [5.6]	1. Express the numerator as a single fraction. 2. Express the denominator as a single fraction. 3. Divide the numerator by the denominator. 4. Simplify, if possible.	See Example 2, Activity 5.6, page 533.
Radical functions [5.7]	The function defined by $y = \sqrt{g(x)}$ has domain all real x such that $g(x) \geq 0$.	$y = \sqrt{2x + 1}$ 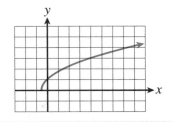
Radical functions [5.7]	The function defined by $y = -\sqrt{g(x)}$ has domain all real x such that $g(x) \geq 0$.	$y = -\sqrt{2x + 1}$ 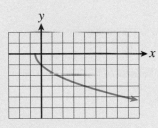
Solving an equation involving one radical expression [5.8]	1. Isolate the radical term on one side of the equation. 2. Square both sides of the equation. 3. Solve the resulting equation. 4. Check all solutions in the original equation.	See Appendix A.
Solving an equation involving more than one radical expression [5.8]	1. Isolate one radical term on one side of the equation. 2. Square both sides of the equation. 3. Solve the resulting equation. If a radical remains, repeat steps 1 and 2. 4. Check all solutions in the original equation.	See Example 3, Activity 5.8, pages 556–557.

CONCEPT/SKILL	DESCRIPTION	EXAMPLE
Domain of a function defined by an equation of the form $y = \sqrt[n]{g(x)}$, where n is a positive integer and $g(x)$ is a polynomial [5.9]	The domain of $y = \sqrt[n]{g(x)}$ is all real numbers if n is an odd positive integer. The domain of $y = \sqrt[n]{g(x)}$ is all real numbers for which $g(x) \geq 0$ if n is an even positive integer.	$y = \sqrt[3]{x + 3}$ 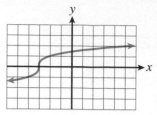
Solving radical equations that contain radical expressions with an index other than 2 [5.9]	1. Isolate a radical term on one side of the equation. 2. Raise each side of the equation to the power that matches the index of the radical. 3. If a radical remains, repeat steps 1 and 2. Solve the resulting equation. 4. Check all solutions in the original equation.	See Example 1, Activity 5.9, page 566.

1. According to the blueprint, the floor area of the stage in the new auditorium at your college must be rectangular and equal to 1200 square feet. The width of the stage is key to all of the theater productions. Therefore, in this situation, the stage's depth is a function of its width.

   a. Let d represent the depth (in feet) and w represent the width (in feet). Write an equation that expresses d as a function of w.

   b. Complete the following table using the equation from part a.

w (feet)	30	35	40	50	60
d (feet)					

   c. What happens to the depth as the width increases?

   d. What happens if the width is 100 feet? Is this realistic? Explain.

   e. Can the width be zero? Explain.

   f. What do you think is the practical domain for this function?

   g. What type of a function do you have in this situation?

   h. What is the domain of the general function?

   i. What is the vertical asymptote?

   j. What is the horizontal asymptote? Explain in words how you determined it.

2. Sketch the following graphs without using your graphing calculator.

   a. $f(x) = \dfrac{1}{x^2}$

   b. $g(x) = \dfrac{-1}{x^3}$

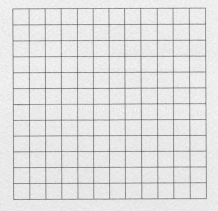
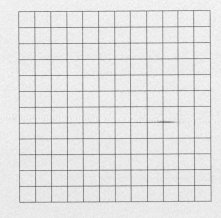

nswers to all Gateway exercises are included in the Selected Answers appendix.

577

c. Describe how the graphs are similar and how they are different.

3. a. If y varies inversely as x, and $x = 10$ when $y = 12$, then determine the value of y when $x = 30$.

b. The loudness, in decibels, of a stereo varies inversely to the square of the distance from the speaker to the person listening. If the loudness is 32 decibels at a distance of 4 feet, then what is the loudness when the listener is 10 feet from the speaker?

c. When the volume of a circular cylinder is constant, the height varies inversely as the square of the radius. If the radius is 2 inches when the height is 8 inches, determine the height when the radius is 5 inches.

4. Determine the horizontal and vertical asymptotes and the intercepts of each of the following. Then sketch a graph of each function. Verify using a graphing calculator.

a. $f(x) = \dfrac{10}{x}$ 　　　　　　**b.** $g(x) = \dfrac{4}{x - 3}$ 　　　　　　**c.** $f(x) = \dfrac{2x}{x + 2}$

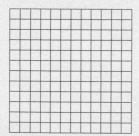　　　　

5. Students from the local community college plan to celebrate their 10-year reunion. They reserve a restaurant for an evening of entertainment. The fee for the band is $600, and food will cost each person $45.

a. If $f(n)$ represents the total cost for n people to participate in the reunion, write an equation for the total cost.

b. If 100 people attend, what will be the total cost of the event?

c. Determine a function, A in terms of n, that will represent the average cost per person to attend the event.

d. If 100 people attend the reunion, how much should each person pay?

e. Use your graphing calculator to complete the following table.

n (no. of people attending)	50	100	150	200	250
$A(n)$ (cost per person) ($)					

f. If the committee thinks that each person should pay at most $50, how many would have to attend for the cost per person to be $50? Show your answer algebraically, and check it using your graphing calculator.

g. Determine the practical domain of the function.

h. From the graph, determine the vertical asymptote. Is there a practical meaning of this asymptote in this situation?

i. From the graph of function A, determine the horizontal asymptote. Is there a practical meaning of this asymptote in this situation?

6. a. Solve the equation $\frac{4}{x-2} + 3 = 9$ using an algebraic approach. Verify your answer graphically.

b. When you graph the function $f(x) = \frac{4}{x-2} - 6$, what do you discover about the solution to the equation in part a and the x-intercept of the graph of the function f? Explain.

7. Solve each of the following equations using an algebraic approach. Verify your solutions graphically.

a. $\dfrac{3}{x+2} = 5$

b. $\dfrac{-2x}{3x-4} = 2$

8. The local grocery store has just hired you and your friend. You can stock a shelf in 15 minutes. Your friend will take 20 minutes to do the shelf. How long will it take to stock the shelf if you work together?

9. You work in the admissions office at your community college. You must assemble all of the packets for the placement test sessions. You work with a friend who takes twice as long as you do to assemble the packets. If you work together, the packets can be completed in 45 minutes. On the day you must assemble the packets, you have a big exam. How many hours does it take your friend to do the job alone?

10. Solve each of the following equations algebraically. Verify your answers graphically.

a. $\dfrac{1}{6} - \dfrac{3}{2x} = \dfrac{1}{5x}$

b. $\dfrac{2}{x} + \dfrac{3}{4x} = \dfrac{1}{12}$

11. Solve each of the following equations for the indicated variable. Express your answer as a single fraction.

a. Solve $S = \dfrac{C}{1 - r}$ for r.

b. Solve $\dfrac{1}{a} + \dfrac{3}{b} = \dfrac{4}{c}$ for b.

12. Simplify each of the following complex expressions.

a. $\dfrac{\dfrac{1}{a} + \dfrac{2}{b}}{\dfrac{2}{a} + \dfrac{1}{b}}$

b. $\dfrac{1 + \dfrac{1}{x - 2}}{1 - \dfrac{3}{x + 2}}$

3. A camera lens possesses a measurement called the focal length, f. When an object is in focus, the focal length is related to the distance of the object from the lens, p, and the image distance from the lens, q, by the formula

$$f = \frac{1}{\dfrac{1}{p} + \dfrac{1}{q}}.$$

a. Determine f if p is 4 meters and q is 3 meters.

b. Simplify the complex fraction on the right side of the original formula.

c. Redo part a using the new formula from part b. How do the answers compare?

4. a. What is the domain of the function defined by $f(x) = \sqrt{x + 4}$.

b. Sketch the graph of the function f.

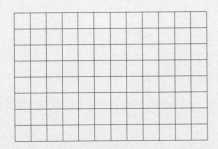

c. As the input increases, what happens to the output values?

d. What is the range of this function?

e. Are there any intercepts? If so, what are they?

f. How is the function f similar to the function $g(x) = \sqrt{x - 4}$?

g. How is the graph of the function f similar to $h(x) = -\sqrt{x + 4}$?

15. a. Draw the graph of $f(x) = \sqrt{x}$.

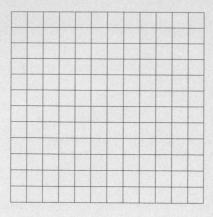

b. Determine the equation of the inverse of the function f.

c. Sketch the graph of the inverse on the same axes as the graph of f in part a from $x \geq 0$.

d. From the graphs, describe how you know that they are inverses.

e. Show that f and f^{-1} are inverses using composition of functions.

16. For each of the given functions,

  i. Determine the domain and range.

  ii. Determine the x- and y-intercepts.

  iii. Sketch a graph.

a. $f(x) = \sqrt{x} + 4$ **b.** $f(x) = \sqrt{x + 4}$

    i.   **i.**

    ii.   **ii.**

    iii.   **iii.**

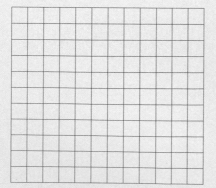
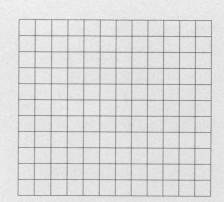

17. Solve each of the following equations using an algebraic approach. Verify your solutions graphically.

a. $\sqrt{3x - 2} - 6 = -2$

b. $\sqrt{2x + 1} - \sqrt{x + 7} = 0$

c. $\sqrt[3]{5x + 4} = 3$

d. $\sqrt{4x + 8} - 3 = -5$

e. $x^{4/3} = 81$

18. What is the domain of each function?

a. $y = \sqrt[3]{x + 8}$

b. $y = \sqrt[4]{x - 6}$

c. $y = (x + 1)^{1/6}$

19. A submarine periscope must be a certain distance above the water for it to be used to locate a ship a certain number of miles away. The model for the distance (in miles) that the submarine periscope can see is the formula $d = \sqrt{1.5h}$, where h represents the height (in feet) above the surface of the water.

How far above the surface of the water would the periscope have to be to see a ship that is 6 miles away?

20. A ring is dropped from the American span of the Thousand Island Bridge and hits the water 3.1 seconds later. What is the height of the bridge? Use the formula $T = \sqrt{\frac{d}{16}}$, where d represents the distance in feet and T represents the time in seconds.

Introduction to the Trigonometric Functions

Cluster 1   Introducing the Sine, Cosine, and Tangent Functions

Activity 6.1

The Leaning Tower of Pisa

Objectives

1. Identify the sides and corresponding angles of a right triangle.

2. Determine the length of the sides of similar right triangles using proportions.

3. Determine the sine, cosine, and tangent of an angle using a right triangle.

4. Determine the sine, cosine, and tangent of an acute angle by using the graphing calculator.

During a trip to Italy, you visit a wonder of the world, the Leaning Tower of Pisa. Your guidebook explains that the tower now makes an 85-degree angle with the ground and measures 179 feet in length. If you drop a stone straight down from the top of the tower, how far from the base will it land?

To answer this question accurately, you need a branch of mathematics called **trigonometry**. Although the development of trigonometry is generally credited to the ancient Greeks, there is evidence that the ancient Egyptian cultures used trigonometry in constructing the pyramids.

Right Triangles

As you will see, the accurate answer to the Leaning Tower question requires some knowledge of **right triangles**. Consider the following right triangle with angles A, B, and C and sides a, b, and c.

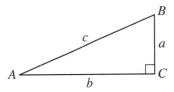

Definition

Angle C is the right angle, the angle measuring 90°. The side opposite the right angle, c, is called the **hypotenuse**. Side a is said to be **opposite** angle A because it is not part of angle A. Side b is said to be **adjacent** to angle A because it and the hypotenuse form angle A. Similarly, side b is the side opposite angle B, and side a is the side adjacent to angle B. The sides a and b are called **legs**.

Note that in any right triangle, the lengths of the sides are related by the Pythagorean theorem.

$$c^2 = a^2 + b^2$$

In words, the square of the hypotenuse is equal to the sum of the squares of the other two sides.

Example 1   *Consider the following right triangle.*

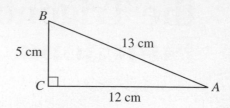

The side opposite angle B is 12 centimeters long. The side opposite angle A is 5 centimeters long. The side adjacent to angle B is 5 centimeters long. The side adjacent to angle A 12 centimeters long. The hypotenuse is 13 centimeters long. Note that the lengths of the sides of the triangle satisfy the Pythagorean Theorem; $13^2 = 5^2 + 12^2$.

1. Consider the following right triangle.

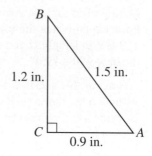

Determine the length of each of the following.

a. the side opposite angle B

b. the side adjacent to angle B

c. the side opposite angle A

d. the side adjacent to angle A

e. the hypotenuse

f. Demonstrate that the length of the sides satisfies the Pythagorean theorem.

Similar Triangles

Consider the following right triangles whose angles are the same but whose sides are different lengths. These three triangles are called **similar triangles**.

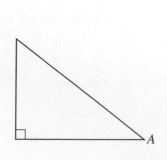

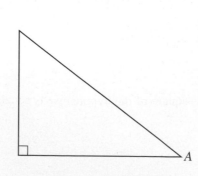

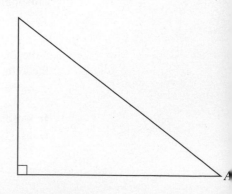

2. a. Using a protractor, estimate the measure of angle A to the nearest degree. (For a review of angles measured in degrees, see Appendix B.) Because $0° \leq A < 90°$, angle A is called an **acute angle**.

b. Use a metric or English ruler to complete the following table.

	LENGTH OF THE HYPOTENUSE	LENGTH OF THE SIDE OPPOSITE A	LENGTH OF THE SIDE ADJACENT TO A
Small Triangle			
Midsize Triangle			
Large Triangle			

3. a. Use the information in the table in Problem 2b to complete the ratios in the following table with respect to angle A. Write each ratio as a decimal rounded to the nearest tenth.

	LENGTH OF THE SIDE OPPOSITE ANGLE A / LENGTH OF THE HYPOTENUSE	LENGTH OF THE SIDE ADJACENT TO ANGLE A / LENGTH OF THE HYPOTENUSE	LENGTH OF THE SIDE OPPOSITE ANGLE A / LENGTH OF THE SIDE ADJACENT TO ANGLE A
Small Triangle	$\frac{1}{1.7} \approx 0.6$	$\frac{1.3}{1.7} \approx 0.8$	$\frac{1}{1.3} \approx 0.8$
Midsize Triangle			
Large Triangle			

b. What do you observe about the ratio $\dfrac{\text{length of the side opposite angle } A}{\text{length of the hypotenuse}}$ for each of the three right triangles?

The table in Problem 3a illustrates the geometric principle that **corresponding sides of similar triangles are proportional**. Recall that the ratios $\dfrac{a}{b}$ and $\dfrac{c}{d}$ are proportional if

$$\frac{a}{b} = \frac{c}{d}.$$

4. Consider another right triangle in which the measure of angle A is not the same as the measure of angle A in the three similar triangles from Problems 2 and 3.

a. Using a protractor, estimate the measure of angle A to the nearest degree.

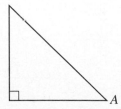

b. For this new triangle, measure the lengths of its sides and then complete the follow
ing table with respect to angle A. Write each as a ratio and then as a decimal
rounded to the nearest tenth.

LENGTH OF THE SIDE OPPOSITE ANGLE A / LENGTH OF THE HYPOTENUSE	LENGTH OF THE SIDE ADJACENT TO ANGLE A / LENGTH OF THE HYPOTENUSE	LENGTH OF THE SIDE OPPOSITE ANGLE A / LENGTH OF THE SIDE ADJACENT TO ANGLE A
New Triangle		

c. Are the ratios for the new triangle the same as the ratios for the three similar triangles?

d. What changed from the similar triangles to the new triangle to make the ratios change?

Sine, Cosine, and Tangent Functions

As indicated in Problem 4, the ratios of the sides of a right triangle are dependent on the size
the angle A. If the angle changes, the ratios change. This fact is fundamental to trigonometry.

The ratios of the sides of a right triangle with acute angle A are a function of the size of A.

The ratios are given special names and are defined as follows.

Definition

Let A be an acute angle (less than 90°) of a right triangle. The **sine**, **cosine**, and **tangent**
of angle A are defined by

$$\text{sine of } A = \sin A = \frac{\text{length of the side opposite } A}{\text{length of the hypotenuse}},$$

$$\text{cosine of } A = \cos A = \frac{\text{length of the side adjacent to } A}{\text{length of the hypotenuse}},$$

$$\text{tangent of } A = \tan A = \frac{\text{length of the side opposite } A}{\text{length of the side adjacent to } A},$$

where sin, cos, and tan are the standard abbreviations for sine, cosine, and tangent,
respectively.

* Sine, cosine, and tangent are called **trigonometric functions**.
* Note that the input values for the trigonometric functions are angles and the output
values are ratios.
* A mnemonic device often used to help remember the trigonometric relationships is
SOH CAH TOA, where SOH indicates that the sine function is the length of the side
opposite divided by the length of the hypotenuse, CAH indicates the cosine
definition, and TOA indicates the definition of the tangent function.

Example 2   *Consider the following right triangle (the large triangle in Problems 2 and 3) with the given approximate dimensions.*

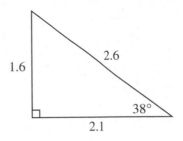

$$\sin 38° = \frac{1.6}{2.6} \approx 0.615 \text{ or } 0.6 \text{ (nearest tenth)}$$

$$\cos 38° = \frac{2.1}{2.6} \approx 0.808 \text{ or } 0.8 \text{ (nearest tenth)}$$

$$\tan 38° = \frac{1.6}{2.1} \approx 0.762 \text{ or } 0.8 \text{ (nearest tenth)}$$

Therefore, for any size right triangle with a 38° angle as one of its acute angles, the sine of 38° is always approximately 0.615, the cosine of 38° is always approximately 0.808, and the tangent of 38° is always approximately 0.762.

You can use your graphing calculator to evaluate sin 38°. Make sure the calculator is in degree mode. Press the (SIN) key, followed by (3), (8), (1) and (ENTER).

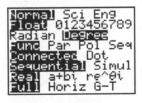   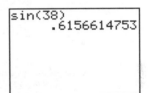

5. a. The sum of the three angles in a triangle is 180°. If one angle measures 90°, the sum of the measures of the other two angles must be 90°. If one of the acute angles measures 38°, the other acute angle is 52°. Use the accompanying figure to determine each of the following.

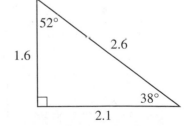

 i. sin 52°

 ii. cos 52°

 iii. tan 52°

b. Verify your rounded answers in part a using your graphing calculator.

Example 3   *Consider the following right triangle, where A and B are acute angles*

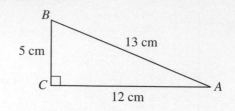

$$\sin A = \frac{5}{13}, \cos A = \frac{12}{13}, \tan A = \frac{5}{12}, \sin B = \frac{12}{13}, \cos B = \frac{5}{13}, \tan B = \frac{12}{5}$$

6. Consider the following right triangle.

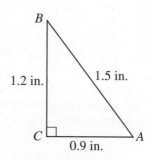

Calculate the following (round to four decimal places).

    a. $\sin A$ **b.** $\cos A$ **c.** $\tan A$

    d. $\sin B$ **e.** $\cos B$ **f.** $\tan B$

Example 4   *Solve the following equations. Round your answers to the nearest tenth.*

    a. $\sin 56° = \dfrac{x}{15}$ **b.** $\cos 13° = \dfrac{24}{x}$ **c.** $\tan 72° = \dfrac{x}{24.7}$

SOLUTION

Use the calculator in degree mode to evaluate the function values,

  a. $\qquad 0.829 = \dfrac{x}{15}$ **b.** $\qquad 0.974 = \dfrac{24}{x}$ **c.** $\qquad 3.078 = \dfrac{x}{24.7}$

    $15(0.829) = x$ $(0.974)x = 24$ $(3.078)(24.7) = x$

        $12.4 \approx x$ $x = \dfrac{24}{0.974}$ $76.0 \approx x$

                                      $x \approx 24.6$

7. Solve the following equations. Round your answers to the nearest tenth.

    a. $\sin 24° = \dfrac{x}{10}$ **b.** $\cos 63° = \dfrac{x}{23.5}$ **c.** $\tan 48° = \dfrac{16}{x}$

Trigonometric Values of Special Angles

To determine the trigonometric function values for 30° and 60°, start with an equilateral triangle, a triangle with three equal sides. Assume for now that all three sides are 2 units in length. Note that all three angles must measure 60°.

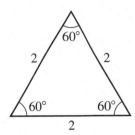

Now bisect one of the angles to form two congruent right triangles with angles measuring 30°, 60°, and 90°.

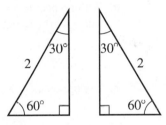

8. a. Considering one of these two new triangles, what are the lengths of the two legs of the right triangle?

b. With respect to the 30° angle, what are the lengths of the opposite side, the adjacent side, and the hypotenuse, respectively?

c. With respect to the 60° angle, what are the lengths of the opposite side, the adjacent side, and the hypotenuse, respectively?

d. Use the results of parts b and c to complete the following table.

θ	sin θ	cos θ	tan θ
30°			
60°			

To determine the trigonometric function values of 45°, start with a 45-45-90 isosceles rig⟩ triangle.

9. For convenience, now assume that the equal legs have length 1.

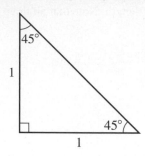

a. Determine the length of the hypotenuse.

b. With respect to the 45° angle, what are the lengths of the opposite side, the adjacent side, and the hypotenuse, respectively?

c. Use the results of part b to complete the following table.

θ	$\sin \theta$	$\cos \theta$	$\tan \theta$
45°			

Knowing the trigonometric function values for 30°, 45°, and 60° can be very helpful i⟩ understanding the behavior of these functions. Keeping these values handy or memorizin⟩ them is a good idea.

Example 5   *Consider the following two right triangles.*

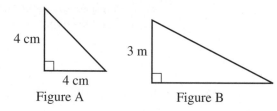

           Figure A   Figure B

a. Given that the length of one leg of a 45-45-90 triangle is 4 centimeters, determine the exact length of the other two sides. (See Figure A.)

SOLUTION

In Figure A, the two acute angles both measure 45° and the two legs are the same length. S⟩ the other leg is also 4 centimeters long. To determine the length of the hypotenuse, use th⟩ sine function.

$$\sin 45° = \frac{4}{h}; \quad \frac{1}{\sqrt{2}} = \frac{4}{h}; \quad h = 4\sqrt{2} \text{ centimeters}$$

b. Given that the length of the shortest side of a 30-60-90 triangle is 3 meters, determine the lengths of the other two sides. (See Figure B.)

SOLUTION

In Figure B, the smallest angle measures 30°. With respect to the 30° angle, the opposite side is given. To determine the length of the adjacent side, the tangent function could be used.

$$\tan 30° = \frac{3}{a}; \quad \frac{1}{\sqrt{3}} = \frac{3}{a}; \quad a = 3\sqrt{3} \text{ meters}$$

To determine the length of the hypotenuse, the sine function could be used.

$$\sin 30° = \frac{3}{h}; \quad \frac{1}{2} = \frac{3}{h}; \quad h = 6 \text{ meters}$$

Tower of Pisa Problem Revisited

You are now ready to answer the original Tower of Pisa problem.

10. a. Recall that the tower makes an 85° angle with the ground and the tower is 179 feet in length. Construct a right triangle that satisfies these conditions.

b. With respect to the 85° angle, is the height of the tower represented by the length of an opposite side, the length of an adjacent side, or the length of the hypotenuse of your triangle?

c. You want to determine how far from the base of the tower the stone hits the ground. Therefore, you want to determine the length of which side of the triangle with respect to the 85° angle?

d. Which trigonometric function relates the side with the length you know and the side with the length you want to know?

e. Write an equation using the information in parts a–d.

f. Using your calculator to evaluate cos 85°, solve the equation in part e.

SUMMARY ACTIVITY 6.1

1. The **trigonometric functions** are functions whose inputs are measures of the acute angles of a right triangle and whose outputs are ratios of the lengths of the sides of the right triangle.

2. The three sides of a right triangle are the **adjacent** side, the **opposite** side, and the **hypotenuse**. The hypotenuse is always the side opposite the right (90°) angle. The other two sides vary, depending on which angle is used as the input.

3. The **sine**, **cosine**, and **tangent** of the acute angle A of a right triangle are defined by

$$\sin A = \frac{\textit{length of the side opposite A}}{\textit{length of the hypotenuse}}$$

$$\cos A = \frac{\textit{length of the side adjacent to A}}{\textit{length of the hypotenuse}}$$

$$\tan A = \frac{\textit{length of the side opposite A}}{\textit{length of the side adjacent to A}}.$$

For additional practice working with the trigonometric functions in right triangles and special triangles, see Appendix B.

EXERCISES: ACTIVITY 6.1

1. Triangle ABC is a right triangle.

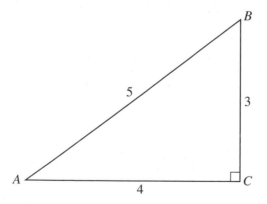

Determine each of the following.

a. $\sin A$ **b.** $\sin B$ **c.** $\cos A$

d. $\cos B$ **e.** $\tan A$ **f.** $\tan B$

2. In a certain right triangle, $\sin A = \frac{24}{25}$.

    a. Determine possible lengths of the three sides of a right triangle.
      Hint: Use the Pythagorean theorem, $c^2 = a^2 + b^2$, to determine the length of any
      unknown side.

    b. Determine $\cos A$.

    c. Determine $\tan A$.

3. In a certain right triangle, $\tan B = \frac{7}{4}$.

    a. Determine possible lengths of the three sides of the right triangle.

    b. Determine $\sin B$.

    c. Determine $\cos B$.

4. Consider the accompanying right triangle.

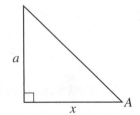

    a. Which of the trigonometric functions relates angle A and sides a and x?

    b. What equation involving angle A and side a would you solve to determine the value
      of x?

5. Given angle B and side c in the accompanying diagram, answer the questions in parts a and b.

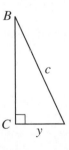

    a. Which of the trigonometric functions relates angle B and sides c and y?

    b. What equation involving angle B and side c would you solve to determine the value of y?

6. Consider the accompanying right triangle.

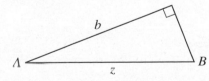

    a. Which of the trigonometric functions relates angle A and sides b and z?

    b. What equation involving angle A and side b would you solve to determine the value of z?

7. Solve the following equations. Round your answer to the nearest tenth.

   a. $\sin 49° = \dfrac{x}{12}$ b. $\tan 84° = \dfrac{x}{9}$ c. $\sin 22° = \dfrac{23}{x}$

8. a. Given that the length of the hypotenuse of a 45-45-90 triangle is 7 centimeters, determine the exact length of the two legs. Sketch and label a diagram before calculating.

   b. Given that the length of the longer leg of 30-60-90 triangle is 20 feet, determine the exact length of the short leg and the hypotenuse. Sketch and label a diagram before calculating.

9. A friend asks you to help build a ramp at his mother-in-law's house. There are three 7-inch-high steps leading to the front door. Another friend donates a 15-foot ramp. The building inspector informs you that any access ramp for people with disabilities can have an inclination no greater than 5°.

   a. Sketch a diagram that assumes the land in front of the steps is level and that the ramp makes a 5° angle with the top of the steps.

   b. What is the increase in height from one end of the ramp to the other?

   c. Would the donated ramp be long enough to meet the code? Explain.

Activity 6.2

A Gasoline Problem

Objectives

1. Identify complementary angles.

2. Demonstrate that the sine of one of the complementary angles equals the cosine of the other.

You and a friend take a camping trip to the American Southwest. On the way home, you realize that your Jeep is running low on gas, so you stop at a filling station. Unfortunately, the attendant informs you that his station has been without gas for several days. However, he is certain that there is a station on a side road up ahead that should have gas to sell.

A quick phone call confirms this information, and you begin to ask about the exact location of the station. The attendant's map of the area indicates that the station is about 10 miles away at an angle of 35° with the road you are currently traveling.

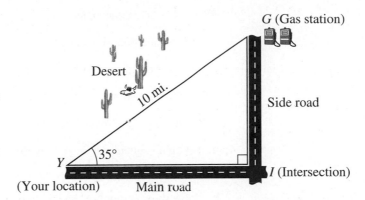

Your Jeep is equipped with a GPS, so driving through the desert is not a great problem, although you would prefer to stay on the road. You estimate that you have enough gas for 15 miles. Do you have enough gas to make it to the other station without going through the desert?

This problem can be solved using the triangle above and some trigonometry to determine the distance from you to the intersection, side *YI*, and from the intersection to the gas station, side *IG*.

1. With respect to angle *Y*, select the correct response in parts a and b.

   a. the side *YI* is known as

   i. the side opposite angle *Y*,

   ii. the side adjacent to angle *Y*, or

   iii. the hypotenuse

   b. the side *YG* is known as

   i. the side opposite angle *Y*,

   ii. the side adjacent to angle *Y*, or

   iii. the hypotenuse

   c. the function that relates *YI*, *YG*, and angle *Y* is the _____.

2. **a.** Set up an equation indicated by Problem 1c.

   b. Solve the equation for the length of *YI*.

3. With respect to angle Y, select the correct response in parts a and b.

    a. the side IG is known as

        i. the side opposite angle Y,

        ii. the side adjacent to angle Y, or

        iii. the hypotenuse

    b. the side YG is known as

        i. the side opposite angle Y,

        ii. the side adjacent to angle Y, or

        iii. the hypotenuse

    c. the function that relates IG, YG, and angle Y is the _____.

4. a. Set up an equation indicated by Problem 3c.

    b. Solve the equation for the length of IG.

5. Do you have enough gas to make it without journeying through the desert? Explain.

Meanwhile at the second service station, the attendant is also making some calculations. H
is aware of your situation and is trying to anticipate which option you will choose in case h
has to go look for you.

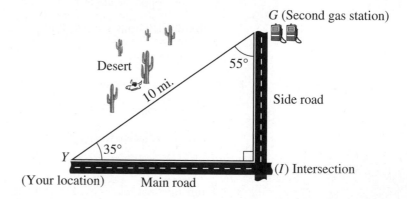

He knows that the line to the first station makes a 55° angle with his road.

6. With respect to angle G, select the correct response in parts a and b.

    a. the side YI is known as

        i. the side opposite angle G,

        ii. the side adjacent to angle G, or

        iii. the hypotenuse

    b. the side *YG* is known as

        i. the side opposite angle *G*,

        ii. the side adjacent to angle *G*, or

        iii. the hypotenuse

    c. the function that relates *YI*, *YG*, and angle *G* is the _____.

7. a. Set up an equation indicated by Problem 6c.

    b. Solve the equation for the length of *YI*.

    c. How does the answer in part b compare to the answer in Problem 2b?

8. With respect to angle *G*, select the correct response in parts a and b.

    a. the side *IG* is known as

        i. the side opposite angle *G*,

        ii. the side adjacent to angle *G*, or

        iii. the hypotenuse

    b. the side *YG* is known as

        i. the side opposite angle *G*,

        ii. the side adjacent to angle *G*, or

        iii. the hypotenuse

    c. the function that relates *IG*, *YG*, and angle *G* is the _____.

9. a. Set up an equation indicated by Problem 8c.

    b. Solve the equation for the length of *IG*.

    c. How does the answer in part b compare to the answer in Problem 4b?

Complementary Angles

Although you used different angles of the triangle *YGI* in Problems 1–9, your results for the lengths of *YI* and *IG* should have been the same. The angles involved in these calculations were 35° and 55°. Because their sum is 90°, they are called **complementary angles**.

   Definition

> Two acute angles *A* and *B* whose measures sum to 90° are called **complementary angles**. If *x* represents the measure of an acute angle, then $90 - x$ represents the measure of its complementary angle.

Since the measures of the two acute angles in a right triangle sum to 90°, they are complementary.

Example 1

x, ACUTE ANGLE	$90 - x$, COMPLEMENTARY ANGLE
35	55
55	35
72	18
18	72

10. a. The solution to Problem 2b involved cos 35° and the solution to Problem 7b involved sin 55°. Compare the values of cos 35° and sin 55°.

b. The solution to Problem 4b involved sin 35° and the solution to Problem 9b involve cos 55°. Compare the values of sin 35° and cos 55°.

This situation demonstrates another fundamental principle of trigonometry.

Cofunctions of complementary angles are equal. Symbolically, if x is an acute angle, then

$$\sin x = \cos (90° - x)$$

and

$$\cos x = \sin (90° - x).$$

In fact, the name *cosine* is derived from the words *sine* and *complement*.

Example 2

a. $\sin 35° = \cos 55°$ **b.** $\sin 55° = \cos 35°$

c. $\sin 72° = \cos 18°$ **d.** $\sin 18° = \cos 72°$

11. Complete the following table, where x represents the measure of an angle in degrees Use your graphing calculator to determine values of $\sin x$ and $\cos (90 - x)$ to fou decimal places.

x	$(90 - x)°$	$\sin x$	$\cos (90 - x)°$
0°			
15°			
30°			
45°			
60°			
75°			
90°			

SUMMARY: ACTIVITY 6.2

1. Complementary angles are two acute angles whose measures sum to 90°.

2. The two acute angles in a right triangle are complementary.

3. The *co* in cosine is from the word *complement*.

4. Cofunctions of complementary angles are equal.

EXERCISES: ACTIVITY 6.2

1. You are in a rowboat on Devil Lake in Ontario, Canada. Your lakeside cabin has no running water, so you sometimes go to a fresh spring at a different point on the lakeshore. You row in a direction 60° north of east, for half a mile. You are not yet tired, and it is a lovely day.

a. How far would you now have to row if you were to return to your cabin by rowing directly south and then directly west? Note that the resulting figure is a 30-60-90 triangle. Round your answer to the nearest hundredth.

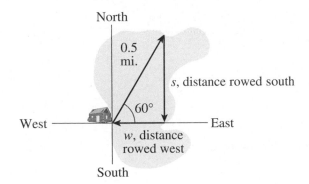

b. Check your solution to part a, using the complementary angle of the angle used in part a.

2. One afternoon you don't pay enough attention while rowing. You row 20° off course, too far north, instead of directly west as you had intended. You row off course for 300 meters.

a. Draw a diagram of this situation.

b. How far west have you gone? Round your answer to the nearest meter.

c. How far north are you of where you had originally planned to be?

d. Check your solutions to parts b and c, using the complementary angle of the angle used in parts b and c.

3. a. Complete the following table, where x represents the measure of an angle in degrees. Use your graphing calculator to determine values of $\sin x$ and $\cos (90 - x)$ to four decimal places.

x	$(90 - x)$	$\sin x$	$\cos (90 - x)$
7°			
17°			
24°			
33°			
48°			
67°			
77°			

b. What trigonometric property does this table illustrate?

4. a. Complete the following table, where x represents the measure of an angle in degrees. For the special values in the table, determine the exact values of $\cos x$ and $\sin (90 - x)$.

x	$(90 - x)$	$\cos (x)$	$\sin (90 - x)$
30°			
45°			
60°			

b. What trigonometric property does this table illustrate?

Activity 6.3

The Sidewalks of New York

Objectives

1. Determine the inverse tangent of a number.

2. Determine the inverse sine and cosine of a number using the graphing calculator.

3. Identify the domain and range of the inverse sine, cosine, and tangent functions.

A friend of yours is having a party in her Manhattan apartment, which borders Central Park. She gives you the following directions from your place, which also borders the park.

i. If you are coming after dark, head east for three blocks, going around the park, and then go north for two blocks.

ii. If you can come early, you can cut through the park, a shorter route.

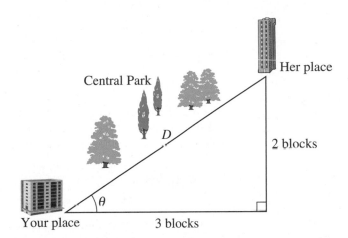

Central Park

Her place

D

2 blocks

θ

Your place

3 blocks

You leave your place early and decide to cut through the park. You want to compute the shortest distance, D (in blocks), between your apartments.

1. Using the Pythagorean theorem, $c^2 = a^2 + b^2$, determine the distance, D, in blocks, across the park.

Now you want to determine the direction (angle) you need to go to get to your friend's place. You can represent the angle by θ, the Greek letter *theta*. Because you now have the lengths of all three sides of the triangle, you can use any of the trigonometric functions to help determine θ. Begin with the tangent function.

Inverse Tangent Function

Recall that the input for the tangent function is an angle and the output is the ratio of the length of the side opposite to the length of the side adjacent in a right triangle.

$$tangent\ of\ an\ angle = \frac{length\ of\ the\ opposite\ side}{length\ of\ the\ adjacent\ side}$$

2. Determine the value of the tangent of θ, written tan θ, for the Central Park triangle.

If you want to determine the tangent of a known angle, you can use the $\boxed{\text{TAN}}$ key on your calculator. From Problem 2, you know the value of the tangent of the angle θ, but do not know the value of the θ. That is, you know the output for the tangent function, but you do not know the input.

3. Use the table feature on your calculator to approximate θ from Problem 2. You should compute the tangent of several possible values of θ to get as close as possible to the desired answer. Complete the following table. Round your answers to five decimal places.

θ	$\tan \theta$
30°	
	0.83910
35°	
	0.67451
33.7°	
	0.66667

You don't have to experiment every time to determine θ when you know the tangent of θ. There is a more direct method using the inverse tangent function. Recall that with inverse functions, the inputs and outputs are interchanged.

Definition

The input of the **inverse tangent** function is the ratio $\dfrac{length\ of\ the\ opposite\ side}{length\ of\ the\ adjacent\ side}$ for the angle. The output is the acute angle. The inverse tangent function is denoted by $\tan^{-1}$ or **arctan** and defined by

$$\tan^{-1} x = \theta, \text{ where } \theta \text{ is the acute angle whose tangent is } x.$$

Thus the input, x, represents the ratio of the length of the opposite side to the length of the adjacent side for the angle θ.

Just as $\log x = y$ has the equivalent exponential form $10^y = x$, $\tan^{-1} x = \theta$ is equivalent to $\tan \theta = x$.

Example 1 *Consider the following triangle.*

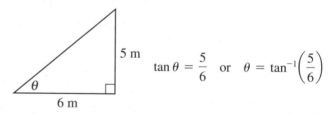

$$\tan \theta = \frac{5}{6} \quad \text{or} \quad \theta = \tan^{-1}\!\left(\frac{5}{6}\right)$$

Using your calculator, you can determine that $\theta \approx 39.8°$. See the following TI-83/TI-84 Plus screen. Be careful; your calculator must be in degree mode.

The inverse tangent function is located on your calculator as second function to tangent. That is, you will need to press the (2nd) key before you press the tangent key. Note that your calculator uses the notation $\tan^{-1}$ rather than arctan.

4. Use your calculator to determine the inverse tangent of the answer to Problem 2. (Make sure your calculator is in degree mode.) Compare this answer with your approximation from Problem 3.

$$Remember: \tan^{-1}(\text{ratio}) = \text{angle}$$

$$\tan(\text{angle}) = \text{ratio}$$

Inverse Sine and Cosine Function

There are similar definitions for the inverse sine and inverse cosine functions.

Definition

1. The input of the **inverse sine** function is the ratio $\dfrac{length\ of\ the\ opposite\ side}{length\ of\ the\ hypotenuse}$ for the angle. The output is the acute angle. The inverse sine function is denoted by **sin**$^{-1}$ or **arcsin** and defined by

$$\sin^{-1} x = \theta, \text{ where } \theta \text{ is the acute angle whose sine is } x.$$

Thus the input, x, represents the ratio of the length of the opposite side to the length of the hypotenuse for the angle θ.

2. The input of the **inverse cosine** function is the ratio $\dfrac{length\ of\ the\ adjacent\ side}{length\ of\ the\ hypotenuse}$ for the angle. The output is the acute angle. The inverse cosine function is denoted by **cos**$^{-1}$ or **arccos** and defined by

$$\cos^{-1} x = \theta, \text{ where } \theta \text{ is the acute angle whose cosine is } x.$$

Thus the input, x, represents the ratio of the length of the adjacent side to the length of the hypotenuse for the angle θ.

Again, $\sin^{-1} x = \theta$ has the equivalent form $\sin\theta = x$ and $\cos^{-1} x = \theta$ has the equivalent form $\cos\theta = x$.

Example 2 *Consider the following triangle.*

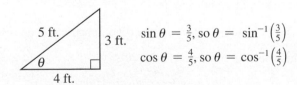

$\sin\theta = \frac{3}{5}$, so $\theta = \sin^{-1}\left(\frac{3}{5}\right)$

$\cos\theta = \frac{4}{5}$, so $\theta = \cos^{-1}\left(\frac{4}{5}\right)$

Using either the inverse sine or the inverse cosine function, you determine that $\theta \approx 36.9°$. See the following TI-83/TI-84 Plus screen.

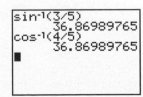

5. a. In Problems 2 and 4 you determined the values of θ in the Central Park situation using the inverse tangent function. Now use the inverse sine function to determine the value of θ in the Central Park situation.

b. Use the inverse cosine function to determine the value of θ in the Central Park situation.

6. The term used by highway departments when describing the steepness of a hill i **percent grade**. For example, a hill with a 5% grade possesses a slope of $\frac{5}{100}$ or $\frac{1}{20}$. Thi means there will be a 5-foot vertical change for every 100 feet of horizontal change o 1 foot of vertical change for every 20 feet of horizontal change as a car ascends o descends the hill.

a. You are driving along Route 17B in the Catskill Mountains of New York State. Just before coming to the top of a hill, you spot a sign that reads "7% Grade Next 3 Miles. Trucks Use Lower Gear." Draw a triangle, and label the appropriate parts to model this situation.

b. Use an inverse trigonometric function to determine the angle that the base of the hil makes with the horizontal.

c. Use trigonometry to determine how many feet of elevation you will lose from the top of the hill to the bottom.

SUMMARY: ACTIVITY 6.3

1. The domain (set of inputs) of each **inverse trigonometric function** is a set of ratios of the lengths of the sides of a right triangle.

2. The range (set of outputs) of each inverse trigonometric functions is the set of angles.

3. For inverse trigonometric functions

$$\sin^{-1} x = \theta \text{ is equivalent to } \sin \theta = x,$$

$$\cos^{-1} x = \theta \text{ is equivalent to } \cos \theta = x, \text{ and}$$

$$\tan^{-1} x = \theta \text{ is equivalent to } \tan \theta = x.$$

EXERCISES: ACTIVITY 6.3

1. For each of the following, use your calculator to determine θ to the nearest $0.01°$.

   a. $\theta = \arcsin\left(\frac{1}{2}\right)$ **b.** $\theta = \cos^{-1}\left(\frac{3}{7}\right)$ **c.** $\theta = \arctan(2.36)$

   d. $\theta = \sin^{-1}(0.8974)$ **e.** $\tan\theta = \frac{7}{3}$ **f.** $\cos\theta = \frac{3}{7}$

   g. $\sin\theta = 0.3791$ **h.** $\tan\theta = 0.3791$

2. For each of the following, determine θ without using your calculator.

   a. $\tan\theta = 1$ **b.** $\sin\theta = 0.5$ **c.** $\cos\theta = \frac{\sqrt{3}}{2}$

3. Complete the accompanying table, which refers to right triangles, labeled as in the figure below.

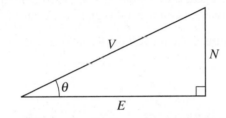

V	θ	E	N
32	65°	13.5	29.0
		12	20
4.1		2.4	
26			18
4.5	45°		

4. While hiking, you see an interesting rock formation on the side of a vertical cliff. You want to describe for a friend how he might see it when he walks down the path. If you stand on the path at a certain place, 50 feet from the base of the cliff, the rock formation is visible about 30 feet up the cliff. At what angle should you tell your friend to look?

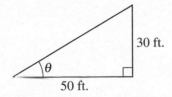

30 ft.

50 ft.

5. A warehouse access ramp claims to have a 10% grade. The ramp is 15 feet long.

   a. Draw a diagram of this situation.

   b. What angle does the ramp make with the horizontal?

   c. How much does the elevation change from one end of the ramp to the other?

6. You are at a hot air balloon festival and enjoying watching the balloons inflate and rise into the atmosphere. You are 200 feet away from one particular balloon when it begins its ascent.

   a. Draw a diagram of this situation. Let θ represent the angle formed by the balloon, you, and the balloon's lift-off point.

   b. Determine the value of θ when the balloon is 200 feet in the air.

   c. When the balloon begins its ascent, it is 400 feet away from you, what is the value of θ?

   d. Did you recognize the angles in parts b and c as special angles?

Activity 6.4

Solving a Murder

Objective

1. Determine the measure of all sides and all angles of a right triangle.

There has been a fatal shooting 110 feet from the base of a 25-story building. Each story measures approximately 12 feet. Two suspects live in the building, one on the 7th floor, the other on the 20th. Both suspects were in their apartments at the time of the murder. Forensic specialists report that the bullet was fired from somewhere in the building and entered the body at an angle of approximately 58° with the ground.

1. Draw a diagram of this situation.

Based on the diagram, the question is now: Is α or is θ equal to 58°?

2. Use the appropriate trigonometric function to determine which of the two suspects could not have committed the murder.

3. On what other floors of the building should the police question additional possible suspects? Explain.

As demonstrated in Problems 2 and 3, determining the measure of the sides and angles of a right triangle can be useful.

Definition

To **solve a triangle** means to determine the measure of all sides and all angles. This process can be especially useful for architects, surveyors, and navigators.

Example 1   *Solve the right triangle ABC, with A = 33.0°, C = 90.0°, and c = 12.2 inches.*

SOLUTION

Consider the following diagram.

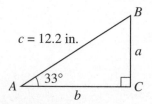

You need to determine the measurements of the remaining sides and angles; angle B, side a and side b. Because $A + B = 90°$ and $A = 33°$,

$$33 + B = 90, \text{ or}$$

$$B = 90° - 33° = 57°.$$

To determine the length of side a, you can use angle A, side c, and the sine function.

$$\sin A = \frac{a}{c}$$

$$\sin 33° = \frac{a}{12.2}$$

So $a = 12.2 \sin 33° \approx 6.6$ inches (nearest tenth). See the following calculator screen.

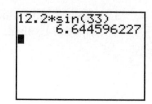

To determine the length of side b, you can use angle A, side c, and the cosine function.

$$\cos A = \frac{b}{c}$$

$$\cos 33° = \frac{b}{12.2}$$

$$b = 12.2 \cos 33° \approx 12.2(0.8387) \approx 10.2 \text{ inches}$$

4. Use the given information to solve each of the following right triangles.

a.

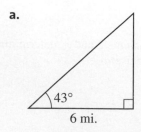

b.

c.

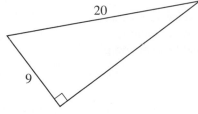

SUMMARY: ACTIVITY 6.4

1. Many trigonometric problems involve solving right triangles; that is, determining the measures of all sides and angles.

2. The following is a trigonometric problem-solving strategy:

   a. Draw a diagram of the situation using right triangles.

   b. Identify all known sides and angles.

   c. Identify sides and/or angles you want to know.

   d. Identify functions that relate the known and unknown.

   e. Write and solve the appropriate trigonometric equation(s).

EXERCISES: ACTIVITY 6.4

1. Use the given information to solve each of the following right triangles.

   a.

     6.5 ft.

     57°

   b.

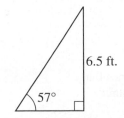

c.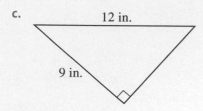

12 in.

9 in.

2. You need to construct new steps for your deck and read that stringers are on sale at the local lumber company. Stringers are precut side supports to which you nail the steps; they are made in three-, four-, five-, six-, or seven-step sizes. Each step on the stringers is 7 inches high and 12 inches deep.

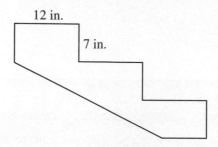

12 in.

7 in.

a. If the vertical rise of your deck measures 3.5 feet, which size stringer should you buy? Explain.

b. How far out will your steps extend from the porch? Explain.

c. What angle will a line from the lower right end of the stringer to the edge of the deck make with the ground? Explain.

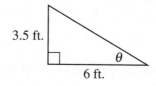

3.5 ft.

θ

6 ft.

d. What angle will the line in part c make with the vertical? Explain.

3. Some application problems involve a horizontal line of sight, which is used as a reference line. An angle measured *above* the sight line is called an **angle of elevation**. An angle measured *below* the sight line is called an **angle of depression**.

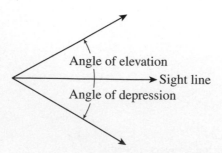

Angle of elevation

Sight line

Angle of depression

You and some friends take a trip to Colton Point State Park in the Grand Canyon of Pennsylvania. Some of your group go white-water rafting, while some of your friends join you for a hike. You reach the observation deck in time to see the rest of your party battling the white water. Someone in the group asks you how close you think the rafts actually get to you as they float by. You have no idea, but you ask a nearby park ranger.

She doesn't know either, but she does tell you that the canyon is approximately 800 feet deep at Colton Point and that the angle of depression to the creek is about 22°.

a. Draw a diagram of this situation.

b. Use trigonometry to estimate how close the rafters get to you on the observation deck.

Project Activity 6.5

How Stable Is That Tower?

Objectives

1. Solve problems using right-triangle trigonometry.

2. Solve optimization problems using right-triangle trigonometry with a graphing approach.

Situation 1: Stabilizing a Tower

You are considering buying property near a cell phone tower and are concerned about you property values. You decide to do some reading about issues involving towers, such as ae thetics, safety, and stability. Of course, anyone living near the tower would like a guarante that it could not blow down. Guy wires are part of that guarantee.

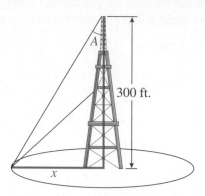

The tower rises 300 feet and is supported by several pairs of guy wires all attached on th ground at the same distance from the base of the tower. In each pair, one guy wire extend from the ground to the top of the tower, and the other attaches halfway up. The accompany ing diagram illustrates one pair of wires.

1. New guidelines for stability recommend that the angle (*A*) the guy wires make with th line through the center of the tower (not the angle of elevation) must be at least 40° Therefore, the existing guy wires may need to be replaced. Because you are concerne about how close the wires will come to your property, you need to compute the shortes distance from the tower at which the guy wires may be attached. Use trigonometry t compute this distance.

2. To improve stability, the authors of the guidelines propose increasing the minimun angle the guy wire makes with the tower from 40° to 50°. What is the effect on the short est distance from the tower at which the guy wires may be attached?

Situation 2: Climbing a Mountain

You are camping in the Adirondacks and decide to climb a mountain that dominates the loca area. You are curious about the vertical rise of the mountain and recall from your mathemat ics course that surveyors can measure the angle of elevation of a mountain summit with theodolite. You are able to borrow a theodolite from the local community college to gathe some pertinent data. You find a level field and take a first reading of an angle of elevation o 23°. Then you walk 100 feet toward the mountain summit and take a second reading o 24° angle of elevation (see accompanying illustration).

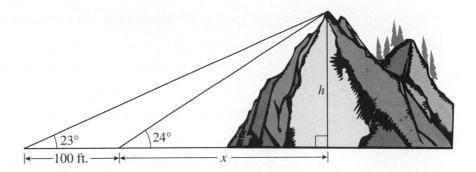

3. Use h to represent the vertical rise of the mountain, and write expressions for tan 23° and tan 24° in terms of h and x.

   a.   **b.**

4. **a.** Solve the system of equations in Problem 3 to determine x. Round your answer to the nearest tenth.

   b. Use the value of x from part a to determine the vertical rise, h, of the mountain.

Situation 3: Seeing Abraham Lincoln

You are traveling to South Dakota and plan to see Mount Rushmore. In preparation for your trip, you do some research and discover that from the observation center, the vertical rise of the mountain is approximately 500 feet and the height of Abraham Lincoln's face is 60 feet (see accompanying diagram). To get the best view, you want to position yourself so that your viewing angle of Lincoln's face is as large as possible.

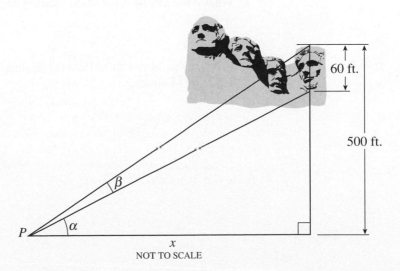

NOT TO SCALE

5. a. Determine from the diagram which angle is the viewing angle.

b. Write an appropriate trigonometric equation for the angle $(\alpha + \beta)$ in terms of x.

c. Write an appropriate trigonometric equation for α in terms of x.

d. Write the equivalent inverse function expression for the equations in parts b and c.

e. Write an equation that defines the viewing angle as a function of the distance, x, tha you are standing from the base.

6. Enter the equation you determined in Problem 5e into Y1 of your graphing calculato Choose a window that corresponds to the graph below. How does your graph compar to the graph below?

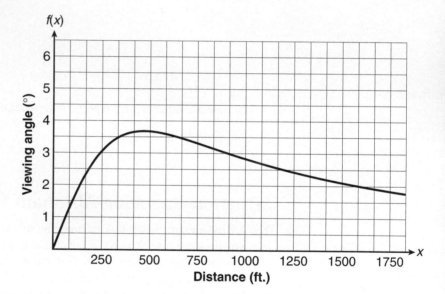

7. What is the largest value of the viewing angle? Justify your conclusion.

8. How far should you stand from the mountain to obtain this maximum value of the view- ing angle? Explain.

EXERCISES: ACTIVITY 6.5

1. You are driving on a straight highway at sea level and begin to climb a hill with a 5% grade. (Recall that a grade is given in a percent but may be expressed as a fraction. In that way you can view grade as a slope.)

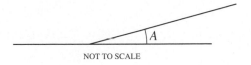

NOT TO SCALE

a. What is the slope of the highway with a grade of 5%?

b. What is the **angle of elevation**, A? That is, what angle does the highway make with the horizontal? Explain.

c. If you walk along the highway for one mile, how many feet above sea level are you? Explain.

2. You are standing 92 meters from the base of the CN Tower in Toronto, Canada. You are able to measure the angle of elevation to the top of the tower as 80.6°.

a. Draw a diagram and indicate the angle of elevation.

b. What is the height of the tower? Round your answer to the nearest tenth.

3. You are in a spy satellite, equipped with a measuring device like a theodolite, orbiting 5 miles above Earth. Your mission is to discover the length of a secret airport runway. You measure the angles of depression to each end of the runway as 30° and 25°, respectively. What is the length of the runway?

4. The Empire State Building rises 1414 feet above the ground, and you are standing across 34th Street, approximately 80 feet from the base of the building.

   a. If you look up to the top of the building, what angle of elevation do you make with the ground? Round your answer to the nearest degree.

   b. How far from the building must you be to make your angle of elevation with the ground 85°? Round your answer to the nearest foot.

5. Consider the function defined by $f(x) = \arctan \frac{20}{x} - \arctan \frac{10}{x}$ defined for $0.01 \le x \le 100$.

   a. Use your graphing calculator to sketch a graph of this function.

   b. Over the given domain, what is the maximum value of the function, and where does it occur?

   c. Over the given domain, what is the minimum value, and where does it occur?

6. You are interested in constructing a feeding trough for your cattle that can hold the largest amount of feed. You buy a 15-by-50-foot piece of aluminum to construct a 50-foot-long trapezoidal trough with a base of 5 feet. You bend up the two 5-foot sides through an angle of $t°$ with the horizontal. Each cross section is a trapezoid (see accompanying diagram). You need to determine the angle, t, that produces the largest volume for the trough.

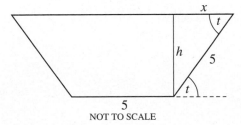

50 ft.

5 ft.   5 ft.

5 ft.
NOT TO SCALE

   a. Recall the formula for the area of a trapezoid (see inside back cover of the textbook, if necessary), and write the area of the trapezoidal cross section in terms of h and x (see accompanying diagram).

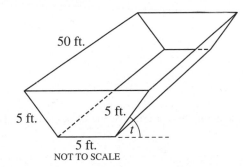

x

t

h

5

t

5
NOT TO SCALE

b. Using right-triangle trigonometry, write h in terms of t. In a similar way, write x in terms of t. Using this information, write an equation for the area of the trapezoid as a function of t.

c. Use your graphing calculator to graph the area function you constructed in part b with $0° < t < 90°$. Label your axes, and remember to indicate units.

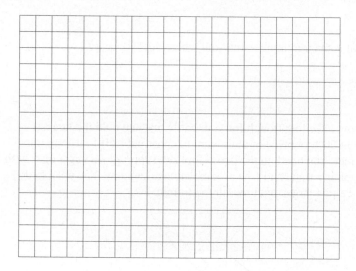

d. Use the graph to determine what angle produces the largest area for the trapezoid? Explain.

e. What is the maximum area? Explain.

f. Write the volume of the trough as a function of the angle t.

g. Use your graphing calculator to graph the volume function you constructed in part f with
$0° < t < 90°$. Label axes and show your units.

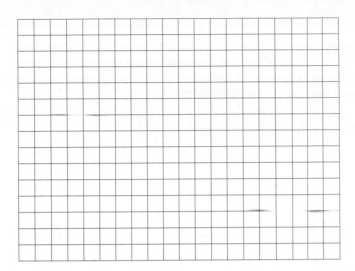

h. Use the graph to determine what angle produces the largest volume, and what is that
volume? Justify your conclusion.

i. What do you conclude about the angle that produces the largest cross-sectional area and the
angle that produces the largest volume for the trough?

1. Classical right-triangle trigonometry was developed by the ancient Greeks to solve problems in surveying, astronomy, and navigation. For purposes of computation, the side opposite the angle θ, side O, is called the opposite; the side opposite the right angle, side H, is called the hypotenuse; and the third side, side A, is called the adjacent side to angle θ.

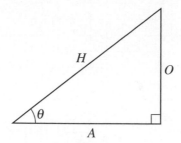

Define the three major trigonometric functions—$\sin \theta$, $\cos \theta$, and $\tan \theta$—in terms of H, A, and O.

2. a. Given any right triangle, which trigonometric function would you use to determine the length of the opposite side if you knew the angle measure and the length of the hypotenuse?

b. Which trigonometric function would you use to determine the length of the adjacent side if you knew the angle measure and the length of the hypotenuse?

c. Which trigonometric function would you use to determine the length of the adjacent side if you knew the angle measure and the length of the opposite side?

d. Which trigonometric function would you use to determine the length of the opposite side if you knew the angle measure and the length of the adjacent side?

3. a. Suppose for a right triangle you know the length of the side opposite an angle and you know the length of the hypotenuse. How can you determine the angle?

b. There is another way to solve part a. Describe this alternative technique.

4. If you know the lengths of two sides of a right triangle, how can you determine all the angles in the triangle? Make up an example, and determine all the angles. Remember that all the interior angles of a triangle add up to 180°.

5. Consider the following right triangle.

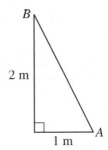

Determine the exact value of each of the following.

a. $\sin A$

b. $\cos A$

c. $\tan A$

d. $\sin B$

e. $\cos B$

f. $\tan B$

g. What trigonometric property is illustrated by parts a–f? Explain.

6. Consider the following two calculator screens.

 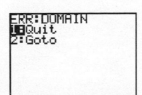

a. The second screen indicates that 1.257 is not in the domain of the inverse sine function. Do you agree? Explain.

b. The following screen indicates that you do not have the same problem for the inverse tangent function. Why not? Explain.

7. Using diagrams, explain the difference between angle of depression and angle of elevation.

Cluster 1) How Can I Practice?

1. Triangle *ABC* is a right triangle.

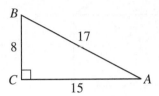

Determine each of the following. Write your answer as a ratio.

a. tan *A*   **b.** tan *B*   **c.** cos *A*

d. cos *B*   **e.** sin *A*   **f.** sin *B*

2. Use your graphing calculator to determine the values of each of the following. Round your answers to the nearest thousandth.

a. sin 47° =   **b.** cos 55° =

c. tan 31° =   **d.** tan 80° =

3. Given sin $A = \frac{5}{13}$, determine cos *A* and tan *A* exactly.

4. Given tan $B = \frac{7}{4}$, determine sin *B* and cos *B* exactly.

5. Use your calculator to determine θ, where $0° \leq \theta \leq 90°$. Round to the nearest tenth of a degree.

a. sin $\theta = \frac{3}{4}$ **b.** cos $\theta = 0.9172$ **c.** $\theta = \arctan \frac{7}{2}$

d. $\theta = \sin^{-1}\frac{2}{7}$ **e.** $\theta = \tan^{-1} 0.9714$ **f.** $\theta = \arccos 0.9714$

6. Solve the following right triangle. That is, determine all the missing sides and angles.

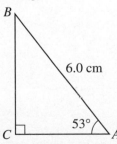

7. You are building a new garage to be attached to your home and investigate several 30-foot-wide trusses to support the roof. You narrow down your choices to three: One has an angle of 45° with the horizontal, and the others have angles of 35° and 25° with the horizontal. You determine that the walls of the garage must be 10 feet high. To match the height of the rest of the house, the peak of the garage should be approximately 20 feet high. Which truss should you buy?

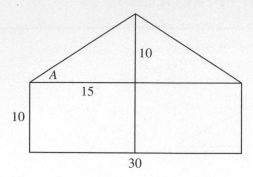

8. The side view of your swimming pool is shown here. The dimensions are in feet.

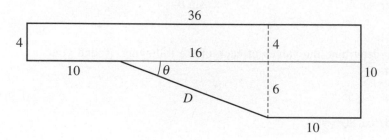

   a. What is the angle of depression, θ?

   b. What is the length of the inclined side, D?

9. a. As part of your summer vacation, you rent a cottage on a large lake. One day, you decide to visit a small island that is 6 miles east and $2\frac{1}{2}$ miles north of your cottage. Draw a diagram for this situation. How far from your cottage is the island?

   b. At what angle with respect to due east should you direct your boat to make the trip from your cottage to the island as short as possible?

Cluster 2

Why Are the Trigonometric Functions Called Circular Functions?

Activity 6.6

Learn Trig or Crash!

Objectives

1. Determine the coordinates of points on a unit circle using sine and cosine functions.

2. Sketch the graph of $y = \sin x$ and $y = \cos x$.

3. Identify the properties of the graphs of the sine and cosine functions.

You are piloting a small plane and want to land at the local airport. Due to an emergency on the ground, the air traffic controller places you in a circular holding pattern at a constant altitude with a radius of 1 mile. Because your fuel is low, you are concerned with the distance traveled in the holding pattern. Of course, you communicate with air traffic control about your coordinates so that you do not collide with another airplane.

The given diagram shows your path in the air. The airport is located at the center $(0, 0)$ of the circle on the ground. The radius of the circle is 1 mile. A circle centered at $(0, 0)$ and having radius 1 is called a **unit circle**. You begin your holding pattern at $(1, 0)$.

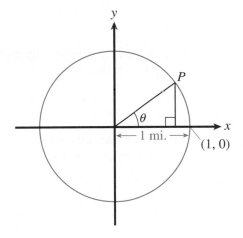

Let's examine the beginning of your first circular loop.

1. What distance (in miles) do you fly in one loop?

2. Let P represent your position after flying only one-tenth of a loop. (See preceding diagram.)

   a. Compute the distance traveled from $(1, 0)$ to P.

   b. Determine the number of degrees of the central angle, θ, when you fly one-tenth of a loop (see preceding diagram). Recall that there are $360°$ in a circle.

The angle in Problem 2b is called a **central angle**.

Definition

A **central angle** is an angle with its vertex at the center of a circle.

3. Now refer back to the right triangle in the preceding diagram. The measure of the central angle, θ, is $36°$.

   a. What is the length of the hypotenuse of the right triangle?

b. If (x, y) represents the coordinates of point P, then which coordinate, x or y, represents the length of the side opposite the 36° angle? Which letter represents the length of the side adjacent to the 36° angle?

c. Use the appropriate trigonometric function to determine the value of x.

d. Use the appropriate trigonometric function to determine the value of y.

e. What are the coordinates of point P?

Problem 3 demonstrates that if

i. an object is moving a distance d counterclockwise on the unit circle from the starting point $(1, 0)$, and

ii. $P(x, y)$ represents the position of the object on the unit circle after it has moved distance d, and

iii. θ represents the corresponding central angle, then the coordinates of P are given by $(\cos \theta, \sin \theta)$.

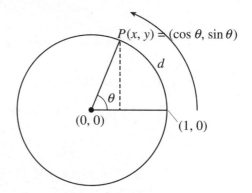

4. Repeat the procedure demonstrated in Problem 3 for the following fractions of a loop and record your results in the table.

LOOP	CENTRAL ANGLE, θ	DISTANCE TRAVELED	POSITION ON THE CIRCLE $(\cos \theta, \sin \theta)$
$\frac{1}{5}$	72°	$\frac{2\pi}{5}$ mi. or 1.26 mi.	(0.31, 0.95)
$\frac{1}{8}$			
$\frac{1}{20}$			

Trigonometric Functions of Angles Greater than 90°

5. Your instruments tell you that you have traveled 2 miles in the holding pattern. What is the central angle, θ? (*Hint:* What fraction of the loop have you traveled?)

For $\theta > 90°$, the coordinates of P are now *defined* to be $(\cos \theta, \sin \theta)$. This extends the idea of Problem 3 to larger angles, θ. Depending on where you are on the circle, these coordinates may be positive or negative.

6. a. From Problem 5, you know that the central angle is 114.6° (see accompanying diagram). What is the measure of the central angle, θ', contained within the right triangle?

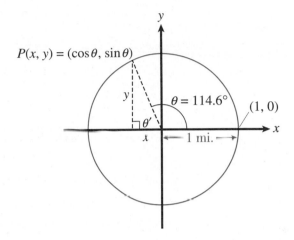

b. What is the length of the hypotenuse?

c. Using the sine and cosine functions, determine the lengths of the remaining two sides of the right triangle.

d. Using the results from part c, what are the coordinates of point P? Remember, the point is in quadrant II.

7. Your graphing calculator can be used to determine the coordinates of a point in a direct manner. Doing so involves calculating the sine and cosine of the central angle θ.

a. Determine the value of each of the following.

cos 114.6°

sin 114.6°

b. How do the results in part a compare to the coordinates of point P in Problem 6d?

In general, the position $P(x, y)$ of an object moving on a unit circle is given by $x = \cos \theta$ and $y = \sin \theta$, where θ is a central angle with initial side the positive x-axis and terminal side OP, where O is the origin, the center of the circle.

8. Give an example of where on the unit circle (circle of radius 1) both coordinates of are negative. Place θ and the coordinates of P on the following diagram.

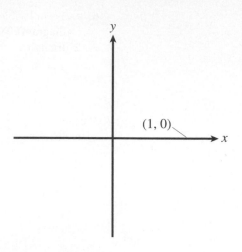

9. a. By the time you have traveled 5 miles in the holding pattern, what is the central angle, θ?

b. What are the coordinates of point P?

Graphs of Sine and Cosine Functions

10. Complete the following table. Be sure your calculator is in degree mode. Round th first column to the nearest hundredth and the third and fourth columns to the neare thousandth.

DISTANCE TRAVELED	CENTRAL ANGLE θ	cos θ	sin θ
0	0°		
	30°		
	45°		
	60°		
1.57			
	120°		
3.14			
	220°		
6			
	360°		

11. Plot the data pairs $(\theta, \sin\theta)$ on the following grid. Draw a smooth curve through these points. Verify the results with your graphing calculator by sketching a graph of $y = \sin x$. Use the window Xmin = 0, Xmax = 360, Ymin = -1, and Ymax = 1 and degree mode.

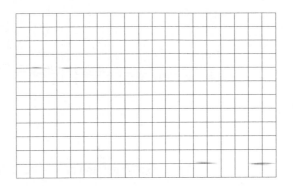

12. Repeat Problem 11 for data pairs $(\theta, \cos\theta)$.

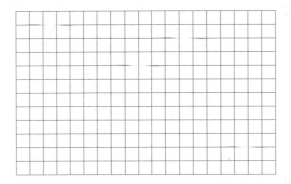

13. You notice your instruments read 11.28 miles from the start of the holding pattern. What is the central angle, θ, and what are the coordinates of point P?

14. Suppose you are at some particular point in the holding pattern. How many miles will you travel in the loop to return to the same coordinates?

15. a. Complete the following table.

θ (degrees)	0°	90°	180°	270°	360°	450°	540°	630°	720°
sin θ									
cos θ									

   b. Sketch a graph of $y = \sin \theta$ for $0° \leq \theta \leq 720°$. Verify using your graphing calculator.

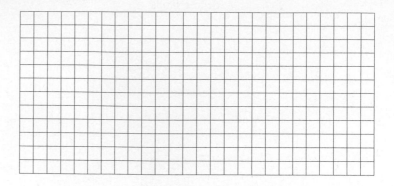

   c. Sketch a graph of $y = \cos \theta$ for $0° \leq \theta \leq 720°$.

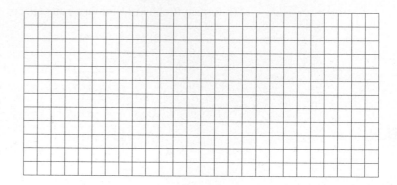

   d. What pattern do you observe in each of the graphs?

Note that these repeating graphs in Problem 15 show the **periodic** or **cyclic** behavior of the trigonometric functions. Because many real-world phenomena involve this repeating behavior, the trigonometric functions are very useful in modeling these phenomena.

Definition

The shortest horizontal distance it takes for one cycle to be completed is called the **period**. The period is 360° for $y = \sin x$ and $y = \cos x$.

16. A negative angle is used to indicate that the object is moving along the circumference of a unit circle in a clockwise direction.

    a. Locate the point P that has a central angle of $-30°$ on the following diagram.

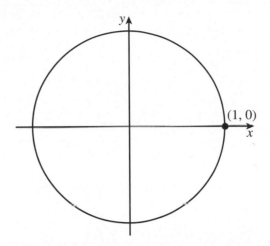

    b. Construct the appropriate right triangle. What is the measure of the central angle within this right triangle? What is the length of the hypotenuse?

    c. Use the sine and cosine functions to determine the lengths of the other two sides.

    d. What are the coordinates of the point P?

    e. Use your graphing calculator to determine $\cos(-30°)$ and $\sin(-30°)$. How do these results compare to the coordinates of point P in part d?

17. a. Complete the following table.

θ (Degrees)	$\cos\theta$	$\sin\theta$
0°		
$-30°$		
$-90°$		
$-180°$		
$-270°$		
$-360°$		

b. Graph the points of the form $(\theta, \sin \theta)$ using the values from the table in part a.

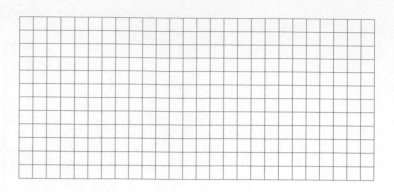

c. Graph the points of the form $(\theta, \cos \theta)$ using the values from the table in part a.

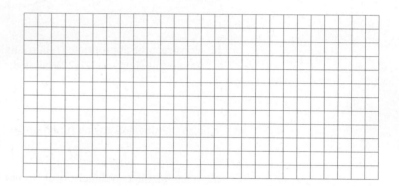

d. Use your graphing calculator to graph $y = \sin x$ and $y = \cos x$ for $-720° \le \theta \le 720°$.

18. a. What is the domain of the sine and cosine functions?

b. What is the range of the sine and cosine functions?

SUMMARY: ACTIVITY 6.6

1. A **central angle** is an angle whose vertex is the center of a circle.

2. The position $P(x, y)$ of an object moving on the unit circle from the point $(0, 1)$ defines the sine and cosine functions by the rules $x = \cos \theta$, $y = \sin \theta$, where θ is a central angle formed by the positive x-axis and the line segment OP. Because of this connection to the unit circle, the sine and cosine functions are often called **circular functions**.

3. The **domain** of the sine and cosine functions is all angles, both positive and negative.

4. The **range** of the sine and cosine functions is all values of N such that $-1 \le N \le 1$.

5. The graphs (one cycle) of $y = \sin x$ and $y = \cos x$ look like the following.

sin x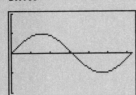

cos x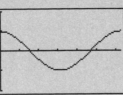

6. The **period** is 360° for $y = \sin x$ and $y = \cos x$.

EXERCISES: ACTIVITY 6.6

1. Determine the coordinates of the point on the unit circle corresponding to the following central angles. Round to the nearest hundredth.

  a. 72°   **b.** 310°   **c.** 270°

  d. 111°   **e.** 212°   **f.** 435°

  g. −70°

2. For each of the points on the unit circle determined in Exercise 1, determine the distance traveled from $(1, 0)$ to the point along the circle. Round your answers to the nearest hundredth.

  a.   **b.**   **c.**

  d.   **e.**   **f.**

  g.

3. Consider the graph of the following function.

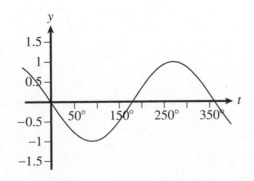

  a. Compare this graph to graphs studied in this activity.

b. What is the motion along the unit circle described by the graph?

4. Consider the graph of the following function.

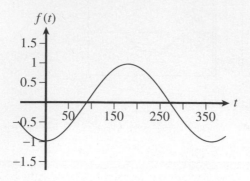

a. How does this graph compare to the graphs studied in this activity?

b. What is the motion around the unit circle described by the graph?

5. The following table represents the number of daylight hours for a certain city in the Western Hemisphere on the dates indicated.

MAR. 21	APR. 21	MAY 21	JUNE 21	JULY 21	AUG. 21	SEPT. 21	OCT. 21	NOV. 21	DEC. 21	JAN. 21	FEB. 21	MAR
11.9	10.5	9.6	8.7	9.7	10.6	12.1	13.5	14.7	15.7	14.7	13.4	11.

a. Plot the data on the following grid.

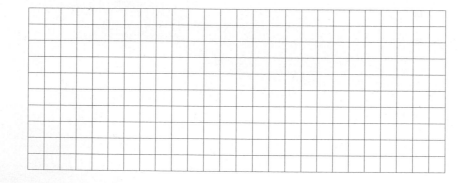

b. Do the data indicate any similarities to a circular function (a function defined by points on the unit circle)? Explain.

c. How does this function compare with the others in this activity?

d. Is the city in question north or south of the equator? Explain.

Objectives

. Convert between degree and radian measure.

. Identify the period and frequency of a function defined by $y = a \sin(bx)$ or $y = a \cos(bx)$ using the graph.

Household electric current is called alternating current, or AC, because it changes magnitude and direction with time. The household current through a 60-watt lightbulb is given by the equation

$$A = 2 \sin(120\pi t),$$

where A is the current in amperes and t is time in seconds.

Note that the input of the sine function in this activity is time measured by real numbers (seconds) and not angles measured in degrees. In order for this to make sense, an alternate real-number method for measuring angles must be introduced. This method is called **radian measure**.

Degree and radian measure correspond in the following way:

180° is the same as 1π radians, or about 3.14 radians.

Therefore, $180° = 1\pi$ radians. Dividing each side by 180 gives you $1° = \frac{\pi}{180}$ radians. Dividing each side by π gives you 1 radian $= \frac{180°}{\pi}$.

Whenever you see an angle measure without a degree symbol, assume that the angle is measured in radians.

Example 1 *Dividing each side of the equality* **180°** $=$ π *radians by 2 shows that 90 degrees is the same as* $\frac{\pi}{2}$ *radians. You can also convert 90 degrees to radians by multiplying* $1°$ *by 90.*

$$90° = 90 \cdot 1° = 90 \cdot \frac{\pi}{180} \text{ radians} = \frac{90\pi}{180} \text{ radians} = \frac{\pi}{2} \text{ radians}$$

Similarly, $10° = 10 \cdot \frac{\pi}{180}$ radians $= \frac{\pi}{18}$ radians

1. In the following table, convert degree measures to radian measure. Round to the nearest thousandth.

DEGREE MEASURE	RADIAN MEASURE
10°	
20°	
30°	
60°	
120°	
360°	

2. **a.** If you divide the equality $180° = \pi$ radians by π, you find that 1 radian $= \frac{180°}{\pi}$. Then 2 radians would be $2\left(\frac{180°}{\pi}\right)$ and so on.

Generalize this to describe a procedure to convert radians to degrees.

   b. How many degrees are there in 1.5π radians?

   c. How many degrees are there in 2π radians?

   d. How many degrees are there in $\frac{\pi}{10}$ radians?

Appendix

For more practice converting degree measure to radian measure and vice versa, se[e] Appendix B.

Periodic Behavior of Graphs of Sine and Cosine Functions

As stated, the function defined by $A = 2 \sin(120\pi t)$, where A is the current in amperes an[d] t is time in seconds, gives the household current through a 60-watt lightbulb.

3. Graph the equation $A = 2 \sin(120\pi t)$ using your calculator. Your calculator must b[e] in radian mode instead of degree mode. Using the indicated window, the graph shoul[d] appear as follows:

   Xmin = 0   Ymin = -3
   Xmax = $\frac{1}{20}$ Ymax = 3
   Xscl = $\frac{1}{240}$ Yscl = 1

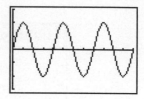

4. Using the graph, determine the maximum current. Explain.

5. What do you think is happening to the current when the graph drops below the horizon[tal] tal axis? (Reread the description of alternating current.)

There is a pattern on the graph of $A = 2 \sin(120\pi t)$ starting at the 0 level, going up to $+2$[,] down to -2, and returning to the 0 level. The pattern repeats. This pattern is called a **cycle**.

Recall from Activity 6.6 that the shortest horizontal distance (on the input axis) it takes fo[r] one cycle to be completed is called the **period**.

Example 2   *Determine the period of each of the following using its graph.*

a. $y = \sin(2x)$ **b.** $y = \sin\left(\frac{1}{2}x\right)$

SOLUTION

a.
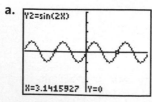

b.
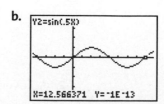

The screen in part a above indicates that the graph of $y = \sin(2x)$ completes one cycle from[m] $(0, 0)$ to $(\pi, 0)$. Therefore, the period is π. The period in part b is 4π.

6. What is the *period* of the electric current function from Problem 3? (The tick marks on the horizontal axis are at $\frac{1}{240}$-second intervals.)

7. Use the graph of each of the following functions to determine the period of the following functions. Use the window Xmin $= 0$, Xmax $= 6\pi$, Xscl $= \frac{\pi}{4}$, Ymin $= -2$, Ymax $= 2$, and Yscl $= 1$.

a. $y = \sin x$
b. $y = \sin\left(\frac{1}{2}x\right)$

c. $y = \cos(3x)$
d. $y = \cos\left(\frac{2}{3}x\right)$

Definition

The **frequency** of $y = a \sin(bx)$ or $y = a \cos(bx)$ is the number of cycles completed when the input, x, increases 2π units.

Example 3   *Determine the frequency of each of the following using its graph.*

a. $y = \sin(2x)$
b. $y = \sin\left(\frac{1}{2}x\right)$

SOLUTION

a.
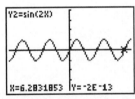

The frequency is 2 because the graph completes two complete cycles in 2π units.

b.
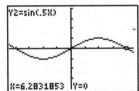

The frequency is 0.5 because the graph completes half of one cycle in 2π units.

8. Determine the frequency, the number of cycles completed in 2π units, for each of the following functions.

a. $y = \sin x$
b. $y = \cos \frac{1}{2}x$
c. $y = \cos 2x$

In general, the frequency of $y = \cos(bx)$ or $y = \sin(bx)$ is b.

---

9. For normal household current described by $A = 2 \sin(120\pi t)$, how many cycles occur in 1 second?

SUMMARY: ACTIVITY 6.7

1. **Radian** measure is used when the input of a repeating function is better defined by real numbers than angles measured in degrees.

2. Since $1°$ corresponds to $\frac{\pi}{180}$ radians, to convert degree measure to radian measure, multiply the number of degrees by $\dfrac{\pi \text{ radians}}{180°}$.

3. To convert radian measure to degree measure, multiply the number of radians by $\dfrac{180°}{\pi}$.

4. Radian and degree equivalent

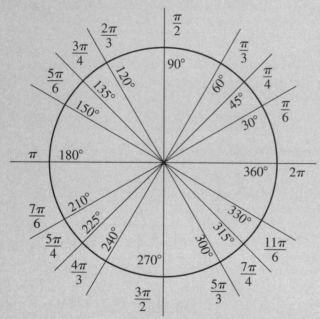

5. The pattern of a graph that covers all y-values once and is repeated is called a **cycle**.

6. The smallest interval of input necessary for the graph of a function to complete one cycle is called the **period**.

   Note: A formula for period will be developed in Activity 6.8.

7. The **frequency** of a periodic or cyclic function is the number of cycles completed when the input is increased by 2π units.

EXERCISES: ACTIVITY 6.7

1. Convert the following degree measures to radian measures.

   a. $45°$ **b.** $140°$

   c. $330°$ **d.** $-36°$

2. Convert the following radian measures to degree measures.

a. $\frac{3\pi}{4}$

b. 2.5π

c. 6π

d. 1.8π

3. Complete the following table.

Degree Measure	0°	30°		60°	90°		180°	210°		360°
Radian Measure	0		$\frac{\pi}{4}$		$\frac{3\pi}{4}$				$\frac{3\pi}{2}$	

For Exercises 4–9, be sure your calculator is in radian mode.

4. a. Complete the following table.

Radian Measure, x	0	$\frac{\pi}{4}$	$\frac{\pi}{2}$	$\frac{3\pi}{4}$	π	$\frac{5\pi}{4}$	$\frac{3\pi}{2}$	$\frac{7\pi}{4}$	2π
$\sin x$									

b. Sketch a graph of $f(x) = \sin x$ using the table in part a.

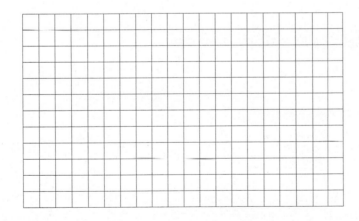

c. Use your graphing calculator to sketch a graph of $f(x) = \sin x$ for $0 \le x \le 2\pi$.

5. How do the graphs of each pair of the following functions compare? Use your graphing calculator.

a. $y = 2 \cos x, y = \cos 2x$

b. $y = \cos \frac{1}{3} x, y = \cos 3x$

For Exercises 6–9, graph each function, and then determine the following for each function:

a. *The largest (maximum) value of the function*

b. *The smallest (minimum) value of the function*

c. *The period (the shortest interval for which the graph completes one cycle)*

6. $y = 0.5 \sin 2x$ **7.** $y = -3 \sin 3x$ **8.** $f(x) = 2.3 \cos(0.5x)$

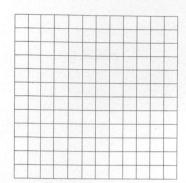
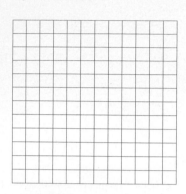
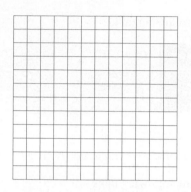

9. You are setting up a budget for the new year. Your utility bill for natural gas and electric usage is a large part of your budget. To help determine the amount you might need to spend on gas and electricity, you examine your previous 3 years' bills. Because you live in a rural area, you are billed every 2 months instead of every month. The data you obtain appears in the following graph.

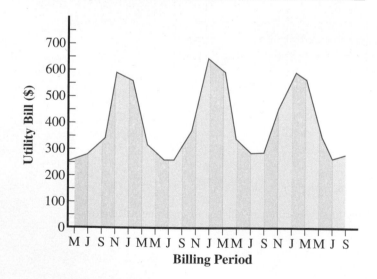

You notice that a pattern develops, which is repeated. This function, though not exactly periodic, models a periodic function for practical purposes. Label the horizontal axis with bimonthly periods, beginning with the June and July bill from 3 years ago.

a. For what months of the year is the utility bill the highest? How much is this bill?

b. What months of the year is the utility bill the lowest? How much is this bill?

c. What is the largest value for the function whose graph is given here?

d. What is the period of the graph?

e. Your power company announces that its rates will increase 5% beginning in April of the coming year. How will this change affect the graph of this function? Will it affect the periodic nature of the function?

f. You heat your house and your water with natural gas and use electricity for all other purposes. You do not currently have air-conditioning in your house. If you were to install a central air-conditioning unit next summer, what changes do you think might occur in the shape of the graph?

Activity 6.8

Get in Shape

Objectives

1. Determine the amplitude of the graph of
$y = a \sin(bx)$ and
$y = a \cos(bx)$.

2. Determine the period of the graph of
$y = a \sin(bx)$ and
$y = a \cos(bx)$ using a formula.

You decide to try jogging to shape up. You are fortunate to have a large neighborhood par nearby that has a circular track with a radius of 100 meters.

1. If you run one lap around the track, how many meters have you traveled? Explain.

2. You start off averaging a relatively slow rate of approximately 100 meters per minute How long does it take you to complete one lap?

You want to improve your speed. Gathering data describing your position on the trac as a function of time may be useful. You sketch the track on a coordinate system with a cente at the origin. Assume that the starting line has coordinates (100, 0) and that you ru counterclockwise.

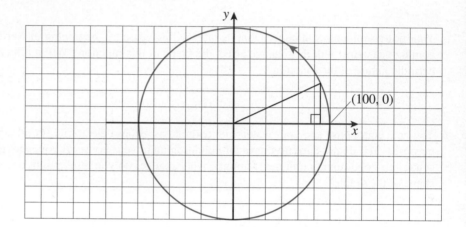

3. You begin to gather data about your position at various times. Using the precedin; diagram and the results from Problem 2, complete the following table to locate you coordinates at selected times along your path.

t (min)	YOUR x-COORDINATE	YOUR y-COORDINATE
0		
$\frac{\pi}{2}$		
π		
$\frac{3\pi}{2}$		
2π		
$\frac{5\pi}{2}$		
3π		

4. What patterns do you notice about the numerical data in Problem 3? Predict your coor dinates as your time increases.

5. To better analyze your position at times other than those listed in the previous table, you decide to make some educated guesses. Let's consider $t = \frac{\pi}{4}$ minutes. Use the graph to approximate the coordinates of your position when $t = \frac{\pi}{4}$, and label these coordinates on the graph.

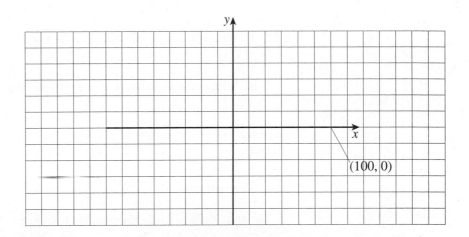

6. To check your guess, you recall that right-triangle trigonometry gives some useful information about special right triangles. (If you are not familiar with special right triangles, see Appendix B.) First, however, you need to compute θ, given in the graph of Problem 5. Calculate θ and explain how you arrive at your answer.

7. On the following grid, use the data from Problems 3, 5, and 6 to plot (t, y), show in your y-coordinate as a function of t. Connect your data pairs to make a smooth graph, and predict what will happen to the graph for values of t before and after the values of t in the tables.

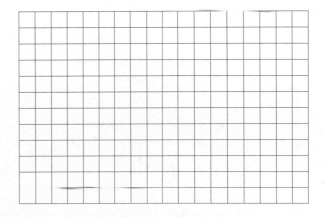

8. Use your graphing calculator to plot $y = 100 \sin(t)$. Make sure your graphing calcula tor is in radian mode. Note that the name of the input t will need to be changed to x t conform with the calculator.

9. Compare your answers to Problems 7 and 8.

10. a. What is the maximum value of the sine function in Problem 8?

b. What is the minimum value of the sine function in Problem 8?

Another important feature of the graphs of the sine and cosine functions is called th **amplitude**.

Definition

The **amplitude** of a periodic function equals

$$\tfrac{1}{2}(M - m),$$

where M is the maximum output value of the function and m is the minimum output value of the function.

> **Example 1**   *The amplitude of the function defined by $y = \sin x$ is 1. The maximum output value is 1. The minimum output value is -1. The amplitude is $\tfrac{1}{2}(1 - (-1)) = \tfrac{1}{2}(1 + 1) = \tfrac{1}{2}(2) = 1$.*

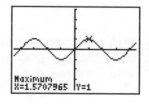 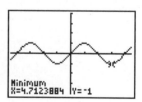

11. a. What is the amplitude of the sine function in the equation $y = 100 \sin x$?

b. Is there a relationship between the amplitude of $y = 100 \sin x$ and the coefficient 100? Explain.

12. Determine by inspection the amplitude of the following functions. Then verify you answers using your graphing calculator.

a. $y = 1.5 \sin x$ **b.** $f(x) = 15 \sin(2x)$ **c.** $y = 3 \cos\left(\tfrac{1}{3}x\right)$

13. a. Is -2 the amplitude of $y = -2 \sin x$? Explain.

b. What is the amplitude of $y = -2 \sin x$?

c. Use your graphing calculator to sketch a graph of $y = -2 \sin x$.

d. How does the graph of $y = -2 \sin x$ compare to the graph of $y = 2 \sin x$?

e. What is the general effect of the negative sign of the coefficient a in $y = a \sin x$?

> In general, the amplitude of the functions defined by $y = a \sin (bx)$ and $y = a \cos (bx)$ is $|a|$.

Period of the Sine and Cosine Functions

After a lot of practice, you begin to speed up on your circular track of radius 100 meters.

You finally achieve your personal goal of 200 meters per minute.

14. a. If you run 200 meters per minute, how long does it take you to complete one lap? Note that this amount of time to complete one lap will be important in defining the key concept of period for the trigonometric functions in the following problems.

b. Complete the following table to give your coordinates at selected special points along your path. Since one lap takes π min, each $\frac{\pi}{4}$ of a loop covers $\frac{1}{4}$ of the circular track.

t (min)	YOUR x-COORDINATE	YOUR y-COORDINATE
0		
$\frac{\pi}{2}$		
π		
2π		

c. On the following grid, use the data from Problem 14b to plot (t, y), your y-coordinate, as a function of t. Connect your data pairs to make a smooth graph, as in the previous activity, and predict what will happen to the graph for values of t before and after the values of t in the table.

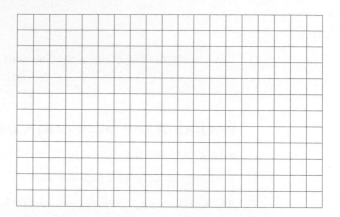

d. What is the amount of time it takes for the graph to complete one full cycle (that is, for you to complete one full lap)?

e. What effect does doubling your speed have on the amount of time to complete one cycle? Explain.

15. a. Use your graphing calculator to plot $y = 100 \sin (2t)$. Note that the name of the input t will need to be changed to x to conform with your calculator.

b. Compare the graph in part a to the graphics in Problem 14c.

16. Recall that the period of a trigonometric function is the shortest interval of inputs needed to complete one full cycle. Determine the period of each of the following using the graph.

a. $y = 100 \sin x$ **b.** $y = 100 \sin (2x)$

The period of the graph of a sine function can be determined directly from the equation that defines the function. For example, the period of $y = 100 \sin (2x)$ is π units, which is half of the period of $y = 100 \sin (x)$. It appears that the coefficient of x in the function affects the period of the function.

In general, if $y = a \sin (bx)$ and $b > 0$, then the period is $\dfrac{2\pi}{b}$.

17. On the following grid, repeat Problem 14c to plot the data pairs (t, x).

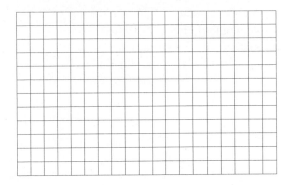

18. a. What equation should you enter into your calculator to produce the graph from Problem 17?

b. Enter the equation from part a into your calculator, and obtain a graph.

19. a. If you were to triple your speed on the track from the original 100 meters per minute, what effect would this have on the amount of time to complete one lap?

b. What equations would describe the x- and y-coordinates of your position?

20. Determine the period of each of the following functions.

a. $y = 100 \cos x$

b. $g(x) = 100 \cos (2x)$

c. $h(x) = 100 \cos (3x)$

SUMMARY: ACTIVITY 6.8

1. The **amplitude** of trigonometric functions defined by $y = a \sin(bx)$ or $y = a \cos(bx)$ is defined by

$$\tfrac{1}{2}(M - m),$$

where M represents the maximum function value and m represents the minimum function value. The amplitude is equivalent to $|a|$, the absolute value of the coefficient of the functions defined by $y = a \sin(bx)$ and $y = a \cos(bx)$. Therefore,

$$|a| = \tfrac{1}{2}(M - m).$$

2. For the trigonometric functions defined by $y = a \sin(bx)$ or $y = a \cos(bx)$, the **period** is $\dfrac{2\pi}{b}$, where $b > 0$.

3. The **frequency** of trigonometric functions defined by $y = a \sin(bx)$ or $y = a \cos(bx)$ is the number of cycles completed by the graphs over intervals of length 2π. The **frequency** of trigonometric functions defined by $y = a \sin(bx)$ or $y = a \cos(bx)$ is b.

EXERCISES: ACTIVITY 6.8

1. On the following grid, repeat Problem 17 to plot the data pairs (t, x).

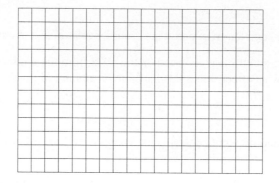

2. Use your graphing calculator to plot $x = 100 \cos(t)$. Make sure your graphing calculator is in radian mode. Note that the names of both the input t and output (your x-coordinate) will need to be changed to x and y, respectively, to conform with your calculator.

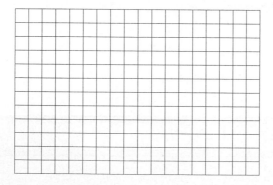

3. a. What is the maximum function value of the cosine function in Problem 2?

   b. What is the minimum function value of the cosine function in Problem 2?

4. What is the connection among the function values in Problem 3, the 100-meter radius of the circle, and the coefficient 100 of the function?

5. If you are forced to run on a larger circular track of radius 150 meters and you increase your speed to 150 meters per minute, predict the equations of the functions describing the *x*- and *y*-coordinates of your position. Explain.

6. Determine by inspection the amplitude of the following functions. Then verify the results with your graphing calculator. (*Remember:* Amplitude *cannot* be negative.)

a. $y = 3 \sin x$

b. $y = 0.4 \cos x$

c. $f(x) = -2 \cos x$

d. $g(x) = -2.3 \sin x$

e. $y = 2 \sin (3x)$

f. $h(x) = -4 \cos (x)$

7. Is there any relationship between the amplitude and the period of a sine function?

8. For each of the following tables, identify a function of the form $y = a \sin bx$ or $y = a \cos bx$ that approximately satisfies the table.

a.

x	0	0.7854	1.5708	2.3562	3.1416
y	0	− 15	0	15	0

b.

x	0	2.244	4.488	6.732	8.976
y	1.3	0	− 1.3	0	1.3

9. For each of the following graphs, identify a function of the form $y = a \sin bx$ or $y = a \cos bx$ that the graph approximates.

a.

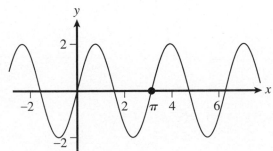

b.

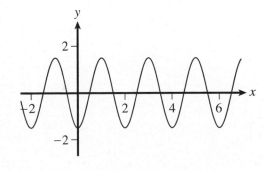

10. Match the given equation to one of the graphs that follow. Assume that Xscl $= 1$ and Yscl $= 1$.

a. $y = 2 \cos 0.5x$

b. $y = -0.5 \sin 2x$

c. $y = 0.5 \cos 2x$

d. $y = 2 \sin 0.5x$

i.

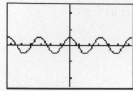

ii.

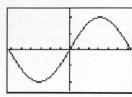

iii.

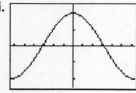

iv.

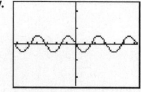

Objective

1. Determine the
   displacement of
   $y = a \sin(bx + c)$ and
   $y = a \cos(bx + c)$
   using a formula.

You have developed a keen interest in carousels. You developed enough of an interest to visit six carousel parks on your vacation.

At one of the parks you are watching a carousel and notice a young girl and her brother as they rush to pick out their special horses for their rides. The girl chooses one right in front of you. Her brother chooses one in the same row only $\frac{1}{4}$ of the way around the carousel, and of course ahead of his sister. You decide to investigate their relative positions as they enjoy their rides.

Each child is 20 feet from the center of the carousel. Assume the girl starts at the point $(20, 0)$ and her brother starts at $(0, 20)$.

1. Complete the following table to give coordinates for both the girl and her brother at selected special points on their ride.

t (AMOUNT OF ROTATION IN RADIANS)	THE GIRL'S x-COORDINATE	THE GIRL'S y-COORDINATE	HER BROTHER'S x-COORDINATE	HER BROTHER'S y-COORDINATE
0	20	0	0	20
$\frac{\pi}{2}$				
π				
$\frac{3\pi}{2}$				
2π				
$\frac{5\pi}{2}$				

2. On the following grid, use the data from Problem 1 to plot (t, y) for both the girl and her brother. Connect the points to smooth out your graphs.

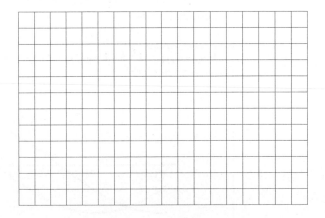

3. What is the relationship between the two graphs?

Definition

The **displacement**, or **horizontal shift**, of the graph of $y = a \sin (bx + c)$ is the smallest movement (left or right) necessary for the graph of $y = a \sin (bx)$ to match the graph of $y = a \sin (bx + c)$ exactly.

Example 1   *Consider the graphs of $y = \sin x$ and $y = \sin (x + 1)$.*

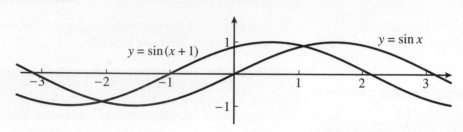

The graph of $y = \sin x$ must be moved 1 unit to the left to match the graph of $y = \sin (x + 1$
exactly, so the displacement, or horizontal shift, is -1.

4. a. Which graph is displaced in Problem 2?

b. What is the displacement?

5. Would you expect the same type of relationship between the two graphs representing the
x-coordinates? Explain.

Displacement, or horizontal shift, is defined for the cosine function in the same manner it i
defined for sine.

Example 2   *Consider the graphs of $y = 3 \cos 2x$ and $y = 3 \cos (2x - 1)$.*

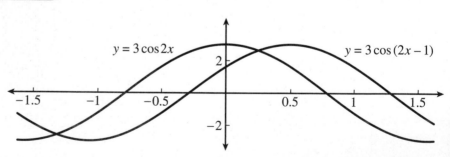

The graph of $y = 3 \cos (2x - 1)$ appears to be about $\frac{1}{2}$ unit to the right of the graph of
$y = 3 \cos 2x$, so the displacement is approximately $\frac{1}{2}$.

6. a. If the girl's x-coordinate is given by $20 \cos t$, predict the defining equation of the boy's x-coordinate.

b. Use your graphing calculator to test your prediction.

> In general, in the functions defined by $y = a \sin (bx + c)$ and $y = a \cos (bx + c)$, the values for b and c affect the displacement, or horizontal shift, of the function. The horizontal shift is given by $-\frac{c}{b}$. Note that if $-\frac{c}{b}$ is negative, the shift is to the left. If $-\frac{c}{b}$ is positive, the shift is to the right.

7. a. Using the expression $-\frac{c}{b}$, what is the displacement of $y = 3 \cos (2x - 1)$?

b. Is your result consistent with the graph in Example 2?

8. In this activity, the boy's y-coordinate is given by $y = a \sin (bx + c)$, where $a = 20$ and $b = 1$. Calculate c.

SUMMARY: ACTIVITY 6.9

1. The **displacement**, or **horizontal shift**, of the graph of $y = a \sin (bx + c), b > 0$, is the smallest movement (left or right) necessary for the graph of $y = a \sin bx$ to match the graph of $y = a \sin (bx + c)$ exactly.

2. The **displacement**, or **horizontal shift**, of the graph of $y = a \cos (bx + c), b > 0$, is the smallest movement (left or right) necessary for the graph of $y = a \cos (bx)$ to match the graph of $y = a \cos (bx + c)$ exactly.

3. For the functions defined by $y = a \sin (bx + c)$ and $y = a \cos (bx + c), b > 0$, the horizontal shift is given by $-\frac{c}{b}$.

4. If $-\frac{c}{b}$ is negative, the shift is to the left. If $-\frac{c}{b}$ is positive, the shift is to the right.

EXERCISES: ACTIVITY 6.9

1. For each following equations, identify the amplitude, period, and displacement of its graph.

a. $y = 0.7 \cos \left(2x + \frac{\pi}{2}\right)$

b. $y = 3 \sin (x - 1)$

c. $f(x) = -2.5 \sin\left(0.4x + \frac{\pi}{3}\right)$ **d.** $g(x) = 15 \sin\left(2\pi x - 0.3\right)$

2. For each of the following tables, identify a function of the form $y = a \sin(bx + c)$ or $y = a \cos(bx + c)$ that approximately satisfies the table.

a.

x	−0.7854	0.7854	2.3562	3.927	5.4978
y	0	3	0	−3	0

b.

x	1	2.5708	4.1416	5.7124	7.2832
y	−0.5	0	0.5	0	−0.5

3. Sketch one cycle of the graph of the function defined by $f(x) = 2 \sin\left(x + \frac{\pi}{2}\right)$.

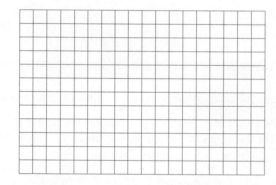

4. Determine an equation for the function defined by the following graph.

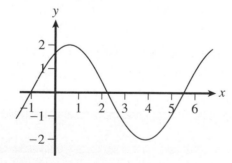

5. Match the given equation to one of the graphs that follow. (Assume that Xscl = 1 and Yscl = 1.)

a. $y = 2 \cos (x - 1)$

b. $y = 2 \sin (x - 1)$

c. $y = 2 \cos (x + 2)$

d. $y = 2 \sin (x + 2)$

i.
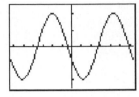

ii.
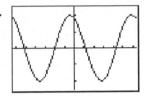

iii.
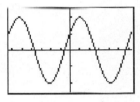

iv.
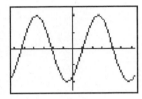

Activity 6.10

Texas Temperatures

Objectives

1. Determine the equation of a sine function that best fits the given data.

2. Make predictions using a sine regression equation.

According to the National Weather Service, the average monthly high temperature in th Midland-Odessa, Texas, area between 1971 and 2000 is given by the following table.

MONTH	AVER. MAX. (°F)
Jan	56.8
Feb	63.0
Mar	70.9
Apr	78.8
May	86.8
Jun	92.7
Jul	94.3
Aug	92.8
Sep	86.1
Oct	77.4
Nov	65.9
Dec	58.4

1. To get a feel for the relationship between the month and the average high temperatures let x represent the month, where 1 = Jan of the first year, 12 = Dec of the first yea 13 = Jan of the second year, and 24 = Dec of the second year. Plot the average maxi mum monthly temperatures for Midland-Odessa over a 2-year period on the grid below Use the same output data for the 1st and 2nd year.

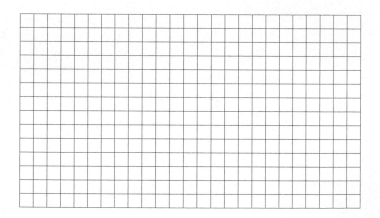

2. Enter the data plotted in Problem 1 into your calculator, and verify your scatterplot by creating a stat plot. Your plot should resemble the screen below.

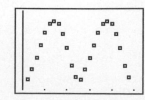

Notice the wavelike and repetitive nature of the data. This is typical of many natural periodi phenomena.

To produce a sine function that models the data, set your calculator to radian mode and use the Stat Calc menu and Option C: SinReg.

Note that the sine regression model is of the form

$$y = a\sin(bx + c) + d.$$

3. Enter your regression model here, rounding the values of a, b, c, and d to the nearest 0.001.

4. Add the graph of this function to your stat plot to verify that it is a good model. Your screen should appear as follows.

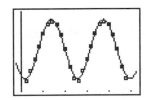

SUMMARY: ACTIVITY 6.10

1. The **sine function** can be used to model periodic wavelike behavior.

2. To determine the sine regression using the TI-83/TI-84 Plus calculator,

   a. set your MODE to radian, and

   b. use the SinReg option of the Stat Calc menu.

EXERCISES: ACTIVITY 6.10

The phases of the moon are periodic and repetitive even though the names of the phases can be confusing.

  i. A new moon occurs when no moon is visible to an observer on Earth.

 ii. During the first quarter, half of the side of the moon facing Earth is visible.

iii. During a full moon, the entire side of the moon is visible.

iv. During the last quarter, the other half of the side of the moon facing Earth is visible.

The U.S. Naval Observatory lists the following dates for phases of the moon in a recent year.

Date	Jan 29	Feb 5	Feb 13	Feb 21	Feb 28	Mar 6	Mar 14	Mar 22	Mar 29
Moon Phase	New	1st quarter	Full	Last quarter	New	1st quarter	Full	Last quarter	New

1. **a.** Let *x* represent the number of days since Jan 29 (be careful here) and let *y* represent the amount of moon visible, where New = 0, the 1st quarter = 0.5, Full = 1, and the Last quarter = 0.5. Complete the following table.

Date	Jan 29	Feb 5	Feb 13	Feb 21	Feb 28	Mar 6	Mar 14	Mar 22	Mar 29
x; Days Since Jan 29									
y; The Amount of Moon Visible									

b. Plot the data points generated in the table in part a.

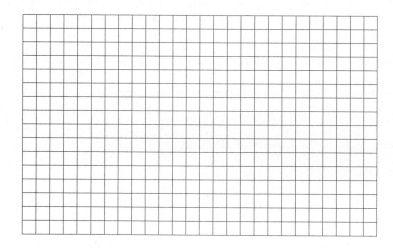

c. Does this data indicate that a sine regression may be appropriate in this case? Explain.

d. Use your TI-83/TI-84 Plus calculator to produce a sine regression model for the data in part a. Round *a*, *b*, *c*, and *d* to the nearest 0.0001 and record the model below.

e. Use your model to predict approximately how much of the moon will be visible 95 days after Jan 29th, provided the sky is clear.

2. Tides are the regular rising or falling of the ocean's surface. This is due in large part to the gravitational forces of the moon. The following table represents the water level of the tide off the coast of Pensacola, Florida, for a 24-hour period, January 2 and 3 of a recent year.

Hour #	0	2	4	6	8	10	12	14	16	18	20	22	24
Measurement in Feet Above Average Low Tide	−.24	−.07	.29	.66	1.02	1.27	1.32	1.07	.66	.2	−.16	−.47	−.53

Data source: NOAA tidesonline.nos.noaa.gov

a. Plot the data points given in the table above using hour as the independent variable.

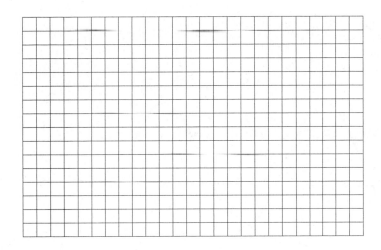

b. Does this data indicate that a sine regression may be appropriate in this case? Explain.

c. Use your TI-83/TI-84 Plus calculator to produce a sine regression model for the data in the table. Round a, b, c, and d to the nearest 0.001 and record the model below.

d. Use your model to predict the height of the tide 30 hours after the original observation.

e. Do you expect your model from part c to be a good predictor of tide height on a year-round basis?

Cluster 2   What Have I Learned?

1. Explain why the trigonometric functions could be called circular functions.

2. Sometimes the difference between the trigonometric functions and the circular functions is explained by the difference in inputs. The input values for the trigonometric functions are angle measurements, and the input values for the circular functions are real numbers. How does your knowledge of radian measure relate to this?

3. **a.** Estimate the amplitude of the function defined by the following graph.

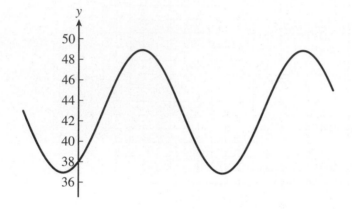

  b. Use the definition of amplitude to answer part a.

4. The period of $y = \sin x$ is 2π and the period of $y = \cos x$ is 2π. Explain why this makes sense when sine and cosine are viewed as circular functions.

5. Given a function defined by $y = a \sin(bx + c)$ and $b > 0$, and given the fact that b and c possess opposite signs, determine whether the graph of the function is displaced to the right or to the left. Explain.

Cluster 2 How Can I Practice?

1. Determine the coordinates of the point on the unit circle corresponding to the following central angles. If necessary, round your results to the nearest hundredth.

   a. $36°$

   b. $210°$

   c. $-90°$

   d. $317°$

   e. $-144°$

   f. $450°$

2. For each of the points on the unit circle determined in Exercise 1, determine the distance traveled along the circle to the point from $(1, 0)$.

   a.

   b.

   c.

   d.

   e.

   f.

3. Convert the following degree measures to radian measures in terms of π.

   a. $18°$

   b. $150°$

   c. $390°$

   d. $-72°$

4. Convert the following radian measures to degree measures.

   a. $\frac{5\pi}{6}$

   b. 1.7π

   c. -3π

   d. 0.9π

Obtain the following information about each of the functions defined by the equations in Exercises 5–9.

a. *Use your graphing calculator to sketch a graph.*

b. *From the defining equation, determine the amplitude. Then use your graph to verify that your amplitude is correct.*

c. *From the defining equation, determine the period. Then use your graph to verify that your period is correct.*

d. *From the defining equation, determine the displacement. Then use your graph to verify that your displacement is correct.*

5. $y = 4 \cos 3x$

6. $y = -2 \sin (x - 1)$

Answers to all How Can I Practice exercises are included in the Selected Answers appendix.

7. $s = 3.2 \sin(-2x)$

8. $f(x) = -\cos\left(\frac{x}{2} + 1\right)$

9. $g(x) = -3 \cos(4x - 1)$

10. You rent a cottage on the ocean for a week one summer and notice that the tide comes in twice a day with approximate regularity. Remembering that the trigonometric functions model repetitive behavior, you place a meterstick in the water to measure water height every hour from 6:00 A.M. to midnight. At low tide the height of the water is 0 centimeter, and at high tide the height is 80 centimeters.

   a. Explain why a sine or a cosine function models this relationship between height of water in centimeters and time in hours.

   b. What is the amplitude of this function?

   c. Approximate the period of this function. Explain.

   d. Determine a reasonable defining equation for this function. Explain.

The bracketed numbers following each concept indicate the activity in which the concept is discussed.

CONCEPT/SKILL	DESCRIPTION	EXAMPLE
The sine function of the acute angle A of a right triangle [6.1]	$\sin A = \dfrac{\text{length of the side opposite } A}{\text{length of the hypotenuse}}$	Examples 2 and 3, Activity 6.1, pages 589–590
The cosine function of the acute angle A of a right triangle [6.1]	$\cos A = \dfrac{\text{length of the side adjacent to } A}{\text{length of the hypotenuse}}$	Examples 2 and 3, Activity 6.1, pages 589–590
The tangent function of the acute angle A of a right triangle [6.1]	$\tan A = \dfrac{\text{length of the side opposite } A}{\text{length of the side adjacent to } A}$	Examples 2 and 3, Activity 6.1, pages 589–590
Complementary angles [6.2]	Complementary angles are two acute angles, the sum of whose measures is 90°.	Angles of 30° and 60° are complementary angles.
Cofunctions related to complementary angles [6.2]	Cofunctions of complementary angles are equal.	$\sin 35° = \cos 55°$
Inverse trigonometric functions [6.3]	The inverse trigonometric functions are defined by $\sin^{-1} x = \theta$ is equivalent to $\sin \theta = x$ $\cos^{-1} x = \theta$ is equivalent to $\cos \theta = x$ $\tan^{-1} x = \theta$ is equivalent to $\tan \theta = x$	Examples 1 and 2, Activity 6.3, pages 604–605
The domain of the inverse trigonometric functions [6.3]	The domain (set of inputs) of the inverse trigonometric functions is the set of ratios of the lengths of the sides of a right triangle.	The domain of the inverse sine function is all real numbers from −1 to 1, including both −1 and 1.
The range of the inverse trigonometric functions [6.3]	The range (set of outputs) of the inverse trigonometric functions is the set of angles.	The range of the inverse tangent function is all angles from −90° to 90°.
Solving right triangles [6.4]	When trigonometric problems involve solving right triangles, employ the following trigonometric problem-solving strategy. 1. Draw a diagram of the situation using right triangles. 2. Identify all known sides and angles. 3. Identify sides and/or angles you want to know. 4. Identify functions that relate the known and unknown. 5. Write and solve the appropriate trigonometric equation(s).	Example 1, Activity 6.4, page 610

CONCEPT/SKILL	DESCRIPTION	EXAMPLE
Central angle [6.6]	A central angle is an angle whose vertex is the center of a circle.	Problem 2b, Activity 6.6, page 627
The domain of the sine and cosine functions [6.6]	The domain of each of the sine and cosine functions is all angles, both positive and negative.	The domains are all real numbers.
The range of the sine and cosine functions [6.6]	The range of the sine and cosine functions is all values of N such that $-1 \leq N \leq 1$.	The range is all real numbers from -1 to 1 inclusive.
The graph of $y = \sin x$ [6.6]	The graph of $y = \sin x$ is a periodic wave. One cycle is shown in the Example.	
The graph of $y = \cos x$ [6.6]	The graph of $y = \cos x$ is a periodic wave. One cycle is shown in the Example.	
The period of the sine and cosine functions [6.6]	The period is the number of units required to complete one cycle of the graph of a function.	The period is 360° for $y = \sin x$ and $y = \cos x$.
Radian measure [6.7]	Radian measure is used when the input of a repeating function is better defined by real numbers than angles measured in degrees.	360° is equivalent to 2π radians; the circumference of a unit circle.
Converting from radian measure to degree measure [6.7]	To convert degree measure to radian measure, multiply the degree measure by $\dfrac{\pi \text{ radians}}{180°}$.	$30° = 30 \cdot \dfrac{\pi}{180} = \dfrac{\pi}{6}$ radians
Converting from degree measure to radian measure [6.7]	To convert radian measure to degree measure, multiply the radian measure by $\dfrac{180°}{\pi \text{ radians}}$.	$\dfrac{2\pi}{3} = \dfrac{2\pi}{3} \cdot \dfrac{180°}{\pi} = \dfrac{360°}{3} = 120°$

CONCEPT/SKILL	DESCRIPTION	EXAMPLE
The amplitude of trigonometric functions [6.8]	The amplitude of trigonometric functions defined by $y = a\sin(bx)$ or $y = a\cos(bx)$ is defined by $$\frac{1}{2}(M - m),$$ where M represents the maximum function value and m represents the minimum function value.	Example 1, Activity 6.8, page 646
The frequency of sine and cosine [6.8]	The frequency of trigonometric functions defined by $y = a\sin(bx)$ or $y = a\cos(bx)$, $b > 0$, is the number of cycles completed by the graphs over intervals of length 2π. The frequency of trigonometric functions defined by $y = a\sin(bx)$ or $y = a\cos(bx)$, $b > 0$, is b.	The frequency of $y = \sin(2x)$ is 2; two complete cycles occur in 2π units.
The period of sine and cosine [6.8]	For the trigonometric functions defined by $y = a\sin(bx)$ or $y = a\cos(bx)$, $b > 0$, the period is $\frac{2\pi}{b}$.	The period of $y = \sin(2x)$ is $\frac{2\pi}{2} = \pi$ units.
The displacement, or horizontal shift, of the graph of sine or cosine [6.9]	The displacement, or horizontal shift, of the graph of $y = a\sin(bx + c)$, $b > 0$, is the smallest movement (left or right) necessary for the graph of $y = a\sin bx$ to match the graph of $y = a\sin(bx + c)$ exactly. Similarly for the cosine.	Examples 1 and 2, Activity 6.9 page 654
The displacement, or horizontal shift, of the graph of sine or cosine [6.9]	For the functions $y = a\sin(bx + c)$ and $y = a\cos(bx + c)$, $b > 0$, the horizontal shift is given by $-\frac{c}{b}$.	The displacement of the function $y = 2\sin\left(3x + \frac{\pi}{2}\right)$ is given by $-\frac{\frac{\pi}{2}}{3} = -\frac{\pi}{6}$. The shift is $\frac{\pi}{6}$ units to the left.
The sine function can be used to model periodic wavelike behavior [6.10]	Using the TI-83/TI-84 Plus calculator, set your MODE to radian and use the SinReg option on the Stat Calc menu.	Problems 1–4, Activity 6.10 pages 658–659

1. You walk 7 miles in a straight line 63° north of east.

   a. Determine how far north you have traveled.

   b. Determine how far east you have traveled.

2. Solve the following triangles.

   a.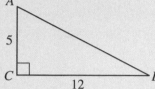

   b.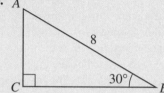

   c.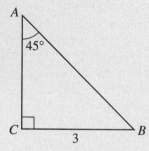

   d.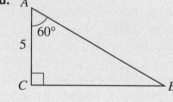

3. **a.** Given $\sin \theta = \frac{6}{10}$, determine $\cos \theta$ and $\tan \theta$ without using your calculator.

   b. Given $\cos \theta = \frac{\sqrt{3}}{2}$, determine $\sin \theta$, $\tan \theta$, and θ without using your calculator.

   c. Given $\tan \theta = \frac{8}{5}$, determine $\sin \theta$ and $\cos \theta$ without using your calculator.

Answers to all Gateway exercises are included in the Selected Answers appendix.

668

4. You are taking your nephew to see the Empire State Building. When you are 100 feet away from the building, you and your nephew look up to see the top. You are 6 feet tall, your nephew is 3 feet tall, and the Empire State Building is 1414 feet high. You notice that even though he is only half your height, your nephew does not have to tilt his head any more than you do. Is your observation correct? Is the angle always independent of people's heights? Explain.

5. In the diagram below, determine the lengths of *a*, *b*, *c*, and *h* to the nearest tenth.

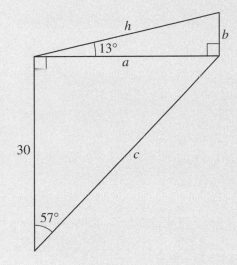

6. In the following diagram, determine *x* and *h*. (*Hint:* See Exercises 2c and d.)

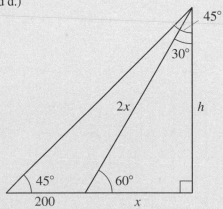

7. Using the following triangles, complete the table without using your calculator.

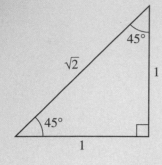

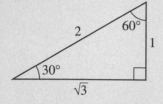

ANGLE θ	sin θ	cos θ	tan θ
120°			
135°			
150°			
180°			
210°			
225°			
240°			
270°			
300°			
315°			
330°			
360°			

8. Determine the amplitude and period of the given functions, and then sketch their graphs. Use your graphing calculator to verify your results.

a. $y = 2 \sin x$ **b.** $y = -2 \sin x$ **c.** $y = \cos 2x$

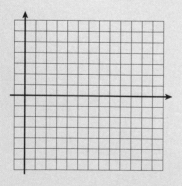

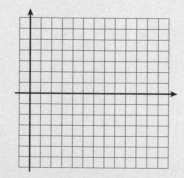

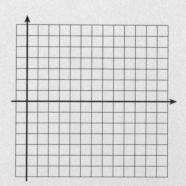

d. $y = \cos 2\pi x$

e. $y = \sin \frac{x}{2}$

f. $y = \sin \frac{\pi x}{2}$

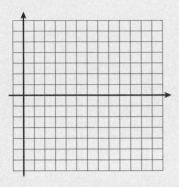

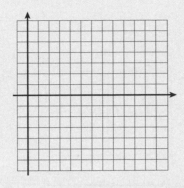

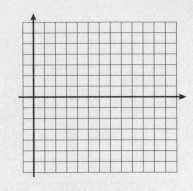

g. $y = \cos x$

h. $y = \frac{2}{3} \cos (2x)$

i. $y = \sin (2\pi x - 3\pi)$

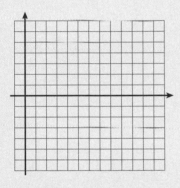

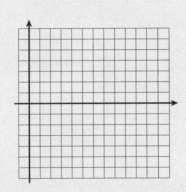

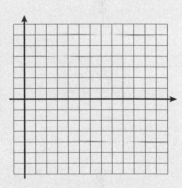

j. $y = 3 \sin (2\pi x - 3\pi)$

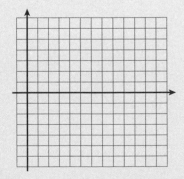

9. Match the given equation to one of the accompanying graphs. (Assume that Xscl = 1 and Yscl = 1.)

a. $y = -3 \sin x$

b. $y = 2 \sin \left(\frac{\pi x}{2} + \frac{\pi}{2} \right)$

c. $y = 2 \sin \left(\frac{\pi x}{2} - \frac{\pi}{2} \right)$

d. $y = -3 \cos x$

e. $y = -2 \cos \left(\frac{\pi x}{2} + \frac{\pi}{2} \right)$

f. $y = -\cos \left(\pi x - \frac{\pi}{2} \right)$

i.

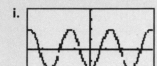

ii.

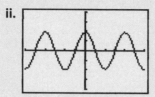

iii.

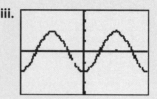

iv.

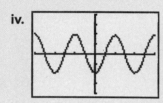

v.

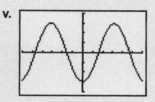

vi.

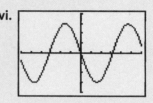

iv.

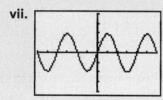

v.

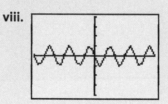

vii.

viii.

Concept Review

Properties of Exponents

The basic properties of exponents are summarized as follows.

> If a is a nonzero real number and n and m are rational numbers, then
>
> **1.** $a^n a^m = a^{n+m}$ **2.** $\dfrac{a^n}{a^m} = a^{n-m}$ **3.** $(a^n)^m = a^{nm}$
>
> **4.** $(ab)^n = a^n b^n$ **5.** $a^0 = 1$ **6.** $a^{-n} = \left(\dfrac{1}{a}\right)^n = \dfrac{1}{a^n}$

Property 1: $a^n a^m = a^{n+m}$ If you are multiplying two powers of the same base, keep the base the same and add the exponents.

Example 1: $3^4 \cdot 3^7 = 3^{4+7} = 3^{11}$

Note that the exponents were added and the base did not change.

Property 2: $\dfrac{a^n}{a^m} = a^{n-m}$ If you are dividing two powers of the same base, keep the base the same and subtract the exponents.

Example 2: $\dfrac{6^6}{6^4} = 6^{6-4} = 6^2 = 36$

Note that the exponents were subtracted and the base did not change.

Property 3: $(a^n)^m = a^{nm}$ If an exponent is applied to a power, multiply the exponents.

Example 3: $(y^3)^4 = y^{12}$

The exponents were multiplied. The base did not change.

Property 4: $(ab)^n = a^n b^n$ If an exponent is applied to a product, the exponent is distributed to each factor.

Example 4: $(2x^2 y^3)^3 = 2^3 \cdot (x^2)^3 \cdot (y^3)^3 = 8x^6 y^9$

Note that since the base contained three factors, each of those was raised to the third power. The common mistake in an expansion such as this is not to raise the coefficient to the power.

Property 5: $a^0 = 1, a \neq 0$. Often presented as a definition, Property 5 states that a▮ nonzero base raised to the zero power is 1. This property or definition is a result of Property▮ of exponents as follows.

Consider $\dfrac{x^5}{x^5}$. Using Property 2, $x^{5-5} = x^0$. However, you know that any fraction in which t▮ numerator and the denominator are equal is equivalent to 1. Therefore, $x^0 = 1$.

Example 5: $\left(\dfrac{2x^3}{3yz^5}\right)^0 = 1$

Given a nonzero base, if the exponent is 0, the value is 1.

Property 6: $a^{-n} = \left(\dfrac{1}{a}\right)^n = \dfrac{1}{a^n}$. Sometimes presented as a definition, Property 6 states th▮ any base

raised to a negative power is equivalent to the reciprocal of the base raised to the positiv▮ power. Note that the negative exponent does not have any effect on the sign of the base. Th▮ property could also be viewed as a result of the second property of exponents as follows.

Consider $\dfrac{x^3}{x^5}$. Using Property 2, $x^{3-5} = x^{-2}$. If you view this expression algebraically, yo▮ have three factors of x in the numerator and five in the denominator. If you divide out th▮ three common factors, you are left with $\dfrac{1}{x^2}$. Therefore, if Property 2 is true, then $x^{-2} = \dfrac{1}{x^2}$.

Example 6: Write each of the following without negative exponents.

$$\textbf{a. } 3^{-2} \qquad\qquad\qquad \textbf{b. } \dfrac{2}{x^{-3}}$$

Solution: $\textbf{a. } 3^{-2} = \left(\dfrac{1}{3}\right)^2 = \dfrac{1}{3^2} = \dfrac{1}{9}$ $\textbf{b. } \dfrac{2}{x^{-3}} = \dfrac{2}{\dfrac{1}{x^3}} = 2 \div \dfrac{1}{x^3} = 2 \cdot x^3 = 2x^3$

A **factor** of the form b^x can be moved from a numerator to a denominator or from a denominator to a numerator by changing the *sign of the exponent.*

Example 7: Simplify and express your results with positive exponents only.

$$\left(\dfrac{x^3 y^{-4}}{2x^{-3} y^{-2} z}\right) \cdot \left(\dfrac{4x^3 y^2 z}{x^5 y^{-3} z^3}\right)$$

Solution: Simplify each factor by writing it with positive exponents only.

$$\left(\dfrac{x^6}{2y^2 z}\right) \cdot \left(\dfrac{4y^5}{x^2 z^2}\right)$$

Now multiply and simplify.

$$\dfrac{4x^6 y^5}{2x^2 y^2 z^3} = \dfrac{2x^4 y^3}{z^3}$$

Exercises

Simplify and express your results with positive exponents only. Assume all variables represent nonzero real numbers.

1. 5^{-3}

2. $\dfrac{1}{x^{-5}}$

3. $\dfrac{3x}{y^{-2}}$

4. $\dfrac{10x^2y^5}{2x^{-3}}$

5. $\dfrac{5^{-1}z}{x^{-1}z^{-2}}$

6. $5x^0$

7. $(a + b)^0$

8. $-3(x^0 - 4y^0)$

9. $x^6 \cdot x^{-3}$

10. $\dfrac{4^{-2}}{4^{-3}}$

11. $(4x^2y^3) \cdot (3x^{-3}y^{-2})$

12. $\dfrac{24x^{-2}y^3}{6x^3y^{-1}}$

13. $\dfrac{(4x^{-2}y^{-3}) \cdot (5x^3y^{-2})}{6x^2y^{-3}z^{-3}}$

14. $\left(\dfrac{2x^{-2}y^{-3}}{z^2}\right) \cdot \left(\dfrac{x^5y^3}{z^{-3}}\right)$

15. $\dfrac{(6x^4y^{-3}z^{-2})(3x^{-3}y^4)}{15x^{-3}y^{-3}z^2}$

Solving 3 × 3 Linear Systems Algebraically

Linear equations such as $3x + 2y - z = 4$ involve three variables, x, y, and z. A solution ⊙ such an equation is an ordered triple (x, y, z) such that if the values of x, y, and z in th ordered triple are substituted into the equation, the result is a true statement.

A system of three linear equations in three variables (a 3 × 3 system) such as

$$-x + y + z = -3$$
$$3x + 9y + 5z = 5$$
$$x + 3y + 2z = 4$$

has as its solution all ordered triples (x, y, z) that will make all three equations true.

Solving a 3 × 3 Linear System Algebraically

1. Eliminate one variable using any two of the given three equations to obtain an equatic in two variables (or less).

2. Eliminate the same variable using the third equation not used in step 1 and either of th other two equations to obtain a second equation in two variables (or less).

3. Solve the system consisting of the two equations found in steps 1 and 2.

4. Substitute the values obtained in step 3 into any equation involving all three variables ▸ determine the value of the third variable.

5. Check your solution in all three equations.

Example 1: Determine all solutions of

$$-x + y + z = -3$$
$$3x + 9y + 5z = 5$$
$$x + 3y + 2z = 4.$$

Solution: You may choose to eliminate x using the addition method.

Step 1.   Multiply both sides of the first equation by 3 and add the results to the second equation.

$$3(-x + y + z = -3) \quad \text{or} \quad -3x + 3y + 3z = -9$$
$$3x + 9y + 5z = 5 \qquad\qquad 3x + 9y + 5z = 5$$

The sum is $12y + 8z = -4$.

Step 2.   Multiply the third equation by -3 and add the results to the second.

$$3x + 9y + 5z = 5 \qquad\qquad 3x + 9y + 5z = 5$$
$$-3(x + 3y + 2z = 4) \quad \text{or} \quad -3x - 9y - 6z = -12$$

The sum is $-z = -7$.

Step 3.   The 2 × 2 system resulting from steps 1 and 2 is

$$12y + 8z = -4$$
$$-z = -7.$$

The second equation is equivalent to $z = 7$. Substituting this value into the first equation of the new system yields

$$12y + 8(7) = -4$$
$$12y + 56 = -4$$
$$12y = -60$$
$$y = -5.$$

Step 4. Using the values of y and z in the third equation of the original system,

$$x + 3(-5) + 2(7) = 4$$
$$x - 15 + 14 = 4$$
$$x - 1 = 4$$
$$x = 5.$$

The potential solution is $(5, -5, 7)$ and should be checked in all three of the original equations.

Not every 3×3 linear system has unique solutions; some have multiple solutions. These are called **dependent systems**. Some systems have no solution and are called **inconsistent systems**.

Example 2: Solve the following system.

$$x + 2y + 3z = 5$$
$$-x + y - z = -6$$
$$2x + y + 4z = 4$$

Solution: Again, you may eliminate x using the addition method.

Step 1. Sum the first two equations to eliminate x.

$$x + 2y + 3z = 5$$
$$-x + y - z = -6$$

The sum is $3y + 2z = -1$.

Step 2. Multiply the second equation by 2 and add the results to the third.

$$-2x + 2y - 2z = -12$$
$$2x + y + 4z = 4$$

The sum is $3y + 2z = -8$. The new system is

$$3y + 2z = -1$$
$$3y + 2z = -8.$$

Step 3. To solve the new system, multiply the first equation by -1 and add the results to the second.

$$-3y - 2z = 1$$
$$3y + 2z = -8$$

The sum is $0 = -7$.

Since $0 = -7$ is a false statement, the conclusion is there is no solution. The original system was an inconsistent system.

Had the sum of the equations in step 3 resulted in a true statement such as $0 = 0$, the conclusion would have been that there were an infinite number of solutions. That is, the system would have been *dependent*.

Exercises

Solve the following systems algebraically. If the system is dependent or inconsistent, sta
this as the answer.

1. $x + y - z = 9$
$x + y + z = 5$
$x - y + 2z = 1$

2. $-2x + y + 4z = 3$
$x + y - 3z = 2$
$x - y + 2z = 1$

3. $x + 2y + 3z = 5$
$-x + y - z = -6$
$2x + y + 4z = 4$

4. $3x - 2y + 3z = 11$
$2x + 3y - 2z = -5$
$x + 4y - z = -5$

5. $x - 4y + z = -5$

$3x - 12y + 3z = -15$

$-2x + 8y - 2z = 10$

6. $2x + 3y + 4z = 3$

$6x - 6y + 8z = 3$

$4x + 3y - 4z = 2$

7. $x + 2y = 10$

$-x + 3z = -23$

$4y - z = 9$

Inequalities Involving Absolute Value

The key to solving absolute value inequalities algebraically is to rewrite them using the following properties.

> **Absolute Value Properties**
>
> For any real number x and $a > 0$,
>
> $$|x| < a \text{ is equivalent to } -a < x < a.$$
>
> For any real number x and $a > 0$,
>
> $$|x| > a \text{ is equivalent to the statement } x > a \text{ or } x < -a.$$

> **Solving Absolute Value Inequalities**
>
> **1.** Rewrite the inequality with the absolute value isolated.
>
> **2.** Rewrite the inequality as a compound inequality or pair of inequalities.
>
> **3.** Solve the resulting inequality or inequalities.

Example 1: Solve $|2x - 3| + 3 \le 8$.

Solution: Subtract 3 from both sides.

$$|2x - 3| + 3 - 3 \le 8 - 3 \quad \text{or} \quad |2x - 3| \le 5$$

Using the first property,

$$-5 \le 2x - 3 \le 5$$

Add 3 to each part.

$$-5 + 3 \le 2x - 3 + 3 \le 5 + 3$$
$$-2 \le 2x \le 8$$

Divide each part by 2.

$$\frac{-2}{2} \le \frac{2x}{2} \le \frac{8}{2} \quad \text{or} \quad -1 \le x \le 4$$

Example 2: Solve $|4x + 3| - 4 > 7$.

Solution: Add 4 to both sides to isolate the absolute value.

$$|4x + 3| - 4 + 4 > 7 + 4$$
$$|4x + 3| > 11$$

Using the second absolute value property,

$$4x + 3 > 11 \quad \text{or} \quad 4x + 3 < -11.$$

Solving these inequalities,

$$4x + 3 - 3 > 11 - 3 \quad \text{or} \quad 4x + 3 - 3 < -11 - 3$$
$$4x > 8 \qquad\qquad\qquad 4x < -14$$
$$x > 2 \qquad\qquad\qquad x < -\tfrac{7}{2}.$$

Exercises

Solve the following inequalities.

1. $|3x - 5| < 5$

2. $|x - 3| - 2 \leq 3$

3. $|4x - 1| > 3$

4. $|2x - 1| - 4 \geq 7$

Solving Equations by Factoring

Some quadratic and higher-order polynomial equations can be solved by using factoring and the zero-product property.

The process is as follows.

> **Solving an Equation by Factoring**
>
> **1.** Use the addition principle to remove all terms from one side of the equation. This results in the equation having one side equal to zero.
>
> **2.** Combine like terms and then factor.
>
> **3.** Use the zero-product rule to set each factor containing a variable equal to zero and then solve the equations.
>
> **4.** Check your solutions in the original equation.

Example 1: Solve the equation $x(x + 5) = 0$.

Solution: This equation already satisfies the first two steps in our process, so we simply start at step 3.

$$x = 0 \quad x + 5 = 0$$
$$x = -5$$

Thus, we have two solutions $x = 0$ and $x = -5$. The check is left to the reader.

Example 2: Solve the equation $6x^2 = 16x$.

Solution: Making one side of the equation equal to zero, $6x^2 - 16x = 0$. Since there are no like terms, factor the binomial.

$$2x(3x - 8) = 0$$

Using the zero-product principle,

$$2x = 0 \quad \text{or} \quad 3x - 8 = 0$$
$$x = 0 \quad \text{or} \quad 3x = 8$$
$$x = \tfrac{8}{3}.$$

The two potential solutions are $x = 0$ and $x = \tfrac{8}{3}$. The check is left to the reader.

Example 3: Solve $3x^2 - 2 = -x$.

Solution: Making one side of the equation equal to zero, $3x^2 + x - 2 = 0$. Since there are no like terms, factor the trinomial.

$$(3x - 2)(x + 1) = 0$$

Using the zero-product principle,

$$3x - 2 = 0 \quad \text{or} \quad x + 1 = 0$$
$$3x = 2 \quad \text{or} \quad x = -1$$
$$x = \tfrac{2}{3}.$$

The two potential solutions are $x = \tfrac{2}{3}$ and $x = -1$. The check is left to the reader.

Example 4: Solve $3x^3 - 8x^2 = 3x$.

Solution: Making one side of the equation equal to zero, $3x^3 - 8x^2 - 3x = 0$. Since there are no like terms, factor the trinomial.

$$x(3x^2 - 8x - 3) = 0$$
$$x(3x + 1)(x - 3) = 0$$

Using the zero-product principle,

$$x = 0 \quad \text{or} \quad 3x + 1 = 0 \quad \text{or} \quad x - 3 = 0$$
$$3x = -1 \quad \text{or} \quad x = 3$$
$$x = -\tfrac{1}{3}.$$

The three potential solutions are $x = 0$, $x = -\tfrac{1}{3}$, and $x = 3$. The check is left to the reader.

Exercises

Solve each of the following equations.

1. $x(x + 7) = 0$

2. $3(x - 5)(2x + 1) = 0$

3. $12x = x^2$

4. $x^2 + 5x = 0$

5. $x^2 - 2x - 63 = 0$

6. $3x^2 - 9x - 30 = 0$

7. $-7x + 6x^2 = 10$

8. $3y^2 = 2 - y$

9. $-28x^2 + 15x - 2 = 0$

10. $4x^2 - 25 = 0$

11. $(x + 4)^2 - 16 = 0$

12. $(x + 1)^2 - 3x = 7$

13. $2(x + 2)(x - 2) = (x - 2)(x + 3) - 2$

14. $18x^3 = 15x^2 + 12x$

Solving Quadratic Equations by Completing the Square

The square root property can be used to solve equations of the form $B^2 = a$.

> **Square Root Property**
>
> If $B^2 = a$, where a is a real number, then $B = \pm\sqrt{a}$.

Example 1:   Solve the equation $(x + 3)^2 = 9$.

Solution: This equation fits the form of the hypothesis of the square root property, where $B = x + 3$. Therefore,

$$x + 3 = \pm\sqrt{9} = \pm 3.$$

You now have two equations to solve, $x + 3 = 3$ and $x + 3 = -3$. The solutions a $x = 0, -6$. Both of these values make the original statement true. Hence, both are solutions

The next example illustrates an algebraic technique of solving quadratic equations known *completing the square*. The strategy is to rewrite the quadratic equation $ax^2 + bx + c =$ $a \neq 0$, in the form $(x + h)^2 = k$ and solve as in Example 1. This requires an algebra process known as completing the square.

Consider the binomial $x^2 + 6x$. What term must be added to the binomial to produce a trin mial that is a perfect square? The answer is one-half the coefficient of the linear term square In this case, one-half of 6 is 3, and $3^2 = 9$, and

$$x^2 + 6x + 9 = (x + 3)^2.$$

This process can be helpful in solving quadratic equations as follows.

Solving a Quadratic Equation by Completing the Square

1. Use the multiplication principle to make the coefficient of x^2 equal to 1.

2. Rewrite the equation with the constant term isolated on one side.

3. Use the addition principle to add the square of one-half the coefficient of the linear term to both sides of the equation.

4. Replace the trinomial with its factored form, a perfect square.

5. Apply the square root property.

6. Solve the resulting linear equations.

7. Check your solutions in the original equation.

Example 2:   Solve $x^2 - 4x - 5 = 0$ by the completing the square method.

Since the coefficient of x^2 is one, step 1 is not necessary.

Step 2.   Adding 5 to both sides to isolate the constant term yields

$$x^2 - 4x = 5.$$

Step 3.   The value needed to complete the square is $\left(\frac{1}{2} \cdot (-4)\right)^2 = (-2)^2 = 4$. Using the addition principle to add this to both sides produces

$$x^2 - 4x + 4 = 5 + 4$$
$$x^2 - 4x + 4 = 9.$$

Step 4. Replacing the trinomial with its perfect square form,

$$(x - 2)^2 = 9.$$

Steps 5 and 6. Applying the square root principle and solving,

$$x - 2 = \pm 3$$
$$x = 2 \pm 3$$
$$x = -1, 5.$$

The checking of the solutions is left to the reader.

Example 3: Solve $6x + 6 = -x^2$ by the completing the square method.

Step 1. Multiply each term by -1 to make the coefficient of x^2 equal to 1.

$$-6x - 6 = x^2$$

Step 2. Using the addition principle to isolate the constant term,

$$-6 = x^2 + 6x.$$

Step 3. The value necessary to complete the square is $\left(\frac{1}{2} \cdot 6\right)^2 = 9$. Completing the square yields

$$-6 + 9 = x^2 + 6x + 9.$$

Step 4. Factoring and simplifying,

$$3 = (x + 3)^2.$$

Steps 5 and 6. Taking the square root of both sides and solving,

$$\pm\sqrt{3} = x + 3$$
$$x = -3 \pm \sqrt{3}.$$

The check is left to the reader. Note that the solutions in this case are real, but not rational.

Exercises

Solve the following quadratic equations using the completing the square.

1. $x^2 - 6x + 8 = 0$ **2.** $x^2 - 9x + 14 = 0$

3. $-4x = -x^2 + 12$ **4.** $2x^2 + 2x - 24 = 0$

5. $3x^2 + 2x = 1$

6. $-\frac{1}{2}x^2 - x + \frac{3}{2} = 0$

7. $10x^2 + 6x = 5$

8. $15x^2 - 10x - 3 = 0$

Derivation of the Quadratic Formula

The quadratic formula results from applying the completing the square method to the genera
quadratic equation $ax^2 + bx + c = 0$, where $a > 0$.

Step 1. Make the coefficient of x^2 equal 1, by multiplying both sides of the equation by $\frac{1}{a}$.

$$x^2 + \frac{b}{a}x + \frac{c}{a} = 0$$

Step 2. Use the addition principle to isolate the constant term on one side of the equal sign

$$x^2 + \frac{b}{a}x = -\frac{c}{a}$$

Step 3. Complete the square of the binomial. The coefficient of the linear term is $\frac{b}{a}$. The

term needed to complete the square is $\left(\frac{1}{2} \cdot \frac{b}{a}\right)^2 = \frac{b^2}{4a^2}$. Use the addition principl

to add this term to both sides.

$$x^2 + \frac{b}{a}x + \frac{b^2}{4a^2} = \frac{b^2}{4a^2} - \frac{c}{a}$$

Rewrite the right-hand side as a single fraction.

$$x^2 + \frac{b}{a}x + \frac{b^2}{4a^2} = \frac{b^2 - 4ac}{4a^2}$$

Step 4. Express the left-hand side in factored form.

$$\left(x + \frac{b}{2a}\right)^2 = \frac{b^2 - 4ac}{4a^2}$$

Step 5. Apply the square root property to the resulting equation.

$$x + \frac{b}{2a} = \pm \frac{\sqrt{b^2 - 4ac}}{2a}$$

Step 6. Solve for x.

$$x = -\frac{b}{2a} \pm \frac{\sqrt{b^2 - 4ac}}{2a} \quad \text{or} \quad x = \frac{-b \pm \sqrt{b^2 - 4ac}}{2a}$$

This formula can be used to solve any quadratic equation in standard form $ax^2 + bx + c = 0, a \neq 0$, and is called the **quadratic formula**.

Rational Expressions

You may need to practice skills relating to rational functions to enhance your understanding of these functions.

Simplifying Rational Expressions

1. Factor the numerator and the denominator.

2. Divide the numerator and the denominator by the common factors.

Example 1: Simplify $\dfrac{x^2 - 10x + 24}{x^2 - 5x + 4}$.

Solution:

Step 1. Factor the numerator and the denominator.

$$\frac{x^2 - 10x + 24}{x^2 - 5x + 4} = \frac{(x - 4)(x - 6)}{(x - 1)(x - 4)}$$

Step 2. Divide the numerator and denominator by the common factor.

$$\frac{\dfrac{(x - 4)(x - 6)}{x - 4}}{\dfrac{(x - 1)(x - 4)}{x - 4}} = \frac{x - 6}{x - 1}$$

Multiplying or Dividing Rational Expressions

1. Factor the numerator and denominator of each fraction completely.

2. Divide out the common factors (cancel).

3. Multiply remaining factors.

4. In division, proceed as above after inverting the divisor (the fraction after the division sign).

Example 2: Divide and simplify $\dfrac{x^2 + 3x - 10}{2x} \div \dfrac{x^2 - 5x + 6}{x^2 - 3x}$.

Solution:

Step 1.   Rewrite as multiplication.

$$\frac{x^2 + 3x - 10}{2x} \div \frac{x^2 - 5x + 6}{x^2 - 3x} = \frac{x^2 + 3x - 10}{2x} \cdot \frac{x^2 - 3x}{x^2 - 5x + 6}$$

Step 2.   Factor each fraction completely.

$$\frac{x^2 + 3x - 10}{2x} \cdot \frac{x^2 - 3x}{x^2 - 5x + 6} = \frac{(x + 5)(x - 2)}{2 \cdot x} \cdot \frac{x \cdot (x - 3)}{(x - 2)(x - 3)}$$

Step 3.   Cancel common factors.

$$\frac{(x + 5)(x - 2)}{2 \cdot x} \cdot \frac{x \cdot (x - 3)}{(x - 2)(x - 3)} = \frac{x + 5}{2}$$

Step 4.   The simplified expression is

$$\frac{x + 5}{2}.$$

Adding or Subtracting Rational Expressions

1. Find the LCD (least common denominator).

2. Build each fraction to have the LCD.

3. Add or subtract numerators.

4. Place the numerator over the LCD, and simplify if necessary.

Example 3: Add and simplify $\dfrac{x}{x + 1} + \dfrac{3}{(x + 1)^2}$.

Solution:

Step 1.   Since the denominators are already factored, it is clear that the LCD is $(x + 1)^2$

Step 2.   Build each fraction to have the LCD.

$$\frac{x}{x + 1} + \frac{3}{(x + 1)^2} = \frac{x(x + 1)}{(x + 1)(x + 1)} + \frac{3}{(x + 1)^2} = \frac{x^2 + x}{(x + 1)^2} + \frac{3}{(x + 1)^2}$$

Step 3.   Add or subtract the numerators.

$$\frac{x^2 + x}{(x + 1)^2} + \frac{3}{(x + 1)^2} = \frac{x^2 + x + 3}{(x + 1)^2}$$

Step 4.   Since the numerator can't be factored, we are finished.

Solving Rational Equations

1. Find the LCD of all fractions in the equation.

2. Multiply both sides of the equation by $\frac{LCD}{1}$ (clear all denominators).

3. Solve the resulting equation.

4. Check for extraneous roots.

Example 4: Solve $3 - \dfrac{4}{x} = \dfrac{5}{2}$.

Solution:

Step 1. The LCD is $2x$.

Step 2. Multiply both sides of the equation by $\dfrac{2x}{1}$.

$$\frac{2x}{1}\left(3 - \frac{4}{x}\right) = \frac{2x}{1}\left(\frac{5}{2}\right)$$

This is equivalent to $6x - \dfrac{8x}{x} = \dfrac{10x}{2}$ or $6x - 8 = 5x$.

Step 3. Solving the resulting equation,

$$6x - 6x - 8 = 5x - 6x \quad \text{or} \quad -8 = -x, x = 8.$$

Step 4. The check is left to the reader.

Exercises

Simplify the following.

1. $\dfrac{3x^2 - 6x}{x^2 + x - 6}$

2. $\dfrac{2x^3 + 2x^2 - 4x}{2x + 4}$

3. $\dfrac{x^2 + 2x - 15}{3 - x}$

Perform the indicated operations and simplify.

4. $\dfrac{4x^2y}{5xz} \cdot \dfrac{15x^6}{8xy^2}$

5. $\dfrac{x^2 + 2x - 15}{3x + 15} \div \dfrac{x - 3}{3}$

6. $\dfrac{3}{x^2} + \dfrac{5}{6x}$

7. $\dfrac{2}{x - 5} - \dfrac{3}{x + 3}$

8. $\dfrac{3}{x - 3} + \dfrac{x - 2}{x^2 - 9}$

9. $\dfrac{5}{x^2 - x - 2} - \dfrac{2}{x^2 + 4x + 3}$

10. $\dfrac{x - 3}{x^2 - 3x + 2} - \dfrac{x + 1}{x^2 - 4}$

Solve the following equations.

11. $\dfrac{x}{3} + \dfrac{2x}{7} = 10$

12. $\dfrac{-2}{x} + \dfrac{8}{3} = \dfrac{2}{x}$

13. $\dfrac{x-2}{x-4} = \dfrac{x}{x-1}$

14. $\dfrac{1}{x-4} + x = \dfrac{-3}{x-4}$

15. Solve $\dfrac{1}{R_1} + \dfrac{1}{R_2} = \dfrac{1}{R}$ for R.

Complex Fractions

Complex fractions are fractions with a fractional expression in the numerator, the denominator, or both. Examples include

$$\dfrac{\dfrac{2}{5} + \dfrac{1}{3}}{7}, \qquad \dfrac{x+3}{\dfrac{x}{x+1} - 2} \qquad \text{and} \qquad \dfrac{x + \dfrac{1}{x} - 3}{x^3 - x - \dfrac{2}{x^2}}.$$

There are two methods commonly used to simplify complex fractions. The first is to express the numerator and denominator as single fractions and then divide.

> **Simplifying a Complex Fraction by Simplifying the Numerator and Denominator**
>
> **1.** Express the numerator as a single fraction.
>
> **2.** Express the denominator as a single fraction.
>
> **3.** Divide the numerator by the denominator.
>
> **4.** Simplify, if possible.

Example 1: Simplify $\dfrac{1 - \dfrac{7}{16}}{3 - \dfrac{2}{5}}.$

Solution:

Step 1.   Simplify the numerator: $1 - \dfrac{7}{16} = \dfrac{16}{16} - \dfrac{7}{16} = \dfrac{9}{16}.$

Step 2.   Simplify the denominator: $3 - \dfrac{2}{5} = \dfrac{3}{1} - \dfrac{2}{5} = \dfrac{15}{5} - \dfrac{2}{5} = \dfrac{13}{5}.$

Step 3.   Divide the numerator by the denominator: $\dfrac{9}{16} \div \dfrac{13}{5} = \dfrac{9}{16} \cdot \dfrac{5}{13} = \dfrac{45}{208}.$

Step 4.   Since the fraction cannot be simplified, the simplified result is $\dfrac{45}{208}.$

Example 2: Simplify $\dfrac{\dfrac{1}{x} + \dfrac{2}{x^2}}{2 + \dfrac{1}{x^2}}$.

Solution:

Step 1.   Simplify the numerator: $\dfrac{1}{x} + \dfrac{2}{x^2} = \dfrac{x}{x^2} + \dfrac{2}{x^2} = \dfrac{x + 2}{x^2}$.

Step 2.   Simplify the denominator: $2 + \dfrac{1}{x^2} = \dfrac{2}{1} + \dfrac{1}{x^2} = \dfrac{2x^2}{x^2} + \dfrac{1}{x^2} = \dfrac{2x^2 + 1}{x^2}$.

Step 3.   Divide the numerator by the denominator.

$$\frac{x + 2}{x^2} \div \frac{2x^2 + 1}{x^2} = \frac{x + 2}{x^2} \cdot \frac{x^2}{2x^2 + 1} = \frac{x + 2}{2x^2 + 1}$$

Step 4.   The result in step 3 is simplified.

The second method of simplifying a complex fraction is to multiply the numerator and denominator by the LCD of the entire fraction.

> **Simplifying a Complex Fraction by Multiplying by the LCD**
>
> **1.** Determine the LCD of the numerator fractions and denominator fractions.
>
> **2.** Multiply the fraction by 1 in the form $\frac{\text{LCD}}{\text{LCD}}$.
>
> **3.** Simplify, if possible.

Example 3: Simplify $\dfrac{1 - \frac{7}{16}}{3 - \frac{2}{5}}$.

Solution:

Step 1.   The only denominators are 16 and 5. Since there are no common factors, the LCD is 80. Multiply by $\dfrac{80}{80}$.

Step 2. $\dfrac{80\left(1 - \frac{7}{16}\right)}{80\left(3 - \frac{2}{5}\right)} = \dfrac{80 - \frac{80 \cdot 7}{16}}{240 - \frac{80 \cdot 2}{5}} = \dfrac{80 - 5 \cdot 7}{240 - 16 \cdot 2} = \dfrac{80 - 35}{240 - 32} = \dfrac{45}{208}$

Step 3.   Since the fraction is simplified, $\dfrac{45}{208}$ is the desired result.

Example 4: Simplify $\dfrac{\dfrac{3}{n - 5} - 2}{1 - \dfrac{4}{n - 5}}$.

Step 1.   The LCD is $n - 5$.

Step 2.   Multiply the original fraction by $\dfrac{(n - 5)}{(n - 5)}$.

$$\frac{(n - 5) \cdot \left(\dfrac{3}{n - 5} - 2\right)}{(n - 5) \cdot \left(1 - \dfrac{4}{n - 5}\right)} = \frac{\dfrac{3(n - 5)}{n - 5} - 2(n - 5)}{(n - 5) - \dfrac{4(n - 5)}{n - 5}} = \frac{3 - 2n + 10}{n - 5 - 4} = \frac{-2n + 13}{n - 9}$$

Step 3.   Since the numerator and denominator have no common factors, the simplified result is $\dfrac{-2n + 13}{n - 9}$.

Exercises

Simplify the following complex fractions.

1. $\dfrac{\frac{1}{2} - \frac{1}{4}}{\frac{5}{8} + \frac{3}{4}}$

2. $\dfrac{\frac{5}{6y}}{\frac{10}{3xy}}$

3. $\dfrac{\dfrac{8x^2y}{3z^3}}{\dfrac{4xy}{9z^5}}$

4. $\dfrac{3 - \frac{1}{x}}{1 - \frac{1}{x}}$

5. $\dfrac{\dfrac{x^2}{y} - y}{\dfrac{y^2}{x} - x}$

6. $\dfrac{4 + \dfrac{6}{n+1}}{7 - \dfrac{4}{n+1}}$

7. $\dfrac{\dfrac{1}{y-2} + \dfrac{3}{x}}{\dfrac{5}{x} - \dfrac{4}{xy-2x}}$

8. $\dfrac{\dfrac{x}{x+1} - 1}{\dfrac{x+1}{x-1}}$

9. $\dfrac{1 + \dfrac{x}{x+1}}{\dfrac{2x+1}{x-1}}$

10. $\dfrac{\dfrac{x+1}{x-1} + \dfrac{x-1}{x+1}}{\dfrac{x+1}{x-1} - \dfrac{x-1}{x+1}}$

Radicals and Fractional Exponents

You may need to practice skills relating to radical functions to enhance your understanding of these functions.

Translating Radical Expressions to Expressions Using Rational Exponents

1. The integer power on the base, b, becomes the numerator of the exponent on b.

2. The index (root) becomes the denominator of the exponent.

For example, $\sqrt[3]{x^2} = x^{2/3}$. In reverse, $a^{5/4} = \sqrt[4]{a^5}$.

When a $\sqrt{}$ is written without an index, the index is assumed to be 2.

Fractional exponents also obey the laws of exponents as outlined on page A-1.

1. $a^n a^m = a^{n+m}$ **2.** $\dfrac{a^n}{a^m} = a^{n-m}$ **3.** $(a^n)^m = a^{nm}$

4. $(ab)^n = a^n b^n$ **5.** $a^0 = 1, a \neq 0$ **6.** $a^{-n} = \dfrac{1}{a^n}$

In Examples 1–6, the following steps are used:

1. Write each expression using rational exponents.

2. Apply the appropriate property of exponents.

3. Write the expression using radical notation.

Example 1: $\sqrt[3]{x^2} \cdot \sqrt[4]{x} = x^{2/3} \cdot x^{1/4} = x^{2/3+1/4}$
$$= x^{(8/12)+(3/12)} = x^{11/12} = \sqrt[12]{x^{11}}$$

Example 2: $\dfrac{\sqrt[5]{x^4}}{\sqrt[10]{x^7}} = \dfrac{x^{4/5}}{x^{7/10}} = x^{4/5-7/10} = x^{1/10} = \sqrt[10]{x}$

Example 3: $\left(\sqrt[3]{\sqrt{x^5}}\right) = (x^{5/2})^{1/3} = x^{(5/2)\cdot(1/3)} = x^{5/6} = \sqrt[6]{x^5}$

Example 4: $\sqrt[3]{ab^2} = (ab^2)^{1/3} = a^{1/3} \cdot b^{2/3} = \sqrt[3]{a} \cdot \sqrt[3]{b^2}$

Example 5: $(\sqrt[4]{x})^0 = (x^{1/4})^0 = x^0 = 1$

Example 6: $x^{-2/3} = \dfrac{1}{x^{2/3}} = \dfrac{1}{\sqrt[3]{x^2}}$

Exercises

Rewrite the following using exponents.

1. a. $\sqrt[5]{x^4}$ **b.** $\sqrt[6]{x^3}$

c. $\sqrt[3]{(x+y)^2}$ **d.** $\sqrt[3]{(a-b)^3}$

Rewrite these expressions using a radical.

2. a. $9x^{3/2}$ **b.** $(9x)^{3/2}$

c. $(4x^2 - 9y^2)^{1/2}$ **d.** $(x-y)^{4/5}$

Simplify and express your results in radical form, if appropriate. Assume variables represen[t] nonzero values.

3. a. $\sqrt{x^{12}}$

b. $\sqrt[5]{6.87^5}$

c. $\sqrt{\sqrt{\sqrt{a^2b}}}$

d. $(\sqrt[3]{a^4bc^3})^{30}$

e. $x^{1/4} \cdot x^{3/8}$

f. $\dfrac{x^{1/2}}{x^{1/3}}$

g. $(x^{-2/5})^{1/4}$

h. $(2x^{1/3})^0$

Trigonometry

In *degrees*, a protractor measures angles from 0 to 180. In *radians*, the angles range in value from 0 to π:

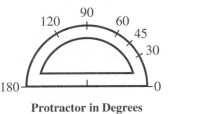

Protractor in Degrees

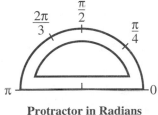

Protractor in Radians

The fundamental idea is that the measure of a straight angle can be taken to be either 180 degrees or π radians.

$$180° \equiv \pi \text{ radians} \qquad (1)$$

All other angles are done proportionately. The following table gives some examples. The last line of the table is useful for converting *any* angle from degrees to radians.

ANGLE (in degrees)	REASONING	CALCULATIONS	ANGLE (in radians)
90	90 is *one-half* of 180	$\frac{1}{2} \cdot \pi$	$\frac{\pi}{2}$
60	60 is *one-third* of 180	$\frac{1}{3} \cdot \pi$	$\frac{\pi}{3}$
45	45 is *one-fourth* of 180	$\frac{1}{4} \cdot \pi$	$\frac{\pi}{4}$
30	30 is *one-sixth* of 180	$\frac{1}{6} \cdot \pi$	$\frac{\pi}{6}$
120	120 is *two-thirds* of 180	$\frac{2}{3} \cdot \pi$	$\frac{2\pi}{3}$
1	1 is *one one-hundred-eightieth* of 180	$\frac{1}{180} \cdot \pi$	$\frac{\pi}{180}$

$$1° \equiv \frac{\pi}{180} \text{ radians} \qquad (2)$$

The following table gives examples of the use of formula (2).

ANGLE (in degrees)	REASONING	CALCULATIONS	ANGLE (in radians)
12°	12 is *twelve* times 1	$12 \cdot \frac{\pi}{180}$	$\frac{\pi}{15}$
7°	7 is *seven* times 1	$7 \cdot \frac{\pi}{180}$	$\frac{7\pi}{180}$
345°	345 is *345* times 1	$345 \cdot \frac{\pi}{180}$	$\frac{23\pi}{12}$

The angle to wrap around a *full circle* is *twice* a straight angle of 180°, so it is 360°, or 2π radians. You can also have angles that wrap around a circle more than once! (Think of a fishing reel or spool of wire, with the string or wire wrapped around many times.)

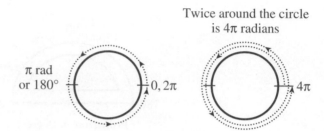

Equivalence (1) also enables us to convert angles from radians to degrees. Study these examples.

ANGLE (in radians)	REASONING	CALCULATIONS	ANGLE (in degrees)
$\frac{2\pi}{3}$	*Two-thirds* of π	$\frac{2}{3} \cdot 180$	120°
7π	*Seven* times π	$7 \cdot 180$	1260°
1	From (1), $\pi \equiv 180°$. Divide both sides of this equivalence by π.	$\frac{180}{\pi}$	$\frac{180}{\pi} \approx 57.3°$

The last line of the preceding table gives us an equivalence useful in converting from radians to degrees.

$$1 \text{ radian} = \frac{180°}{\pi} \qquad (3)$$

Since $\pi \approx 3.14$, equivalence (3) shows that 1 radian $\approx$ 57.3°. This is worth seeing on a protractor.

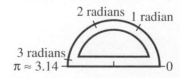

To convert 5 radians: $5 \cdot \dfrac{180}{\pi} = \dfrac{900}{\pi}$ degrees.

To convert 0.6 radian: $0.6 \cdot \dfrac{180}{\pi} = \dfrac{108}{\pi}$ degrees.

The answers just given can be written approximately: 5 radians $\approx 286.5°$ and 0.6 radian $\approx 34.4°$.

Exercises

In Exercises 1–9, convert the given angle from degrees to radians or vice versa.

1. $30°$

2. $135°$

3. $\dfrac{2\pi}{5}$ radians

4. $150°$

5. $\dfrac{5\pi}{3}$ radians

6. 1.5 radians

7. $27°$

8. $\dfrac{2}{3}$ radian

9. $450°$

10. How many times would you have to wrap a length of string around a circle to mark off an angle of 4π radians? 12π radians? 15π radians? 7 radians? 2000 radians?

11. Recall that the circumference of a circle, C, is given by the formula $C = 2\pi r$, where r is the radius of the circle. A **unit circle** is one whose radius is 1. Explain why the circumference of a unit circle equals the radian measure of the angle needed to wrap once around the circle.

12. a. Explain why a central angle of 1 radian in a unit circle subtends an arc whose length is 1 unit. (*Hint*: See Exercise 11.)

b. Explain why a central angle of t radians in a unit circle subtends an arc whose length is t units.

c. Explain why a central angle of *t* radians in a circle of radius *r* subtends an arc whose length is *tr* units.

13. Label the following radian measures on the circle:

$$\frac{\pi}{4}, \frac{\pi}{2}, \frac{5\pi}{4}, \frac{3\pi}{2}, 2\pi.$$

14. Label the following radian measures on the circle:

$$\frac{\pi}{3}, \frac{\pi}{2}, \frac{2\pi}{3}, \frac{4\pi}{3}, \pi.$$

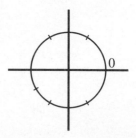

15. Label the following radian measures on the circle:

$$\frac{\pi}{6}, \frac{\pi}{2}, \frac{7\pi}{6}, \frac{4\pi}{3}, \frac{11\pi}{6}.$$

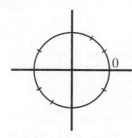

16. Locate approximately the following radian measures on the circle:

$$1, \frac{\pi}{6}, 2, 0.6, 5, 3, \frac{3\pi}{4}, \frac{3\pi}{2}, 1.4.$$

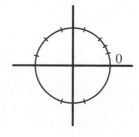

Trigonometric Functions in Right Triangles

For an angle θ in a right triangle (as pictured), the basic trigonometric functions (sine, cosine, tangent) are defined by

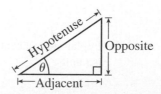

$$\sin\theta = \frac{\text{opposite}}{\text{hypotenuse}} \qquad \cos\theta = \frac{\text{adjacent}}{\text{hypotenuse}} \qquad \tan\theta = \frac{\text{opposite}}{\text{adjacent}}.$$

The acronym **SOH CAH TOA** summarizes this; for example, **SOH** tells you that **s**ine equals **o**pposite over the **h**ypotenuse. Using the accompanying triangle,

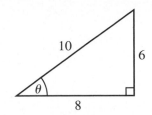

$$\sin \theta = \frac{\text{opp}}{\text{hyp}} = \frac{6}{10} = \frac{3}{5}$$

$$\cos \theta = \frac{\text{adj}}{\text{hyp}} = \frac{8}{10} = \frac{4}{5}$$

$$\tan \theta = \frac{\text{opp}}{\text{adj}} = \frac{6}{8} = \frac{3}{4}.$$

Frequently, you must use the **Pythagorean theorem** for right triangles.

As you recall, the theorem says that in a right triangle, $c^2 = a^2 + b^2$. For example, to determine the values of trigonometric functions in the second triangle to the right, you first use the Pythagorean theorem to determine the missing side:

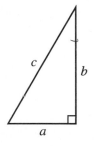

$$2^2 = 1^2 + x^2 \quad \text{or} \quad 4 = 1 + x^2 \quad \text{or} \quad x^2 = 3$$

$$x = \sqrt{3}.$$

Note: Since x represents the length of the side of a triangle we take the positive root.

Then, as before,

$$\sin \theta = \frac{\text{opp}}{\text{hyp}} = \frac{\sqrt{3}}{2}$$

$$\cos \theta = \frac{\text{adj}}{\text{hyp}} = \frac{1}{2}$$

$$\tan \theta = \frac{\text{opp}}{\text{adj}} = \frac{\sqrt{3}}{1} = \sqrt{3}.$$

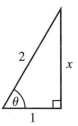

Exercises

In Exercises 1–6, find sin θ, cos θ, and tan θ.

1.

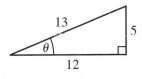

2.

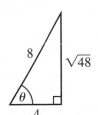

3.

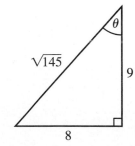

4.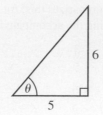

5.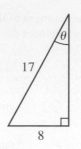

6.

7. Using the Pythagorean theorem and SOH CAH TOA, show that $\sin^2 \theta + \cos^2 \theta = 1$ fo any angle θ in a right triangle.

Trigonometric Function Values of Special Angles

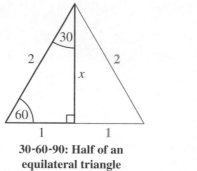

30-60-90: Half of an equilateral triangle

45-45-90: Isosceles; two equal legs

The third side in each triangle can be found using the Pythagorean theorem. The proportions are

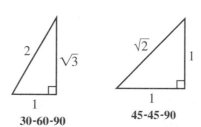

30-60-90   **45-45-90**

From the triangles, you obtain these important values for trigonometric functions.

ANGLE (in degrees)	sin θ	cos θ	tan θ
30	$\dfrac{1}{2}$	$\dfrac{\sqrt{3}}{2}$	$\dfrac{1}{\sqrt{3}}$
45	$\dfrac{\sqrt{2}}{2}$	$\dfrac{\sqrt{2}}{2}$	1
60	$\dfrac{\sqrt{3}}{2}$	$\dfrac{1}{2}$	$\sqrt{3}$

Trigonometric Functions for More General Angles

When an angle θ is larger than 90° (in radians, $\theta > \frac{\pi}{2}$), you can still evaluate sine, cosine, and tangent. You work in an xy plane, make the positive x-axis the initial side of the angle, and make a *reference triangle* by dropping a perpendicular to the x-axis from a point on the terminal side of the angle. (See the figure below.) For positive angles, you rotate *counterclockwise* to find the terminal side.

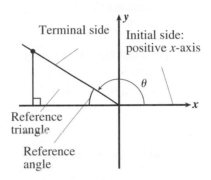

You then use SOH CAH TOA on the *reference triangle*. **Warning:** The adjacent and oppo-site sides may be *negative* in this situation, depending on the quadrant in which the terminal side of the angle lies.

For example, in the figure shown, note the negative sign for the adjacent side. (It lies on the *negative x*-axis.) You have

$$\sin \theta = \frac{\text{opp}}{\text{hyp}} = \frac{4}{5}$$

$$\cos \theta = \frac{\text{adj}}{\text{hyp}} = \frac{-3}{5}$$

$$\tan \theta = \frac{\text{opp}}{\text{adj}} = \frac{4}{-3} = \frac{-4}{3}.$$

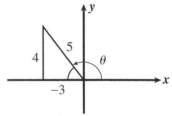

Note: The hypotenuse is ALWAYS positive.

You also need to use the Pythagorean theorem at times. In the next figure, you have $6^2 = (-2)^2 + y^2$ or $y^2 = 32$. Since y must be *negative* (do you see why?), $y = -\sqrt{32} = -4\sqrt{2}$. You now find

$$\sin \theta = \frac{\text{opp}}{\text{hyp}} = \frac{-4\sqrt{2}}{6} = \frac{-2\sqrt{2}}{3}$$

$$\cos \theta = \frac{\text{adj}}{\text{hyp}} = \frac{-2}{6} = \frac{-1}{3}$$

$$\tan \theta = \frac{\text{opp}}{\text{adj}} = \frac{-4\sqrt{2}}{-2} = 2\sqrt{2}.$$

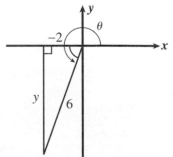

As noted previously, reference triangles with angles of 30°, 60°, or 45° show up frequently because of the sym-metry involved. To find the values of the three trigono-metric functions for $\theta = 5\pi/3$ radians, we convert to degrees and sketch the angle and the reference triangle.

$$\frac{5\pi}{3} = \frac{5}{3} \cdot 180° = 300°$$

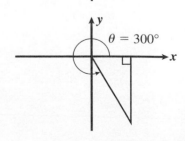

To help remember the $\pm$ signs, the phrase *All Students Take Calculus* is useful. The four words go in the four quadrants of the *xy* plane.

The All means *all* trigonometric functions are positive in quadrant I; **S** for **Students** means S*ine* is positive in quadrant II; **T** for **Take** means T*angent* is positive in quadrant III; **C** for **Calculus** means C*osine* is positive in quadrant IV.

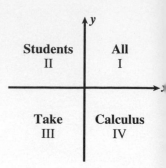

Note that for *negative* angles we locate the terminal side by rotating *clockwise* from the positive *x*-axis. The figure at the right shows that

$$\sin\left(\frac{-3\pi}{4}\right) = \frac{-1}{\sqrt{2}} = \frac{-\sqrt{2}}{2},$$

and by similar calculations

$$\cos\left(\frac{-3\pi}{4}\right) = \frac{-\sqrt{2}}{2} \quad \text{and} \quad \tan\left(\frac{-3\pi}{4}\right) = 1.$$

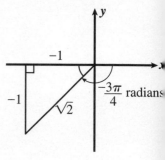

Exercises

In Exercises 1–4, find the remaining side of the reference triangle and evaluate $\sin\theta$, $\cos\theta$, and $\tan\theta$.

1.

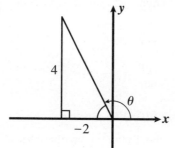

2.

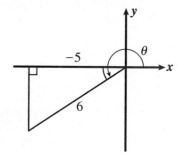

3.

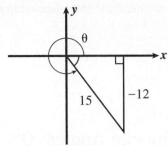

4.

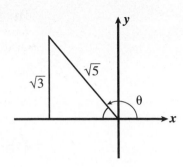

In Exercises 5–10, sketch the angle and reference triangle and then work out the values of the three trigonometric functions for that angle.

5. $\theta = \frac{\pi}{2}$ radians

6. $\theta = 225°$

7. $\theta = -150°$

8. $\theta = 330°$

9. θ is fourth quadrant angle whose sine is $-\frac{6}{8}$.

10. θ is fourth quadrant angle whose tangent is $-\frac{3}{5}$.

11. Show that for θ in *any* quadrant, the relationship $\sin^2\theta + \cos^2\theta = 1$ holds.

Dealing with Special Angles: $0°, \pm 90°, \pm 180°$

For multiples of 90° (equivalently, multiples of $\frac{\pi}{2}$ radians), the reference triangle degenerates to a straight line segment. Either the adjacent or opposite side degenerates to 0:

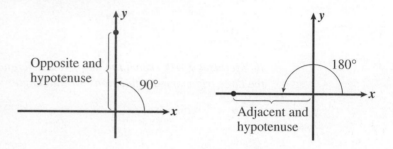

For simplicity, notice that for ordinary angles, when we mark a point on the terminal side of an angle θ, the *x-value* gives the value of the *adjacent* side, and the *y*-value gives the *opposite* side. Let us also use r for the length of the hypotenuse. Therefore, we could have defined

$$\sin\theta = \frac{\text{opp}}{\text{hyp}} = \frac{y}{\text{hyp}} = \frac{y}{r} \qquad \cos\theta = \frac{\text{adj}}{\text{hyp}} = \frac{x}{\text{hyp}} = \frac{x}{r} \qquad \tan\theta = \frac{\text{opp}}{\text{adj}} = \frac{y}{x}. \qquad (4)$$

We use $\frac{y}{r}, \frac{x}{r}$, and $\frac{y}{x}$ for the special angles to find the values of the trigonometric functions. When we have a multiple of 90°, we mark a point on the terminal side and label it with its x and y numbers. Use 0 and ± 1 for simplicity; then use formula (4). Here are the figures for $-90°$ and 180°. Remember that r is always positive.

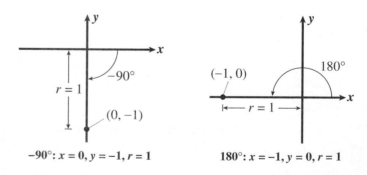

$-90°: x = 0, y = -1, r = 1$ $180°: x = -1, y = 0, r = 1$

Now the trigonometric values are

$$\sin(-90°) = \frac{\text{opp}}{\text{hyp}} = \frac{y}{r} = \frac{-1}{1} = -1 \qquad \sin 180° = \frac{\text{opp}}{\text{hyp}} = \frac{y}{r} = \frac{0}{1} = 0$$

$$\cos(-90°) = \frac{\text{adj}}{\text{hyp}} = \frac{x}{r} = \frac{0}{1} = 0 \qquad \cos 180° = \frac{\text{adj}}{\text{hyp}} = \frac{x}{r} = -\frac{1}{1} = -1$$

$$\tan(-90°) = \frac{\text{opp}}{\text{adj}} = \frac{y}{x} = \frac{-1}{0} = \textit{undefined.} \qquad \tan 180° = \frac{\text{opp}}{\text{adj}} = \frac{y}{x} = \frac{0}{-1} = 0$$

If you use only 0 and ±1 for x- and y-values of these special angles, r always equals 1. Also, whenever the x value is 0, the tangent is *undefined*, since the formula then involves division by 0.

Exercises

In Exercises 1–6, for each given value of θ draw the angle, label a point on the terminal side, and use formulas to get the values of sin θ, cos θ, and tan θ.

1. $\theta = 90°$

2. $\theta = -\pi$ radians

3. $\theta = 720°$

4. $\theta = -630°$

5. $\theta = \dfrac{7\pi}{2}$ radians

6. $\theta = 23\pi$ radians

Getting Started with the TI-83/TI-84 Plus Family of Calculators

ON-OFF

To turn on the calculator, press the ⊂ON⊃ key. To turn off the calculator, press ⊂2nd⊃ and then ⊂ON⊃.

Most keys on the calculator have multiple purposes. The number or symbolic function/command written directly on the key is accessed by simply pressing the key. The symbolic function/commands written above each key are accessed with the aid of the ⊂2nd⊃ and ⊂ALPHA⊃ keys. The command above and to the left is color-coded to match the ⊂2nd⊃ key. That command is accessed by first pressing the ⊂2nd⊃ key and then pressing the key itself. Similarly, the command above and to the right is color-coded to match the ⊂ALPHA⊃ key and is accessed by first pressing the ⊂ALPHA⊃ key and then pressing the key itself.

Contrast

To adjust the contrast on your screen, press and release the ⊂2nd⊃ key and hold ⊂▲⊃ to darken and ⊂▼⊃ to lighten.

Mode

The ⊂MODE⊃ key controls many calculator settings. The activated settings are highlighted. For most of your work in this course, the settings in the left-hand column should be highlighted.

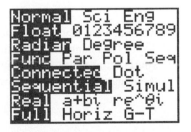

To change a setting, move the cursor to the desired setting and press ⊂ENTER⊃.

The Home Screen

The home screen is used for calculations.

You may return to the home screen at any time by using the QUIT command. This command is accessed by pressing (2nd) (MODE). All calculations in the home screen are subject to the order of operations convention.

Enter all expressions as you would write them. Always observe the order of operations. Once you have typed the expression, press (ENTER) to obtain the simplified result. Before you press (ENTER), you may edit your expression by using the arrow keys, the delete command (DEL), and the insert command (2nd) (DEL).

Three keys of special note are the reciprocal key (X⁻¹), the caret (^) key, and the negative key (─).

Typing a number and then pressing the reciprocal command key (X⁻¹) give the reciprocal of the number. The reciprocal of a nonzero number, n, is $\frac{1}{n}$. As noted in the screen below, when performing an operation on a fraction, the fraction MUST be enclosed in parentheses before accessing this command.

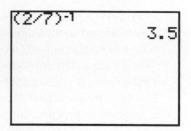

The caret key (^) is used to raise numbers to powers.

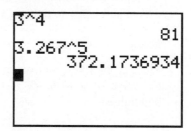

The negative key (─) on the bottom of the keyboard is different from the subtraction key (─). They cannot be used interchangeably. The negative key is used to change the sign of a single number or symbol; it will not perform a subtraction operation. If you mistakenly use the negative key in attempting to subtract, you will likely obtain an ERROR message.

A table of some frequently used keys and their functions follows.

KEY	FUNCTION DESCRIPTION
ON	Turns calculator on or off.
CLEAR	Clears the line you are currently typing. If cursor is on a blank line when (CLEAR) is pressed, it clears the entire home screen.
ENTER	Executes a command.
(−)	Calculates the additive inverse.
MODE	Displays current operating settings.
DEL	Deletes the character at the cursor.
^	Symbol used for exponentiation.
ANS	Storage location of the result of the most recent calculation.
ENTRY	Retrieves the previously executed expression so that you may edit it.

ANS and ENTRY

The last two commands in the table can be real time savers. The result of your last calculation is always stored in a memory location known as ANS. It is accessed by pressing (2nd) ((−)) or it can be automatically accessed by pressing any operation button.

Suppose you want to evaluate $12.5\sqrt{1 + 0.5 \cdot (0.55)^2}$. It could be evaluated in one expression and checked with a series of calculations using ANS.

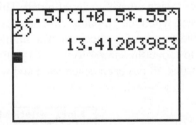
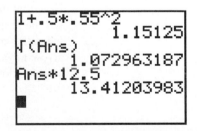

After you have keyed in an expression and pressed (ENTER), you cannot move the cursor back up to edit or recalculate this expression. This is where the ENTRY ((2nd) (ENTER)) command is used. The ENTRY command retrieves the previous expression and places the cursor at the end of the expression. You can use the left and right arrow keys to move the cursor to any location in the expression that you wish to modify.

Suppose you want to evaluate the compound interest expression $P\left(1 + \frac{r}{n}\right)^{nt}$, where P is the principal, r is the interest rate, n is the number of compounding periods annually, and t is the number of years, when $P = \$1000$, $r = 6.5\%$, $n = 1$, and $t = 2, 5,$ and 15 years.

Using the ENTRY command, this expression would be entered once and edited twice.

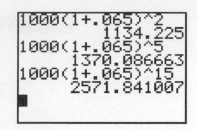

Note that there are many last expressions stored in the ENTRY memory location. You ca
repeat the ENTRY command as many times as you want to retrieve a previously entere
expression.

Functions and Graphing with the TI-83/TI-84 Plus Family of Calculators

"Y =" Menu

Functions of the form $y = f(x)$ can be entered into the TI-83/TI-84 Plus using the "Y=
menu. To access the "Y =" menu, press the (Y=) key. Type the expression for $f(x)$ after Y
using the (X,T,θ,*n*) key for the variable x and press (ENTER).

For example, enter the function $f(x) = 3x^5 - 4x + 1$.

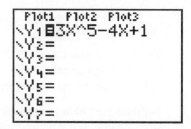

Note that the = sign after Y_1 is highlighted. This indicates that the function Y_1 is active an
will be graphed when the graphing command is executed and will be included in your tabl
when the table command is executed. The highlighting may be turned on or off by using th
arrow keys to move the cursor to the = symbol and then pressing (ENTER). Notice in th
screen below that Y_1 has been deactivated and will not be graphed nor appear in a table.

Once the function is entered in the "Y =" menu, function values may be evaluated in th
home screen.

For example, given $f(x) = 3x^5 - 4x + 1$, evaluate $f(4)$. In the home screen, press $\boxed{\text{VARS}}$.

Move the cursor to Y-VARS and press $\boxed{\text{ENTER}}$.

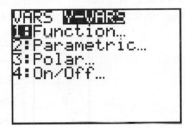 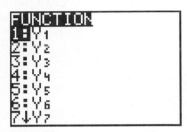

Press $\boxed{\text{ENTER}}$ again to select Y_1. Y_1 now appears in the home screen.

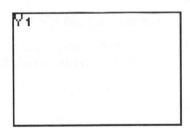

To evaluate $f(4)$, press $\boxed{(}$ $\boxed{4}$ $\boxed{)}$ after Y_1 and press $\boxed{\text{ENTER}}$.

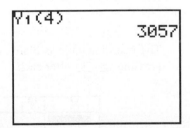

Tables of Values

If you are interested in viewing several function values for the same function, you may want to construct a table.

Before constructing the table, make sure the function appears in the "Y =" menu with its "=" highlighted. You may also want to deactivate or clear any functions that you do not need to see in your table. Next, you will need to check the settings in the Table Setup menu. To do this, use the TBLSET command ($\boxed{\text{2nd}}$ $\boxed{\text{WINDOW}}$).

As shown in the screen above, the default setting for the table highlights the Auto options for both the independent (x) and dependent (y) variables. Choosing this option will display ordered pairs of the function with equally spaced x-values. TblStart is the first x-value to be displayed, and here is assigned the value −2. ΔTbl represents the equal spacing between consecutive x-values, and here is assigned the value 0.5. The TABLE command (2nd GRAPH) brings up the table displayed in the screen below.

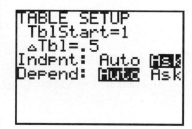

Use the ▲ and ▼ keys to view additional ordered pairs of the function.

If the input values of interest are not evenly spaced, you may want to choose the Ask mode for the independent variable from the Table Setup menu.

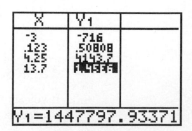

The resulting table is blank, but you can fill it by choosing any values you like for x and pressing ENTER after each.

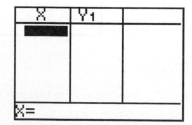

Note that the number of digits shown in the output is limited by the table width, but if you want more digits, move the cursor to the desired output and more digits appear at the bottom of the screen.

Graphing a Function

Once a function is entered in the "Y =" menu and activated, it can be displayed and analyzed. For this discussion we will use the function $f(x) = -x^2 + 10x + 12$. Enter this as Y_1, making sure to use the negation key $(\,(-)\,)$ and not the subtraction key $(\,-\,)$.

The Viewing Window

The viewing window is the portion of the rectangular coordinate system that is displayed when you graph a function.

Xmin defines the left edge of the window.

Xmax defines the right edge of the window.

Xscl defines the distance between horizontal tick marks.

Ymin defines the bottom edge of the window.

Ymax defines the top edge of the window.

Yscl defines the distance between vertical tick marks.

In the standard viewing window, Xmin $= -10$, Xmax $= 10$, Xscl $= 1$, Ymin $= -10$, Ymax $= 10$, and Yscl $= 1$.

To select the standard viewing window, press $(\,ZOOM\,)$ $(\,6\,)$.

You will view the following:

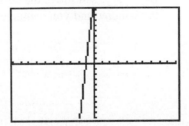

Is this an accurate and/or complete picture of your function, or is the window giving you a misleading impression? You may want to use your table function to view the output values that correspond to the input values from -10 to 10.

X	Y₁		X	Y₁		X	Y₁	
-10	-188		-3	-27		4	36	
-9	-159		-2	-12		5	37	
-8	-132		-1	1		6	36	
-7	-107		0	12		7	33	
-6	-84		1	21		8	28	
-5	-63		2	28		9	21	
-4	-44		3	33		10	12	
X=-10			X=3			X=10		

The table indicates that the minimum output value on the interval from $x = -10$ to $x = 10$ −188, occurring at $x = -10$, and the maximum output value is 37 occurring at $x =$ Press (WINDOW) and reset the settings to the following:

$$Xmin = -10, Xmax = 10, Xscl = 1,$$
$$Ymin = -190, Ymax = 40, Yscl = 10$$

Press (GRAPH) to view the graph with these new settings.

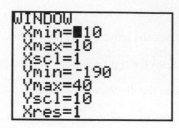 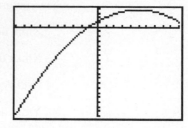

The new graph gives us a much more complete picture of the behavior of the function on th interval $[-10, 10]$.

The coordinates of specific points on the curve can be viewed by activating the trace featur While in the graph window, press (TRACE). The function equation will be displayed at th top of the screen, a flashing cursor will appear on the curve at the middle of the screen, an the coordinates of the cursor location will be displayed at the bottom of the screen.

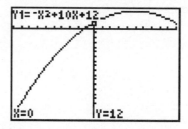

The left arrow key, (◄), will move the cursor toward smaller input values. The right arrov key, (►), will move the cursor toward larger input values. If the cursor reaches the edge o the window and you continue to move the cursor, the window will adjust automatically.

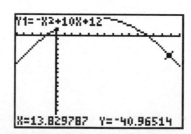

Zoom Menu

The Zoom menu offers several options for changing the window very quickly.

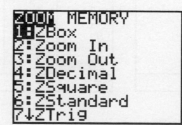 

The features of each of the commands are summarized in the following table.

ZOOM COMMAND	DESCRIPTION
1: ZBox	Draws a box to define the viewing window.
2: Zoom In	Magnifies the graph near the cursor.
3: Zoom Out	Increases the viewing window around the cursor.
4: ZDecimal	Sets a window so that Xscl and Yscl are 0.1.
5: ZSquare	Sets equal size pixels on the x- and y-axes.
6: ZStandard	Sets the window to standard settings.
7: ZTrig	Sets built-in trig window variables.
8: ZInteger	Sets integer values on the x- and y-axes.
9: ZoomStat	Sets window based on the current values in the stat lists.
0: ZoomFit	Replots graph to include the max and min output values for the current Xmin and Xmax.

Solving Equations Graphically Using the TI-83/TI-84 Plus Family of Calculators

The Intersection Method

This method is based on the fact that solutions to the equation $f(x) = g(x)$ are input values of x that produce the same output for the functions f and g. Graphically, these are the x-coordinates of the intersection points of $y = f(x)$ and $y = g(x)$.

The following procedure illustrates how to use the intersection method to solve $x^3 + 3 = 3x$ graphically.

Step 1. Enter the left-hand side of the equation as Y_1 in the "Y =" editor and the right-hand side as Y_2. Select the standard viewing window.

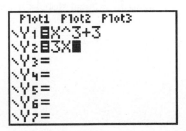

Step 2. Examine the graphs to determine the number of intersection points.

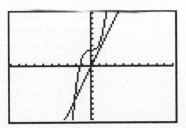

You may need to examine several windows to be certain of the number of intersectio[...] points.

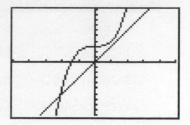

Step 3. Access the Calculate menu by pushing 〔2nd〕 〔TRACE〕, then choose option 5: intersec[...]

 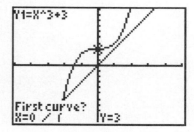

The cursor will appear on the first curve in the center of the window.

Step 4. Move the cursor close to the desired intersection point and press 〔ENTER〕.

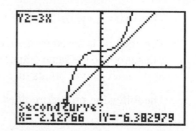

The cursor will now jump vertically to the other curve.

Step 5. Repeat step 4 for the second curve.

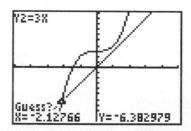

Step 6. To use the cursor's current location as your guess, press 〔ENTER〕 in response to the question on the screen that asks Guess? If you want to move to a better guess value do so before you press 〔ENTER〕. You can also enter a Guess using the calculator keypad and then pressing 〔ENTER〕.

The coordinates of the intersection point appear below the word Intersection.

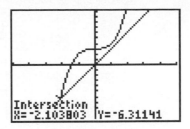

The *x*-coordinate is a solution to the equation.

If there are other intersection points, repeat the process as necessary.

Using the TI-83/TI-84 Plus Family of Calculators to Determine the Linear Regression Equation for a Set of Paired Data Values

Example 1:

INPUT	OUTPUT
2	2
3	5
4	3
5	7
6	9

Enter the data into the calculator as follows:

1. Press (STAT) and choose EDIT.

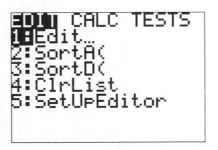

2. The calculator has six built-in lists, L1, L2, . . . , L6. If there is data in L1, clear the list as follows:

   a. Use the arrows to place the cursor on L1 at the top of the list. Press (CLEAR) followed by (ENTER), followed by the down arrow.

   b. Follow the same procedure to clear L2 if necessary.

c. Enter the input values into L1 and the corresponding output values into L2.

L1	L2	L3	2
2	2	------	
3	5		
4	3		
5	7		
6	9		

L2(6) =

To see a scatterplot of the data, proceed as follows.

1. STAT PLOT is the 2nd function of the (Y=) key. You must press (2nd) before pressing (Y=) to access the STAT PLOT menu.

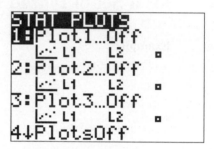

2. Select Plot 1 and make sure that Plots 2 and 3 are Off. The screen shown below will appear. Select On and then choose the scatterplot option (first icon) on the Type line. Confirm that your x and y values are stored, respectively, in L_1 and L_2. The symbols L_1 and L_2 are 2nd functions of the (1) and (2) keys, respectively. Finally, select the small square as the mark that will be used to plot each point.

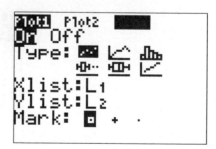

3. Press (Y=) and clear or deselect any functions currently stored.

Plot1	Plot2	Plot3
\Y1=		
\Y2=		
\Y3=		
\Y4=		
\Y5=		
\Y6=		
\Y7=		

4. To display the scatterplot, have the calculator determine an appropriate window by pressing ZOOM and then 9 (ZoomStat).

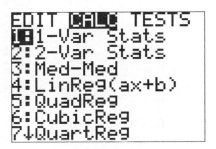

The following instructions will calculate the linear regression equation and store it in Y_1.

1. Press STAT and right arrow to highlight CALC.

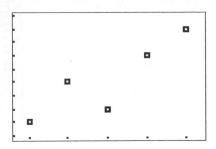

2. Choose 4: LinReg (ax + b). LinReg (ax + b) will be pasted to the home screen. To tell the calculator where the data is, press 2nd and 1 (for L1), then , then 2nd and 2 (for L2) because the Xlist and Ylist are stored in L_1 and L_2, respectively. The display should look like this:

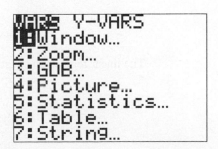

3. Press , and then press VARS .

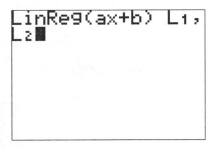

4. Right arrow to highlight Y-VARS.

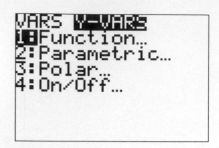

5. Choose 1, FUNCTION.

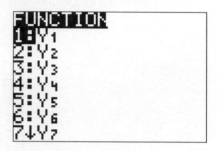

6. Choose 1 for Y_1 (or 2 for Y_2, etc., if you prefer to store the regression equation in another location).

```
LinReg(ax+b) L₁,
L₂,Y₁
```

7. Press ⎡ENTER⎤.

```
LinReg
 y=ax+b
 a=1.6
 b=-1.2
```

The linear regression equation for this data is $y = 1.6x - 1.2$.

8. To display the regression line on the scatterplot screen, press $\boxed{\text{GRAPH}}$.

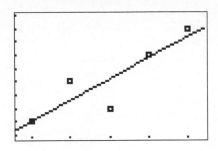

9. Press the $\boxed{\text{Y=}}$ key to view the equation.

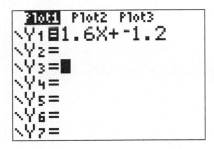

Selected Answers

Chapter 1

Activity 1.1 Exercises: 1. a. Weight is input, and height output. **c.** Height is input, and weight is output. **. a.** Yes, for each numerical grade there will correspond only one letter grade. **b.** No, for each letter grade there may correspond several different numerical grades. For example, an A may be based on 92% or 94%. **c.** This is not a function. For each number of hours studied for the exam, there may correspond several different scores. For example, a study time of 3 hours could result in a score of 1, 82, or 94. **d.** True. For any specific area, the corresponding number of tiles required is unique. This is assuming that no tiles are damaged in the installation of the tiles. **.** This situation represents a function. For any given final selling price, there is just one corresponding amount of tax. **. a.** Yes, in this table each elevation is paired with only one amount of snowfall. **b.** Yes, in this table each quantity of snow is paired with one elevation. **7. a.** Yes, each input value has only one output value. **b.** No, the one input 5 is paired with four different outputs. **9. a.** The input is x, the output is $g(x)$ or y. The function name is g, y equals g of x, **.** The input is 6. The output is 3.527. The name is f, f of 6 equals 3.527, **e.** The input is price. The output is sales tax. The name is T. Sales tax is a function of price. **10. a.** 1600; 400, **b.** $f(6) = 2400$

Activity 1.2 Exercises: 1. a. The independent (input) variable is the price of an item. The dependent (output) variable is the sales tax. **b.** $h(x) = 0.08x$. **.** $f(2) = -1, f(-3.2) = -11.4, f(a) = 2a - 5$; **.** $f(2) = 4, f(-3.2) = 4, f(a) = 4$;

x	$h(x)$
10	0.1
20	0.05
30	0.033
40	0.025

9. a. The distance traveled is 3 times the number of hours I have hiked. **b.** The input is hours. The output is distance. **d.** $h(t)$ is the dependent variable since distance is the output. **f.** $h(7) = 3(7)$ $h(7) = 3(7) = 21. (7, 21).$ If I hike for 7 hours, I expect to travel 21 miles. **h.** The practical domain depends on the individual and in this situation is probably real numbers from 0 to about 8. Using this domain, the range is real numbers from 0 to 24.

10. a. domain $\{-2, 0, 5, 8\}$, range $\{3, 4, 8, 11\}$;
11. a. $0 \le C \le 100$; **b.** $32 \le F \le 212$

Activity 1.3 Exercises:

1. a.

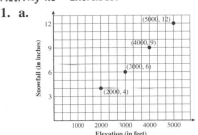

b. Continuous, because the amount of snowfall is defined at any elevation. **3.** Graph d; **4.** Graph b; **5.** Graph a; **6.** Graph c; **7. a.** No graph appears on the screen. **b.** 24, 19, 16, 15, 16, 19, 24, **c.** The y-values are increasing from a minimum of 15. Therefore, you need to have Ymax of at least 30. **d.**

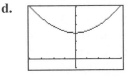

Activity 1.4 Exercises: 1. a. 3.7, 6.4, 9.1, 11.8, **b.** $f(2500) = 7.75$ inches of snow, **c.** $f(-2000) = -4.4$ It has no meaning in this context. -2000 would mean 2000 feet below sea level, but -4.4 inches of snow is not possible. **d.** $\text{Xmin} = 0, \text{Xmax} = 5000, \text{Ymin} = 0, \text{Ymax} = 15$, **e.** Yes, any vertical line will intersect the graph no more than once, **f.** increasing, **g.** It is the same, 7.75; **3. a.** What is the value of the truck after a certain number of years? **b.** The value of the truck and the number of years of ownership. **c.** The independent variable is the number of years you own the truck. The dependent variable is the value of the truck. **d.** 11,580, 10,760, 9940, 9120, 8300, **e.** The value of the truck is obtained by subtracting the product of 820 and the number of years from 12,400. **f.** Let v represent the value of the truck and t represent the number of years. $v = 12,400 - 820t$ **g.** Using the equation from part d, you could subtract 820 twice from the value of the truck after five years. **h.** The rate of depreciation was a constant 820 dollars per year. No, the truck would depreciate more when it's newer and has greater value.

4. a. 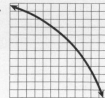   **b.** The graph is a horizontal line;
5. a. This is a function.
b. This is not a function.
6. Xmin $= -9$, Xmax $= 9$,
Ymin $= -54,000$,
Ymax $= 54,000$,
answers may vary

Activity 1.5   Exercises: 2. a. number of attempts, time in seconds **b.** The time required to complete a task decreases as the number of attempts increases. As a person attempts a task more times, the task takes less time to complete. After five attempts, no further improvement is made.   **3. a.** Selling price, number of units sold **b.** As the selling price increases, the number of units sold increases slightly at first, reaches a maximum, and then declines until none are sold.

How Can I Practice? 1. a. Yes, because for each point total there is only one grade. **b.** Yes. For each numerical grade in the table, there is one value of total points.
c. $f = \{(432, 86.4), (394, 78.8), (495, 99), (330, 66), (213, 42.6)\}$,
d.

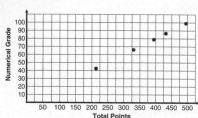

e. $f(394) = 78.8$,

f. A student with 394 points has a 78.8 average,
g. $f(213) = 42.6$, **h.** A student with 213 points has a 42.6 average, **i.** $n = 330$;   **2.** This is a function.   **3.** This could be a function, depending on how activity level is measured.   **4.** This is a function.   **5.** This is not a function.   **6.** This is a function.   **7.** This is not a function. The input -3 has two different outputs.   **8.** This is a function.   **9.** This is a function.   **10.** This is not a function.   **11. a.** $c = 120h$, **b.** $f(h) = 120h$, **c.** 240, 480, 840, 960, 1320, **d.** $360, (3,360)$, **e.** $h = 5$ hours, **f.** $f(h)$ or c is the output variable. This is the variable that depends on the number of credits taken. **g.** h is the independent variable. It is the input variable. **h.** For each value of input, there is one value of output, **i.** Assuming there are no half-credit courses, the practical domain is all whole numbers from 0 to 11, depending on the college, **j.** The horizontal axis represents the input.

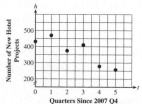

k. $f(h)$ is a function because the graph passes the vertical line test, **m.** 1080;   **12. a.** $p(3) = 13$, **b.** $p(-4) = -1$, **c.** $p\left(\frac{1}{2}\right) = 8$, **d.** $p(0) = 7$;   **13. a.** $t(2) = -3$,

b. $t(-3) = 18 + 9 - 5 = 22$;   **14. a.** 64.8, 66.6, 68.4, 72.9, 73.8, 75.6, **b.** The life expectancy for a male born in 1985 will be 71.1 years. **c.** Xmin $= 0$, Xmax $= 60$, Ymin $= 60$, Ymax $= 80$, **d.** increasing, **e.** They are the same.   **15. a.** domain $\{3, 4, 5, 6\}$, range $\{5, 8, 10\}$, **b.** domain $\{0, 50, 100, 150, 200\}$, range $\{19.95, 23.45, 26.95, 30.45, 33.95\}$, **c.** domain $-3 \leq x \leq 4$, range $-1 \leq y \leq 3$, **d.** domain $-3 \leq x \leq 3$, range $0 \leq y \leq 4$, **e.** Domain is all real numbers, range is all real numbers;   **16. a.** The net profit increases during the first two quarters of 2009. The net profit then decreases for about 2 quarters, and then it increases through the final quarter of 2010. **b.** The annual income rises rather steadily for 3 years; in the fourth year, it rises sharply. Then it suffers a sharp decline during the next year. During the last year, the income recovers to about the point it was originally.
17.

Hours of Daylight

Dec 21   March 21   June 21   Oct 21
Time

Activity 1.6   Exercises: 1. a. $\frac{27.5 - 22.8}{2010 - 1950} = \frac{4.7}{60} = 0.0783$ years of age/year, **b.** The median age of a man at the time of his first marriage is increasing at an average rate of 0.0783 years/year.   **3.** $\frac{27.5 - 25.1}{100} = \frac{2.4}{100} \approx 0.024$ years/year;
5. a. It means that the median age of a man at the time of his first marriage is decreasing. **b.** 1910–1920, 1920–1930, or 1940–1950, **c.** The graph would fall to the right.
7. a.

Number of New Hotel Projects

Quarters Since 2007 Q4

b. $\frac{\Delta h}{\Delta t} = \frac{459 - 439}{1 - 0} = 20$ new hotels/quarter.
c. $\frac{\Delta h}{\Delta t} = \frac{257 - 403}{5 - 3} = \frac{-146}{2} = -73$ new hotels/quarter.
8. a. $\frac{760 - 668}{1970 - 1960} = \frac{92}{10} = 9.2$ gallons/year,
b. $\frac{520 - 668}{1990 - 1960} = \frac{-148}{30} \approx -4.93$ gallons/year,
c. $\frac{547 - 530}{2007 - 1995} = \frac{17}{12} \approx 1.42$ gallons/year,
d. $\frac{547 - 668}{2007 - 1960} = \frac{-121}{47} \approx -2.57$ gallons/year,
e. It means that from 1960 to 2007, the average fuel consumption per year of a passenger car in the United States decreased by about 2.6 gallons/year.

Activity 1.7   Exercises: 1. a. yes, linear; the constant rate of change is 10, **b.** no, not linear, **c.** yes, linear; the constant rate of change is $\frac{-9}{4}$;   **3. a.** Yes, the average rate of change is a constant -3. **b.** No, between weeks 1 and 2 the slope is -5. Between weeks 2 and 3, the slope is -4. **c.** Yes, the slope is 0 for all pairs of points.
5. a. $m = \frac{5 - (-7)}{0 - 2} = \frac{12}{-2} = -6$, **b.** $(0, 5)$,
c. $f(x) = -6x + 5$, **d.** $\left(\frac{5}{6}, 0\right)$;   **7. a.** Yes, the slope is a constant. **b.** $m = \frac{3000 - 3500}{20 - 0} = \frac{-500}{20} = -25$ ft/sec,

The jet is losing altitude, **d.** (0, 3500),
$h = -25t + 3500$, **f.** (140, 0); The jet lands in
40 seconds.   **8. a.** $\left(\frac{-1}{2}, 0\right)$, **b.** (6, 0)

ctivity 1.8   Exercises: 1. a. $y = \frac{1}{2}x - 1$,
. $y = -\frac{4}{3}x + 1$, **c.** $y = -3x - 5$

. $m = \dfrac{6 - (-3)}{2 - (-4)} = \dfrac{6 + 3}{2 + 4} = \dfrac{9}{6} = \dfrac{3}{2}, y = \dfrac{3}{2}x + 3$,

. $y = 3x - 11$;   **2. a.** (0, 35); The vertical intercept
ccurs where the input $x = 0$.

. $m = \dfrac{40 - 35}{100 - 0} = \dfrac{5}{100} = 0.05$; The mileage charges are
0.05 per mile.   **c.** $c = 0.05x + 35$;

. a. $m = \dfrac{145 - 75}{4 - 2} = \dfrac{70}{2} = 35$ mph. This represents
e average rate of change or the average speed of the boat
om $t = 2$ to $t = 4$. **b.** $d = 35t + 5$;

. a. i. $m = 1$, **ii.** (0, −2), **iii.** $y = x - 2$,
i. $m = -2$, **ii.** (0, 6), **iii.** $y = -2x + 6$;

. a. $m = \dfrac{31.42 - 0}{5 - 0} = 6.28$, **b.** (0, 0), **c.** $C = 6.28r$,

. $C = 2\pi r$, **e.** Yes, π is approximately 3.14, so 2π is
pproximately 6.28.   **9. a.** (0, 2000); the vertical intercept
ccurs where the input value is 0.

. $\dfrac{\Delta c}{\Delta h} = \dfrac{311000 - 2000}{3000} = \dfrac{309000}{3000} = 103$, building costs are
103 per square foot.   **c.** $c = 103h + 2000$.

. $c = 103(2500) + 2000 = 259{,}500$.

1. a. The slope $\dfrac{2226 - 1952}{95 - 75} = \dfrac{274}{20} = 13.7$.   **b.** For each
dditional 1 kilogram increase in weight, a 20-year-old,
90.5-centimeter-tall male has an increase of 13.7
dditional calories in his basal energy requirement.

. $B = m(w - h) + k, B = 13.7(w - 75) + 1952$,
$= 13.7w - 1027.5 + 1952, B = 13.7w + 924.5$, the
mbolic rule $B = 13.7w + 924.5$ expresses the basal en-
gy rate B for a 20-year-old, 190.5-centimeter-tall male in
rms of his weight, w. **d.** The B-intercept has no practical
eaning in this situation because it would indicate a weight
f 0 kilograms. A possible practical domain is a set of
eights from 55 to 182 kilograms.

ctivity 1.9   Exercises: 1. a. $y = 2x - 3, m = 2, (0, -3)$,
. $y = -x - 2, m = -1, (0, -2)$, **c.** $y = \frac{2}{3}x - \frac{7}{3}, m = \frac{2}{3}$,
$), -\frac{7}{3})$, **d.** $y = \frac{1}{2}x + 2, m = \frac{1}{2}, (0, 2)$
. $y = 4, m = 0, (0, 4)$;

. a.

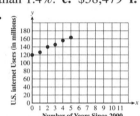

b. This is not a function. It
does not pass the vertical line
test; **c.** $x = -2$; **d.** The slope
is undefined.
e. vertical: none; horizontal:
(−2, 0);   **5. a.** $f(x) = 2000$,
b. 2000, 2000, 2000;

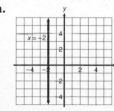

d. The slope is zero. This means the fee does not change.
e. The graph is a horizontal line through (0, 2000).
6. a. $250w$, **b.** $200d$, **c.** $250w + 200d = 10{,}000$,
d. $d = \dfrac{10{,}000 - 250w}{200} = 50 - \frac{5}{4}w$, **e.** (40, 0); The maximum
number of washers I can purchase is 40. **f.** (0, 50); The
maximum number of dryers I can purchase is 50.

Activity 1.10   Exercises:

1. a.

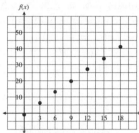

b. (Answers will vary.)
Yes, the points are very
close to a line. **c.** $f(x) = 2.299x - 0.761$,
d. 22.229, **e.** 56.714,

f. $f(10)$ is more accurate; 10 is within the given data. 25 is
not. $f(10)$ uses interpolation. $f(25)$ uses extrapolation;
2. b. $f(t) = 2.30t + 8.579$, **c.** The slope of the line is
2.30. This means that the average debt per person is
increasing at an average rate of $2300 per year. **d.** The
regression line predicts an average debt of $13,179 in 2004.
This is 179 above the actual, 13,000, an error of slightly
less than 1.4%. **e.** $38,479 **f.** extrapolation
4. a.

b. $y = 8.66x + 120.52$

c. The slope of the line is 8.66, indicating that for each
year after 2000, the number of Internet users increased by
approximately 8.66 million. **d.** According to the model,
the number of Internet users in the United States will reach
200 million in 2009. I am somewhat confident of this prediction because the graph of the data from 1996 to 2002 indicates that the relationship has been very linear and 2009
is only four years from the latest data shown.

How Can I Practice? 1. a. $C = 50 + 10t$, where the
independent variable, t, represents the number of months
and the dependent variable, C, represents the total cost in
dollars. The vertical intercept is called the C-intercept and
represents the $50 down payment. The slope represents the
month charge of $10. **b.** $V = 16{,}000 - 1500x$, where the
independent variable, x, represents the number of years
since the car was purchased and the dependent variable, V,
represents the value of the car in dollars. The vertical
intercept is called the V-intercept and represents the value
of the car on the day of purchase. The slope represents
the car's yearly depreciation.   **2. a.** $g(x) = 2x - 3$,
b. $h(x) = -2x - 3$, **c.** $x = 2$, **d.** none,
e. $f(x) = -2x + 3$, **f.** $y = -2x$, **g.** $y = 2$, **h.** none,
i. none;   **3. a.** 28, 40, 52, 60, 68, **b.** yes, **c.** 4,
d. $c = f(m) = 4m + 20$,

e.

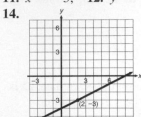

f. This represents the monthly charge. **g.** $(0, 20)$; It indicates that the initial rental cost is \$20. **h.** $(-5, 0)$. It has no practical meaning in this case.

i. $65 = 4m + 20$ or $45 = 4m$ or $m = 11.25$. I can keep the graphing calculator for 11 months. **4. a.** 1.5, **b.** 1.5, **c.** $s(t)$ is a linear function because the rate of change is constant. **5.** $m = \frac{12 - 8}{-5 - 3} = -\frac{1}{2}$; **6.** $m = -4$; **7.** $m = \frac{2}{5}$; **8.** $y = -7x + 4$; **9.** $y = 2x + 10$; **10.** $y = 5$; **11.** $x = -3$; **12.** $y = -\frac{1}{2}x - 2$; **13.** $y = \frac{1}{3}x - 3$;

14.

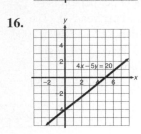

15.

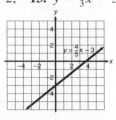

16.

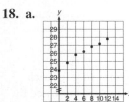

17.

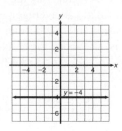

18. a.

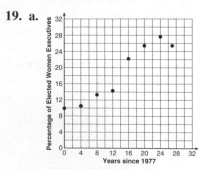

b. $y = 0.322x + 24.156$,
c. $y = 27.054$,
d. $y = 30.596$;

19. a.

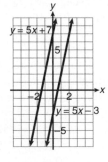

b. $y = 0.728x + 8.472$, **c.** In 1986, $x = 9$; $f(9) = 15.0\%$. In 2010, $x = 33$; $f(33) = 32.5\%$, **d.** 1986 because it is interpolated from the data. The 2010 percentage was a result of extrapolation, which is usually less accurate.

Activity 1.11   Exercises:
1. a. Numerically   Graphically

x	y_1	y_2
−2	−1	8
−1	1	7
0	3	6
1	5	5
2	7	4
3	9	3

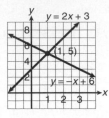

Algebraically (substitution method)

$y = 2x + 3 \quad y = -x + 6$

$$2x + 3 = -x + 6 \quad y = -1 + 6$$
$$\underline{+\ x \qquad\quad +\ x} \qquad y = 5$$
$$3x + 3 = 6 \qquad \text{The answer is } (1, 5)$$
$$\underline{-\ 3 - 3}$$
$$\frac{3x}{3} = \frac{3}{3}$$
$$x = 1$$

c. Numerically   Graphically

x	y_1	y_1
0	−3	7
1	2	12
2	7	17
3	12	22
4	17	27

Algebraically (substitution method)
$y = 5x - 3 \quad y = 5x + 7$

$$5x - 3 = 5x + 7 \qquad \text{There is no solution.}$$
$$\underline{-5x \qquad\quad -5x}$$
$$-3 = 7 \text{ false}$$

2. a. $s = 17.2 + 1.5n$, **b.** $s = 9.6 + 2.3n$, **c.** in the year 2020;   **3. a.** $c = 3560 + 15n$, **b.** $c = 2850 + 28n$, **c.** $n = 54.6$ or 55 months, **d.** dealer 1's system;

a.

t, NUMBER OF YEARS SINCE 1975	LIFE EXPECTANCY FOR WOMEN	LIFE EXPECTANCY FOR MEN
0	77.42	69.80
25	80.30	75.10
50	83.17	80.40
100	88.92	91.00
75	86.05	85.70
80	86.62	86.76
79	88.51	86.55

. 79 years after 1980, in the year 2059, the life expectancy or both men and women will be 86.5 years. **c.** (78.56, 86.45), **.** $x = 78.56$, $E = 86.45$

ctivity 1.12 Exercises: 1. a. $x = -5$, **b.** $x = -2$,
. $\frac{4}{7} = x$; **3. a.** $(1, 5)$, **b.** $(2, -3)$;
. a. $8x + 5y = 106$
$x + 6y = 24$,
. $x = -6y + 24$
$8(-6y + 24) + 5y = 106$
$-48y + 192 + 5y = 106$
$-43y = -86$
$y = 2$
$= -6(2) + 24 = 12$

4 2, 2) Each centerpiece costs $12, and each glass costs $2.

$$8x + 5y = 106 \qquad 8x + 5y = 106$$
$$-8(x + 6y) = -8(24) \qquad -8x - 48y = -192$$
$$\overline{\qquad\qquad\qquad -43y = -86}$$
$$y = 2$$
$$x + 6(2) = 24$$
$$x + 12 = 24$$
$$x = 12,$$

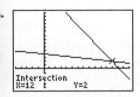

Intersection
X=12   Y=2

ctivity 1.13 Exercises: 1. $(0, -3, 5)$; **3.** $(-5, 3, 1)$;
. a. dependent, **b.** inconsistent;

ctivity 1.14 Exercises:

. $\begin{bmatrix} 4 & 3 & -1 \\ 2 & -1 & -13 \end{bmatrix}$; **3.** $\begin{bmatrix} 4 & -2 & 1 & 15 \\ 3 & 2 & -2 & -4 \\ 1 & 0 & 1 & 5 \end{bmatrix}$;

. $4x + 3y = 15$
$2x - 5y = 1$;
. $(x, y, z) = (1, -1, 2)$; **9.** Thirty-four 2-pointers and leven 3-pointers were made for a total of 101 points.
1. 6 luxury sedans, 14 small hatchbacks, 8 hybrids.

Activity 1.15 Exercises: 1. $l + w + d \le 61$;
3. $C(A) < C(B)$; **5.** $24{,}650 < i \le 59{,}750$;
7. $x > -2$ **9.** $x < 5$

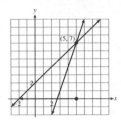
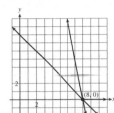

11. $x \ge 8$ **13.** $x \ge -0.4$

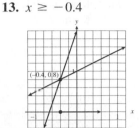

15. $1 < x < 2$ **17. a.** $-9.90t + 529.54 < 200$,
b. $t > 33.29$, 34 years after 1990, the year 2024;
18. a. $C = 19.99 + 0.79n$, **b.** $C = 29.99 + 0.59n$,
c. $29.99 + 0.59n < 19.99 + 0.79n$, **d.** $50 < n$.
19. a. $150 + 60n$, **b.** $150 + 60n \le 1200$, **c.** $n \le 17.5$
The maximum number of boxes that can be placed in the elevator is 17. **20.** $57.5 \le w \le 70$;
21. a. $-79.8 \le F \le 134$, **b.** $-79.8 \le 1.8C + 32 \le 134$
c. $-62.1 \le C \le 56.7$

Activity 1.16 Exercises:

1. f. $1.20 + .90(11) = \$11.10$;
3. a.

$$f(x) = \begin{cases} 2.5x & x \le 15{,}000 \\ 37{,}500 + 3(x - 15{,}000) & 15{,}000 < x \le 21{,}000 \\ 55{,}500 + 4(x - 21{,}000) & x > 21{,}000 \end{cases}$$

c. 23,375 books;
5. a. $f(x)$: 8, 7, 6, 5, 4, 3, 2, 1, 0, 1, 2;
$g(x)$: 2, 1, 0, 1, 2, 3, 4, 5, 6, 7, 8;
b.

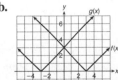

c. f is the same as $y = |x|$, only it is shifted 3 units to the right. g is the same as $y = |x|$, only it is shifted 3 units to the left.
d. $f(x) = \begin{cases} -x + 3 & if \ x \le 3 \\ x - 3 & if \ x > 3 \end{cases}$

e. The domains of f and g are all real numbers.
f. The ranges of f and g are $y \ge 0$.

How Can I Practice?
1. a. $(1, -4)$ **b.** $(3, 2)$

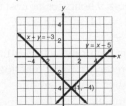
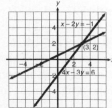

c. $(-4, -5)$

d. $-2 = 6$ inconsistent; no solution

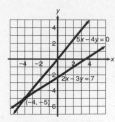

2. a. $(1, -4)$, **b.** $(3, 2)$, **c.** $(-4, -5)$, **d.** inconsistent;
3. $(1, 5, -2)$; **4. a.** $x \geq 1.8$, **b.** $x > -6$,
c. $1 \leq x < 5$; **5. a.** $t + d = 80, 0.50t + 0.75d = 52$,
b. $d = 48, t = 32$, **c.**

6. a. $y = 80 + 30x, y = 50 + 40x$,
b.

COLUMN 2	COLUMN 3
140	130
200	210
260	290
320	370

c.

d. 3 hours; the total cost is $170,
e. $x = 3, y = 80 + 90 = 170$, **f.** I will use Towne Truck;
its graph is below World Transport for $x = 6$;
7. a. $y = 2x$, **b.** $x = 7, y = 14, z = 6$, **c.** It checks;
8. a. $-5 < x \leq 6$, **b.** $-3 \leq x < 4$; **9. a.** $x \geq 1367$,
b. $1034 \leq x \leq 1500$

Gateway Review 1. a. Yes, it is a function. **b.** No, it is not
a function. There are two different outputs paired with 2.
c. Yes, it is a function.

2. 20, 36, 44, 60, 76, **a.** Yes, for each input there is one
output. **b.** The input is x, the number of hours worked.
c. The dependent variable is $f(x)$, the total cost. **d.** Negative
values would not be realistic domain values. A negative number of hours worked does not make sense. **e.** The rate of
change is $8 per hour. **f.** The rate of change is $8 per
hour. **g.** The rate of change between any two points is $8
per hour. **h.** The relationship is linear. **i.** $f(x) = 8x + 20$,
j. The slope is the hourly rate I charge, $8 per hour,
k. $(0, 20)$ is the vertical intercept. The 20 represents the fertilizer cost. **l.** $f(4) = 8(4) + 20 = 52$, **m.** $8x + 20 = 92$
or $8x = 72$ or $x = 9$; I need to work 9 hours for the cost to
equal exactly $92. **3. a.** $f(-2) = 14, g(-2) = 10$,
b. $-6 + (-5) = -11$, **c.** $24 - 16 = 8$,
d. $36(-2) = -72$; **4. a.** This represents a linear function. The slope is 4. **b.** This represents a linear function,
$m = \frac{-2}{5}$. **c.** This does not represent a linear function.
d. This represents a linear function, $m = 10.7$.

5. a. $m = \dfrac{9 + 3}{-4 - 5} = \dfrac{12}{-9} = \dfrac{-4}{3}$, **b.** $m = \frac{3}{7}$,

c. $m = \frac{1}{2}$;

6. a. $y = 4$, **b.** $y = 2x + 5$,
c. $-14 = -3(6) + b, b = 4$
$y = -3x + 4$,
d. $-2 = 2(7) + b, b = -16$
$y = 2x - 16$,

e. $x = 2$, **f.** $0 = -5(4) + b, b = 20, y = -5x + 20$,
g. $16 = 4(2) + b, b = 8, y = 4x + 8$, **h.** $y = \frac{-1}{2}x + 5$
7. $y = \frac{-2}{5}x + 2$; **8. a.** $f(x) = 300,000 - 10,000x$,
b. $m = -10,000$. The building depreciates $10,000 per
year. **c.** $(0,300,000)$; the original value is $300,000,
d. $(30, 0)$; it takes 30 years for the building to fully depreciate; **9. a.** $(0, -3)$, **b.** $(0, -3)$, **c.** $(0, -3)$,
d. The graphs all intersect at the point $(0, -3)$.
e. The results are the same. **10. a.** $m = -2; (0, 1)$,
b. $m = -2; (0, -1)$, **c.** $m = -2; (0, -3)$, **d.** The
graphs are parallel lines. **e.** The results are the same.
11. a. $m = -3; (0, 2)$, **b.** $m = -3; (0, 2)$ **c.** $m = -3$
$(0, 2)$, **d.** The graphs are all the same. **e.** the slopes,
f. the slopes and the y-intercepts,
g. The results are the same. **12. a.** $(0, 150), (75, 0)$,
b.

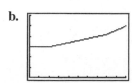

c. The domain and range are all real
numbers. **d.** $w(t) = -2t + 150$,
e. They are essentially the same.
f. The vertical intercept is $(0, 150)$.
It indicates the person's initial
weight of 150 pounds. The horizontal intercept $(75, 0)$ indicates that
after 75 weeks of weight loss, the
person weighs nothing.

g. The practical domain is $0 \leq t \leq 15$. The practical range
is $120 \leq w(t) \leq 150$. **13. a.** $f(x) = 25$, horizontal line
through $(0, 25)$,
b.

c. The slope is 0.

14. a.

$$f(x) = \begin{cases} 1500 & \text{if} \quad x \leq 10,000 \\ 1500 + 0.02(x - 10,000) & \text{if} \quad 10,000 < x \leq \\ 2100 + 0.04(x - 40,000) & \text{if} \quad x > 40,000 \end{cases}$$

b.

c. $f(25,000) = 1500 + 0.02(15,000) = 1800$,
d. $x = 66,250$; **15. a.** $y = 1040x + 7900$ or
$t(n) = 1040n + 7900$, **b.** 1040; the number of finishers
increased at a rate of 1040 per year, **c.** $(0, 7900)$; the
model indicates that there were 7900 finishers in 2006,
d. pretty well, **e.** 14,140, **f.** I used extrapolation because
am predicting outside the original data. **g.** No, 2024 is

...rther from the data than 2000. The farther removed ...e are from the data, the more likely our prediction is ...correct.   **16. a.** $(3, -1)$, **b.** $(-1, 6)$,

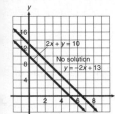

d. $4 = 4$; this is a dependent system. Any pair of numbers that satisfies one equation satisfies both equations.

...**7.** $x = 4.50$, $y = 0.75$;   **18. a.** $(0, 1, 2)$, **b.** $(-3, 1, 0)$, ...$(0.5, 0.25, -0.5)$, **d.** $(12, 7, 9)$;   **19. a.** It is a good ...eal. **b.** Answers will vary. I would not take advantage of ...is. I don't give away many pictures.
...**0. a.**

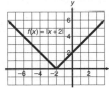

b. increasing $x > -2$, decreasing $x < -2$, **c.** The domain is all real numbers. **d.** The range is $y \geq 0$. **e.** g is the reflection through the x-axis. **f.** f shifts the graph of $y = |x|$ 2 units to the left. h shifts the graph of $y = |x|$ 2 units up.

Chapter 2

...ctivity 2.1   Exercises:

. a.

COLUMN 2	COLUMN 3	COLUMN 4
100	250	350
100	500	600
100	750	850
100	1000	1100
100	1250	1350

. $C(x) = 12.50x + 100$, **c.** $0.15(20) = 3$,
. $3(750) = \$2250$, **e.** $0.13(2250) = \$292.50$,

COLUMN 2	COLUMN 3	COLUMN 4
3	2,250	292.50
6	4,500	585.00
9	6,750	877.50
12	9,000	1,170.00
15	11,250	1,462.50

. $T(x) = 0.13(750)(0.15x) = 14.625x$,
. $P(x) = 2.125x - 100$, **i.** $0 \leq x \leq$ the number of ...eople the banquet room will accommodate.
... Set $P(x) = 0$ and solve for x, $x = 47.06$; 48 people ...ust attend. **k.** Set the profit equal to 500 and solve for x;
...00 $= 2.125x - 100$; $x = 282.35$; 283 must attend.

3. $f(x) + g(x)$: $4, -6, 1, 6, 2, 8$; $f(x) - g(x)$:
$2, -4, -1, 8, -4, 0$;   **5. a.** $5x - 2$, **b.** $x^2 + 3x - 8$,
c. $-x + 30$, **d.** $-3x^2 + 8x - 3$, **e.** $15x - 15$, **f.** 11,
g. $5x^2 + 5x - 9$, **h.** $-6x^2 + 15x - 4$, **i.** $-5x + 25$,
j. $33x - 10$;   **7. a.** $4x^2 - x$, **b.** $4x^2 - 5x - 9$,
c. $3 + 4x$, **d.** $4x^2 - 5x + 3$;   **9. a.** $1, -9, -11, -5$,
$9, 31$, **b.** $f(x) - g(x) = x^2 + 5x - 5$, **c.** The answers
check.

Activity 2.2   Exercises: 1. a. $(3x)(2x) = 6x^2$, **b.** $2x - 3$,
c. $(2x - 3)(3x) = 6x^2 - 9x$;   **3. a.** $P(x) = 40 - x$,
b. $N(x) = 3000 + 100x$, **c.** $R(x) = (40 - x)$
$(3000 + 100x)$. **d.** The domain is $0 \leq x \leq 10$.
e.

PRICE PER TICKET, $P(x)$	NUMBER OF TICKETS SOLD, $N(x)$	TOTAL REVENUE, $R(x)$
40	3000	120,000
38	3200	121600
36	3400	122400
34	3600	122400
32	3800	121600
30	4000	120,000

f. The values in the fourth row are the product of the values in the second and third rows. **g.**

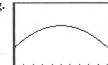

$Xmin = 0$, $Xmax = 10$, $Ymin = 117,000$, $Ymax = 125,000$. **h.** $R(x) = -100x^2 + 1000x + 120,000$.
i. There is only one graph. The graphs are the same.
5.

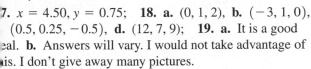

$x^3 + 6x^2 + 4x - 15$;   **7. a.** $6x^2 + 19x + 10$,
b. $6x^2 - 19x + 10$, **c.** $4x^2 + 5x - 6$, **d.** $4x^2 - 5x - 6$;
9. a. $4x^2 + 12x + 9$, **b.** $9x^2 - 12x + 4$, **c.** $25x^2 - 4$,
d. $x^4 - 25$, **e.** The outer product and inner product are opposites. Their sum is 0. **10. a.** $f(x) \cdot g(x) = 2x^2 - x - 3$,
b.

x	$f(x)$	$g(x)$	$f(x) \cdot g(x)$
0	1	-3	-3
1	2	-1	-2
2	3	1	3
3	4	3	12
4	5	5	25

c. Answers may vary depending on the choices of x.

Activity 2.3   Exercises: 1. 8.532×10^{11};   **3. a.** 3×10^{21},
b. 4.5×10^{16}, **c.** 9,000,000,000,000,000,000,000,000,000,000,
d. $\frac{9 \times 10^{27}}{4.5 \times 10^{16}} = 2 \times 10^{11}$; **5. a.** $\frac{6.35 \times 10^{8}}{2.27 \times 10^{9}}$, **b.** 2.797×10^{-1};
7. 1; **9.** 10; **11.** $\frac{4}{x^{4}}$; **13.** $2x^{7}$; **15.** $\frac{3y^{4}}{5x^{4}}$; **17.** $3x^{-3} = \frac{3}{x^{3}}$;
19. $\frac{-1}{2a^{2}b^{3}}$

Activity 2.4   Exercises: 1. a. x^{18}, **b.** $4x^{10}$, **c.** $-27x^{6}$,
d. $\frac{1}{x^{12}}$, **e.** $\frac{16}{a^{10}}$, **f.** $\frac{b^{8}}{a^{8}c^{12}}$, **g.** $-x^{18}$, **h.** x^{18}; **3. a.** $x^{13/12}$,
b. $x^{1/6}$, **c.** $x^{7/15}$; **5. a.** 5, **b.** 0, **c.** not a real number;
7. a. $t = f(L) = 2\pi\left(\frac{L}{32}\right)^{1/2}$, **b.** $t = 2.22$ seconds

How Can I Practice? 1. a. $x = 30$, **b.** $N = f(t) = 30 + t$,
c. $C = g(t) = 20 - 0.5t$, **d.** $N = f(t)$: 30, 32, 34, 36,
38, 40; $C = g(t)$: 20, 19, 18, 17, 16, 15, **e.** $R(t)$: 600, 608,
612, 612, 608, 600, **f.** $R(t) = f(t) \cdot g(t) = (30 + t)$
$(20 - 0.5t) = -0.5t^{2} + 5t + 600$,

g.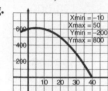
Xmin = -10
Xmax = 50
Ymin = -200
Ymax = 800

h. $612.50 is the maximum
revenue if 35 couples attend.
i. 35 tickets must be sold to
obtain the maximum revenue.

2. a. $3x - 1$, **b.** $-x + 5$, **c.** $2x^{2} + x - 6$, **d.** 2, **e.** 0,
f. $3x + 6$; **3. a.** $-3x + 5$, **b.** $x^{4} - x^{3} - 5x^{2} + 9x - 4$,
c. 0, **d.** $-x^{2} - 7x + 14$; **4. a.** $5x - 2$,
b. $2x^{2} - 2x - 8$, **c.** $-2x + 12$, **d.** $2x^{2} - 13x - 8$,
e. $5x^{2} - x + 2$; **5. a.** x^{4}, **b.** x^{9}, **c.** $6x^{8}$, **d.** $x^{5}y^{6}z$,
e. $10x^{6}y^{5}z^{8}$, **f.** $-30a^{5}b^{3}$; **6. a.** $x^{2} - 7x + 10$,
b. $4x^{2} + 25x - 21$, **c.** $4x^{2} - 9$, **d.** $x^{3} + x^{2} - 11x + 10$,
e. $2x^{3} - x^{2} + 3x + 2$, **f.** $-2x^{2} - 5x - 21$,
g. $11x^{2} - 2x$, **h.** $-x^{5} - x^{3} + 3x^{2} + 2x - 1$,
i. $9x^{2} + 30x + 25$, **j.** $4x^{2} - 28x + 49$,
k. $x^{3} + 12x^{2} + 48x + 64$, **l.** $25x^{2} - 49$;
7. a. 3350, 4050, 4750, 5450, 6150,
b. $f(t) = 140t + 3350$, **c.** 5250, 7000, 8750, 10,500,
12,250, **d.** $g(t) = 350t + 5250$, **e.** 400, 400, 400, 400,
400, **f.** $h(t) = 400$, **g.** $k(t) = 490t + 9000$,

h.

3	3770	6300	400	10,470
12	5030	9450	400	14,880
18	5870	11550	400	17,820
25	6850	14000	400	21,250

i. It will exceed $25,000 33 years after 2005 or in 2038.
8. a. 17, -3, -7, 5, 33, 77,
b. $f(x) - g(x) = 2x^{2} + 2x - 7$, **c.** The answers check.
9. a. $\frac{2}{x^{3}}$, **b.** $9x^{2}$, **c.** $\frac{1}{3^{4}}$, **d.** 1, **e.** $2x^{7}$, **f.** $-6x^{4}y^{8}$, **g.** 4,
h. $\frac{2}{3x^{4}}$, **i.** $\frac{5y^{2}}{x^{4}}$, **j.** $\frac{-15}{x^{5}}$, **k.** $-2x^{6}$, **l.** $\frac{a^{4}}{b^{4}c^{5}}$; **10. a.** x^{8},
b. $x^{3}y^{3}$, **c.** $32x^{20}y^{5}$, **d.** $432x^{10}y^{4}$, **e.** 25, **f.** -25, **g.** $4x^{2}$,
h. $-8x^{3}$, **i.** $-8x^{22}$, **j.** 125, **k.** 16, **l.** -3, **m.** $\frac{27}{64}$,
n. 1.59, **o.** $x^{2}y^{4}$, **p.** $x^{7/6}$, **q.** $2x^{1/3}y^{1/3}$;
11. a. 10,080,000, **b.** I own approximately 0.2 of a
square mile. **c.** 3.4339×10^{14} cubic mile;
12. 3.10233×10^{8}, 2.8×10^{8};

13. $\frac{2.26 \times 10^{10}}{7.57 \times 10^{9}} \approx 0.3 \times 10^{1} = 3$. To divide powers with th[e]
same base, keep the base and subtract the exponents.
14. a. $A = 35 + 12x + x^{2}$, **b.** $(7 + 4)(5 + 4) = 99$,
or $35 + 12(4) + 4^{2} = 99$; $99 - 35 = 64$ square feet,
c. $A = \pi r^{2}$, for the umbrella $A = \pi(r + 2)^{2}$, or
$A = \pi r^{2} + 4\pi r + 4\pi$

Activity 2.5   Exercises: 1. a. $g(2) = 200$. The radius of th[e]
slick is 200 ft. 2 hr. after the spill.
b. $f(g(2)) = f(200) = \pi(200)^{2} = 125,664$. The area o[f]
the oil slick is 125,664 square feet 2 hours after the spill.
c. $f(g(10)) = f(1000) = \pi(1000)^{2} = 1,000,000\pi$.
This is the area of the slick after 10 hours.
d. $f(g(t)) = f(100t) = \pi(100t)^{2} = 10,000\pi t^{2}$,
e. $f(g(10)) = 10,000\pi(10)^{2} = 1,000,000\pi$. The results
are the same. **3. a.** $-18x^{2} + 18x - 3$,
b. $-6t^{2} + 6t + 2$; **5. a.** $L(x) = 0.99x$,
b. $D(x) = 0.90x$,
c. $S(x) = L(D(x)) = L(0.90x) = 0.99(0.90x) = 0.891$[x]
d. $S(500) = 0.891(500) \approx 446$, **e.** $D(x)$ needs to
be improved because the air bags fail 10% of the time,
whereas the seat belts fail only 1% of the time.
6. a. $f(x) = x - 6$, **b.** $g(x) = 0.99x$,
c. $g(f(10,000)) = g(9994) = 0.99(9994) = 9894$.
If 10,000 bottles are processed, 9894 of them should
be properly labeled and capped.

Activity 2.6   Exercises: 1. a. The clerk was thinking
25% + 40% = 65%. **b.** Yes, because I am getting a larg[e]
discount than I should. **c.** No, because I should be getting
a 55% discount. **3. a.** $f(x) = x - 1500$, **b.** $g(x) = 0.9$[x]
c. $g(f(20,000)) = g(20,000 - 1500) =$
$0.9(18,500) = 16,650$. The price of a $20,000 car with a
$1500 rebate and a 10% discount is $16,650;
4. a. $g(f(t)) = g(0.5t) = \pi(0.5t)^{2} = 0.25\pi t^{2}$,
b. The input is the number of seconds after the pebble hits th[e]
water. The output is the area of the outer ripple in square fee[t]

Activity 2.7   Exercises: 1. a. 7, **b.** 4, **c.** x, **d.** x;
2. a. $h^{-1} = \{(3, 2), (4, 3), (5, 4), (6, 5)\}$,
b. $h(3) = 4$ $h^{-1}(h(3)) = 3$,
c. $h^{-1}(5) = 4$ $h(h^{-1}(5)) = 5$; **3.** No, because the inp[ut]
value 2 is paired with two different output values, 0 and 3.
5. a. $m = \frac{112.75 - 56.37}{100 - 50} = 1.1276$, **b.** You have 1.1276
Canadian dollars for every U.S. dollar. **c.** $f(x) = 1.1276$[x]
d. $f(3000) = 1.1276(3000) = 3382.80$. If I have $3000
U.S., then I can exchange it for $3382.80 Canadian.
e. $m = \frac{100 - 50}{112.75 - 56.37} \approx 0.8868$, **f.** You have 0.8868.
U.S. dollar for 1 Canadian dollar. **g.** $g(x) = 0.8868x$,
h. $g(6000) = 0.8868(6000) = 5320.80$.
If I have $6000 Canadian dollars, then I can exchange
them for $5320.80 in U.S. dollars.
i. $f(g(x)) = f(0.8868x) = 1.1276(0.8868x) = x$,
$g(f(x)) = g(1.1276x) = 0.8868(1.1276x) = x$

Activity 2.8   Exercises: 1. a. $P = f(S) = 0.05S + 250$,
b. $f(6000) = 0.05(6000) + 250 = 550$. The weekly
salary for $6000 worth of sales is $550.

$S = g(P) = \frac{P - 250}{0.05}$, **d.** $g(400) = \frac{400 - 250}{0.05} = 3000$. weekly salary of $400 means I sold $3000 worth of merchandise. **e.** $g(f(8000)) = 8000$;
a. $y + 4 = 3x$ or $x = \frac{y + 4}{3}$ or $y = f^{-1}(x) = \frac{x + 4}{3}$,
$2w = z - 4$ or $z = 2w + 4$ or $g^{-1}(z) = 2z + 4$
$t = \frac{5}{s}$ or $s = \frac{5}{t}$;

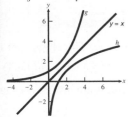

g and h are inverses.
The graphs of g and h are symmetric with respect to the line $y = x$.

a. $g^{-1}(x) = \frac{3x - 6}{4}$, **b.**
Yes, because the graphs are reflections in the line $y = x$.

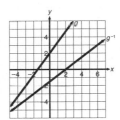

$g^{-1}\left(\frac{6 + 4x}{3}\right) = \frac{3\left(\frac{6 + 4x}{3}\right) - 6}{4} = \frac{4x}{4} = x$.

Yes, because $g^{-1}(g(x)) = x$.
a. 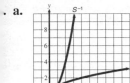 **b.**

The area of the square is the input. The length of the side of the square is the output. **d.** The length of the side of the square is the input. The area of the square is the output.
Given the length of the side, we can determine the area of the granite top.

How Can I Practice? 1. a. $f(-1) = -3$, **b.** $g(5) = 7$,
$f(x + 2) = (x + 2)^2 - 4 = x^2 + 4x$,
d. $g(x^2 - 4) = x^2 - 4 + 2 = x^2 - 2$,
$f(x^2 - 4) = (x^2 - 4)^2 - 4 = x^4 - 8x^2 + 12$,
$x = y + 2$ or $y = g^{-1}(x) = x - 2$;
a. $f(-2) = -6$, **b.** $g(-1) = 2$, **c.** $g(-2) = -2$,
d. $f(4 + x - x^2) = (4 + x - x^2) - 4 = x - x^2$,
$= -x^2 + 9x - 16$, **f.** $x = y - 4$ or
$= f^{-1}(x) = x + 4$; **3.** $-1, 3, 2, 1, 0$;
a. $g(3x^2) = -2(3x^2)^3 = -54x^6$,
b. $f(-2x^3) = 3(-2x^3)^2 = 12x^6$,
$g(48) = -2(48)^3 = -221,184$;
a. $s(4x - 1) = (4x - 1)^2 + 4(4x - 1) - 1 = 16x^2 - 8x + 1 + 16x - 4 - 1 = 16x^2 + 8x - 4$,
b. $t(x^2 + 4x - 1) = 4(x^2 + 4x - 1) - 1 = x^2 + 16x - 5$, **c.** $x = 4y - 1$ or $y = t^{-1}(x) = \frac{x + 1}{4}$;
a. $p(\sqrt{x + 2}) = \frac{1}{\sqrt{x + 2}}$, **b.** $c\left(\frac{1}{x}\right) = \sqrt{\frac{1}{x} + 2}$,

c. $x = \frac{1}{y}$ or $y = p^{-1}(x) = \frac{1}{x}$; **7.** $\{(6, 4), (-9, 7), (1, -2), (0, 0)\}$; **8.** $f(g(x)) = f\left(\frac{x + 3}{2}\right) = 2\left(\frac{x + 3}{2}\right) - 3 = x$
$g(f(x)) = g(2x - 3) = \frac{(2x - 3) + 3}{2} = x$. Because
$f(g(x)) = g(f(x)) = x$, f, and g are inverse functions.
9. a. $y = 4x + 3$; $\frac{y - 3}{4} = x$; $y = f^{-1}(x) = \frac{x - 3}{4}$,
b.

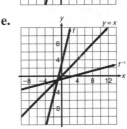

c. The intercepts of f are $(0, 3)$ and $\left(-\frac{3}{4}, 0\right)$. The intercepts of f^{-1} are $\left(0, -\frac{3}{4}\right)$ and $(3, 0)$.
d. The slope of the graph of f is 4. The slope of the graph of the inverse is $\frac{1}{4}$.

e.

10. a. 41, 42.8, 44.6, 47,
b.

41	42.8	44.6	47
0	3	6	10

c. $f^{-1}(x) = \frac{x - 41}{0.6}$, **d.** The population will be 46 million in 1998. **e.** The graphs are symmetrical about the line $y = x$.
f. The horizontal intercept of the function interchanged, is the vertical intercept of its inverse. The vertical intercept of the function interchanged, is the horizontal intercept of its inverse.

HORIZONTAL INTERCEPT	VERTICAL INTERCEPT
$(-68.3, 0)$	$(0, 41)$
$(41, 0)$	$(0, -68.3)$

g. Function: $m = 0.6$; inverse function: $m = \frac{1}{0.6}$ or $\frac{5}{3}$.
The slopes of the two functions are reciprocals.
h. The population will be 50 million in 2005.
11. a. 444.96 euros, **b.** $487.62,
c. $g(f(x)) = g(0.007416x) = 1.39320(0.007416x) = 0.010332x$, **d.** $g(f(60,000)) = 0.010332(60,000) = 619.92; **12.** $x = \frac{4}{3}$; **13. a.** $b = f(x) = x^2$, $V = g(b) = 10b$, **b.** $V = g(f(x)) = g(x^2) = 10x^2$.

Gateway Review 1. a. $2x^2 - 2x - 1$, **b.** $-x^2 + 5x - 4$,
c. $4x^2 - 13x + 3$, **d.** $x^3 - 7x^2 + 13x - 15$,
e. $-11x + 11$, **f.** $2x^4 - 5x^3 + 4x^2 + 7x - 4$;
2. a. $6x^8$, **b.** $16x^6y^2$, **c.** $-2x^5y^3$, **d.** $-10x^5y^5z^4$, **e.** $9x^6y^2$,
f. $-125x^3y^3$, **g.** $2x^3$, **h.** 2, **i.** 1, **j.** $\frac{3y^3}{2z^3}$, **k.** $-5x^{-8} = \frac{-5}{x^8}$,
l. $-4x^2$, **m.** $(-5)^3(x^{-3})^3 = -125x^{-9} = \frac{-125}{x^9}$,
n. $x^{4/5+1/2} = x^{8/10+5/10} = x^{13/10}$, **o.** $x^{(2/3)\cdot3} = x^2$;
3. a. -20, **b.** $4x + 1$,
c. $f(3) - g(3) = 16 - (-3) = 19$,

d. $(6x - 2)(-2x + 3) = -12x^2 + 22x - 6$,
e. $-12x + 16$, **f.** $g(10) = -2(10) + 3 = -17$,
g. $y = 6x - 2$ or $y = f^{-1}(x) = \frac{x + 2}{6}$;
4. a. $x^2 - 4x + 5$, **b.** $= 3x^3 - 5x^2 + 11x - 6$,
c. $= 9x^2 - 15x + 9$, **d.** $g(5) = 3(5) - 2 = 13$;
5. a. 7, **b.** 4, **c.** 81, **d.** 3.214, **e.** 9, **f.** 32, **g.** $\frac{1}{16}$;
6. a. $f(x) = 2(0.01)x^2 = 0.02x^2$,
b. $g(x) = 4(0.004)(x)(3x) = 0.048x^2$,
c. $(f + g)(x) = 0.02x^2 + 0.048x^2 = 0.068x^2$,
d.

f(x)	g	T(x) = f(x) + g(x)
0.08	0.19	0.27
0.32	0.77	1.09
0.72	1.73	2.45
1.28	3.07	4.35
2.00	4.80	6.80

7. a. $C(x) = 12x + 300$, **b.** $R(x) = 25.95x$,
c. $p(x) = 25.95x - (12x + 300) = 13.95x - 300$,
d. 22 hats must be sold, because 21 hats is not quite
enough. The solution was obtained graphically.
e. $C(50) = 900$. The cost of producing 50 hats is $900.
$R(50) = 1297.50$. The revenue from 50 hats is $1297.50.
$p(50) = 397.50$. The profit from selling 50 hats is $397.50.
f. The profit is the difference between the revenue and cost
functions.   **8. a.** $f(x) = 60(110) + x(110 - 2x)$,
b. $f(x) = 6600 + 110x - 2x^2$, **c.** integers $0 \le x \le 30$,
d. $f(15) = 7800$. At regular price the cost is $8250, so the
savings are $450.   **9. a.** -10, **b.** 41, **c.** 2, **d.** 5;
10. a. integers $0 \le x \le 30$, **b.** $f(22) = \$2864.40$.
If 22 snowboards are produced the cost is $2864.40.
c. $f(3.75t) = 150(3.75t) - 0.9(3.75t)^2 =$
$562.50t - 12.65625t^2$, **d.** The input variable is t,
e. $f(g(4)) = \$2047.50$, **f.** $3500 = 562.50t - 12.65625t^2$,
t is about 7.5 hours (determined graphically);
11. a. $x = \frac{2y - 3}{5}; y = \frac{5x + 3}{5}$; or $f^{-1}(x) = \frac{5x + 3}{2}$.
b. The slope of f is $\frac{2}{5}$. The slope of f^{-1} is $\frac{5}{2}$.
The slopes are reciprocals.
12. a. $f(g(x)) = f\left(\frac{1 - x}{2}\right) = -2\left(\frac{1 - x}{2}\right) + 1 = x$;
$g(f(x)) = g(-2x + 1) = \frac{1 - (-2x + 1)}{2} = \frac{2x}{2} = x$.
Since $f(g(x)) = g(f(x)) = x$, f, and g are inverses.
b.

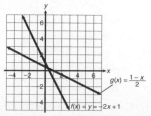

c. f and g are symmetric
with respect to the line
$y = x$;

13. a. Yes, the ratio (change in cost)/(change in number of
tickets) is constant. **b.** $f(n) = 5.5n$, **c.** The cost of one
ticket is $5.50. This represents the slope.

d.

TOTAL COST	NUMBER OF TICKETS
$11.00	2
$27.50	5
$38.50	7
$66.00	12

e. $n = g(c) = \frac{2}{11}c$, **f.** The slope is $\frac{2}{11}$. The slopes are re
ciprocals. **g.** $f(g(c)) = f\left(\frac{2}{11}\right)c = 5.5\left(\frac{2}{11}\right)c = c$; $g(f(n)) =$
$g(5.5n) = \frac{2}{11}(5.5n) = n$. The functions are inverses becaus
they undo one another.

Chapter 3

Activity 3.1   Exercises: 1. a.   top table: 0.008, 0.04, 0.2, 1,
5, 25, 125; bottom table: 0.0537, 0.1424, 0.3774, 1, 2.65,
7.0225, 18.61 **b.**

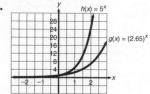

c.

BASE, b	GROWTH FACTOR	x-INTERCEPT	y-INTERCEPT	HORIZONTAL ASYMPTOTE	INCREA DECR
5	5	none	(0, 1)	y = 0	incre
2.65	2.65	none	(0, 1)	y = 0	incre

3. a. This data is exponential with a growth factor of 3.
b. This data is linear with a slope of 0.5. **c.** This data is
exponential with a growth factor of 2.   **5. a.** 2, 4, 8, 16, 32,
b. 256, **c.** The data is exponential with a growth factor
of 2. **d.** (Answers may vary.) The practical domain is the
set of nonnegative integers from 0 to 10. The practical
range is the set of whole number powers of 2 up to 2^{10}.

Activity 3.2   Exercises: 1. a. first table 23.32, 8.16, 2.86, 1
0.35, 0.1225, 0.043, second table 125, 25, 5, 1, 0.2, 0.04,
0.008, **b.**

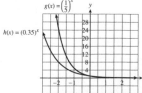

c. first row 0.35, decay, none, (0, 1) $y = 0$, decreasing,
second row $\frac{1}{5}$, decay, none, (0, 1), $y = 0$, decreasing;
3. a. This data is linear with a slope of 0.5, **b.** This data
is exponential with a growth factor of 4, **c.** This data is
exponential with a decay factor of 0.4.   **5. a.** The functio
f will increase slower than g because its growth factor is
smaller, **b.** The function f will decrease faster than g
because its decay factor is smaller.

Activity 3.3   Exercises: 1. a. $P = 148.0(0.997)^t$, **b.** Close; substituting 14 for t yields $P = 148.0(0.997)^{14} \approx 141.9$; $P = 148.0(0.997)^{20} \approx 139.4$ million;   **2. a.** ii, **b.** i; **a.** The function is increasing. The base 5 is greater than and is a growth factor. **b.** The function is decreasing. The base $\frac{1}{2}$ is less than 1 and is a decay factor. **c.** The function is increasing. The base 1.5 is a growth factor (greater than 1). **d.** The function is decreasing. The base 0.2 is a decay factor (less than 1).   **6. a.** $f(-2) = \frac{3}{16}$, **b.** $f\left(\frac{1}{2}\right) = 6$, **c.** $f(2) = 48$, **d.** $f(1.3) \approx 18.1886$; **a.** 2.5 ppm, **b.** 2.5, 1.75, 1.225, 0.8575, 0.6003, 0.4202,

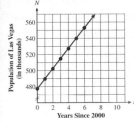

d. $A(3) = 0.8575$ ppm, **e.** Chlorine should be added in 1.4 days.   **9. a.** The data is exponential; the consecutive ratios are approximately constant. **b.** The growth factor is $b \approx 1.024$. **c.** $N = 478.4(1.024)^t$,

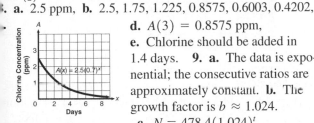

. The vertical intercept is $(0, 478.4)$. 478.4 is the population of Las Vegas in 2000 measured in thousands. **f.** In 2020, $= 20$, so $N = 478.4 \cdot (1.024)^{20} \approx 768.8$ thousands. This is probably not a good estimate. 20 years is extrapolating 4 years into the future and probably too far from the data to be accurate. **g.** According to the model, the population of Las Vegas will double in 29.2 years.

Activity 3.4   Exercises:

GROWTH FACTOR	GROWTH RATE	DECAY FACTOR	DECAY RATE
1.02	2%	0.77	23%
1.029	2.9%	.32	68%
2.23	123%	0.953	4.7%
1.34	34%	.803	19.7%
1.0002	.02%	0.9948	.52%

. a. Helena: $P = 27,885 \cdot (1.0113)^t$, Butte: $P = 30,752 \cdot (0.99)^t$, **.** Helena: $P = 27,885 \cdot (1.0113)^4 = 29,167$, Butte: $P = 30,752 \cdot (0.99)^4 = 29,540$. **.** The population of Helena will be 55,770 when $t \approx 61.7$. **.** The populations will be equal when $t \approx 4.6$.

4. a.

b. The growth factor is $b = 1.0125$; the growth rate is $0.0125 = 1.25\%$, **c.** $P(70) = 120.6(1.0125)^{70} \approx 287.7$ million. The prediction is a little higher. **5. a.** $V(t) = 20,000(0.85)^t$, **b.** The decay rate is $15\% = 0.15$, **c.** The decay factor is 0.85, **d.** $V(5) = 20,000(0.85)^5 \approx \$8874.11$, **e.**

f. The value will be \$10,000 when $t \approx 4.3$ years.

Activity 3.5   Exercises: 1. a. $A = 25,000\left(1 + \frac{0.045}{4}\right)^{4t}$, **c.** approximately 21.4 years, **e.** approximately 21.2 years; **3. a.** $A = 1900e^{0.06 \cdot 2} \approx \$2142.24$, **b.** approximately 11.6 years; **5. a.** $b - \left(1 + \frac{0.048}{12}\right)^{12}$, **b.** $b = 1.04907$, **c.** $r_e = 4.907\%$.

Activity 3.6   Exercises: 1. a. The function is increasing because in the model, $b > 1$. **b.** It is a growth rate: $r = b - 1$; $r = 1.03 - 1 = 0.03 = 3\%$. **c.** The initial value is 537.6 thousand. According to the model, it is the population of Charlotte, North Carolina in 2000. **d.** $k = 0.0296$, **e.** $y = 537.6e^{0.0296t}$, **f.** The population of Charlotte will be approximately 838.1 thousand in 2015 if the population grows at the same rate.   **3. a.** The initial value is 33 and is increasing at the rate of 9.7%. **b.** The initial value is 97.8 and is decreasing at the rate of 23%. **c.** The initial value is 3250 and is decreasing at the rate of 27%. **d.** The initial value is 0.987 and is increasing at the rate of 7.6%. **5. a.** $y = 20e^{-0.0244(20)} \approx 12.277$ grams, **b.** The half-life is approximately 28.4 years, **c.** The decay factor is $b = e^{-0.0244} \approx 0.9759$. The decay rate is 0.0241 or 2.41%.   **7. a.** In 1996, $x = 9$, so $A = 36.2e^{0.14(9)} \approx 127.62$ billion dollars, **b.**

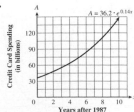

c. The vertical intercept is $(0, 36.2)$. There was \$36,200,000,000 in holiday credit-card spending in 1987. **d.** 1992 $(x = 5)$, **e.** approximately 4.95 years.

Activity 3.7   Exercises: 1. a. The data is not linear. The output seems to be increasing exponentially.

b.

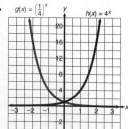

c. Yes; the scatterplot looks like an exponential function. **d.** $C = 282.6 \cdot 1.083^t$, **e.** $C = \$2,074,400,000,000$ **f.** The growth factor, b, is 1.083,

g. 0.083 or 8.3%, **h.** 2001 $(t = 21)$, **i.** approximately 8.7 years; **3. a.** the set of all real numbers, **b.** the set of all positive real numbers, **c.** $y = a \cdot b^x$ is positive for all values of x. **d.** $y = a \cdot b^x$ is never negative. **e.** $(0, a)$

How Can I Practice? 1. a. $C = 17,000(1.04)^t$, **b.** The growth rate is 0.04; the growth factor is 1.04. **c.** $C = 17,000(1.04)^3 \approx \$19,122.69$, **d.** 14.5 years; **2. a.** Graph i is function g, **b.** Graph ii is function h; **3.** Graph i is function g because it is decreasing with a decay factor of 0.47 which is between 0 and 1. Graph ii is function h because it is increasing with the growth factor of 1.47. **4. a.** 13.01, 33.18, 84.61, **b.** 1.26, 0.76, 0.46, **c.** 216, 7776, 279,936; **5. a.** $y = 2(2.55)^x$, **b.** $y = 3.5(0.6)^x$, **c.** $y = \frac{1}{6}(36)^x$; **6. a.** $f(0) = 1.3$; decreasing, **b.** $f(0) = 0.6$; increasing, **c.** $f(0) = 3$; decreasing; **7. a.** Yes, **b.** The constant ratio is 2.5 or $\frac{5}{2}$, **c.** $y = 2(2.5)^x$; **8. a.** Plan 1: $S = 22,000 + 1000x$; Plan 2: $S = 22,000(1.04)^x$, **b.** Plan 1: $22,000, $23,000, $25,000, $27,000, $32,000, $37,000; Plan 2: $22,000, $22,880, $24,747, $26,766, $32,565, $39,621, **c.** It depends. If I plan to be with the company less than 8 years, I would take plan 1, because it takes plan 2 about 7 years to catch up. If I expect to be with the company for a long time, say 20 years, I would choose plan 2, because by then I would be better off by more than $6000 per year. **9. a.** $N = f(t) = 2e^{0.075t}$, **b.** $N = f(8) = 2e^{0.075(8)} \approx 3.6442$ thousand,

c.

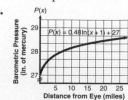

d. 14.6482 or about 15 weeks;

10. a. top table: $\frac{1}{64}, \frac{1}{16}, \frac{1}{4}, 1, 4, 16, 64$; bottom table: 64, 16, 4, 1, $\frac{1}{4}, \frac{1}{16}, \frac{1}{64}$, **b.**

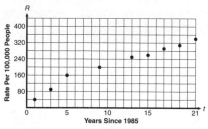

c. Row 1 : 4, growth, none, $(0, 1)$, x-axis, increasing; row 2: $\frac{1}{4}$, decay, none, $(0, 1)$, x-axis, decreasing; **11. a.** $415, **b.** 0.0118 or 1.18% per month, **c.** 1.0118, **d.** $f(x) = 415(1.0118)^x$, **e.** $(0, 415)$, **f.** This represents the initial balance on the card. **g.** $466.65, **h.** With no payments, I exceed my credit limit during the sixteenth month;

12. a. The ratios are all approximately 1.02, **b.** 1.02, **c.** $w(t) = 12.50(1.02)^x$, **d.** 2%, **e.** $15.24, **f.** about 35 years; **13. a.** $A = 10,000(1.01)^{12t}$, **b.** approximately $18,166.97, **c.** approximately 6 years, **d.** $A = 10,000e^{0.12t}$ **e.** $18,221.19, $54.22 more than in part b; **14. a.**

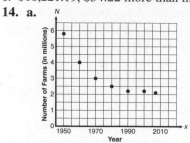

b. An exponential decay model would better model the data. The data is decreasing, but not at a constant rate. **c.** $N = 4.81(0.9832)^x$, **d.** $4.81(0.9832)^{65} \approx 1.60$ million farms, **e.** 0.9832, **f.** $0.9832 = 1 - r$, $r = 1 - 0.9832 = 0.0168$ or 1.68%, **g.** The number of farms is decreasing a a rate of 1.68% per year, **h.** The original amount in the mode is 4.81; half is 2.405. The time is approximately 41 years.

Activity 3.8 Exercises: 1. a. 5, **b.** 3, **c.** -1, **d.** -6, **e.** 0, **f.** 2, **g.** $\frac{1}{2}$, **h.** $\frac{1}{2}$, **i.** 0, **j.** 0, **k.** 5, **l.** -2, **m.** 0; **3. a.** $\log_3 9 = 2$, **b.** $\log_{121} 11 = \frac{1}{2}$, **c.** $\log_4 27 = t$, **d.** $\log_b 19 = 3$, **5. a.** $x = 0.512$, **b.** $x = 2.771$, **c.** $x = -5.347$

Activity 3.9 Exercises: 1. a. $x > 0$, **b.** the set of all real numbers, **c.** $x > 1$, **d.** $0 < x < 1$, **e.** $x = 1$, **f.** $x = 10$ **3.** $y = \log_2 x$; **4. a.** $x > 0$, **b.** the set of all real numbers, **c.** $x > 1$, **d.** $0 < x < 1$, **e.** $x = 1$, **f.** $x = e$; **5.** $n = \dfrac{\log(2500) - \log(45,000)}{\log(1 - 0.40)} \approx 5.7$ years.

Activity 3.10 Exercises: 1. a.

b. It could very well be logarithmic. It tends to increase more slowly as the input increases. **c.** $R = 9.433 + 98.109\ln(t)$ **d.** Yes, it is a very good fit, **e.** 2015 is 30 years after 1985, so evaluate R when $t = 30$. $R = 9.433 + 98.109\ln(30) \approx 343$ per 100,000 population. **3. a.** $f(0) = 27$ inches. This is the pressure in the eye of the storm.

b.

$P(x)$

$P(x) = 0.48\ln(x + 1) + 27$

Barometric Pressure (in. of mercury)

Distance from Eye (miles)

c. As you move away from the hurricane's eye, the pressure increases quickly at first and then more slowly.

a.

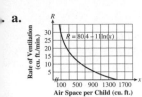

$R = 80.4 - 11 \ln(300) \approx 17.66$ cubic feet;

a. $t = \dfrac{\ln 2}{4 \ln\left(1 + \frac{0.055}{4}\right)} \approx 12.689$ years,

$t = \dfrac{\ln 3}{4 \ln\left(1 + \frac{0.055}{4}\right)} \approx 20.11$ years,

$t = \dfrac{\ln 2}{12 \ln\left(1 + \frac{0.055}{12}\right)} \approx 12.632$ years

Activity 3.11   Exercises: 1. a. $\log_b 3 + \log_b 7$,

$\log_3 3 + \log_3 13 = 1 + \log_3 13$, **c.** $\log_7 13 - \log_7 17$,

$\log_3 x + \log_3 y - \log_3 3 = \log_3 x + \log_3 y - 1$;

a.   **b.** The graphs are the same. This is not surprising because the log of a product is the sum of the logs.

a. $7.4 \log(15) = 8.7$ or 9 cars, **b.** 9, 13, 21, **c.** The sum of the sales from the smaller ads exceeds the sales from the larger ad by 1. **d.** Pretty close. 15 times 50 equals 750, so I would have expected the sum of the sales from the smaller ads to equal the sales from the largest. The error is due to rounding. **e.** Forget about the giant ad. It is a waste of money.   **7. a.** $\log_2 245$, **b.** $\log \sqrt[4]{\dfrac{x^3}{z^5}}$,

$\ln \dfrac{2^2 5^2 z^4}{5^3} = \ln \dfrac{4 z^4}{5}$, **d.** $\log_5 \dfrac{x^2 + 3x + 2}{x^2 + 6x + 9}$;   **9. a.** 0.8271,

b. 0.7557, **c.** 1.5011, **d.** 0.7112

Activity 3.12   Exercises: 1. a. $N(5) = 49.6(0.91)^5 \approx 31.0$; 31,000 arrests in the year 2005, **b.** $49.6(0.91)^t = 20$; $\ln(0.91) = \ln(20/49.6)$, $t \approx 9.6$; there will be 20,000 arrests 10 years after 2000 or in the year 2010.

a. Yes 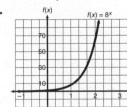   **b.** $A(t) = 352.65(1.006)^t$,

$t = \dfrac{\ln\left(\frac{705.3}{352.65}\right)}{\ln(1.006)} \approx 116$; 116 years after 1990 would be the year 2106.

4. $x = \dfrac{\ln 14}{\ln 2} \approx 3.81$;

6. $t = \dfrac{\ln 2}{\ln(1.04)} \approx 17.7$;

$x \approx 1.881$;   **10. a.** $t \approx 8$ days, **b.** $\frac{1}{5}P_0 = P_0 e^{-0.086t}$ or $t = \dfrac{\ln(0.2)}{-0.086} \approx 19$ days

How Can I Practice? 1. a. $\log_4 16 = 2$,

b. $\log_{10}(0.0001) = -4$, **c.** $\log_3\left(\frac{1}{81}\right) = -4$;

a. $2^5 = 32$, **b.** $5^0 = 1$, **c.** $10^{-3} = .001$, **d.** $e^1 = e$;

a. $x = 4^{-3} = \frac{1}{64}$, **b.** $b = 2$, **c.** $y = 3$;

4. a. $-1; -0.5; 0; 1; 2; 3$, **b.**

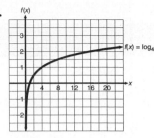

d. $(1, 0)$, **e.** $x > 0$, **f.** all real numbers, **g.** increasing, **h.** The y-axis ($x = 0$) is a vertical asymptote, **i.** $f(32) = 2.5$, **j.** $x \approx 90.5$;

5. a. $\log_b x + 2 \log_b y - \log_b z$,

b. $\frac{3}{2} \log_3 x + \frac{1}{2} \log_3 y - \log_3 z$, **c.** $\log_5 x + \frac{1}{2} \log_5 (x^2 + 4)$,

d. $\frac{1}{3} \log_4 x + \frac{2}{3} \log_4 y - \frac{2}{3} \log_4 z$;   **6. a.** $\log \dfrac{x \sqrt[3]{y}}{\sqrt{z}}$,

b. $\log_3 (x + 3)^3 z^2$, **c.** $\log_3 \sqrt[3]{\dfrac{x}{y^2 z^4}}$;   **7. a.** $\dfrac{\log 17}{\log 5} = 1.76$,

b. $\frac{1}{3} \cdot \dfrac{\log 41}{\log 13} = 0.4826$;   **8. a.** $x \approx 0.0067$,

b. $x = 646.08$;   **9. a.** $x = \dfrac{\log 17}{\log 3} = 2.5789$,

b. $x = \dfrac{\ln 14}{1.7} \approx 1.55$;

10. a.

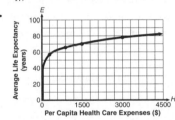

b. $E = 0.035 + 9.669 \ln(1500) \approx 70.7$ years, **c.** $H \approx \$2865$

Gateway Review 1. a. The growth factor is $100\% + 6\% = 106\%$ or 1.06. Let T represent the tuition. $T = 300(1.06)^t$, $T = 300(1.06)^5 \approx 401.47$. In 5 years, the tuition will be $\$401.47$ per credit. $T = 300(1.06)^{10} \approx 537.25$. In 10 years, a credit will cost $\$537.25$. **b.** $\dfrac{\$401.37 - \$300}{5} = \$20.29$; Tuition increases approximately $\$20.29$ per credit per year over the next 5 years. **c.** $\dfrac{\$537.25 - \$300}{10} \approx \$23.73$; Tuition increases approximately $\$23.73$ per credit per year. **d.** Tuition will double in about 12 years if inflation stays at 6%.   **2. a.** $\frac{1}{8}, \frac{1}{2}$; 1; 8; 16; 64; 512.

b.

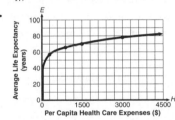

c. The function is increasing, because $b = 8 > 1$, **d.** all real numbers, **e.** $y > 0$, **f.** There is no x-intercept. The y-intercept is $(0, 1)$, **g.** There is one horizontal asymptote, the x-axis,

$y = 0$, **h.** The domain and range are the same. The graphs are reflections in the y-axis. f is increasing; g is decreasing. **i.** f is moved vertically upward 5 units to obtain h. **j.** Solve $y = 8^x$ for x: $x = \log_8 y$. Interchange x and y: $y = \log_8 x$.

3.

BASE, b	GROWTH OR DECAY FACTOR	x-INTERCEPT	y-INTERCEPT	HORIZONTAL ASYMPTOTE	INCREASING OR DECREASING
6	growth	none	(0, 1)	$y = 0$	increasing
$\frac{1}{3}$	decay	none	(0, 1)	$y = 0$	decreasing
2.34	growth	none	(0, 5)	$y = 0$	increasing
0.78	decay	none	(0, 3)	$y = 0$	decreasing
2	growth	(2, 0)	(0, −3)	$y = -4$	increasing

4.

all reals	all reals	all reals	$x > 0$	$x > 3$
$y > 0$	$y > 2$	$y > -5$	all reals	all reals

5. a. The table is approximately exponential. The growth factor is about 1.55. **b.** $y = 10 \cdot 1.55^x$;   **6. a.** 12.48, 25.46, 51.94, **b.** 2.21, 1.55, 1.09, **c.** 64, 1024, 16,384, **d. i.** $y = 3.00(2.04)^x$ **ii.** $y = 4.50(0.7)^x$ **iii.** $y = 0.25(16)^x$ **7. a.** 15,000, 15,225, 15,453, 15,685, 15,920, 16,159, **b.** $y = 15,000(1.015)^x$, **c.** $y \approx \$16,897$; this is reasonable if you assume that 15,000 is a reasonable starting salary and that the 1.5% salary increase per year remains constant.

d. $x = \log_{1.015} 2 = \dfrac{\ln (2)}{\ln (1.015)} \approx 46.6$ years;

8. a. $A = 5000e^{0.065(8)} \approx \$8410.14$, **b.** $t \approx 13.5$ years;
9. a. 125, **b.** 27, **c.** $\frac{1}{32}$, **d.** 25, **e.** −2, **f.** 4, **g.** −3, **h.** 2;
10. a. $\log_6 36 = 2$, **b.** $\log_{10} 0.000001 = -6$,
c. $\log_2 \frac{1}{32} = -5$;   **11. a.** $3^4 = 81$, **b.** $7^0 = 1$,
c. $10^{-4} = 0.0001$, **d.** $e^1 = e$, **e.** $q^b = y$;
12. a. $x = \frac{1}{125}$, **b.** $b = 4$, **c.** $y = 6$, **d.** $x = 8$;
13. a. $-3, -2, -1, 0, 1, 2$
b.

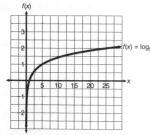

c.

d. $(1, 0)$, **e.** $x > 0$, **f.** all real numbers, **g.** It has a vertical asymptote at $x = 0$. The function gets closer and closer to the y-axis but does not cross it. **h.** $f(23) \approx 1.948$,
i. $x \approx 52.416$;   **14. a.** $\frac{\log 21}{\log 7} \approx 1.56$,

b. $\dfrac{\log \left(\frac{8}{9}\right)}{\log 15} \approx -0.0435$;

15. a. $3 \log_2 x + \log_2 y - \left(\frac{1}{2}\right) \log_2 z$,

b. $\left(\frac{1}{3}\right)(4 \log x + 3 \log y - \log z)$;   **16. a.** $\log \dfrac{x \sqrt[4]{y}}{z^3}$,

b. $\log \sqrt[3]{\dfrac{x}{y^2 z}}$;   **17. a.** $3 + x = \frac{\log 7}{\log 3}$; $x \approx -1.23$,

b. $4x + 9 = 2^4$; $x = 1.75$, **c.** $x \approx 341.5$;

18. a.

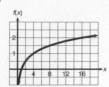

c. 2.87744,
d. $2.319 = \frac{\log x}{2 \log 2}$; $x \approx 24.9$;

19. a. New York, 19.43 million; Florida, 18.20 million,
b.

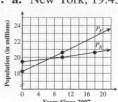

c. Florida's population will equal that of New York 7 years after 2007, or in the year 2014.

d. $25 = 18.20e^{0.0126t}$, $\ln\left(\dfrac{25}{18.20}\right) = 0.0126t$

$t \approx 25$ years. Florida's population will exceed 25 million in the year 2032. **e.** $18.20e^{0.0126t} > 19.43e^{0.0031t}$, $\ln 18.20 + 0.0126t > \ln 19.43 + 0.0031t$, $0.0095t > 0.0654$, $t > 6.88$ years. At current rates, the population of Florida will exceed the population of New York sometime in the year 2014.
20. a. $y = 45.786 - 6.903 \ln x$
b. $y = 45.786 - 6.903 \ln (500) \approx 2.887$ kilometers

Chapter 4

Activity 4.1 Exercises: 1. a. 9, 4, 1, 0, 1, 4, 9,
b.

c. 1, **d.** The domain is all real numbers. The range is all real numbers greater than or equal to 0. **e.** $-9, -4, -1, 0, -1, -4, -9$ **g.** -1, **h.** The graph of $y = -x^2$ is a reflection of $y = x^2$ over the x-axis.
2. a. $a = -2, b = 0, c = 0$, **b.** $a = \frac{2}{5}, b = 0, c = 3$,
c. $a = -1, b = 5, c = 0$, **d.** $a = 5, b = 2, c = -1$;
4. a. f opens upward; g opens downward; both pass through $(0, 0)$, **b.** Both f and h open upward. h is wider than f. **c.** h is g shifted up 2 units; both open upward. **d.** Both f and g open upward. The low point of f is 3 units below the x-axis; the low point of g is 3 units above the x-axis. **e.** f opens upward with a vertical intercept at $(0, 1)$; h opens downward with a vertical intercept at $(0, -1)$; both are symmetric with respect to the y-axis.   **6. a.** downward, **b.** $(0, -4)$;   **8. a.** upward, **b.** $(0, 3)$;   **10. a.** downward, **b.** $(0, -7)$;   **12. a.** The graph of $y = \frac{3}{5}x^2$ is wider than the graph of $y = x^2$, **b.** The graph of $y = x^2$ would have a greater output value for any input x except $x = 0$.

Activity 4.2 Exercises: 1. a. upward, **b.** $x = 0$,
c. $(0, -3)$, minimum, **d.** $(0, -3)$;   **3. a.** upward,
b. $x = -2$, **c.** $(-2, -7)$, minimum, **d.** $(0, -3)$;
5. a. upward, **b.** $x = -1.5$, **c.** $(-1.5, 1.75)$, minimum, **d.** $(0, 4)$;   **7. a.** upward, **b.** $x = 0.25$,

. $(0.25, -3.125)$, minimum, **d.** $(0, -3)$; **9. a.** $(1, 0)$, $(6, 0)$, **b.** D: all real numbers; R: $g(x) \leq 6.25$, **c.** $x < 3.5$, . $x > 3.5$; **11. a.** $(3.46, 0), (-3.46, 0)$, . D: all real numbers; R: $y \geq -12$, **c.** $x > 0$, **d.** $x < 0$; **3. a.** $(-1, 0), (3, 0)$, **b.** D: all real numbers; : $g(x) \leq 4$, **c.** $x < 1$, **d.** $x > 1$; **15. a.** $(0.2, 0), (1, 0)$, . D: all real numbers; R: $y \leq 0.8$, **c.** $x < 0.6$, . $x > 0.6$; **17. a.** 149 feet, **b.** 6.05 seconds, **c.** It indicates the height of the arrow when it is shot, **d.** The practical domain is 0 second $\leq x \leq 6.05$ seconds. The practical range is 0 feet $\leq h(x) \leq 149$ feet, **e.** $(-0.05, 0), (6.05, 0)$; he first has no meaning; the second indicates the time in seconds it takes for the arrow to hit the ground.

9. a. $(30, 200)$,

. $x = -\frac{b}{2a} = \frac{120}{4} = 30$, $C(30) = 200$; the vertex is $30, 200)$, **c.** They are the same. **d.** minimum point, . The cost of production is minimized when 30 oil lamps re produced. **f.** $(0, 2000)$; it costs $2000, even if no il lamps are produced.

Activity 4.3 Exercises: 1. $x = 6$ or $x = -2$; **3.** $x = 9$ or $= -5$; **5.** $x = -11$ or $x = -1$; **7.** $x = \pm 5$; . $x = -3$ or $x = 1$; **11.** $x = 7$ or $x = -4$; **3. a.** $-2 < x < 6$, **b.** $x < -2$ or $x > 6$; **5. a.** $d(55) = 181.5$ ft., **b.** $0.04v^2 + 1.1v = 200$; ≈ 58 mph

Activity 4.4 Exercises: 1. $6x^5(2 - 3x^3)$; . $2x(x^2 - 7x + 13)$; **5.** $(x + 3)(x - 2)$; . $(x + 5y)(x + 2y)$; **9.** $(6 + x)(2 + x)$; **1.** $(3x - 2)(x + 7)$; **13.** $5b^2(4b + 3)(b - 4)$; **4.** $x = 3$ or $x = 2$; **16.** $x = 3$ or $x = -2$; **8.** $x - \frac{1}{3}$ or $x = -4$; **20.** $x = 9$ or $x = -2$; **2. a.**

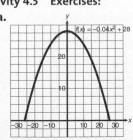

. $A = (20 + 2x)(15 + x) - 15(20) = 00 + 20x + 30x + 2x^2 - 300 = 2x^2 + 50x$, . $2x^2 + 50x = 168$, **d.** $x + 28 = 0$ or $x - 3 = 0$, $= -28$ or $x = 3$; the solution is 3 ft. -28 ft. makes no ense in this situation.

Activity 4.5 Exercises: . **a.**

b. $(0, 28)$ represents the vertex or turning point of the arch. **c.** $x \approx \pm 26.5$; the intercepts are $(26.5, 0)$ and $(-26.5, 0)$, **d.** The intercepts are the same. **e.** The river is approximately $2(26.5)$ or 53 feet wide. **f.** No; the highest point of the arch is 28 feet above the water. **g.** $-0.04x^2 + 28 = 20$, **h.** $x = \pm 14.14$ ft. Place the pole 14.14 feet to the right or left of the center. **3.** $x = -\frac{1}{2}$; **5.** $x = \frac{6 \pm \sqrt{12}}{4} = 2.37$ or 0.63; **7.** $x = \frac{-3 \pm \sqrt{33}}{4} = 0.69$, or -2.19; **9.** $(0, 0)$ and $(-2, 0)$; **11.** $\left(\frac{1 + \sqrt{41}}{4}, 0\right)$ and $\left(\frac{1 - \sqrt{41}}{4}, 0\right)$ or approximately $(1.851, 0)$ and $(-1.351, 0)$; **13. a.** $d = 2.5$ million particles per ft.^{3} **b.** The minimum occurs at the vertex. $r = \frac{-b}{2a} = \frac{16}{4} = 4$ or 400 rpm; $d = 2(4)^2 - 16(4) + 34 = 2$ million particles per ft.^{3} **c.** $r = 11$; 1100 rpm is the speed of the engine. $r = -3$ is not in the practical domain $r > 0$.

Activity 4.6 Exercises: 1. a.

b. $h(t) = -15.9752t^2 + 52.8875t + 2.5536$, **c.** Yes; the curve touches nearly every data point. **d.** all real numbers from 0 to 3.36 seconds, **e.** real numbers from 0 to 46.33 feet, **f.** The ball reaches 35 feet on the way up after 0.81 second. It reaches 35 feet again on the way down, approximately 2.50 seconds after it was struck. **g.** There are only two solutions, so I got them all. **3. a.** $y = 0.086x^2 - 0.842x + 32.487$, **b.** approx. 650 feet, **c.** $0 = 0.086x^2 - 0.842x - 247.513$; using the quadratic formula, a speed of 58.8 mph requires a stopping distance of 280 feet.

Activity 4.7 Exercises: 1. $5i$; **3.** $6i$; **5.** $4i\sqrt{3}$; **7.** $\frac{3}{4}i$; **9.** $-5 + 10i$; **11.** $3 - 2i$; **13.** $10 + 5i$; **15.** $x = \frac{1}{3} \pm \frac{\sqrt{80}}{6}i$ or $\frac{1}{3} + \frac{2\sqrt{5}}{8}i$, no x-intercepts; **17.** $x = 1$, 3.5; **19.** 2 real solutions; **21.** 1 real solution; **23.** 2 complex solutions

How Can I Practice?
1.

VALUE OF a	VALUE OF b	VALUE OF c
5	0	0
$\frac{1}{3}$	3	-1
-2	1	0

2. a. downward, **b.** $x = 0$, **c.** $(0, 4)$, maximum, **d.** $(0, 4)$; **3. a.** upward, **b.** $x = 0$, **c.** $(0, 0)$, maximum **d.** $(0, 0)$; **4. a.** downward, **b.** $x = 1$, **c.** $(1, 10)$, maximum, **d.** $(0, 7)$; **5. a.** upward, **b.** $x = \frac{1}{2}$, **c.** $\left(\frac{1}{2}, -1\right)$, minimum, **d.** $(0, 0)$; **6. a.** upward, **b.** $x = -3$, **c.** $(-3, 0)$, minimum, **d.** $(0, 9)$; **7. a.** upward, **b.** $x = \frac{1}{2}$, **c.** $\left(\frac{1}{2}, \frac{3}{4}\right)$, minimum, **d.** $(0, 1)$; **8. a.** $(-2, 0), (2, 0)$, **b.** D: all real numbers; R: $y \leq 4$, **c.** $x < 0$, **d.** $x > 0$; **9. a.** $(2, 0), (3, 0)$,

b. D: all real numbers; R: $y \geq -0.25$, **c.** $x > 2.5$,
d. $x < 2.5$;   **10. a.** $(0.91, 0), (-2.91, 0)$, **b.** D: all real
numbers; R: $y \leq 11$, **c.** $x < -1$, **d.** $x > -1$;
11. a. none, **b.** D: all real numbers; R: $y \geq 1.427$,
c. $x > 1.61$, **d.** $x < 1.61$;   **12.** $(0.75, 26.125)$;
13. a. $9a^2(a^3 - 3)$, **b.** $6x^2(4x - 1)$,
c. $4x(x - 5)(x + 1)$, **d.** cannot be factored,
e. $(x - 8)(x + 3)$, **f.** $(y + 5)^2$;   **14. a.** $5, 6.05, 7.2,$
$8.45, 9.8, 11.25, x \approx 1.2$, **b.** $6.75, 6.16, 5.59, 5.04, 4.51,$
$-4, x \approx 0.8$, **c.** $0, -2, 2, 12, 28, 50, x = 2$;
15. a. $x = 1.2, -1.2$, **b.** $x = 0.8, 6.2$, **c.** $x = -\frac{1}{3}, 2$;
16. a. $x = 0, 2$, **b.** $x = 9, -2$, **c.** $x = 3, 1$, **d.** $x = 4, 4$,
e. $x = 6, -4$, **f.** $y = 5, -3$, **g.** $a = 3, -2$,
h. $x = \frac{1}{4}, -2$;   **17. a.** $-8 < x < 2$, **b.** $x < -8$ or $x > 2$;

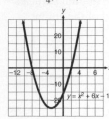

18. a. 105 feet, **b.** Using the calculator to solve $-16t^2 + 80t + 5 = 0$,
$t \approx 5.06$ seconds
c. $-16t^2 + 80t + 5 = 101$,
d. $t = 2, 3$; the ball reaches a height
of 101 feet after 2 seconds. on the way
up and 1 second later on the way down.

19. a. $0 \leq x \leq 100$,
b. $h(50) = 0.01(50)^2 - 50 + 35 = 10$ m;
20. a.

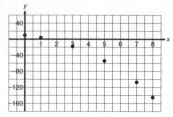

b. $y = -2.096x^2 - 2.25x + 9.038$,
c.

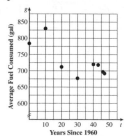

d. Predicted values are very close to the
actual values, **e.** $-33.5, -181$,
f. $x = 4.330, -5.403$;   **21.** $7i$;

22. $3i\sqrt{5}$;   **23.** $11i$;   **24.** $i\sqrt{15}$;   **25.** $4i\sqrt{7}$;   **26.** $5i\sqrt{5}$;
27. $\frac{4}{5}i$;   **28.** $\sqrt{\frac{4}{7}}i = \frac{2}{\sqrt{7}}i$;   **29.** $4 + 3i$;   **30.** $-5 + 6i$;
31. $6 - 4i$;   **32.** $-12 - 24i$;   **33.** $5 + 10i$;
34. i. $a = 3, b = -1, c = -7; b^2 - 4ac = 85$; two real
solutions; $x = 1.70, -1.37$, **ii.** $a = 1, b = -4, c = 10$;
$b^2 - 4ac = -24$; two complex solutions;
$x = 2 \pm i\sqrt{6}$, **iii.** $a = 2, b = -5, c = -3$;
$b^2 - 4ac = 49$; two real solutions; $x = 3, -0.5$,
iv. $a = 9, b = -6, c = 1; b^2 - 4ac = 0$; one real
solution; $x = \frac{1}{3}$;   **35. i.** The discriminant is 0. The graph
only touches the x-axis indicating that there is one, real
solution, **ii.** The discriminant is negative. The graph does not
intersect the x-axis, indicating that there is no real solution,
iii. The discriminant is positive. The graph intersects the
x-axis twice, indicating that there are two real solutions

Activity 4.8   Exercises: 1. a. $2, 32, 64, y = kx$ and
$8 = k1$, so $k = 8$ or $y = 8x$, **b.** $\frac{1}{8}, 27, 216, y = kx^3$ and
$1 = k1^3$, so $k = 1$ or $y = x^3$;   **3.** $y = kx^2$ and $12 = k2^2$,
so $k = 3$. Therefore $y = 3x^2$. When $x = 8, y = 3(8)^2 = 192$;

5. $d = kt^2$ and $20 = k(2)^2$, so $k = 5$. Now $d = 5t^2$, so in
2.5 seconds the skydiver travels $d = 5(2.5)^2$ or 31.25 meters

7. 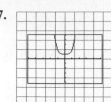   **9.**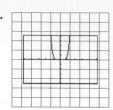

11. $f(x)$ is increasing for $x > 0$;   **13.** $y = x^2$ is rising
more slowly than $y = x^3$ for $x > 1$. Multiplying x^2 by x
gives x^3, and this makes a larger output when $x > 1$;
15. $y = -2x^3$ is decreasing and goes through $(0, 0)$,
whereas $y = 2x^3 + 1$ is increasing and does not pass
through the origin. Both have a similar S-like shape.

Activity 4.9   Exercises: 1. $f(x) = x(x + 2)(x + 1), (0, 0)$
$(-1, 0), (-2, 0)$;   **3.** $h(x) = (x^2 - 4)(x^2 - 9)$,
$h(x) = (x + 2)(x - 2)(x + 3)(x - 3), (2, 0), (-2, 0),$
$(3, 0), (-3, 0)$;

5. a.

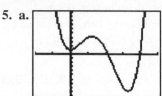

b. The domain is all real numbers.   **c.** The range is
y-values greater than or equal to -8.91.   **d.** $(2.12, 0)$ and
$(3.97, 0)$ **e.** There are two minimum points $(0, 1)$ and
$(3.28, -8.91)$ and one maximum point $(1.22, 4.23)$.   **7.** No
as x increases without bound, y increases without bound.
9. a. increase **b.** decreasing **c.** 1

Activity 4.10   Exercises:
1. a.

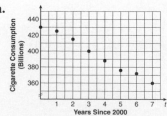

The data does not appear to be linear because as the input
increases, the output increases and decreases. No line
would be close to all of the points.   **b.** quadratic:
$g = 0.0531t^2 - 4.845t + 810.794$; cubic:
$g = 0.0054t^3 - 0.330t^2 + 1.757t + 797.664$; quartic:
$g = -0.00095t^4 + 0.096t^3 - 3.011t^2 + 25.902$
$+ 784.264$

3. a.

. $y = -10.52x + 432.58$;
. $y = 0.083x^2 - 11.11x + 433.17$ **d.** There appears to
be no difference between the two models. Both fit closely
to the data. **e.** The linear model predicts 274.78 and the
quadratic predicts 285.195. **f.** We are predicting in far
outside our practical domain, so I am not very confident in
either model's prediction.

How Can I Practice? 1. $y = kx^2$ and $45 = k3^2$, so $k = 5$
$= 5(6)^2 = 180$; **2. a.** double, **b.** $k = 1080$; k represents the speed at which the sound of thunder travels in feet
per second. **3.** $v = kt$ and $60 = k3$ so $k = 20$
$= 20(4) = 80$ feet per second

. a.

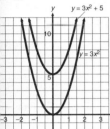

These are the same shape and
size; however, $y = 3x^2 + 5$ is
shifted up 5 units.

.

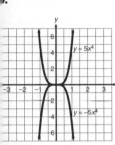

These are the same shape but are
reflections of each other in the
x-axis.

.

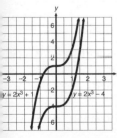

These are the same shape and
size, but $y = 2x^3 - 4$ is shifted
vertically 5 units below
$y = 2x^3 + 1$.

.

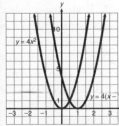

These are the same shape and
size, but $y = 4(x - 1)^2$ is
shifted horizontally 1 unit to the
right of $y = 4x^2$.

5. a. i. $(0, 0)$, **ii.** $(0, 0), (-4, 0), (2, 0)$, **iii.** $(-2.4, 16.9)$,
$(1.1, -5.0)$, **b. i.** $(0, 3)$, **ii.** $(-1, 0), (1.57, 0)$, **iii.** $(0.79, 4.2)$.

6. a.

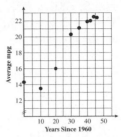

b. Cubic $y = -0.0003638x^3 + 0.0270x^2 - 0.300x$
$+14.251$; Quartic: $y = 0.000006374x^4 - 0.0009609x^3$
$+ 0.0444x^2 - 0.4534x + 14.3314$, **c.** Cubic: 18.9 mpg,
Quartic: 22.2 mpg, **d.** In both cases we are extrapolating.
The cubic goes down, which is not likely. The quartic
remains about the same. I'm not thrilled with that, either,
but it is better than the cubic.

7. a.

$W(t)$ becomes negative after 7.19 minutes. **b.** $W(0) = 10$
gallons **c.** 60.8 gallons, found by using the CALC menu
on the graphing calculator. This is the output value at the
highest point on the graph. **d.** 7.19 minutes

Gateway Review 1. a. up, **b.** $x = 0$, **c.** $(0, 2)$, **d.** $(0, 2)$;
2. a. down, **b.** $x = 0$, **c.** $(0, 0)$, **d.** $(0, 0)$; **3. a.** down,
b. $x = 0$, **c.** $(0, 4)$, **d.** $(0, 4)$; **4. a.** up, **b.** $x = \frac{1}{4}$,
c. $\left(\frac{1}{4}, -\frac{1}{8}\right)$, **d.** $(0, 0)$; **5. a.** up, **b.** $x = -\frac{5}{2}$,
c. $(-2.5, -0.25)$, **d.** $(0, 6)$; **6. a.** up, **b.** $x = \frac{3}{2}$,
c. $(1.5, 1.75)$, **d.** $(0, 4)$; **7. a.** up, **b.** $x = 1$, **c.** $(1, 0)$,
d. $(0, 1)$; **8. a.** down, **b.** $x = 2.5$, **c.** $(2.5, 0.25)$,
d. $(0, -6)$; **9. a.** $(-3, 0), (-1, 0)$, **b.** D: all real
numbers; R: $g(x) \geq -1$, **c.** $x > -2$, **d.** $x < -2$;
10. a. $(-3, 0), (1, 0)$, **b.** D: all real numbers;
R: $f(x) \geq -4$, **c.** $x > -1$, **d.** $x < -1$;
11. a. $(0.382, 0), (2.62, 0)$, **b.** D: all real numbers;
R: $y \geq -1.25$, **c.** $x > 1.5$, **d.** $x < 1.5$;
12. a. $(-3.22, 0), (-0.775, 0)$, **b.** D: all real numbers;
R: $h(x) \geq -3$, **c.** $x > -2$, **d.** $x < -2$; **13. a.** $(2, 0)$,
$(-2, 0)$, **b.** D: all real numbers; R: $y(x) \leq 8$, **c.** $x < 0$,
d. $x > 0$; **14. a.** $\left(\frac{1}{3}, 0\right) (1, 0)$, **b.** D: all real numbers;
R: $f(x) \leq \frac{1}{3}$, **c.** $x < \frac{2}{3}$, **d.** $x > \frac{2}{3}$; **15. a.** none,
b. D: all real numbers; R: $g(x) \geq 5$, **c.** $x > 0$, **d.** $x < 0$;
16. $x = -2$; **17.** $x = 2, 3$; **18.** $x = -0.51, 6.51$;
19. $x = -5, 2$; **20.** $x = \pm 1.1$; **21.** $x = -0.2, -4.8$;
22. a. $9a^2(a^3 - 3)$, **b.** $6x^2(4x - 1)$,
c. $4x(x - 5)(x + 1)$, **d.** cannot be factored,
e. $(x - 8)(x + 3)$, **f.** $(t + 5)^2$; **23.** $x = \pm 3$;
24. $x = \pm 6$; **25.** $x = 3, 4$; **26.** $x = -3, 9$;
27. $x = 0, -1$; **28.** $a = 1, b = 5, c = 3; x = -0.7$,
-4.3; **29.** $a = 2, b = -1, c = 3; x = 0.25 \pm 1.2i$;

30. $a = 1, b = 0, c = -81; x = \pm 9;$ **31.** $a = 3,$
$b = 5, c = -12; x = -3, \frac{4}{3};$ **32.** $a = 2, b = -3,$
$c = -5; x = -1, 2.5;$ **33.** From the graphing
calculator: $(0.42, 0), (3.58, 0)$

$$x = \frac{-(-8) \pm \sqrt{(-8)^2 - 4(2)(3)}}{2(2)} = \frac{8 \pm \sqrt{40}}{4} =$$

$3.58, 0.42;$ **34. a.** $7i,$ **b.** $4i\sqrt{3},$ **c.** $3i,$ **d.** $i\sqrt{23},$
e. $\frac{\sqrt{5}}{3}i,$ **f.** $\frac{\sqrt{17}}{4}i;$ **35. a.** $-5 + 17i,$ **b.** $5 - 16i,$
c. $32 + 12i,$ **d.** $27 + 6i;$ **36.** $b^2 - 4ac = 1;$ two real
solutions; **37.** $b^2 - 4ac = 256;$ two real
solutions; **38.** $b^2 - 4ac = 36;$ two real solutions;
39. $b^2 - 4ac = -20;$ two complex solutions;

$$\textbf{40. } x = \frac{-2 \pm \sqrt{2^2 - 4(3)(2)}}{2(3)} = \frac{-2 \pm \sqrt{-20}}{6} =$$

$\dfrac{-2 \pm 2i\sqrt{5}}{2(3)} = \dfrac{-1 \pm i\sqrt{5}}{3};$ the graph has no x-intercepts,
confirming complex solution; **41. a.** $-2 < x < 3,$
b. $x < -2$ or $x > 3;$ **42. a.** $y = 20,$ **b.** $y = 32,$
c. $y = 40;$ **43. a.** $(2, 0),$ **b.** D: all real numbers; R: all
real numbers, **c.** increasing for all real numbers;
44. a. $(-1, 0),$ **b.** D: all real numbers; R: all real
numbers, **c.** decreasing for all real numbers;
45. a. $(-1.68, 0), (1.68, 0),$ **b.** D: all real numbers;
R: $y \geq -8,$ **c.** inc: $x > 0;$ dec: $x < 0;$
46. a. $(0, 0), (-1.26, 0),$ **b.** D: all real numbers;
R: $y \geq -1.19,$ **c.** inc: $x > -0.8;$ dec: $x < -0.8;$
47. a. none, **b.** D: all real numbers; R: $y \geq 5,$
c. inc: $x > 0;$ dec: $x < 0;$

48. a.

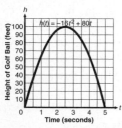

The practical domain is
$0 \leq x \leq 5.$

b. $(2.5, 100);$ the ball reaches its highest level, 100 feet,
2.5 seconds after being struck, **c.** $(0, 0);$ the ball is on the
ground when the club makes contact with it, **d.** $(0, 0),$
$(5, 0);$ the ball is on the ground when the club makes contact,
$t = 0,$ and returns to the ground 5 seconds later, **e.** I am
assuming that the elevations are the same. **49. a.** $(-5, 6),$
b. $(0, 0), (-5, 6), (-10, 0),$ **c.** $y = -0.24x^2 - 2.4x;$
50. a. vertex: $(2.5, h(2.5))$ or $(2.5, 105);$ the maximum
height is 105 feet, **b.** Set $h(t) = 0; t = 5.06$ seconds;
51. a. $s(44) = 122.5$ feet away, **b.** $v \approx -51.78$ or $19.78;$
reject the negative; 19.78 feet per second ≈ 13.5 miles
per hour.

Chapter 5

Activity 5.1 Exercises: 1. a. The average speed
$$= \frac{20 \text{ km}}{1 \text{ hr } 15 \text{ min}} = \frac{20 \text{ km}}{1.25 \text{ hr}} \approx 16 \text{ km/hr.,} \textbf{ b. } 20, 16, 13.33,$$

$11.43, 10, 8.89, 8,$ **c.** $s = f(t) = \frac{20}{t},$ **d. i.** the set of all
nonzero real numbers, **ii.** (Answers will vary.) $1 \leq t \leq 5$
Because 20 kilometers per hour (when $t = 1$) is fast for a
distance runner and 4 kilometers per hour (when $t = 5$) is
slow for a distance runner, most times will fall between
these values.
iii.

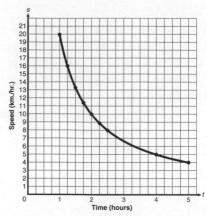

e. The average speed decreases, approaching 0. **f.** The
average speed increases without bound.

3. a. $D = f(N) = \frac{1400 - 200}{N} = \frac{1200}{N},$ **b.** $1200, 600, 400,$
$200, 100, 50,$
c.

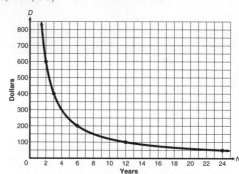

d. Decrease. As N gets larger, D gets smaller.

Activity 5.2 Exercises: 1. a. (Answers will vary.) If a person is 6 feet $= 72$ inches tall and weighs 200 pounds,
$$B = \frac{705(200)}{72^2} = 27.2, \textbf{ b. } B = \frac{119,850}{h^2}, \textbf{ c. } 0 < h < 84$$
This works unless the person is over 7 feet tall, **d.** $33.3,$
$29.3, 25.9, 23.1, 20.8, 18.7,$
e.

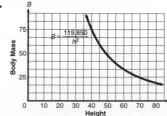

f. Body-mass index B
gets smaller and it ac-
tually approaches 0.
Yes, this makes sense
in this case because
taller persons with the
same weight should be
skinnier.

g. $69.2 < h < 79.4;$
3. i. graph b, **ii.** graph c, **iii.** graph a, **iv.** graph d;
5. a. all nonzero real numbers,

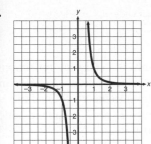

c.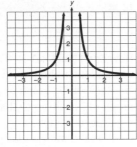

7. a. $y = \frac{2}{x}$; Table: 4, 1, $\frac{1}{3}$, **b.** $y = \frac{8}{x^3}$; Table: 64, 8, $\frac{1}{27}$;

9. $I = \frac{120}{R}$; $I = \frac{120}{15} = 8$ amps;

11. a.

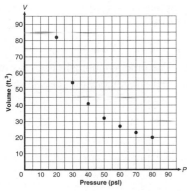

b. No. Using $P = 20$, $V = 82$; $k = 20^2(82) = 32{,}800$.
If $V = \frac{32{,}800}{P^2}$, then $P = 30$ would yield $V = \frac{32{,}800}{30^2} = 36.44$
(not very close), **c.** Yes. Using $P = 20$, $V = 82$;
$k = 20(82) = 1640$; $V = \frac{1640}{30} = 54.67$; $V = \frac{1640}{40} = 41$,
d. Answers will vary, $V \approx 25$ cubic feet.
e. $V = \frac{1640}{P}$; $V = \frac{1640}{65} = 25.2$ cubic feet;

Activity 5.3 Exercises: 1. a.
$$V = \frac{25{,}000 + 55{,}000}{P - 1.5 - 0.40 - 0.60} = \frac{80{,}000}{P - 2.5},$$ **b.** undefined,
160,000, 32,000, 10,667, 3555.56, **c.** V decreases,
d. $V(2) = -160{,}000$. A price of \$2 per cubic meter is not
practical, **e.** $P > 2.5$,
f.

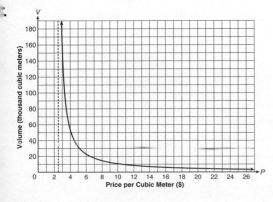

3. a. domain: all real numbers except $x = 7$; vertical asymptote: $x = 7$

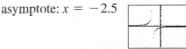

b. domain: all real numbers except $x = 25$; vertical asymptote: $x = 25$

c. domain: all real numbers except $x = 5$; vertical asymptote: $x = 5$

d. domain: all real numbers except $x = 14$; vertical asymptote: $x = 14$

e. domain: all real numbers except $x = -2.5$; vertical asymptote: $x = -2.5$

5. a. $x = 5$, **b.** $f(x)$ gets large, approaching infinity. $g(x)$ gets large in magnitude in a negative direction, approaching negative infinity. **c.** $f(x)$ gets large in magnitude in a negative direction, approaching negative infinity. $g(x)$ gets large, approaching infinity.

Activity 5.4 Exercises:
1. a. $500 + 600 + 500 + 400 = \$2000$, **b.** $2000 + 50n$,
c. $m = \frac{50n + 2000}{n}$, **d.** $m = \frac{50(100) + 2000}{100} = \$70$, **e.** The practical domain is whole numbers from 1 to the size of your class, say 250. **f.** 90, 70, 63.33, 60, 58, **g.** Yes, $m = 50$ is the horizontal asymptote. It makes sense because as the number of attendees increases, the fixed costs attributed to each person get smaller and smaller.
h.

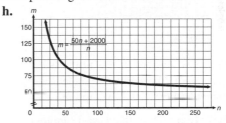

2. a. i. all real numbers except -2,
ii. $x = -2$,
iii.

iv. $y = 4$,
b. i. all real numbers except -1,
ii. $x = -1$,
iii.

iv. $y = -1$,
c. i. all real numbers except 4,
ii. $x = 4$,
iii.

iv. $y = 3$,
d. i. all real numbers except $\frac{1}{2}$,
ii. $x = \frac{1}{2}$,
iii.

iv. $y = 15$;

3. a. $x = 1$ **b.** $x = \frac{7}{9} \approx 0.778$ **c.** $x = -0.859$

d. $x = 0.5$

4. a. $16 = \frac{20}{t}$; $16t = 20$, $t = \frac{20}{16} = 1.25$ hours,
b. $18 = \frac{20}{t}$; $18t = 20$, $t = \frac{20}{18} = 1.11$ hours
5. a. $15d^2 = 1500$; $d^2 = 100$; $d = \pm 10$ feet, but only 10 makes sense, **b.** $8000 = \frac{1500}{d^2}$; $8000d^2 = 1500$; $d^2 = 0.1875$; $d = 0.433$ feet;
7. a.
$$R = \frac{250(312) + 12.5(4009) + 1000(34) - 1250(11) + 6.25(478)}{3(478)}$$
$= 105.5$ (rounded to the nearest tenth)
b.
$$R = \frac{250(371) + 12.5(4002) + 1000(27) - 1250(12) + 6.25(555)}{3(555)}$$
$= 95.0$ (rounded to the nearest tenth)
8. a. $R \approx 7.85$ prey per week, **b.** $61.3 = n$, $n \approx$ 61 prey/sq.mi., **c.** $17.3 \approx n$; putting the two together, $18 \le n \le 61$.
d.

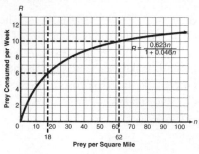

e. The result is negative so discard it. It is not possible for the predator to consume 20 prey/week. Under these conditions, 20 is above the horizontal asymptote.

Activity 5.5   Exercises: 1. a. $5T = 90$, $T = 18$ minutes,
b. $t_2T + t_1T = t_1t_2$, $T = \frac{t_1 t_2}{t_1 + t_2}$,
c. $T = \frac{20(15)}{20 + 15} = \frac{300}{35} = 8.57$ minutes, **d.** $3t_2 = 40$, $t_2 \approx 13.3$ minutes, **e.** $14T = 150$, $T \approx 10.7$ minutes;
3. $t = \frac{72 \pm \sqrt{72^2 - 4(70)}}{2}$, $t \approx 71$ or 1; the only value that makes sense in this situation is $t = 71$ minutes;
5. $21 = 2t$, $t = 10.5$ hours

Activity 5.6   Exercises:
1. a. $h = \dfrac{330}{1 - \frac{40}{770}} \approx 348.08$ hertz; the pitch I hear is higher than the actual pitch, **b.** $h = a \div \frac{770 - s}{770}$; $h = \frac{770a}{770 - s}$; **c.** $h = \frac{770(330)}{770 - 40} \approx 348.08$ hertz; the results are the same, **d.** $h = \frac{770(330)}{770 - 60} \approx 357.89$ hertz;
3. a. $= \frac{x - 4}{2x} \cdot \frac{1}{x - 4} = \frac{1}{2x}$,

b. $= \frac{2 - x}{2x} \cdot \frac{4x^2}{(2 - x)(2 + x)} = \frac{2x}{2 + x}$,
c. $= \frac{3x^2 - 6x}{x^2 - 4x + 3}$, **d.** $= \frac{x}{2x + 4}$;

How Can I Practice? 1. The graphs of f and g are reflections of each other about the x-axis (or the y-axis),   **2.** They are similar, but the graph of g is closer to the x-axis, and the graph of f is closer to the y-axis,   **3. a.** $T =$ time in hours $s =$ speed in mph, $T = \frac{145}{s}$, **b.** $0 < s < 80$, **c.** all real numbers except 0;   **4. a.** domain: all real numbers except $x = -5$; vertical asymptote: $x = -5$, **b.** domain: all real numbers except $x = \frac{13}{2}$; vertical asymptote: $x = \frac{13}{2}$, **c.** domain: all real numbers except $x = \frac{8}{5}$; vertical asymptote: $x = \frac{8}{5}$, **d.** domain: all real numbers except $x \approx 0.5614$; vertical asymptote: $x = 0.5614$;
5. $4000^2 \cdot 100 = k = 1.6 \cdot 10^9$; $w = \frac{1.6 \cdot 10^9}{d^2} = \frac{1.6 \cdot 10^9}{(4500)^2}$ ≈ 79.01 lb.;   **6. a.** The practical domain is all positive integers, with some realistic upper limit, depending on the specific situation.

b.

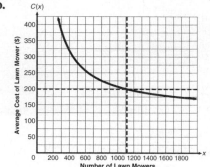

$199x = 132x + 75,250$, $x = 1124$ mowers.

7. a. $t = \dfrac{-(-14) \pm \sqrt{14^2 - 4(0.15)(0.125)}}{2(0.15)} \approx 93.32$ minutes or 0.0089 minutes. Only 93.32 is practical.

b. The drug will be at its highest concentration approximately 0.913 minute after infection.

8. a. $x = -0.25$ **b.** $x = \frac{50}{17} \approx 2.94$

c. $x = \frac{-116}{40} = -2.9$

d.
$$x = \frac{5.8}{2.4 - (0.3)5.8} = \frac{290}{33}$$
≈ 8.788

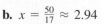

. a. (Answers will vary.) 200-pound man ≈ 90.9
kilograms, $W = \dfrac{90.9}{\left(1 + \frac{15}{6400}\right)^2} \approx 90.475$ kilograms,

. ** $W = \dfrac{70}{\left(1 + \frac{h}{6400}\right)^2}$, **c. 70, 69.78, 67.86, 52.36, 45.94,

-0.64, 10.66, 4.11, **d.** The weight decreases. **e.** 57.69
kilograms, **f.** (Answers will vary at the upper end.) The
domain is $0 \le h \le 40{,}000$.

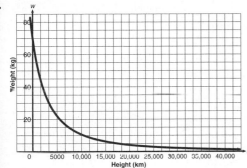

**. ** $h = 2650.9668$ kilometers
0. a. $R_3 = 12$ ohms, **b.** $R = \dfrac{R_1 R_2 R_3}{R_1 R_2 + R_2 R_3 + R_1 R_3}$,

**. ** $R = \dfrac{4(6)(12)}{4(6) + 4(12) + 6(12)} = \dfrac{288}{24 + 48 + 72} = \dfrac{288}{144} = 2$

ohms;
1. a. $x = 1.3$ **b.** $x = -1$

. ** $x = 5$ **d. $x = \frac{1}{4}$

2. a. $s = \dfrac{2(15.3)}{\frac{15.3}{45} + \frac{15.3}{40}} = \dfrac{30.6}{0.34 + 0.3825} \approx 42.4$ mph,

. ** $s = \dfrac{2 d r_1 r_2}{d(r_1 + r_2)} = \dfrac{2 r_1 r_2}{r_1 + r_2}$ **c. $s = \dfrac{2(45)(40)}{(40 + 45)} \approx 42.4$ mph;
the results are the same. **13. a.** $= \dfrac{4x + 2}{x - 3}$,

. ** $= \dfrac{(x + 5)(x - 5)}{(x + 5)} = x - 5$, **c. $= \dfrac{1}{x + 2} \cdot \dfrac{x + 2}{x + 3} = \dfrac{1}{x + 3}$;

Activity 5.7 Exercises: 1. a. 5.48, **b.** 2.45, **c.** 169, **d.** 27;
3. a. (0, 4.04) In 2000–2001 there were approximately
4.04 million undergraduate students receiving Stafford
loans. **b.** 4.04, 4.80, 5.27, 5.64, 5.96, 6.24, 6.49; the equation matches reasonably although it is not perfect.

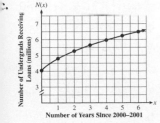

d. $x = 12$, so $N(12) = 1.222\sqrt{12 + 0.24} + 3.442 \approx$
7.72 million.

5. a. 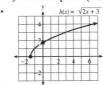 **b.** No, the graph of f is below
the graph of g for $0 < x < 1$.
c. Yes, the graph of f is above
the graph of g for $x > 1$.

7. a. i. The domain is all real numbers such that
$2x + 3 \ge 0$ or $x \ge -\frac{3}{2}$, **ii.** The x-intercept is $\left(-\frac{3}{2}, 0\right)$.
The y-intercept is $(0, \sqrt{3})$.
iii.

b. i. The domain is all real numbers such that $4x + 8 \ge 0$
or $x \ge -2$. **ii.** The x-intercept is $(-2, 0)$; the y-intercept
is $(0, -\sqrt{8})$.
iii.

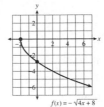

c. i. The domain is all real numbers such that $5 - x \ge 0$
or $x \le 5$. **ii.** The x-intercept is $(5, 0)$. The y-intercept is
$(0, \sqrt{5})$.
iii.

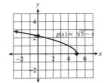

**9. ** $d = \sqrt{12^2 + 24^2 + 17^2} = \sqrt{1009} \approx 31.8$ inches. It
will not fit. **10. a.** $s = \sqrt{30(0.85)l} = \sqrt{25.5l}$,

b. ** $s = \sqrt{25.5(90)} \approx 47.9$ mph, **c. 0 feet $\le l \le 300$
feet is possible,

d.

e. The length of the skid marks is approximately 192 feet.

Activity 5.8 Exercises: 1. a. $x = 4$, **b.** $\sqrt{x + 1} = -4$;
this can't happen; a positive radical can't equal -4. There
is no solution. Equation b has no solution. The left side of
equation b will always be greater than 1, so no solution is
possible. **c.** $x = -2$, but $x = 1$ does not check.

3. a. $x = -4$ checks. **b.** $x = 2$ checks.

c. $x = 1$; $x = -4$ does not check.

5. $L = 32\left(\frac{1.95}{2\pi}\right)^2$, $L \approx 3.08$ feet;

7. a. $V = \sqrt{\frac{1000(10)}{3}} \approx 57.7$ mph,

b. $P = \frac{14700}{1000} = 14.7$ pound per square feet;

9. a. $A = \sqrt{\frac{70 \cdot 200}{3131}} \approx 2.11$ square meters, **b.** $w = \frac{3131A^2}{h}$

Activity 5.9 Exercises: 1. a. 4, **b.** 2, **c.** -3, **d.** 5, **e.** $\frac{1}{6}$,
f. not real, **g.** 10, **h.** not real; **3.** The difference is
$\sqrt[3]{1450} - \sqrt[3]{1280} \approx 0.46$ in.; **5. a.** all real numbers,

b. $x \geq 3$, **c.** all real numbers, **d.** $x \leq 2$; **7. a.** $x = 64$,

b. $x = 81$; **9. a.** $r = \sqrt[3]{\frac{3 \cdot 40}{4\pi}} \approx 2.12$ centimeters,

b. $V = \frac{4\pi(3.5)^3}{3} \approx 179.6$ cubic feet, **c.** $V = \frac{4\pi r^3}{3}$

How Can I Practice? 1. a. $x = 98$, **b.** $x = 41$, **c.** $x = 12$,
d. $x = \pm\sqrt{61}$, **e.** no solution, **f.** $x = 10$, **g.** $x = 10.5$,
h. $x \approx 0.95$; **2. a.** $x \leq 6$, **b.** all real numbers; **c.** $x \geq 2$
or $x \leq -2$; **3.** Length is approximately 7.71 inches.
4. $r = \sqrt[3]{\frac{3V}{4\pi}}$, $V = 620$, $r \approx 5.29$ centimeters;
5. $v = 100$, $100 = \sqrt{64d}$, $d = 156.25$ feet;
6. $x \approx 6.5$ inches. The dimensions of the bottom of the
box are 6.5 inches $\times$ 6.5 inches. **7.** The graphs are
reflections about the line $x = 2$.

Gateway Review 1. a. $d = \frac{1200}{w}$, **b.** 40, 34.286, 30, 24, 20
c. As width increases, the depth decreases. **d.** The depth
is 12 feet, not enough room for most theater sets. **e.** No.
Division by 0 is undefined. **f.** (Answers may vary.)
$30 \leq w \leq 60$, **g.** a rational function, **h.** all real numbers
except 0, **i.** $w = 0$, **j.** $d = 0$; as w increases,
d approaches 0.
2. a. **b.**

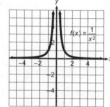
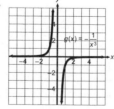

c. The graphs have the same horizontal and vertical asymptotes. $f(x) = \frac{1}{x^2}$ is symmetrical with respect to the y-axis.
$g(x) = -\frac{1}{x^3}$ is symmetrical with respect to the origin in quadrants II and IV. $f(x)$ is always positive. $g(x)$ is both positive
and negative. **3. a.** $y = \frac{k}{x}$; $12 = \frac{k}{10}$; $k = 120$; $y = \frac{120}{30} = 4$,

b. $l = \frac{k}{d}$; $32 = \frac{k}{16}$; $k = 512$; $l = \frac{512}{100} = 5.12$ decibels,

c. $h = \frac{k}{r}$; $8 = \frac{k}{4}$; $32 = k$; $h = \frac{32}{25} = 1.28$ inches;

4. a. H: $y = 0$; V: $x = 0$; no y-intercept, no x-intercept,

b. H: $y = 0$, V: $x = 3$, $\left(0, -\frac{4}{3}\right)$; no x-intercept,

c. H: $y = 2$; V: $x = -2$, $(0, 0)$, $(0, 0)$
4. a.

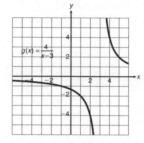

4. b. **4. c.**

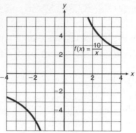

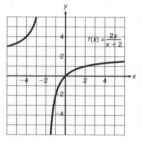

5. a. $f(n) = 45n + 600$, **b.** $f(100) = \$5100$,

c. $A(n) = \frac{45n + 600}{n}$, **d.** $A(100) = \frac{45(100) + 600}{100} = \$51$,

e. 57, 51, 49, 48, 47.40, **f.** $50 = \frac{45n + 600}{n}$; $50n = 45n + 600$
$50n = 45n + 600$; $5n = 600$; $n = 120$ people,

g. $0 < n <$ seating capacity of restaurant, **h.** The vertical asymptote is $n = 0$. You cannot calculate an average
value if no people attend. **i.** The horizontal asymptote is
$A(n) = 45$. As the number of people attending increases,
the average cost approaches \$45 per person.

6. a. $\frac{4}{x - 2} = 6$; $4 = 6x - 12$; $16 = 6x$; $x = \frac{8}{3}$,

b. The solution is the x-coordinate of the x-intercept.

7. a. $x = -\frac{7}{5} = -1.4$

b. $x = 1$

8. $\frac{1}{20} + \frac{1}{15} = \frac{1}{x}$; $60x\left(\frac{1}{20} + \frac{1}{15}\right) = 60x\left(\frac{1}{x}\right)$; $3x + 4x = 60$;
$7x = 60$; $x \approx 8.57$ minutes; **9.** $x = 67.5$, $2x = 135$
minutes, or $2\frac{1}{4}$ hours; **10. a.** $x = 10.2$, **b.** $x = 33$;

11. a. $S = \frac{C}{1 - r}$; $S(1 - r) = C$; $S - Sr = C$; $S - C = Sr$;
$r = \frac{S - C}{S}$, **b.** $bc - 4ab = -3ac$; $b(c - 4a) = -3ac$;
$b = \frac{-3ac}{c - 4a}$; **12. a.** $\frac{b + 2a}{2b + a}$, **b.** $\frac{x + 2}{x - 2}$;

3. a. $f = \dfrac{1}{\frac{1}{4}+\frac{1}{3}} = \dfrac{1}{\frac{3}{12}+\frac{4}{12}} = 1 \div \frac{7}{12} = \frac{12}{7} = 1.71$

meters, **b.** $f = \dfrac{1}{\frac{1}{p}+\frac{1}{q}} = \dfrac{1}{\frac{q}{pq}+\frac{p}{pq}} = \dfrac{1}{\frac{p+q}{pq}} = \dfrac{pq}{p+q}$,

c. $f = \frac{4(3)}{4+3} = \frac{12}{7} = 1.71$ meters; the values are the same.

14. a. $x \geq -4$, **b.**

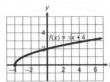

c. The output is increasing. **d.** $y \geq 0$, **e.** The x-intercept is $(-4, 0)$. The y-intercept is $(0, 2)$. **f.** g has the same shape but is shifted 8 units to the right. **g.** The graphs are reflected through the x-axis.

15. a.

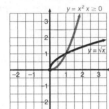

b. $y = \sqrt{x}$; $x = \sqrt{y}$; $y = x^2$; $f^{-1}(x) = x^2$; $x \geq 0$, **d.** The graphs are reflections in $y = x$. **e.** $f(f^{-1}(x)) = f(x^2) = \sqrt{x^2} = x$; $(f^{-1}(f(x))) = f^{-1}(\sqrt{x}) = (\sqrt{x})^2 = x$;

16. a. i. $x \geq 0, y \geq 4$,

i. $(0, 4)$ only,

ii.

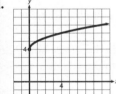

b. i. $x \geq -4, y \geq 0$, **ii.** $(0, 2)$ and $(-4, 0)$,

iii.

17. a. $x = 6$, **b.** $x = 6$, **c.** $x = \frac{23}{5} = 4.6$, **d.** $x = -1$, -1 does not check. There is no solution. **e.** $x = 27$;
18. a. all real numbers, **b.** $x \geq 6$, **c.** $x \geq -1$;
19. $36 = 1.5h$, $h = 24$ feet, **20.** $d \approx 153.76$ feet

Chapter 6

Activity 6.1   Exercises: 1. a. 0.6000, **b.** 0.8000, **c.** 0.8000, **d.** 0.6000, **e.** 0.7500, **f.** 1.3333;   **3. a.** Given $\tan B = \frac{7}{4}$, if the side opposite angle B is 7; then the side adjacent to angle B is 4. Using the Pythagorean theorem, I determine that the

hypotenuse is $\sqrt{65}$. **b.** $\sin B = \frac{7}{\sqrt{65}} = 0.8682$,
c. $\cos B = \frac{4}{\sqrt{65}} = 0.4961$;   **5. a.** the sine function,
b. $\sin B = \frac{y}{c}$;   **7. a.** $x \approx 9.1$, **b.** $x \approx 85.6$, **c.** $x \approx 61.4$;
9. a.

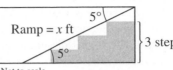

Note: Not to scale

b. $3 \cdot 7 = 21$ inches. The increase in height from one end of the ramp to the top of the stairs is $\frac{21}{12} = 1.75$ feet.
c. $x \approx 20.1$ feet. The ramp needs to be at least 20.1 feet long. Therefore, the donated ramp will not be long enough to meet the code. Alternative approach: $15 \sin 5 = 1.3$ feet. The three steps must measure at most 1.3 feet high for the 15-foot ramp to satisfy the code. Each solution suggests ways to think about modifications to either the ramp or the steps (or both) that could be used to meet the code.

Activity 6.2   Exercises: 1. a. $s + w = 0.68$ miles,
b. These calculations confirm the result in part a.
3. a.

$(90 - x)$	$\sin x$	$\cos(90 - x)$
83°	0.1219	0.1219
73°	0.2924	0.2924
66°	0.4067	0.4067
57°	0.5446	0.5446
42°	0.7431	0.7431
23°	0.9205	0.9205
13°	0.9744	0.9744

b. The table in part a illustrates the property that cofunctions of complementary angles are equal.

Activity 6.3   Exercises: 1. a. $\theta - 30°$, **b.** $\theta - 64.62°$,
c. $\theta = 67.04°$, **d.** $\theta = 63.82°$, **e.** $\theta = 66.80°$,
f. $\theta = 64.62°$, **g.** $\theta = 22.28°$, **h.** $\theta = 20.76°$;

3.

V	θ	E	N
32	65°	13.5	29.0
23.3	59°	12	20
4.1	54°	2.4	3.3
26	43.8°	18.8	18
4.5	45°	3.2	3.2

4. $\theta = \tan^{-1}\left(\frac{30}{50}\right) \approx 31°$. My friend should look up at an angle of approximately 31°.
5. a.

NOT TO SCALE

b. Because grade is rise over run, $\tan \theta = 0.1$.
$\theta = \tan^{-1}(0.1) = 5.7°$. The

ramp makes an angle of 5.7° with the horizontal.
c. $y = 15 \sin(5.7) \approx 1.5$ feet. The elevation changes 1.5 feet from one end of the ramp to the other.

Activity 6.4 Exercises: 1. a. The side adjacent to the 57° angle is 4.2 feet. The hypotenuse is 7.8 feet. The other acute angle is 33°. **b.** The hypotenuse is 19.0. The angle adjacent to side 18 is 18.4°. The other acute angle is 71.6°. **c.** The other leg is 7.9 inches. The angle adjacent to side 9 inches is 41.4°. The other angle is 48.6°.
3. a.

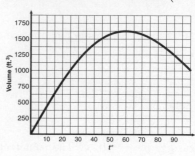

b. The direct distance, d, from the observation deck to the raft is approximately $\frac{800}{\sin(22°)} = 2136$ feet. If you could walk straight down the cliff and straight across at the base of the cliff to the creek, the distance would be approximately $b + c = 2780$ feet, where $b = \frac{800}{\tan 22}$.

Project Activity 6.5 Exercises: 1. a. The slope is $\frac{5}{100}$ or $\frac{1}{20}$ or 0.05. **b.** $A = \tan^{-1}(0.05) = 2.86°$. The highway makes an angle of 2.86° with the horizontal. This angle is called the angle of elevation. **c.** 1 mile is equivalent to 5280 feet. If x represents the number of feet above sea level after walking 1 mile, then $x = 5280 \cdot \sin(2.86°) =$ approximately 263 feet. I would be 263 feet above sea level after 1 mile. **3.** Using the following diagram: The two equations are (a)
$\tan 25° = \frac{5}{x + y}$ and
(b) $\tan 30° = \frac{5}{y}$.
Solving equation (b), $y \approx 8.7$ miles. Then equation (a) becomes $\tan 25° = \frac{5}{x + 8.7}$.
Solve this equation for x.
$(x + 8.7) \tan 25° = 5$, $x \tan 25° = 5 - 8.7 \tan 25°$,
$x = \frac{5 - 8.7 \tan 25°}{\tan 25°}$,
$x \approx 2.0$ miles. The runway is approximately 2 miles long.
6. a. Let A represent the area of the trapezoidal cross section. The height of the cross section is h, and the two bases are 5 and $5 + 2x$, respectively. The area is then determined by the formula $A = \frac{1}{2}h(10 + 2x)$, which, after simplifying, is $A = h(5 + x)$ or $A = 5h + hx$.
b. $h = 5 \sin t$, $x = 5 \cos t$, $A = 5(\sin t)(5 + 5 \cos t)$ or $A = 25 \sin t(1 + \cos t)$ or $A = 25 \sin t + 25 \sin t (\cos t)$,
c.

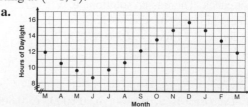

d. The graph in part c indicates that the area of the trapezoidal cross section (output) is the greatest when the angle t is 60°.

e. The area is approximately 32.5 square feet as read from the graph in part c. **f.** Let V represent the volume. Then $V = 50 \cdot A$, where A is the cross-sectional area. In terms of t, the volume is $V = 1250 \sin t (1 + \cos t)$.
g.

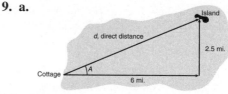

h. The graph indicates the greatest value for the volume between 0° and 90° is approximately 1625 cubic feet when the angle t is 60°. **i.** The angle is the same, namely 60° in this scenario.

How Can I Practice? 1. a. $\frac{8}{15}$, **b.** $\frac{15}{8}$, **c.** $\frac{15}{17}$, **d.** $\frac{8}{17}$, **e.** $\frac{8}{17}$, **f.** $\frac{15}{17}$; **2. a.** 0.731, **b.** 0.574, **c.** 0.601, **d.** 5.671;
3. $\cos A = \frac{12}{13}$ and $\tan A = \frac{5}{12}$; **4.** $\sin B = \frac{7}{\sqrt{65}}$ and
$\cos B = \frac{4}{\sqrt{65}}$; **5. a.** $\theta = 48.6°$, **b.** $\theta = 23.5°$,
c. $\theta = 74.1°$, **d.** $\theta = 16.6°$, **e.** $\theta = 44.2°$, **f.** $\theta = 13.7°$;
6. $\overline{BC} = 4.8$ centimeters, $\overline{AC} = 3.6$ centimeters, $\angle B = 37°$, $\angle C = 90°$; **7.** $A = \arctan\left(\frac{10}{15}\right) = 33.7°$;
therefore, I should buy the 35° trusses. **8. a.** $\theta = 20.6°$,
b. $D = \sqrt{16^2 + 6^2} = 17.1$ feet.
9. a.

The direct distance, d, from the cottage to the island is $d = \sqrt{2.5^2 + 6^2} = 6.5$ miles.
b. $A = \arctan\left(\frac{2.5}{6}\right) = 22.6°$; I should direct my boat 22.6° north of east to get from the cottage to the island in the shortest distance.

Activity 6.6 Exercises: 1. a. $(0.31, 0.95)$, **b.** $(0.64, -0.77)$, **c.** $(0, -1)$, **d.** $(-0.36, 0.93)$, **e.** $(-0.85, -0.53)$,
f. $(0.26, 0.97)$, **g.** $(0.34, -0.94)$; **2. a.** $\frac{72}{360} \cdot 2\pi = 1.26$,
b. $\frac{310}{360} \cdot 2\pi = 5.41$, **c.** $\frac{270}{360} \cdot 2\pi = 4.71$,
d. $\frac{111}{360} \cdot 2\pi = 1.94$, **e.** $\frac{212}{360} \cdot 2\pi = 3.70$,
f. $\frac{435}{360} \cdot 2\pi = 7.59$, **g.** $\frac{-70}{360} \cdot 2\pi = -1.22$; distance is 1.22;
4. a. The graph looks like the cosine function reflected in the t-axis. **b.** The graph indicates clockwise rotation starting at $(-1, 0)$.
5. a.

b. Yes, the number of hours of daylight is cyclical. The graph looks like a shifted and stretched sine graph reflected in the

x-axis. **c.** The graph has the same wavelike shape. **d.** South. The number of hours of daylight is greater there from October to February, summer in the Southern Hemisphere.

Activity 6.7 Exercises: 1. a. $45 \cdot \frac{\pi}{180} = \frac{\pi}{4} \approx 0.785$ radians,
b. $140 \cdot \frac{\pi}{180} = \frac{7\pi}{9} \approx 2.443$ radians,
c. $330 \cdot \frac{\pi}{180} = \frac{11\pi}{6} \approx 5.760$ radians,
d. $-36 \cdot \frac{\pi}{180} = \frac{-\pi}{5} \approx -0.628$ radians;

3.

0°	30°	45°	60°	90°	135°	180°	210°	270°	360°
0	$\pi/6$	$\pi/4$	$\pi/3$	$\pi/2$	$3\pi/4$	π	$7\pi/6$	$3\pi/2$	2π

5. a. $y = 2 \cos x$ is $y = \cos x$ stretched vertically by a factor of 2. $y = \cos(2x)$ is $y = \cos x$ compressed horizontally by a factor of 2. **b.** $y = \cos\left(\frac{1}{3}x\right)$ is $y = \cos x$ stretched horizontally by a factor of 3. $y = \cos(3x)$ is $y = \cos x$ compressed horizontally by a factor of 3.

7.

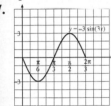

max: 3,
min: -3,
period: $\frac{2\pi}{3}$;

9. a. The bill is highest for December and January. The amount of the bill is approximately $600. **b.** The bill is lowest for May and June. The amount of the bill is approximately $250. **c.** The largest value is $650. **d.** The period is 6 billing periods or 12 months. **e.** The graph will be

stretched vertically by a factor of 1.05. This will not affect the period of the function. **f.** The amount of the bills for the summer months would increase. The graph would flatten out as the monthly charges become more equal.

Activity 6.8 Exercises:

1.

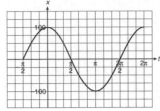

3. a. The maximum value is 100.
b. The minimum value is -100.
5. $x = 150 \cos \theta$
$y = 150 \sin \theta$;

8. a. $y = -15 \sin(2x)$, **b.** $y = 1.3 \cos(0.7x)$;
10. a. Graph is iii, **b.** Graph is iv, **c.** Graph is i, **d.** Graph is ii

Activity 6.9 Exercises: 1. a. amplitude: 0.7, period: π, displacement: $\frac{-\frac{\pi}{2}}{2} = -\frac{\pi}{4}$, **b.** amplitude: 3, period: 2π, displacement: $\frac{-(-1)}{1} = 1$, **c.** amplitude: 2.5, period: 5π, displacement: $\frac{-\frac{\pi}{3}}{0.4} = -\frac{5\pi}{6}$, **d.** amplitude: 15, period: 1, displacement: $\frac{-(-0.3)}{2\pi} = \frac{3}{20\pi} = 0.0477$;

3.

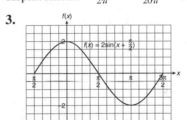

5. a. Graph is iii.
b. Graph is iv.
c. Graph is i,
d. Graph is ii.

Activity 6.10 Exercises: 1. a.

DATE	JAN 29	FEB 5	FEB 13	FEB 21	FEB 28	MAR 6	MAR 14	MAR 22	MAR 29
x; days since Jan 29	0	7	15	23	30	36	44	52	59
y; the amount of moon visible	0	0.5	1	0.5	0	0.5	1	0.5	0

b.

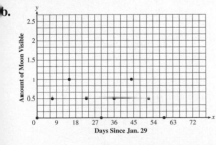

c. Yes. It does show repeated and periodic behavior.
d. $y = 0.5019 \sin(0.2135x - 1.5888) + 0.5306$,
e. $y(95) \approx 0.45$, or about 45% of the side of the moon facing Earth will be visible 95 days after Jan. 29th.

How Can I Practice? 1. a. $(0.81, 0.59)$, **b.** $(-0.87, -0.5)$, **c.** $(0, -1)$, **d.** $(0.73, -0.68)$, **e.** $(-0.81, -0.59)$, **f.** $(0, 1)$; **2. a.** 0.63 units, **b.** 3.67 units, **c.** 1.57 units clockwise, **d.** 5.53 units, **e.** 2.51 units clockwise, **f.** 7.85 units; **3. a.** $18 \cdot \frac{\pi}{180} = \frac{\pi}{10}$, **b.** $150 \cdot \frac{\pi}{180} = \frac{5\pi}{6}$, **c.** $390 \cdot \frac{\pi}{180} = \frac{13\pi}{6}$, **d.** $-72 \cdot \frac{\pi}{180} = \frac{-2\pi}{5}$;
4. a. $\frac{5\pi}{6} \cdot \frac{180}{\pi} = 150°$, **b.** $1.7\pi \cdot \frac{180}{\pi} = 306°$, **c.** $-3\pi \cdot \frac{180}{\pi} = -540°$, **d.** $0.9\pi \cdot \frac{180}{\pi} = 162°$;

5. amplitude: 4
period: $\frac{2\pi}{3}$
displacement: 0

6. amplitude: 2
period: $\frac{2\pi}{1} = 2\pi$
displacement: 1

7. amplitude: 3.2
period: $\frac{2\pi}{2} = \pi$
displacement: 0

8. amplitude: 1
period: $\frac{2\pi}{\frac{1}{2}} = 4\pi$
displacement: -2

9. amplitude: 3
period: $\frac{2\pi}{4} = \frac{\pi}{2}$
displacement: $\frac{-(-1)}{4} = \frac{1}{4}$

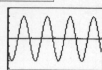

10. a. because of the repetitive nature of the height of the water as a function of time,

b. amplitude $= \dfrac{80 - 0}{2} = 40$,

c. The period is approximately 12 hours, because high tide occurs twice a day. **d.** Let x represent the number of hours since midnight. $y = a \sin(bx + c) + d$, $a = 40$, the amplitude, period: $\frac{2\pi}{b} = 12$, $b = \frac{\pi}{6}$, displacement:

$3 = \dfrac{-c}{b} = \dfrac{-c}{\frac{\pi}{6}}$, $-c = 3 \cdot \left(\frac{\pi}{6}\right)$, $c = \frac{-\pi}{2}$, vertical shift:

$d = 40$, $y = 40 \sin\left(\frac{\pi}{6}x - \frac{\pi}{2}\right) + 40$. Other equations are possible; depends on the choice of displacement.

Gateway Review 1. a. $N = 7 \sin(63°) = 6.24$ miles,

b. $E = 7 \cos(63°) = 3.18$ miles; **2. a.** side $c = 13$, angle $A = 67.4°$, angle $B = 22.6°$, **b.** side $a = 6.93$, side $b = 4$, angle $A = 60°$, **c.** side $b = 3$, side $c = 4.24$, angle $B = 45°$, **d.** side $a = 8.66$, side $c = 10$, angle $B = 30°$;

3. a. $\cos\theta = \frac{8}{10}$, $\tan\theta = \frac{6}{8}$, **b.** $\sin\theta = \frac{1}{2}$, $\tan\theta = \frac{1}{\sqrt{3}}$, $\theta = 30°$, **c.** $c^2 = 5^2 + 8^2 = 89$, $c = \sqrt{89}$, $\sin\theta = \frac{8}{\sqrt{89}}$, $\cos\theta = \frac{5}{\sqrt{89}}$; **4.** No; there is a difference, but it is so small that it is difficult to see. For me: $\theta = \tan^{-1}\left(\frac{1408}{100}\right) = 85.9375°$. For my nephew: $\theta = \tan^{-1}\left(\frac{1411}{100}\right) = 85.9461°$.

5.

$\tan 57° = \frac{a}{30}$
$30 \tan 57° = a$
$a \approx 46.2$
$\cos 57° = \frac{30}{c}$
$c = \frac{30}{\cos 57°}$
$c \approx 55.1$

$\tan 13° = \frac{b}{46.2}$
$b = 46.2 \tan 13°$
$b \approx 10.7$
$\cos 13° = \frac{46.2}{h}$
$h = \frac{46.2}{\cos 13°}$
$h \approx 47.4$

6. So, $(2x)^2 = x^2 + (200 + x)^2$; $x = \dfrac{400 \pm \sqrt{480{,}000}}{4}$.

The negative does not make sense, thus $x = \dfrac{400 + 400\sqrt{3}}{4}$.

$x = 100 + 100\sqrt{3}$; $h = 300 + 100\sqrt{3}$

7.

$\sin\theta$	$\cos\theta$	$\tan\theta$
$\frac{\sqrt{3}}{2}$	$-\frac{1}{2}$	$-\sqrt{3}$
$\frac{1}{\sqrt{2}}$	$-\frac{1}{\sqrt{2}}$	-1
$\frac{1}{2}$	$-\frac{\sqrt{3}}{2}$	$-\frac{1}{\sqrt{3}}$
0	-1	0
$-\frac{1}{2}$	$-\frac{\sqrt{3}}{2}$	$\frac{1}{\sqrt{3}}$
$-\frac{1}{\sqrt{2}}$	$-\frac{1}{\sqrt{2}}$	1
$-\frac{\sqrt{3}}{2}$	$-\frac{1}{2}$	$\sqrt{3}$
-1	0	undef.
$-\frac{\sqrt{3}}{2}$	$\frac{1}{2}$	$-\sqrt{3}$
$-\frac{1}{\sqrt{2}}$	$\frac{1}{\sqrt{2}}$	-1
$-\frac{1}{2}$	$\frac{\sqrt{3}}{2}$	$-\frac{1}{\sqrt{3}}$
0	1	0

8. a. amplitude: 2
period: 2π

b. amplitude: 2
period: 2π

c. amplitude: 1
period: π

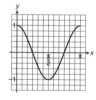

d. amplitude: 1
period: 1

e. amplitude: 1
period: 4π

f. amplitude: 1
period: 4

g. amplitude: 1
   period: 2π

h. amplitude: $\frac{2}{3}$
   period: π

i. amplitude: 1
   period: 1

j. amplitude: 3
   period: 1

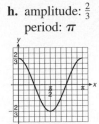

9. a. Graph is vi. **b.** Graph is ii. **c.** Graph is iv.
d. Graph is v. **e.** Graph is vii. **f.** Graph is viii.

Glossary

addition of functions   *See* sum function.

argument   Another name for the input of a function.

average rate of change, or simply, the **rate of change**   The ratio $\dfrac{\Delta y}{\Delta t}$, where Δy represents the change in output and Δt represents the change in input. The average rate of change gives the change in the output for a one-unit increase in the input.

axis of symmetry   A vertical line that separates the graph of a parabola into two mirror images.

change of base formula $\log_b x = \dfrac{\log_a x}{\log_a b}$, where $b > 0, b \neq 1$, is the formula used to change logarithms of one base to logarithms of another base.

coefficient   The numerical multiplier of a variable.

common logarithms   Base 10 logarithms.

complex numbers   Numbers of the form $a + bi$, such that a and b are real numbers and $i = \sqrt{-1}$.

composition function   The function that is created when the output of the function g becomes the input for a second function f. The rule is given symbolically by $y = f(g(x))$.

consistent system of linear equations   A system with exactly one solution.

constant function   A function in which there is no change in the output. The graph of a constant function is a horizontal line.

constant of proportionality   A constant, k, that gives the rate of variation in the direct proportional relationship $y = kx^n$.

constant term   A term that does not change in value.

continuous compounding of an investment   Occurs when the compounding period is so short it is essentially an instant in time. The formula for continuous compounding is $A = Pe^{rt}$.

cubic   A third-degree polynomial function having the general equation $y = ax^3 + bx^2 + cx + d$, where a, b, c, and d are real numbers and $a \neq 0$.

decay factor of an exponential function   The number, b, in the equation $y = a \cdot b^x$, where $0 < b < 1$ and a is the amount when $x = 0$.

decreasing function   A function in which the output decreases in value as the input increases. The graph goes down to the right.

degree of a polynomial function   The exponent of the term with the largest exponent.

dependent variable of a function   The output variable.

difference function, $f - g$ The function that is created from two functions, f and g by the rule $y = f(x) - g(x)$.

direct variation between two variables   A relationship in which as the independent variable (input) increases in value, the dependent variable (output) increases. Also the independent variable decreases as the dependent variable decreases.

discriminant   The expression $b^2 - 4ac$ under the radical of the quadratic formula. The value of the discriminant determines the type of solutions of the equation $ax^2 + bx + c = 0$.

domain of a function   The set of all possible input values of a function.

exponential function   A function of the form $y = a \cdot b^x$ with $b > 0$ and $b \neq 1$ where the independent variable, x, is the exponent.

extraneous solution   A potential solution that is not really a solution to the original equation or problem.

function   A relationship between the input and the output such that for each input value there is exactly one output value.

general form of a linear equation $Ax + By = C$, where A, B, and C are real numbers.

growth factor of the exponential function $y = a \cdot b^x$ The number b, where $b > 1$ and a is the amount when $x = 0$.

horizontal intercepts   All points of the graph of a function whose y-coordinate is 0. (*See* x-intercept.)

identity function   The function in which the output value is always identical to the input value.

imaginary unit $i = \sqrt{-1}$.

inconsistent system of linear equations   A linear system of equations with no solution. (Graphically, two parallel lines represent the system.)

increasing function   A function in which the output increases in value as the input increases. The graph goes up to the right.

input variable   The independent variable.

inverse functions   Two functions f and g related such that $f(g(x)) = x = g(f(x))$. Graphically, these functions are mirror images in the line $y = x$.

inverse variation between two variables   A relationship in which as the independent variable (input) increases in value, the dependent variable (output) decreases. Also, the independent variable decreases as the dependent variable increases.

irrational number   Any real number that cannot be written as a rational number.

linear function   Any function in which the rate of change, or slope, is constant.

linear term   The term of a polynomial function of the form bx, where b is a real number.

logarithm function   A function of the form $y = \log_b x$ where the base $b > 0, b \neq 1$. (The logarithm function is the inverse of the exponential function.)

magnitude   The relative size of a number or quantity, expressed as a distance or absolute value (and is therefore not negative).

mathematical model   A function that best fits the actual data and can be used to predict output values for input values not in the table.

natural logarithm   A logarithm to the base e. The logarithm is written as $\log_e x = \ln x$.

ordered pair   A pair of values, separated by a comma and enclosed in a set of parentheses. The input value is written to the left of the output value.

output variable   The dependent variable.

parabola   The graph of a quadratic function (second-degree polynomial function). The graph is U-shaped, opening either upward or downward.

piecewise function   A function in which the function rule for determining the output is given separately, or in pieces, for different values of the input.

point-slope form of the equation of a line   The equation where m represents the slope of the line and is a fixed point on the line.

polynomial function   Any function defined by a sum of a finite number of terms of the form ax^n, where a is a real number and n is a nonnegative integer.

practical domain   The set of all input values that make sense in a problem situation.

practical range   The set of all output values that make sense in a problem situation.

product function   The function that is created from two functions f and g, by the rule $y = f(x) \cdot g(x)$

profit function   A function that is common in the business world and is defined by Profit = Revenue − Cost.

quadratic formula   The formula $x = \dfrac{-b \pm \sqrt{b^2 - 4ac}}{2a}$ that represents the solutions to the quadratic equation $ax^2 + bx + c = 0$.

quadratic function   A second-degree polynomial function defined by an equation of the form $f(x) = ax^2 + bx + c$, where a, b, and c are real numbers and $a \neq 0$.

quartic function   A fourth-degree polynomial function defined by an equation of the form $f(x) = ax^4 + bx^3 + cx^2 + dx + e$, where a, b, c, d, and e are real numbers and $a \neq 0$.

radical function   Any function involving a radical (square root, cube root, and so on).

radicand   The expression under the radical.

range   The collection of all values of the dependent variable.

rational equation   An equation composed of fractions where the numerators and denominators are polynomials, with the variable appearing in a denominator.

rational function   Any function that can be defined as the ratio of two polynomial functions.

real numbers   All numbers that are either rational or irrational.

slope of a line   The constant rate of change of output to input.

slope-intercept form of the equation of a line   The equation $y = mx + b$, where m represents the slope of the line and $(0, b)$ is the vertical intercept.

solution of a system of equations in two variables   The ordered pair of numbers (x, y) that make both equations true.

sum function, $f + g$ The function that is created from two functions f and g by the rule $y = f(x) + g(x)$.

system of linear equations in two variables   A pair of equations that can be written in the form $y = ax + b$ and $y = cx + d$, respectively where a, b, c, and d are real numbers.

variation   How the dependent variable changes when the independent variable changes. (*See* direct variation or inverse variation.)

vertex   The turning point of the graph of a parabola. It has coordinates $\left(\dfrac{-b}{2a}, f\left(\dfrac{-b}{2a}\right)\right)$, where a and b are determined from the equation $f(x) = ax^2 + bx + c$. The vertex is the highest or lowest point of a parabola.

vertical asymptote of the graph of $y = f(x)$ The vertical line, $x = c$, such that $f(c)$ is undefined and $f(x)$ becomes arbitrarily large in magnitude as x approaches c.

vertical intercept   The point of the graph of the function whose x-coordinate is 0. (*See* y-intercept.)

x-intercepts   All points of the graph of the function at which the y-coordinate is 0. (*See* horizontal intercept.)

Xmax   The largest value of input visible in the window of a graphing calculator.

Xmin   The smallest value of input visible in the window of a graphing calculator.

y-intercept   All points of the graph of the function whose x-coordinate is 0. (*See* vertical intercept.)

Ymax   The largest value of output visible in the window of a graphing calculator.

Ymin   The smallest value of output visible in the window of a graphing calculator.

zero-product principle   The algebraic rule that says if a and b are real numbers such that $a \cdot b = 0$, then either a or b, or both, must be equal to zero.

Index